Mustapha Taleb
Ayoub Ainane
Tarik Ainane

# Practical Nutrition and Dietary Models

Mustapha Taleb
Ayoub Ainane
Tarik Ainane

# Practical Nutrition and Dietary Models

## Cases of nutrition and coronavirus

ScienciaScripts

**Imprint**



Cover image: www.ingimage.com

This book is a translation from the original published under ISBN 978-620-2-54156-5.

Publisher:
Sciencia Scripts
is a trademark of
International Book Market Service Ltd., member of OmniScriptum Publishing Group
17 Meldrum Street, Beau Bassin 71504, Mauritius
Printed at: see last page
**ISBN: 978-620-3-40107-3**

# Authors :

Mustapha TALEB

**Professor at the Faculty of Sciences Dhar El Mahraz, Sidi Mohamed Ben Abdellah University, Fez, Morocco.**

Ayoub AINANE

**Post-doc at the Faculty of Sciences Dhar El Mahraz, Sidi Mohamed Ben Abdellah University, Fez, Morocco.**

Tarik AINANE

**Professor at the Khenifra Higher School of Technology, Sultan Moulay Slimane University, Khenifra, Morocco.**

# Contents

# Chapter 1. The fundamentals of nutrition (nutrients, energy and eating behavior)

**Introduction**

The primary purpose of nutrition is to ensure the coverage of energy needs (macronutrients) and qualitative needs (micronutrients). The main nutrients and their roles are described in this chapter.

The total energy expenditure (TEE) that corresponds to the cost of living is met by drawing on the environment. Food provides the necessary substrates for energy production (energy-rich macronutrients and micronutrients for functioning) which is achieved through the synthesis of ATP (adenosine triphosphate). The transport of energy through redox systems ultimately produces work and heat. Each nutrient has the capacity to produce a certain amount of ATP which is then converted into heat. Calorie is a unit of heat that reflects the energy obtained by the hydrolysis of ATP and the release of phosphorus. Carbohydrates provide approximately 4 kcal/g, as do proteins, lipids provide 9 kcal/g and alcohol 7 kcal/g.

## I. Macronutrients

Macronutrition is the analysis of protein, carbohydrate and fat intake from food. It is these macronutrients that provide the energy essential to the functioning of our body. Macronutrition is therefore essential especially for athletes, patients, or healthy people who follow a diet adapted to his training. The energy provided by these foods is translated into joules or or calorie per gram of food.

### 1. Carbohydrates (CHO)

CHOs (carbohydrates) have in common their chemical structure associating carbon and water molecules in a 1:1 ratio (Cn; H2On). Their energetic interest is considerable since they cover 50 to 70% of energy needs: 1 g of carbohydrates provides 4 kcal. The main metabolic function of carbohydrates is to ensure glycemic homeostasis through adequate intake and the possibility of tissue storage in the form of glycogen, which is to the animal world what starch is to the plant world. Glycogen storage is limited to 300 g, or an energy reserve of 1200 kilocalories. By their sweetening power, monosaccharides and disaccharides contribute to the palatability of the food and thus to its acceptability.

## 1.1 Digestible carbohydrates

### *1.1.1. Classification*

CHOs are very heterogeneous energy nutrients (macronutrients) whose structural and functional classifications are not completely consistent. A distinction is made between non-digestible CHO and digestible CHO (carbohydrates for metabolic use). Among the latter must be distinguished monosaccharides, disaccharides and polysaccharides whose digestion process and metabolic destiny are different.

**a. Monosaccharides :**

Food monosaccharides are the products of the hydrolysis of starch to produce glucose, fructose or galactose. Sorbitol is the alcohol of glucose and xylitol is the alcohol of xyloses; they are of interest only because of their sweetening power. Ribose and deoxyribose are pentoses of endogenous synthesis whose purpose is to produce nucleic acids.

**b. Disaccharides :**

Disaccharides are represented by sucrose (or cooking sugar) which has sweetening power and lactose. Mono- and disaccharides are considered "simple" sugars but their absorption and metabolic fate can be very different from each other. Only glucose and, to a lesser degree, sucrose are to be considered "fast" sugars, implying that they cause early and severe hyperglycemia.

**c. Polysaccharides :**

Not all polysaccharides, which are carbohydrates with a complex structure, are digestible by humans. This is the case of cellulose, while starch, amylopectin and amylose, which are polymers of glucose, are digestible after cooking. Nevertheless, 2 to 5% of the starches in food are resistant to digestive enzymes and are a fermentation substrate for the colonic microflora that transforms them into short-chain fatty acids.

### *1.1.2. Digestion and absorption*

Carbohydrates are absorbed only as monosaccharides obtained by a hydrolysis process that begins upon ingestion under the effect of salivary amylase, which breaks the α 1-4 bonds that link the glucose radicals of amylose, a short-chain glucose polymer. Salivary and pancreatic amylases produce maltose and maltotriose which are hydrolyzed by isomaltase in the intestinal brush border which also breaks the α 1-6 bonds of amylopectin, a long-chain glucose polymer with a structure comparable to that of glycogen. Other intestinal enzymes (disaccharidase, lactase) complete the digestion of the disaccharides formed. The rate of absorption of carbohydrates in the form of glucose depends in principle on the complexity of the carbohydrates ingested. However, although the structure of amylopectin is more complex, blood glucose levels rise more rapidly after ingestion of amylopectin than after amylopectin ingestion, making the distinction between "simple" and "complex" carbohydrates, corollaries of "fast" and "slow" carbohydrates, enshrined in usage, difficult to accept. The presentation of carbohydrates is also important. The conversion of starch to glucose depends on whether or not there is a protein envelope protecting the starch, on the entanglement with structural fibres, or on the combination with other nutrients within a compound meal. Absorption of the dextrorotatory isomer of glucose is rapid due to active absorption. On the other hand, the absorption of the levorotatory isomer of glucose, galactose, sorbitol and xylitol is by passive diffusion with a limiting threshold of about 50g. Beyond this threshold, these carbohydrates have an osmotic effect responsible for digestive discomfort (diarrhea). Fructose is absorbed by a process of facilitated diffusion with a threshold of 100g.

**a. Metabolism :**

The metabolic consequences of carbohydrate intake are not limited to an energy equation but must take into account the importance of the glucose load reaching the general circulation and the kinetics of hyperglycemia. During the first hepatic passage of glucose occurs glycogenogenogenesis which contributes to glucose homeostasis by avoiding excessive post-absorptive hyperglycemia and preventing interprandial hypoglycemia through glycolysis of glycogen stores.

Glycolysis, oxidative phosphorylation and the Krebs tricarboxylic cycle generate ATP. The increase in ATP determines an increase in oxaloacetic acid and acetyl-CoA, which stimulates fatty acid synthesis. Thus, excess carbohydrate intake leads to energy storage in the form of fat, thus avoiding the appearance of hyperglycemia as soon as the glycogen reserves are saturated. The level of intracellular energy reserves acts as a signal to modulate the metabolism. A high level of ATP slows down the Krebs cycle and inhibits glycolysis, while a high level of ADP and AMP induces glycolysis and ATP regeneration. In muscle, the anaerobic metabolism of glucose produces pyruvates that are transformed in situ into CO2 or transported to the liver. In the event of very high exertion, pyruvates produce lactates, the accumulation of which can cause muscle cramps. The insulin secretion induced by hyperglycemia stimulates glycogen formation while glucagon stimulates hepatic and muscle glycogenolysis. Glucose homeostasis is maintained by various procedures designed to circumvent the adverse effects of an inconsistent and discontinuous supply of glucose to the strictly glucose-dependent organs of the brain (140g of glucose per day), the figurative elements of the blood and the renal medulla. When the glycogen reserves are exhausted, a hepatic neoglucogenesis is set up whose main substrates are lactates (Cori cycle), glycerol from free fatty acids and amino acids known as glucoformers whose leader is alanine. Prolonged fasting induces ketogenesis from free fatty acids whose beta-oxidation produces ketone bodies (beta-hydroxybutyrate) that can be used as metabolic fuel by non-glucoid-dependent organs. At the same time, glucose savings are established with a preferential use of fatty acids to provide muscle energy (Randle cycle).

### 1.2. Nutritional approach to carbohydrates

The functional definition of carbohydrates is not superimposed on the biochemical definition. Chemically complex carbohydrates do not necessarily result in lower blood glucose elevations than some simpler carbohydrates. For example, white bread, an important source of starch without sweetening power, does not raise blood sugar levels less than cooking sugar, which is a disaccharide (glucose + fructose). Fructose, a monosaccharide with high sweetening power, raises blood sugar levels less than the starch in white bread. In nutrition, the hyperglycemic power, the metabolic fate and the sweetness count more than the structure. The glycemic index and the glycemic load induced by a carbohydrate are more suitable parameters, although they only provide part of the solution as they depend on the accompanying energy nutrients, the presence or absence of dietary fibre, the speed of gastric emptying and the degree of cooking of the food.

### 1.3 Glycemic index

The glycemic index (GI) of a food is defined in relation to a standard carbohydrate intake. It is defined as the increase in the area under the curve induced by a 50g serving of carbohydrates of a given food expressed as a percentage of the same amount of carbohydrates of a standard food (glucose or 50g of white bread) consumed by the same subject. The GI expresses the degree of hyperglycemia induced by a food. A GI >70% is high while a GI <50% is low. The GI varies according to many factors related to the conditions of ingestion and which cannot be accounted for on the table: proportion of simple carbohydrates, nature of the starches, presence of fiber, cooking method, lipid and protein content of the food or meal in which it is included, industrial process used. There is great variability in the GI within and between individuals and between species of the same food (Fig. 1-1). The time of meal should be taken into consideration, as the GI is higher at breakfast than at lunch. Finally, the notion of GI should be associated with the notion of nutritional profile.

Traditional food and even more so "paleonutrition" are opposed to industrial and modern food with lower GI foods because of a high consumption of starchy foods associated with fiber.

### 1.4. Glycemic load

The glycemic load (GC) appears more relevant than the glycemic index. It is obtained by multiplying the glycemic index by the amount of carbohydrate contained in a serving of a given food. It avoids the pitfall related to the highly variable carbohydrate content of foods which required the consideration of sometimes disproportionate quantities to establish the glycemic index (example: 800g of fruit with a 12% carbohydrate content to reach the equivalent of 100g of glucose). The notion of glycemic load allows comparison of the portions usually consumed and is therefore more adapted to nutritional practice. As an example, the GI of carrots is around 50 but the total amount of carbohydrates provided by a portion of carrots is low since the carbohydrate content is around 12g/100g, hence the low glycemic load of a food with a high nutritional density.

#### *1.4.1. Nutritional implications*

In the general population, a low glycemic load diet appears to be associated with a decrease in cardiometabolic risk related to improved insulin resistance and optimized glucose utilization. It also has a therapeutic purpose in many diseases including overweight, obesity, type 2 diabetes and all high cardiovascular risk situations. Increasing fruit and vegetable and non-digestible fiber intakes is an indirect way to achieve the goal of reducing the carbohydrate load. In addition, it is desirable, except in special cases, to ensure a high carbohydrate intake of about 50 to 55% of the energy ration while maintaining the goal of low glycemic load by using carbohydrates with a low glycemic index in order to reduce energy intakes of lipid origin. Some simple carbohydrates can pose a health risk. Fructose with high sweetening power is a monosaccharide with a low glycemic index (23%) whose absorption by facilitated diffusion allows a hepatic use that is only partially insulin-dependent to produce glucose, lipids and lactates. When ingested in excess (>50 to 75g/d), fructose causes an increase in uric acid, triglycerides and lactates. Its consumption, which has increased sharply over the last few decades (soft drinks, sweets), is associated with the increasing prevalence of obesity. Consumed in large quantities, it stimulates food intake and promotes the production of triglycerides. The consumption of fructose is recommended in the form of fruits but not as added carbohydrates in industrial food. Digestive tolerance to lactose decreases with age. Indeed the intestinal lactase active during childhood is deactivated in adulthood. Adults intolerant to lactose become symptomatic for an intake higher than 5 g/d (i.e. 100mL of milk).

Non-digestible carbohydrates or dietary fibre Dietary fibre is a non-digestible carbohydrate (Table 1). Classified as insoluble and soluble fibers, they do not provide energy and act on gastric emptying, intestinal transit, the balance of intestinal microflora which can partially degrade them. They are associated with foods that are usually low in fat and interact with the absorption of digestible carbohydrates by reducing their GI. Fibre-rich food models offer benefits confirmed by meta-analyses: reduced cardiovascular and metabolic diseases, better weight control and lower prevalence of certain cancers. It is recommended to increase the consumption of fiber (20g/1000kcal) by half in the form of soluble fiber by favoring intakes in natural form: vegetables, legumes, fruits, whole grains. Enrichment of soluble fibre in the form of glucan (from oats) is nevertheless possible, especially at breakfast.

| **Dietary Fibre** | |
|---|---|
| **Soluble fibers** | **Insoluble fibers** |
| Pectins | Cellulose |
| Erasers | Lignines |
| Glucans (oats) | Some hemicelluloses |
| Alginates (algae) | |
| Some hemicelluloses | |

**Table 1:** Dietary Fibre.

## 2. Proteins

Dietary proteins provide the amino acids (AA) needed to cover the body's protein needs. Protein functionalities include growth, function and maintenance of tissues, organs and the immune defence system. They provide 4kcal/g and have the status of an energetic macronutrient. AA are the substrates of all endogenous protein synthesis but also have a complementary metabolic role insofar as proteins constitute an important energy reserve of necessity and some AA contribute to glucose homeostasis by participating in hepatic neoglucogenesis, or even ketogenesis.

### 2.1. Digestion and absorption

Chemically, proteins are defined as AA chains, each of which carries a nitrogen radical. Ingested proteins are digested by gastric pepsin and pancreatic trypsin. Transformed into peptides of a few AA, they are then degraded into AA and dipeptidase by pancreatic and intestinal proteases. The absorbed AA reaches the liver, which is their main site of catabolism.

**a. Metabolism :**

The synthesis of circulating proteins on the one hand and the availability of AA in the general circulation to meet the specific needs of the organs on the other hand depend on the flow of AA. Hepatic protein synthesis is directed according to various metabolic signals. In the case of acute aggression or inflammation, amino substrates are used preferentially for the synthesis of proteins in the acute phase to the detriment of other proteins such as albumin. Outside of acute conditions, the concentrations of albumin and other so-called "nutrition" proteins reflect the state of the protein pool and the nutritional status. The fate of A.A. is diverse. One example is their role in the synthesis of purine and pyrimidine nucleic bases. Protein catabolism provides amino radicals ($NH_2$) that are integrated into the liver in the cycle of ureogenesis, allowing them to be eliminated in the urine. In case of insufficient ureogenesis (hepatocellular deficit), the AA are converted into ammonia ($NH_3$) which has neurotoxic effects. The liver and the kidney are able to synthesize creatine from arginine and glycocol. Creatine is transported to the muscle where it is stored as phosphocreatine and then converted to creatinine which is released into the general circulation before being eliminated in the urine. Urinary creatinine is a good reflection of muscle mass but is also dependent on dietary meat intake. Every day just under 2% of muscle creatine is converted to creatinine.

## 2.2. Biological quality of proteins: essential amino acids

Proteins are not all assimilated equally depending on the nature of the AA they contain (Table 2). A distinction is made between essential AA that cannot be synthesized endogenously by humans. There are nine of them and they must be present to allow the synthesis of other peptides and proteins. It is up to food to ensure their supply. The quality of a protein depends on its AA composition. Formerly defined in relation to animal growth, it is now defined by determining the proportion of AA that can be used without increasing nitrogen losses. The biological value of a protein reflects the quality of the protein. It is calculated by the ratio :

$$\frac{\textbf{N (azote) aliment} - (\textbf{N urinaire} + \textbf{N fécal})}{\textbf{N aliment} - \textbf{N fécal}}$$

The more essential AA in a protein, the better its quality. Proteins of animal origin meet this criterion. Egg white protein ovalbumin has a biological value arbitrarily set at 1. On the contrary, no food of plant origin provides all the essential AA; a combination of several different plant foods according to the principle of complementarity is necessary to meet the needs. All essential AA must be present for protein synthesis, but they can be drawn, if necessary, from the tissue protein pool. On the other hand, not all essential AA may be contained in the same protein food if it is combined with other foods to compensate for the insufficiency of an essential AA that is said to be "limiting. The lower the biological value of a type of protein, the greater the need for that type of protein. Dietary diversification reduces this risk.

| The different amino acids | |
|---|---|
| **Non-essential AA** | **AA Essentials** |
| Valine*, histidine*, leucine*, isoleucine*, lysine*, methionine*, phenylalanine°, threonine*, tryptophan | Arginine, alanine, glutamine, aspartate, asparagine, glycolol, proline serine Cysteine**, tyrosine** ° ° |
| AA aromatic.<br>* AA said plugged in.<br>** These AA become essential if their precursors (methionine and phenylalanine) are present in limited quantities. | |

**Table 2:** The different amino acids.

### 2.3. Nitrogen balance

Nitrogen elimination linked to protein catabolism occurs through several pathways: urine in the form of creatinine, urea, uric acid and ammonia, stools (proteins not absorbed or coming from digestive secretions), epithelia and skin (desquamation and mucous secretions). These losses must be compensated for by an adequate protein intake knowing that proteins contain about 16% nitrogen, which corresponds to 1g of nitrogen for 6.25 g of protein. The nitrogen balance must be in equilibrium or slightly positive. The nitrogen balance is the difference between the nitrogen intake and the sum of the nitrogen expenditure. It varies with overall energy intake. Excessive caloric intake decreases the nitrogen requirement while insufficient energy intake increases it, further negatively affecting the balance.

### 2.4. Protein requirements

They are estimated on the basis of estimated losses. These vary with age, gender, physical activity, physiological condition and health status. Protein requirements are the subject of ANCs that take these particularities into account. The minimum protein requirements are those that ensure good health in adults or normal growth in children. Nitrogen requirements in the case of a diet without protein intake can be compensated by 55mg/kg of nitrogen, i.e. 0.35g/kg of protein, hence the notion that the minimum or indispensable physiological protein requirements are 0.35g/kg. This figure is generally set at 0.55g/kg/day after applying a safety correction factor to take account of individual variation and the fact that protein is less well utilized when intakes are very close to minimum requirements. In order to better take into account the overall needs of a general population, the recommended intakes are set at 0.8g/kg/d in adults.

### 2.5. Effects of cooking on proteins

Some proteins can be denatured and lose their biological value when subjected to high temperatures. Heat decreases the availability of lysine. It inactivates the trypsin inhibitor in soybeans. The phenomenon of meat browning is the consequence of a Maillard reaction with protein glycation and production of glycation end products (EFAs) which are associated with cardiovascular risk. Protein and health

A protein intake of good biological value covering the needs is considered a prerequisite for an optimal state of health. Chronic protein deficiency has formidable consequences: growth disorders in children, skin fragility with delayed healing, altered immune defenses with an increased risk of infection, protein catabolism with sarcopenia and osteopenia. It is particularly feared at the extreme ages of life. However, an excess of protein intake is undesirable, firstly because it is most often to the detriment of carbohydrate intake, which is the preferred source of energy, and secondly because protein foods are most often associated with constitutional fats in which saturated fatty acids dominate. Furthermore, there are epidemiological and experimental arguments showing that a hyperprotidic diet increases glomerular perfusion pressure, which predisposes to chronic renal failure and has a urinary lithogenic effect.

## 3. Lipids

Dietary lipids are multiple. They are important sources of energy (9kcal per 1 gram) that improve the palatability of food and dishes (smoothness). They have a structural (constituents of cell membranes and myelin) and metabolic (precursors of steroid hormones and eicosanoids) role. The usual nomenclature distinguishes between saturated, monounsaturated and polyunsaturated fats. The concept of saturation refers to the presence or absence of double bonds between the carbon atoms that make up the skeleton of fatty acids (FA). These are characterized by the length of the chain (number of carbon atoms), by the number of double bonds and by the isomeric configuration cis (usual) or trans. The nature of the fatty acids ingested has clinical implications.

### 3.1. Lipid digestion and absorption

Hydrophobic lipids in the aqueous phase undergo mechanical and partially chemical digestion (gastric lipase) in the stomach where they are emulsified into fine lipid droplets. This miscellification is maintained thanks to bile salts. The pancreatic lipase cleaves the triglycerides into GA and monoglycerides which are absorbed in the proximal part of the jejunum. The short-chain FAs enter the portal blood where they bind to albumin and reach the liver. Long-chain FAs are re-esterified to triglycerides in the enterocyte and transported to the lymph as chylomicrons. Absorption of the AGs and monoglycerides is almost complete. The longer the GA chain, the longer the absorption is delayed. Medium-chain triglycerides are absorbed rapidly due to more efficient emulsification and greater solubility. They are largely absorbed directly into the portal blood without undergoing re-esterification and are of interest in fat malabsorption situations. Omega-3 (n-3) long-chain FAs are absorbed more rapidly than other long-chain FAs. Cholesterol is absorbed by an active process at a rate of 30-70%, partly as dietary cholesterol and partly as cholesterol in the bile. The balance, sequestered by bile acids, is eliminated in the stool along with phytosterols and stanols. The faecal elimination of fats does not exceed 4 to 6g/d whatever the quantity of lipids ingested. However, the absorption capacity decreases with age. It is reduced by partial gastrectomy and pancreatic insufficiency.

**a. Metabolism :**

Triglyceride FAs are a source of energy that can be used by most organs except the brain, either directly or after storage in fat tissue. Cholesterol and phospholipids are mainly constituents of membranes. GA are either derived from chylomicrons (in the post-prandial phase) and other lipoparticles, or from adipose reserves (fasting) under the action of a lipoprotein lipase stimulated among others by insulin. The chylomicrons that provide most of the transport of triglycerides re-esterified in the enterocyte contain apoprotein (B48) and interact with HDL cholesterol lipoparticles produced in the liver, whose apoprotein C activates lipoprotein lipase and whose apoprotein E facilitates the capture of chylomicron remnants by the liver. The so-called "free" released AGs enter the mitochondria (via carnithine transferase for long-chain AGs) to produce ATP in muscle and adipose tissue. GA not used for energy purposes are re-esterified to triglycerides by glycerol-3-phosphatase, with synthesis requiring glucose and insulin. A high-carbohydrate diet tends to lower the level of GLA and promote the synthesis of reserve triglycerides in case of excessive energy intake. Unused GLA is stored in the form of triglycerides rich in palmitic (saturated) and oleic (monounsaturated) acid. It makes it possible to build up reserves of up to 120,000kcal that can be used after lipolysis favored by insulinopenia during fasting or relative insulinopenia during insulin resistance. The circulating AGs captured by the liver are incorporated into VLDL lipoparticles, which constitute the bulk of the hypertriglyceridemia observed in pathology, particularly in insulin resistance situations. GA from chylomicrons and VLDL can be used for energy purposes by the muscles (especially during prolonged exercise), heart, kidney and platelets. The nature of dietary lipids influences the composition of reserve fat in adipose tissue and the composition of VLDL.

### 3.2. Fatty acids

In chemistry, a fatty acid is a carboxylic acid with an aliphatic chain. Natural fatty acids have a carbon chain of 4 to 36 carbon atoms (rarely more than 28) and typically an even number, because fatty acid biosynthesis, catalyzed by fatty acid synthase, proceeds by iteratively adding groups of two carbon atoms using acetyl-CoA. By extension, the term is sometimes used to refer to all carboxylic acids with a non-cyclic hydrocarbon chain. They are referred to as long-chain fatty acids for a length of 14 to 24 carbon atoms and very long-chain fatty acids if there are more than 24 carbon atoms. Fatty acids are present in animal and vegetable fats, vegetable oils or waxes, in the form of esters.

In biochemistry, fatty acids are a class of lipids that includes aliphatic carboxylic acids and their derivatives (methylated, hydroxylated, hydroperoxylated fatty acids, etc.) and eicosanoids. The latter are derived from eicosapentaenoic acid (omega-3) or arachidonic acid (omega-6) and often act as hormones. Fatty acids play a fundamental structural role in all known life forms through various types of lipids (phosphoglycerides, sphingolipids...) which, in aqueous medium, are organized in two-dimensional networks structuring all biological membranes (cell, plasma, mitochondrial, endoplasmic reticulum, thylacoids, etc...).

They are also important sources of metabolic energy: fatty acids allow living beings to store about 37 kJ of energy (9 kcal) per gram of lipids, compared to about 17 kJ for carbohydrates (4 kcal). They are stored by the body in the form of triglycerides, in which three molecules of fatty acids form an ester with one molecule of glycerol. When they are not bound to other molecules, fatty acids are said to be "free". Their degradation produces large amounts of ATP, the preferred energy molecule of cells: the β-oxidation followed by degradation by the Krebs ring of a saturated fatty acid with $n = 2p$ carbon atoms releases $(10p - 2)$ ATP + $(p - 1) \times$ (FADH2 + NADH+H+), i.e. the energy equivalent of 106 ATP for a palmitic acid molecule $CH_3(-CH_2)_{14}\text{-COOH}$, which contains 16 carbon atoms ($n = 16$, and therefore $p = 8$) (table 3).

Fatty acids can be synthesized by the body through a set of metabolic processes called lipogenesis. They are also provided in large quantities by the diet. On average, the daily energy requirement is 2,000 kcal for a woman and 2,500 kcal for an adult man, of which fat should ideally represent no more than 35%, i.e. 65 g for a woman and 90 g for an adult man.

| Lipidic contribution: the different fatty acids | | | |
|---|---|---|---|
| **Fatty acids (FA)** | **Saturated** | **Monounsaturated** | **Polyunsaturated** |
| Short chains | C4-C8 | - | - |
| Medium chains | C10-C12 | - | - |
| Long chains | C14-C18 | - | - |
| Palmitic acid | C16 | - | - |
| Stearic acid | C18 | - | - |
| Oleic acid | - | C18 : 1 | - |
| Linoleic acid | - | - | C18: 2 n-6* |
| Alpha-linolenic acid | - | - | C18: 3 n-3* |
| Arachidonic acid | - | - | C20: 4 n-6 |
| Eicosapentaenoic acid (EPA) | - | - | C20: 5 n-3 |
| Docosahexaenoic acid (DHA) | - | - | C22: 6 n-3 |
| * GA essential | | | |

**Table 3:** the different fatty acids.

### *3.2.1. Saturated Fatty Acids*

Saturated fatty acids (SFAs) are known to be associated with increased cardiovascular risk while monounsaturated (MUFA) and polyunsaturated (PUFA) are neutral or beneficial. In reality, not all SFAs are deleterious and some are probably neutral or even beneficial (some short-chain SFAs from dairy products and C18 stearic acid). In practice, they are recognizable because they are solid at room temperature.

### *3.2.2. Oleic acid*

Oleic acid (C18: 1 n-9) is the emblematic representative of AGMIS and is associated with the Mediterranean diet.

### *3.2.3. Polyunsaturated fatty acids*

AGPIS are very different from each other. Some are indispensable like linoleic acid (C18: 2 n-6) or alpha-linoleic acid (C18: 3 n-3). These SFAs that cannot be synthesized endogenously have specific effects. Essential AGs are substrates for lipoxygenase and cyclo-oxygenase activity which produce eicosanoids with sometimes opposite functions depending on whether they come from n-3 or n-6 AGPIS.

### *3.2.4. Essential fatty acids n-3*

Essential n-3 AGs are preferentially incorporated into the brain and retina. There is enzymatic competition between the two substrates. It is desirable that the ratio of n-6 AG/n-3 AG should be in the order of 1 to 5 and not more than 10 as it is in the Western diet. The n-3 derivatives have overall favourable health effects with fibrinolytic and anti-inflammatory properties. The alpha-linolenic acid contained in abundance in rapeseed oil, walnuts and soybean and eicosapentaenoic acid (C20 n-3) (EPA) and docohexaenoic acid (C22 n-3) (DHA) provided by marine products (salmon, mackerel, sardines) are n-3 fatty acids that are the source of leukotrienes and thromboxanes with favorable effects. They appear to reduce the risk of sudden death and have a favorable effect on the processes of carcinogenesis, atherogenesis and aging. Supplementation with n-3 FAs limits competition with n-6 FAs, which are metabolized by the same enzymes that then produce compounds that are less favorable to health.

### *3.2.5. Non-essential polyunsaturated fatty acids n-6*

Non-essential n-6 AGPIS are important components of membranes. In excessive intake, n-6 PUFAs are weakened by oxidation, making the lipoparticles rich in them atherogenic.

### *3.2.6. GA essential*

Linoleic acid, a precursor of the n-6 family of GA is limited to 4% in order to respect a linoleic/alpha-linolenic intake < 5 desirable for the prevention of cardiovascular disease and inflammation. Alpha-linolenic acid, a precursor of the n-3 family of fatty acids, whose intake is set at 1%. DHA, due to the low conversion of alpha-linolenic acid whose intake should be 250mg/d.

### *3.2.7. GA not essential*

Oleic acid is the emblematic representative of AGMIS, whose share has been increased to 15 to 20%. The EPA, whose intakes were set at 250 mg/d. Other EFAs, including arachidonic acid, precursor of eicosanoid compounds, some conjugated EFAs (rumenic acid) or trans fatty acids. It is accepted that the dietary lipid ration should represent 35-40% of the energy ration with a theoretical distribution of about 12% for SFAs, 15-20% for MUFAs and 6-8% for SSPFAs. The minimum overall fat intake is set at 20 or 25g/d but the intake of n-3 SIFAs should be at least 2g/d. In situations where carbohydrate intake needs to be reduced (hypertriglycemia), the lipid portion is proportionally increased to the benefit of SIBAs (20%).

### *3.2.8. Special fats: trans fatty acids and conjugated linoleic acid (CLA)*

The metabolism and health impact of GA also depends on their spatial configuration. The majority of FAs are of isomeric "cis" configuration which generates a curvature in the spatial structure of the molecule. Some AGs are of "trans" configuration. These are most often AGMIS. Natural trans fatty acids are present in ruminant products (milk and derivatives, meat). They do not present deleterious effects in terms of cardiovascular risk. On the other hand, trans fatty acids resulting from partial catalytic hydrogenation used in the food industry (pastries, chocolate bars, etc.) are harmful and action is being taken to limit their presence in these manufactured products. The difference between natural and artificial trans fatty acids is due to their composition in isomers, which allows the former to be metabolized. Vaccenic acid can be converted into rumenic acid, a conjugate of linoleic acid (CLA) which has anticarcinogenic and protective properties against cardiovascular risk. On the other hand, the ingestion of trans fatty acids by industrial hydrogenation is associated with an increase in LDL-cholesterol and a decrease in HDL-cholesterol with an increase in cardiovascular risk of approximately 25% for a 2% increase in trans fatty acids. Trans fatty acids are also associated with certain cancers (breast, prostate). AFSSA has proposed to limit trans fatty acids to less than 2% of the energy ration. It remains to use alternatives to industrial hydrogenation. In France, the products marketed have reduced levels of trans fatty acids and consumption is well below the maximum authorized.

#### *3.2.9. Cholesterol*

Cholesterol is only found in foods of animal origin. Plant sterols and stanols interfere with the absorption of cholesterol and can reduce its level. Nevertheless, most of the circulating cholesterol comes from endogenous synthesis by the HMGCoA reductase pathway so that intracellular homeostasis is maintained. Phytosterol enrichment is associated with a decrease in cholesterol levels.

## II. Micronutrients

Various substances from the diet are necessary in small quantities (mg or µg) for the proper functioning of the processes that ensure optimal health. These are vitamins, minerals, trace elements and other microconstituent compounds, which are grouped together under the term microconstituents. Their energy intake is nil or negligible and their role is mainly qualitative. Their total or partial deficiency has repercussions of unequal severity, in principle reversible.

### 1. Organic compounds

#### 1.1. Vitamins

Vitamins include "essential" compounds that are very heterogeneous in their chemical nature and function. They are necessary for the implementation of many enzymatic processes and syntheses. Their endogenous synthesis is either absent or insufficient (vitamin D), or requires a precursor (carotenoids for vitamin A). Their deficiency can be the cause of a specific disease. Vitamins are distinguished according to their functions and according to their hydrosolubility or liposolubility. Fat-soluble vitamins (A, D, E, K) are absorbed with other fats and are stored in the body. Their accumulation in the body following an overdose can be toxic (vit A and D). Water-soluble vitamins (B-complex vitamins and vitamin C) are absorbed more easily and eliminated in the urine when their plasma concentration rises. Their storage is reduced (except for vitamin B12) and they are considered non-toxic (except perhaps vitamin B6).

### 1.2 Vitamin-like substances

Some substances have an interesting, though often ill-defined, qualitative role. Their endogenous synthesis is possible but supplementation improves certain biological processes. Choline is an amino acid considered a key constituent of sphyngomyelin and lecithin, lipids that contribute to the structure of cell membranes and lipoparticles. Its endogenous synthesis from methionine and serine in the presence of vitamin B12 and folate does not fully meet the needs. The taurine involved among other things in neuromodulation is also necessary for the production of bile salts. It plays a role in growth and it could be interesting to complete its endogenous synthesis from cysteine and methionine. Carnithine is a nitrogenous substance synthesized from lysine and methionine that is involved in transesterification reactions and in the transport of long-chain GA to the mitochondria. Its endogenous synthesis is generally insufficient in infants. Known to increase muscle performance, it is provided in abundance by meat and dairy products. Lipoic acid, fat-soluble, is a coenzyme of acetylation reactions like certain B vitamins. Coenzyme Q (ubiquinone), structurally related to vitamin E, is involved as an antioxidant and in the transfer of electrons in the mitochondria. It has potential effects on muscle work and is believed to prevent statin-induced myalgia.

## 2. Microconstituents

Bioflavonoids or polyphenols are a large number of molecules that are believed to have biological effects beneficial to health by acting on the endothelial function and having antioxidant, antithrombogenic and antitumor properties. Fruits and vegetables in general, red wine, green tea, chocolate, are particularly rich in them. Their exact role and recommended intakes are still poorly known in humans but their toxicity is not known. Among the hundreds of molecules identified, some such as resveratrol have acquired notoriety thanks to remarkable specific properties demonstrated experimentally.

## 2.1. Trace elements and minerals

These elements, whose needs are extremely variable, from trace elements (trace elements) to several hundred mg (macrominerals), have in common that they are non-organic. Their content in the tissues where they are stored is commensurate with their needs. Excessive intake or storage leads to toxicity.

### *2.1.1. Trace elements*

Trace elements are involved in many biological and enzymatic processes. The most remarkable are iron (daily needs of 20mg for a stock of 4 g), whose essential role in the transport of oxygen by hemoglobin, copper, zinc, iodine, fluorine, cobalt, selenium, manganese, molybdenum, chromium, nickel, boron is known, Each has one or more more more or less defined function(s), the deficiency of which most often leads to a characterized disease, except perhaps for arsenic and vanadium for which no deficiency has been described in humans.

### *2.1.2. Macrominerals*

**a. Calcium :**

Calcium has a considerable biological role because it is an essential component of the skeleton (1kg of calcium in the body) and is necessary for muscle contraction and many other functions including coagulation. Dairy products are the best providers of calcium.

**b. Phosphorus :**

Intimately bound to bone calcium in the form of hydroxyapatites, phosphorus also acts as a substrate for the synthesis of nucleic acids, phospholipids and in the formation of ATP. Phosphorus deficiency is rare (apart from diabetic ketoacidosis and chronic alcoholism). Protein-rich foods (meat and dairy products) are an excellent source.

**c. Magnesium :**

A component of mitochondrial integrity and a cofactor of more than 300 enzymes, magnesium is supplied by green vegetables, legumes, cereals and marine products. Reserves are of the order of 20 to 30 g for daily needs in excess of 400 mg.

**d. Potassium and sodium :**

Potassium is the main cation in the intracellular space. In addition to its functions on osmotic regulation, it plays an essential role in acidobasic regulation and membrane depolarization, particularly at the level of the cardiomyocyte. Potassium is abundant in vegetables and fruits (especially citrus fruits). Sodium is the main intracellular cation. It plays a major role in the regulation and distribution of water and maintains the transmembrane potential. Its deficiency is responsible for dehydration and functional kidney failure. Its excess can promote arterial hypertension in so-called "salt-sensitive" subjects.

### 2.2. Energy needs

The energy expenditure expressed by the kcals corresponding to the heat expenditure is the sum of several types of expenditure.

### 2.3. Rest energy expenditure

The resting energy expenditure (REE) or basal metabolic rate represents 60 to 65% of the total energy expenditure (TEE) of a sedentary subject. It depends on age, sex and lean body mass (WM) and includes a genetic determinism component (10%). It is estimated at an average of 30kcal/kg of MM and can be calculated more accurately by the Harris and Benedict formula or estimated by nutritional status (Table 5). It is increased in situations of metabolic aggression.

# Chapter 2. Food

## Introduction

Food is often the poor relation of medical knowledge in nutrition. And yet we eat food, not nutrients. This is why the dietary advice we give our patients must be expressed in food if it is to be understandable and accessible. Since dietetics is the art of reconciling the preparation of dishes and food, and the satisfaction of the dietary needs of the healthy or sick person, knowledge of food is indispensable. These dietary needs are themselves threefold: nutritional and energetic needs, psycho-affective and hedonic needs, symbolic and relational needs. However, while there are Recommended Nutrient Intakes (ANC), there are no Recommended Dietary Allowances (AAC), although the National Nutrition and Health Program (PNNS) has endeavored to establish benchmarks. But there are limits to the benchmarks that are given for a population: those related to the dietary habits given in a country or region, which explains the arbitrary nature of these benchmarks. There is no essential food. In fact, only nutrients are indispensable: this is why several means exist to reach the recommended nutritional intakes, which are almost universal. The diversity of edible foods is such that we must indeed consider those that are customary for us. Thus, the definition of food given by Jean Trémolières finds its full meaning: "a food is an edible foodstuff (...) at the same time nourishing, palatable and customary". For not only is there no essential food, but to understand the nature and complexity of food, two other axioms must be added: there is no perfect food (except mother's milk), and therefore diversity and variety are the corollary; there is no bad food (only excesses are), and therefore moderation is the corollary. Alongside this natural complexity is added another complexity, that of variations linked to production (cultivation, breeding...), but also to industrial (process) and domestic processing (preparation, cooking, cooking) which is of great importance in terms of food quality. Although there is no essential food, there are, however, due to the diversity, differences and similarities between foods. It is the notion of food classes or groups that has led to the need to group them together and therefore to a classification. However, even if any classification has a chemical (nutritional) justification, it also has an educational purpose. This is why there are several classifications, all of which are imperfect, but each may have its own interest (and educational limits) depending on the objectives. For example, nutrition education in disadvantaged or developing countries is based on a classification into three groups proposed by the WHO: building foods (those that provide protein, such as meats [and related meats] and pulses); energy foods (including starchy and fatty foods); and protective foods (rich in micronutrients such as fruits and vegetables). We had developed an educationally successful classification into nine groups:

- ✓ Fermented meat, sausages, eggs and cheeses to emphasize a community in terms of saturated fatty acids ;
- ✓ Fish and fish products to emphasize their specific content of long-chain omega 3 fatty acids ;
- ✓ Fruits and vegetables including potatoes to emphasize similarity in terms of phytomicroconstituents, micronutrients and fiber ;
- ✓ Cereals, bread and legumes to enhance their richness in vegetable proteins ;
- ✓ Milk and cheese (except fermented cheese) ;
- ✓ Vegetable fats and butter ;
- ✓ Oleaginous fruits, quite apart ;
- ✓ Sweetened food and beverages to emphasize a high content of simple sugars ;
- ✓ Alcoholic and alcoholic beverages to highlight the specificities of alcohol.

The "official" classifications (PNNS) are simple, even simplifying, but admitted with seven groups :

- ✓ Meat, and related products, eggs, fish ;
- ✓ Dairy products ;
- ✓ Fruits and vegetables ;
- ✓ Starchy foods, potatoes, bread, cereals, pulses ;
- ✓ Fatty substances ;
- ✓ Sweet products ;
- ✓ Beverages.

The classifications will thus lead, through frequency and association benchmarks, to advice in order to achieve what is called dietary balance, a word that hides both our ignorance and a great diversity of formulas to achieve the same goal. Finally, the limits of all these approaches will also be those of the food composition tables, themselves, which are only averages at a given moment, for a given food (even if there are several samples) of a particular origin (botanical for example) resulting from a specific transformation or conservation: the degree of imprecision is great, let's be aware of this. We include a microtable (Table 4), based on our pedagogical classification into nine groups. The science of food is also progressing because food is evolving. It is the state of food science in general that we want to present here. It cannot be ignored.

The micro-table (Table 4), has no other use than to give benchmarks for certain foods or types of foods chosen as examples from each of the food groups. Only energy nutrients (in grams) and energy intake, given in kilocalories, are listed per 100 grams.

The indicated deviation represents the range of values observed for the food in question (it is not the standard deviation in the statistical sense). The mean given provides a mean figure which is not necessarily the median of the deviations but the most frequent usual value (minimum and maximum).

**Group 1 :**

| | | Carbo-hydrates | Lipids | Proteins | Energy |
|---|---|---|---|---|---|
| **Meat** | **gap** | **0–1** | **2–30** | **15–30** | **80–350** |
| | average | 0,1 | 15 | 20 | 200 |
| **Fatty meat** | **gap** | **0** | **15–30** | **15–20** | **260–350** |
| | average | 0 | 20 | 18 | 250 |
| E.g.: mutton chop (raw) | | 0 | 30 | 15 | 330 |
| E.g.: pork chop (raw) | | 0 | 29 | 16 | 330 |
| **Medium-fat meat** | **gap** | **0** | **5–15** | **19–21** | **120–260** |
| | average | 0 | 10 | 20 | average |
| e.g.: beef (stew, raw) | | 0 | 11 | 20 | 180 |
| e.g.: rabbit | | 0 | 8 | 21 | 160 |
| **Lean meat** | **gap** | **0** | **2–5** | **19–22** | **80–120** |
| | average | 0 | 3 | 20 | 110 |
| Ex: chicken (without the skin, raw) | | 0 | 4 | 20 | 120 |
| Ex: horse | | 1 | 2 | 21 | 110 |
| **Delicatessen** | **gap** | **0,1–3** | **10–60** | **10–20** | **150–500** |
| | average | 1 | 30 | 15 | 350 |
| e.g.: white ham | | 1,8 | 30 | 15 | 330 |
| e.g.: garlic sausage | | 0,7 | 28 | 15 | 315 |
| e.g.: bacon | | 5 | 27 | 13 | 310 |
| e.g.: bacon<br>e.g.: Frankfurter sausage | | 1,2 | 29 | 12 | 332 |
| **Eggs** | **average** | **0,9** | **11,5** | **13** | **163** |
| **Cheeses** | **gap** | **0–0,2** | **15–35** | **16–36** | **250–465** |
| | average | 0,1 | 25 | 25 | 300 |
| **Soft cheese** | **gap** | **0–0,2** | **18–28** | **19–25** | **260–370** |
| E.g.: camembert 45%. | | 0,2 | 22 | 21 | 285 |

| | | | | | |
|---|---|---|---|---|---|
| **Hard cheese** | **gap** | **0–0,2** | **26–33** | **27–36** | **380–400** |
| E.g.: Emmental cheese | | 0,2 | 29 | 29 | 380 |
| **Blue-veined cheese** | **gap** | **0** | **29–33** | **18–20** | **340–370** |
| E.g.: blue | | 0 | 29 | 20 | |
| **Firm cheese** | **gap** | **0** | **23–33** | **20–26** | **300–405** |
| E.g.: cantal | | 0 | 30 | 23 | 370 |
| **Goat cheese** | **gap** | **0,1–1,5** | **6–39** | **4,7–20** | **80–465** |
| E.g.: fresh goat | | 1,5 | 6 | 5 | 80 |

## Group 2 :

| | | **Carbohy-drates** | **Lipids** | **Proteins** | **Energy** |
|---|---|---|---|---|---|
| **Fish** | **gap** | **0–0,5** | **0,1–22** | **14–25** | **60–220** |
| | average | 0,1 | 1 | 19 | 115 |
| **Fatty fish** | **gap** | **0–0,5** | **5–22** | **14–23** | **100–220** |
| e.g.: mackerel | | 0 | 12 | 19 | 190 |
| e.g.: sardine in canned oil | | 0 | 15 | 23 | 227 |
| **Lean fish** | **gap** | **0–0,5** | **0,1–5** | **14–23** | **60–100** |
| E.g.: cod | | 0 | 0,3 | 18 | 75 |
| E.g.: canned natural tuna | | 0,7 | 2,1 | 25 | 130 |
| **Crustaceans** | **gap** | **0,5–2** | **0,5–2** | **15–20** | **70–100** |
| | average | 0,5 | 1 | 18 | 85 |
| **Molluscs** | **gap** | **2–6** | **0,3–4** | **8–15** | **65–85** |
| | average | 3 | 1,5 | 10 | 70 |

**Group 3**

| | | Carbohy-drates | Lipids | Proteins | Energy |
|---|---|---|---|---|---|
| **Vegetables** | **gap** | **2–20** | **0,1–1** | **0,4–6** | **20–92** |
| | average | 5 | 0,4 | 2 | 40 |
| Tuber vegetablesEx : Potatoes | | 19 | 0,1 | 2 | 85 |
| Root VegetablesEx : carrot | | 9 | 0,3 | 1,2 | 40 |
| Bulb vegetablesEx : leek | | 11 | 0,3 | 2,2 | 52 |
| Leafy vegetablesEx. : Brussels sprouts | | 7 | 0,6 | 4,5 | 52 |
| Fruit vegetablesEx : pumpkin | | 6 | 0,1 | 1 | 25 |
| Vegetables-LegumesEx : peas | | 16 | 0,4 | 6 | 92 |
| Fresh mushrooms | | 5 | 0,6 | 1,9 | 35 |
| **Fresh Fruits** | **gap** | **4–22** | **0,2–0,6** | **0,2–1** | **30–85** |
| E.g.: apple | | 13,5 | 0,6 | 0,2 | 58 |
| **Dried fruits** | **gap** | **61–76** | **0,2–1,2** | **1,2–4** | **270–300** |
| E.g.: dried grapes | | 76 | 0,2 | 2,5 | 290 |
| E.g.: dried chestnut | | 73 | 5 | 7,4 | 367 |

## Group 4 :

| | | Carbohydrates | Lipids | Proteins | Energy |
|---|---|---|---|---|---|
| **Bread** | **gap** | **50–58** | **0,2–2,5** | **7–9** | **200–260** |
| White bread | | 50 | 1 | 8 | 250 |
| Wholemeal bread | | 42 | 2,5 | 9 | 220 |
| **On dry weight** | **gap** | **66–78** | **1–7** | **8–14** | **330–390** |
| | | 75 | 2 | 7,5 | 350 |
| e.g.: whole grain rice<br>e.g.: whole grain rice<br>e.g.: whole grain rice<br>e.g.: whole grain rice<br>e.g. : regular pasta | | 75 | 1,2 | 12,5 | 370 |
| **Over cooked weight** | **gap** | **20–25** | **0,1–0,5** | **2,5** | **100–120** |
| | | 23 | 0,4 | 3,4 | 111 |
| e.g.: white rice<br>e.g.<br>: regular pasta | | 24 | 0,1 | 2 | 109 |
| **Pulses**<br>**On dry weight** | **gap** | **56–61** | **1,5–5** | **20–25** | **330–340** |
| E.g.: lens | | 56 | 1,1 | 25 | 336 |
| E.g.: white bean | | 57 | 1,6 | 22 | 340 |
| **Over cooked weight** | **gap** | **20–25** | **0,5–1** | **8–12** | **110–130** |
| | | 28 | 0,6 | 12 | 168 |
| E.g.: lens<br>E.g.: white bean | | 20 | 0,6 | 8 | 118 |

## Group 5

| | Carbohy-drates | Lipids | Proteins | Energy |
|---|---|---|---|---|
| **Milk** | 4,5 | 3,5 | 3,2 | 64 |
| e.g.: whole milk e.g.: | | | | |
| semi-skimmed milk e.g. | 4,5 | 1,5 | 3,2 | 44 |
| : skimmed milk | 4,5 | 0,2 | 3,3 | 32 |
| **Yoghurt** | 5 | 3,5 | 4,2 | 71 |
| e.g.: yogurt whole milk | 5 | 1,2 | 4,3 | 50 |
| e.g.: natural yogurt e.g. : skimmed yogurt | 5,2 | 0,3 | 4,5 | 44 |
| **Fresh Cheese** | 3 | 10 | 9,6 | 140 |
| e.g.: little chipmunk 40 % e.g. : little chipmunk 20 %. | 3 | 3,5 | 9 | 79 |
| **White Cheese** | 3,6 | 3,4 | 8,5 | 78 |
| e.g.: 20% fromage blanc e.g. : 0% fromage blanc | 4 | 0,2 | 7,5 | 41 |

## Group 6 :

| | Carbohydrates | Lipids | Proteins | Energy |
|---|---|---|---|---|
| **Butter** | **0** | **82** | **0** | **750** |
| e.g.: half butter e.g.: | 0–3,5 | 41 | 2–7 | 381–411 |
| low-fat butter e.g. | 0–3,5 | 41–65 | 0,1–7 | 369–585 |
| : dairy speciality spread | 0–3,5 | 20–40 | 0,1–7 | 180–392 |
| **Fresh cream** | 4 | 33 | 2,2 | 330 |
| **Margarine** | 0,4 | 82 | 0 | 750 |
| **Oil** | 0 | 100 | 0 | 900 |
| **Lard** | 0 | 94 | 0 | 850670 |
| **Lard** | 0 | 70 10 | | |

## Group 7 :

| | | Carbohydrates | Lipids | Proteins | Energy |
|---|---|---|---|---|---|
| **Oilseed** | **gap** | **5–27** | **11–67** | **1–26** | **115–660** |
| average | 15 | 50 | 15 | 600 | |
| E.g.: roasted peanut | 18 | 50 | 26 | 585 | |
| E.g.: dry almond | 16 | 54 | 18 | 620 | |
| E.g.: dry hazelnut | 15 | 60 | 14 | 650 | |
| E.g.: dry walnut | 15 | 60 | 15 | 660 | |
| E.g.: lawyer | 6 | 20 | 2 | 167 | |
| E.g.: green olive in brine | 5 | 11 | 1 | 116 | |
| E.g.: black olive | 27 | 36 | 2 | 338 | |
| E.g.: coconut | 10 | 36 | 3,9 | 385 | |

**Group 8 :**

| | Carbohy-drates | Lipids | Proteins | Energy |
|---|---|---|---|---|
| Industrial Cake | 64 | 13 | 5 | 398 |
| Milk Chocolate | 57 | 30 | 7,2 | 535 |
| Jam | 70 | 0,1 | 0,5 | 283 |
| Ice Cream | 22 | 8,5 | 3,6 | 180 |
| Bakery croissant | 48 | 19 | 9 | 394 |
| Apple dessert (*fast food*) | 37 | 16 | 3 | 299 |
| Chocolate Eclair | 39 | 9 | 5 | 258 |
| Industrial filled wafer | 66 | 26 | 5 | 516 |
| Industrial Madeleine | 58 | 22 | 6,7 | 457 |
| Gingerbread | 72 | 3 | 8 | 354 |
| Hazelnut spread | 60 | 32 | 7 | 551 |
| Butter biscuit | 77 | 11 | 8 | 438 |
| Sablé | 69 | 19 | 7,8 | 180 |
| Sorbet | 25 | 0,5 | 0,5 | 107 |
| Apple Pie | 28 | 7,6 | 2 | 180 |
| **Sweet drinks** | /L | /L | /L | /L |
| Fruit drink | 110 | 0 | 0 | 440 |
| Cola | 110 | 0 | 0 | 440 |
| Fruit juices | 130 | 0 | 0 | 250 |
| Lemonade | 90 | 0 | 0 | 360 |
| Soda | 130 | 0 | 0 | 520 |
| Tonic | 140 | 0 | 0 | 560 |

**Group 9 :**

| **Alcoholic Beverages and Alcohol** | | **Carbohy-drates** | **Lipids** | **Proteins** | **Energy** |
|---|---|---|---|---|---|
| Beer | gap | 15–40 | 0 | 2–4 | 290–450 |
| Red wine | gap | 2 | 0 | 0 | 560–670 |
| Cider | gap | 30–50 | 0 | 0 | 400 |
| Aniseed aperitif | gap | 0 | 0 | 0 | 2520–2860 |
| Natural sweet wine (Port, Muscat) | 140 | 0 | 3 | 1600 | |

**Table 4:** The micro food table.

# 1. Meat and meat products

## 1.1 Symbolism and history

Meat is the unprocessed flesh (meat) of animals. Carnival precedes a period that used to be without meat, Lent. Man is not a carnivore, but as an omnivore he has been eating meat since ancient times. He was perhaps more of a scavenger than a hunter. Cooking has improved its quality, animal husbandry has increased its consumption, and more recently it is the stalling that has increased its consumption. Vegetarians do not eat meat products, vegans do not eat animal products. Meat has a strong symbolism of strength, virility, but also, through blood, of violence: the animal is killed to be eaten, even if it must be drained of its blood to be religiously edible (kosher or halal).

## 1.2. Meat

### *1.2.1. Nutritional characteristics*

Meat is characterized first of all by its high protein content; all meats have a content of around 20 to 25%. These proteins are said to be of good quality because they have no limiting amino acid (among the essential amino acids). On the other hand, raw meat has a lower digestibility than cooked meat. The other characteristics of meat are a lipid content, variable in quantity (2 to 25%) depending on the species, the cuts and the age of the animal: from the fattest to the least fat, mutton, lamb, pork, beef, poultry, rabbit, veal, horse, the younger animal being less fat, and proportions of fatty acids that vary according to the species with a variable and decreasing content of saturated fatty acids in the following order: mutton, lamb, beef, pork, poultry, rabbit, horse. Within the species themselves, the lipid content is very variable: for beef it decreases from the entrecote, tenderloin, flank steak, sirloin, roast (tenderloin); for pork from chops, roast, small salted meat, tenderloin. It is important to know that most of the lipids of the meat are found in the perimuscular fat, while the muscle of the meat contains very little lipids, present in what is called marbling, with a content of around 2%. The other characteristics of the meat concern the micronutrients: meat is an important source of B vitamins (except B9), iron, zinc, potassium. The iron in meat is of the heminic type, well absorbed; red meat (beef, mutton) contains more than "white" meat (poultry, veal...).

### *1.2.2. Other characteristics according to species*

A distinction must be made, especially with regard to fatty acids, between meats according to whether they are ruminant herbivores (beef, sheep), monogastric herbivores (rabbit, horse), monogastric granivores (pork, poultry), these two species being distinguished by the length of their colon: long (pork), rudimentary (poultry). The existence of the rumen leads to a natural rumen hydrogenation of unsaturated fatty acids resulting in a diversity of saturated fatty acids (about 60%), but also to the appearance of conjugated linoleic acid (CLA) or rumenic acid and natural trans fatty acids (transvacenic), while the content of unsaturated fatty acids (linoleic and oleic) decreases. Monogastric herbivores have no rumen and this leads to a naturally higher alpha-linolenic acid content, with rabbits having the highest content: this is of course related to the high alpha-linolenic acid content of alfalfa and grass. Monogastric animals have a higher content of monounsaturated fatty acids. In addition, the composition of their flesh is very strongly modulated by the nature of their diet: a diet rich (too rich) in corn, sunflower, soya, etc. will increase their linoleic acid content (omega 6 of the animal flesh), while a diet containing a little flax (for example, blue-white), or even hemp or alfalfa, which are rich in alpha-linolenic acid, will result in its incorporation into animal tissues and a reduction in saturated fatty acids and linoleic acid. Modulation of the fatty acid composition of the flesh of grass-fed ruminants (pasture, fodder) also exists, but it is weak.

### *1.2.3. Firing related changes*

Cooking meats independently of the cooking fats reduces the water content and proportionally increases the protein and fat content. But cooking can also lead to a transfer of fat from the surface to the flesh: this is the case for example of chicken, with a transfer of fat from the skin to the flesh: cooking chicken without the skin therefore strongly reduces its lipid content. The overcooking of meat (grill, barbecue) generates the appearance of numerous neoformed compounds (benzo(a)pyrene, heterocyclic amines, polycyclic aromatic hydrocarbons) probably very much involved in carcinogenesis processes. Finally, the cooking of meats can lead to the appearance of compounds derived from cholesterol, oxystéols, which are atherogenic, while dietary cholesterol is little or not associated with increased cardiovascular risk.

### *1.2.4. Offal and related products*

Giblets are the organs outside the muscles of the animal: liver, kidney (kidney), thymus (sweetbreads), heart, cheek, tongue, gizzard, tripe, brain... We often also associate other pieces such as feet (pig's feet...). Their composition is therefore very heterogeneous. Contrary to popular belief, most of them (liver, kidneys, tripe) are lean (2 to 9%). Brain, sweetbreads are rich in cholesterol; others (tripe, foot, cheek...) are rich in collagen, a protein of lower quality; the tongue is fat (15%), but low in cholesterol and iron. The liver and, to a lesser extent, the kidneys, are very rich in B group vitamins (including vitamin B9), and iron, zinc, but also cholesterol. They are also rich in purines. The way they are cooked (added fats) has a lot to do with their final lipid content.

#### *1.2.5. Delicatessen*

The delicatessen also corresponds to a heterogeneous set: ham, sausage, sausage, pâtés, blood sausage, andouillette, snout or head pâté, foie gras. Some charcuteries are lean, such as white ham, mostly rinded and defatted; head pâté is low in fat (7%) and rich in collagen. Sausage, sausage, pâtés generally have a high lipid content (20 to 40%), depending on the amount of pork or poultry fat incorporated; foie gras is very rich in lipids (50%), saturated fatty acids as evidenced by its solidification in cold, and cholesterol. Black pudding (white [25%], black pudding [37%]) and andouillette are very fatty foods; black pudding is the food richest in iron. Raw ham (30%) and smoked ham (35%) are higher in fat than cooked ham (22%) or white ham. Cold meats are an interesting source of protein, iron, zinc, vitamins of the B group. Delicatessen meats are usually high in salt, and therefore in sodium, because salt is an indispensable preservative in delicatessen meats. The addition of nitrites is also essential for food safety reasons and contributes to the nitrite intake, the major source being however the natural nitrates in vegetables and water, leading to the formation of nitrites in the stomach. However, only nitrosamines have a carcinogenic effect, their formation depending on many factors such as digestive flora, the presence of vitamin C and amines, pH... Finally, certain practices such as smoking can generate the appearance of heterocyclic amines or polycyclic aromatic hydrocarbons. But today this stage is very strongly reduced in the manufacturing process, just like salting.

## 2. Fish and fish products

Seafood is broadly defined as all food from the sea and aquatic (freshwater) environments, and could include algae. Fishery products refer to fish and seafood: crustaceans, molluscs and shellfish; in this case, snails would be included in fishery products!

### 2.1. Symbolic and historical

Fish has been eaten since the dawn of time, first raw, then cooked. Probably because this food is a source of fatty acids that are essential for the human brain, human societies developed near fishing grounds, and then civilizations settled near lakes, rivers and seas. It is a food very present in the biblical tradition and in the Christian culture: the first Christians were designated by the symbol of the fish, since ICHTUS (fish in Greek) means Jesus Christ Son of God Savior. Later its consumption was associated with Lent, a time when meat was banned. Little by little it became synonymous with "making lean" and therefore with restriction. For a long time, its symbolism remained relatively non-nutritious and therefore nutritionally negative. Only in the last few decades has modern nutrition taken advantage of it.

## 2.2. Fish: nutritional characteristics

Fish has practically the same nutritional equivalence as meat in terms of protein, both in quantitative terms, with about 20% protein, and in qualitative terms, with excellent biological value and very good protein digestibility. Only cartilaginous fish (dogfish, skate...) have proteins of lower nutritional quality. Fish differ between them mainly by their lipid content. We can classically distinguish lean fish (sole, dab, millet, plaice, haddock, pike, monkfish, hake, cod ...) with less than 3% lipids (often less than 2%), semi-fat fish (whiting, sea bass, sea bream, turbot, red mullet, trout, tuna ...) between 3 and 10% lipids, and fatty fish (sardine, herring, mackerel, salmon, halibut, sprat, toothfish, anchovies ...) with more than 10% lipids. The fatty acids in fish are characterized by a very high content (30 to 50% of lipids) of long-chain omega-3 fatty acids (especially eicosapentaenoic acid [EPA], docosahexaenoic acid [DHA]). The absolute value of omega-3 fatty acids must be related to the lipid content. There are many factors that influence the omega-3 fatty acid content of fish: the species, age of the fish, location and season of fishing, sexual maturity/laying maturity, and of course the diet of the fish. The diet and the species are the most important factors since the fatty acids incorporated in the tissues come from the diet of the fish in particular from phyto- and zooplankton (Krill) which is itself rich in omega 3 fatty acids. The latter is abundant in cold seas: it allows fish living in cold waters to keep their membranes and muscle tissues supple in order to move around in them; it is also used for metabolic adaptation. Farmed fish will have a content of omega 3 fatty acids, depending on its diet, with generally a higher content of both omega 3 and omega 6 fatty acids than wild fish. Marine carnivorous fish must obligatorily receive fish "meal" in order to be enriched in omega 3, herbivorous freshwater fish can be enriched in EPA-DHA if they receive algae (rich in EPA-DHA), but also plants rich in alpha-linolenic acid because they have an effective "desaturasic" activity. Younger fish are less fatty, lean fish have a flesh (muscles) low in lipids but a voluminous and fatty liver (e.g. cod liver), while fatty fish have a small lean liver and a flesh rich in lipids (fish oils are fish flesh oils). But when analyzing the lipid composition of fish, one should also distinguish between the relevant parts of the fish: muscles and subcutaneous parts. Other nutritional characteristics of fish are related to micronutrients. Fish is an important source of B-group vitamins, just like meat. Fish liver is excessively rich in vitamins A and D. Fatty fish are also, independently of liver, a good source of vitamin D. Fish is an important source of iodine, selenium, zinc, phosphorus... Bones contain a lot of calcium used as a dietary supplement and very bioavailable.

## 2.3 Seafood

### *2.3.1. Crustaceans*

Crustaceans, shrimp, lobsters, crayfish, crayfish, lobsters have a very low lipid content and a protein content in the order of 15 to 20%. The percentage of omega 3 fatty acid is close to that of fish but the absolute value is low. The cholesterol content is moderate.

### *2.3.2. Shellfish and molluscs*

Mussels, oysters, and other shellfish are characterized by a moderate protein content (10%) and a low lipid intake, although with a high content of omega-3 fatty acids. Squid and other octopus are low in protein and lipids. Shellfish are particularly rich in vitamins B12, iron, and trace elements: zinc, selenium, iodine.

### *2.3.3. Process modifications*

It is the changes in fatty acids that are the most sensitive. Canned fish causes only a very small loss of fatty acids after 6 to 24 months. However, there may be fatty acid transfers between the cover oil and the fish flesh. For example, canning with sunflower oil increases the omega-6 fatty acid content of the flesh, whereas canning with olive or fish oil does not have the same effect. Freezing, on the other hand, leads to a significant loss of fatty acids at 6 months, which can be as much as 50% of the initial omega 3 fatty acid content. Smoking causes little alteration in omega 3 fatty acids, as does salting or brining. However, these techniques can increase the salt and heterocyclic amine content.

# 3. Eggs

## 3.1 Symbolism and history

They get bad press, and it is believed to be because of their cholesterol content, whereas the egg is first and foremost a symbol of birth and therefore of life. Is it because egg laying naturally resumes around Easter, the day of the Resurrection, that this symbolism is strong. It seems that it already existed beforehand. In ancient times, however, eggs were rarely eaten, because eating an egg meant depriving oneself of the future bird.

### 3.2 Nutritional characteristics

The egg is a concentrate of many nutrients present in both the egg white and the yolk. Whole eggs are an excellent source of protein with 13% protein of remarkable biological value since their amino acid profile has long been considered the reference by the FAO and WHO. These proteins are present in the yolk and the white. It is, on the other hand, the yolk that contains the lipids, i.e. about 11% lipids (per 100 g of eggs). Its fatty acid composition can be partially modulated by the hen's diet, not so much the saturated fatty acids, which are constant, as the unsaturated fatty acids, especially the polyunsaturated fatty acids, whose content can be multiplied by 10 (about 15% of the fatty acids) and whose omega 6/omega 3 proportion depends on the hen's diet. Eggs can thus become an important source of omega-3 fatty acids, particularly alpha-linolenic acid (flaxseed sector), or even an important terrestrial source of EPA and DHA (micro-algae sector), without being able to exceed 4% of total fatty acids. Eggs do not contain carbohydrates. Overall it is low in calories with 160 kcalories per 100 g, well below most other terrestrial sources of animal protein. Its cholesterol content is sometimes expressed per 1 egg, per 100 g egg, per 1 yolk, per 100 g yolk, which leads to distortions. A 60 g egg provides 200mg of cholesterol, a little more than half of the daily recommendations. But it is now known that dietary cholesterol has little or no effect on plasma cholesterol and has no effect on cardiovascular risk below 14 eggs per week. Eggs are also a very significant source of phosphorus, iron, iodine, selenium, vitamins A, D and E, the levels of which are very modulated by the hen's diet and can be multiplied by 10. Eggs are also an interesting but variable source of carotenoids such as lutein and zeaxanthin, depending on the animal's diet. Eggs are the protein feed with the best quality/price ratio.

### *3.3 Process modifications*

The hen's diet can modulate the polyunsaturated fatty acid, certain micronutrients and carotenoid composition of the egg. The "free-range egg" nutritional rearing method increases the risk of dioxin contamination, the risk of salmonellosis, and increases hen mortality as well as the cost of production. The organic method does not guarantee nutritional quality: eggs from hens fed with corn will be richer (too rich) in linoleic acid. The taste of the egg is not very sensitive to the process: on the other hand, fresh eggs taste better unless they are cooked "hard". Cooking does not alter its composition, except when pan-frying (fried, omelette...) because of the possible enrichment in added fats. The color of the egg depends on the presence of lutein in the hen's food.

## 4. Milk and dairy products

They are heterogeneous both from the point of view of their composition and from the point of view of their manufacture. Certain products made from milk do not belong to dairy products in the strict sense, this is the case of (fresh) cream and butter, which belong in their uses to fats, although certain fatty white cheeses are similar to certain low-fat fresh creams. We must therefore distinguish between milk, fresh cheeses (petits suisses and related: fresh squares...) and fermented cheeses, dairy products (yoghurt and cottage cheese) and dairy desserts.

### 4.1. Symbolic and historical

The consumption of mammalian milk dates back to the Neolithic, that is to say to the period of breeding (and cultivation), at least 10000 years BC. The biblical image of milk is associated with a symbolism of wealth and happiness: milk and honey will flow in abundance. Only the association of milk and meat is a food ban of a religious nature among the Jews. The consumption of fresh and fermented cheeses in France is very old, as is the consumption of all fermented dairy products. The consumption of milk, although also ancient, took on an economic dimension after the Second World War with the promotion of milk for both economic and public health purposes. The image of milk remains attached to children and women, while cheese has an image more attached to adults and men. An ideological and irrational "anti-milk" discourse, without foundation, has been spreading like a rumor for some years.

### 4.2 Nutritional characteristics

Whole milk contains 32 ‰ protein (3.2%) 45 ‰ carbohydrates, and 35 ‰ fat. Dairy proteins are complex, including casein and whey rich in globulins (lactoglobulins...) of high biological value. Milk carbohydrates are represented by lactose, which is a disaccharide (galactose-glucose). Milk lipids are characterized by an extremely complex mixture of more than 400 different fatty acids. They can be classified biochemically as saturated (60%), monounsaturated (33%) and polyunsaturated (3%) fatty acids. Although the content of omega-3 polyunsaturated fatty acids is low, dairy products, including butter, account for 25% of the alpha-linolenic acid intake in the French diet. The omega 6/omega 3 ratio of dairy products is also remarkably low (2.4). In addition to cis unsaturated fatty acids, there are natural trans unsaturated fatty acids (transvaccinic acid) and conjugated fatty acids (CLA) in milk, which have interesting biological effects. On the quantitative level, semi-skimmed milk provides 15 ‰ or 1.5% fat, skimmed milk 0%, which leads to the exclusion of interesting and diversified fatty acids. Fresh and fermented cheeses have a real lipid content ranging from 0% (0% fromage blanc) to 33% (Roquefort). The labelling of cheeses in terms of lipids may, by regulation, only concern the percentage on the dry extract, from which, if the wet matter (water content) is known, the real content will be derived. Thus, a 20% fromage blanc containing 80% moisture actually provides 4% fat and a 40% fromage blanc provides 8%, which is about as much as a very light fresh cream. It is the same for fermented cheeses, which are all the less fattening as they are rich in water (fresh goat's cheese, Camembert...) and all the more fattening as they are dry and low in water (old Holland, Parmesan...). Some foods close to yoghurts and enriched with cream ("Greek yoghurt" ...) can be very rich in lipids. As far as carbohydrates are concerned, apart from the added sugars in dairy desserts, drinking yoghurts, puddings, desserts, ice cream and sweetened condensed milk, which contain a lot of them, dairy products contain some carbohydrates in the form of lactose. Lactose is virtually absent from fermented cheeses due to the cheese manufacturing process. Yoghurts are low in lactose, which is transformed into lactic acid by the fermentation obtained from lactic ferments. Heating yoghurts leads to suppress the activity of the lactic ferments and thus to increase the lactose content in the food and thus in the digestive tract. In case of lactose intolerance, the undigested lactose is the object of colonic fermentation and production of hydrogen that can be found in the exhaled air (Breath test). Dairy desserts, most often heated, are no longer "live" fermented products, which alters their properties. The other nutrients in dairy products are (in variable content according to the water content) minerals: calcium (120mg/100mL), phosphorus, iodine, zinc, selenium, iron, magnesium, sodium; vitamins (A, B2, B12 especially). The sodium content is higher in cow's milk than in breast milk. The sodium content of fermented cheeses is high or even very high depending on the method of production. Like many protein foods, dairy products contain functional peptides with multiple biological properties. Yoghurts, fermented milks, fermented cheeses are sources of lactic ferments called probiotics when they settle in the digestive tract. Yoghurts by definition are obtained with two strains: Lactobacillus bulgaricus and Streptococcus thermophilus. Other fermented dairy products related to yoghurts and containing other strains (Bifidobacterium longum, LC1...) exist; this is also the case for whipped milk.

**a. Manufacturing and process effects :**

The stages of milk production are cooling, mixing, skimming and then pasteurization for sale as raw milk, sterilization, and increasingly ultra-high temperature (UHT). The UHT treatment hardly alters the vitamin content, does not alter the mineral content. Skimming reduces lipids and fat-soluble vitamins but not the other constituents (minerals, calcium...). Sterilization or UHT sterilization seems to cause a small denaturation of proteins and the appearance of Maillard compounds in very small quantities. Microfiltration could be a way to better preserve the quality of proteins. Microwave heating reduces the vitamin B2 content of milk. Buttermilk is a whitish liquid resulting from the manufacture of butter during churning. The stages of cheese making are firstly coagulation or curdling of the whole milk (in the presence of rennet) resulting in the formation of a casein gel and whey or whey which is then removed; then draining which results in curdling, maturation of the curd by enzymes, the most active of which are those produced by micro-organisms and finally maturing. The whey is therefore a by-product of cheese making. Each type of fermented cheese has a different manufacturing process resulting in a multitude of cheeses. The type of germs (molds, yeasts, bacteria) will condition a multitude of enzymatic reactions leading to biochemical phenomena of water and nutrient migration, metabolic changes in the nutrients present, determining flavors, rind, mold. This flora depends on the milk, the additions, the maturing cellars. Pressing, cooking, type of fermentation, salting, maturing time, temperature of the stages will lead to modulate other technological, organoleptic and nutritional parameters and to appellations: pressed cheese, firm, soft, hard, blue-veined, bloomy rind, washed rind... In the case of fresh cheeses, the manufacturing process ends after draining. Yoghurts are obtained by inoculating the milk with another yoghurt or with lactic ferments, followed by very gentle and prolonged heating. Other related products (kefir...) are made in a similar way with strains called mothers. Dairy desserts are obtained by heating, allowing longer conservation, to be kept cool or not, depending on the products. They are enriched with sugar, caramel, cream ..., Finally the composition of milk may vary depending on the species, but moderately with some quantitative differences on lipids, proteins, calcium, vitamin C ... For example sheep's milk is richer in protein (5,3 %), in lipids (7,0 %), in calcium (188mg), mare's milk less rich in proteins (2,1 %), in lipids (1,2 %), in calcium (80mg), richer in lactose and vitamin C but free of CLA because the horse is not a ruminant! In spite of prejudices, there is no obvious nutritional advantage to consume the milk of particular species: protein allergies are often crossed.

## 5. Cereals

Cereals are basic "raw" foods such as: wheat, rice, rye, barley, corn, millet, oats, which are grasses. Wheat or wheat also includes hardy or ancient varieties such as triticale, spelt, kamut. By extension we add plants that are not grasses and therefore cereals in the botanical sense such as buckwheat, quinoa, amaranth ...

## 5.1. Symbolic and historical

These foods were introduced into the human diet with agriculture, i.e. in the Neolithic period. They allowed the development of all the great civilizations because, unlike roots, tubers and leafy vegetables, they are good sources of protein, indispensable for the survival of populations; especially when combined with other sources of vegetable proteins, especially legumes. Thus the diet of Amerindian civilizations was based on maize and red or black beans; Asian civilizations had a diet based on the combination of rice-lentil (India), rice-soya (China); the Mediterranean region for a long time had recourse to the classic wheat-chickpea or wheat-bean; in sub-Saharan Africa, it was the combination of millet or sorghum-niebeans (or cornilla, small black-headed bean). In Europe, wheat or, in more northern regions, rye, buckwheat, barley have prevailed. Wheat's letters of nobility pass through bread, a food with considerable and positive symbolism. This sanctity remained ingrained in the minds: the bread was marked with a cross before being cut at the table and one was not to throw some away. It must be said that this was the basis of food at a time when 500 to 1,000 grams a day or more were eaten, which explains why the low price of bread was for a long time a political weapon against revolts; the lack of bread was a factor in the revolt of the people. For a long time the bread of the rich was white bread, the bread of the poor was grey bread. The image of wholemeal bread was inevitably tarnished by this socio-economic aspect.

## 5.2. Cereal foods

### *5.2.1. Nutritional characteristics*

The cereal grain has three parts: the kernel, the germ and the husks. The husks are made up of several layers, the thinnest one attached to the almond is called aleurone or protein base, sometimes still a wonderful layer! The aleurone is surmounted by the outer layers or bran. The almond contains starch, a form of energy reserve in plants, and proteins in particular gluten which contains prolamines: glutenin in wheat, barley, rye, oats but also secalin in rye, hordenin in barley, avenin in oats. The germ is also a source of proteins, but especially lipids, unsaturated fatty acids, especially alpha-linolenic acid, vitamin E. The protein base contains of course proteins, of good biological value, B group vitamins, lignans (the phyto-oestrogens of cereals...). Bran is especially rich in fibres and minerals. Cereal grains have in common on the dry weight a significant protein content 10% to 12%, a high content of carbohydrates, mainly starch with a content of about 70%, a low, but variable, lipid content, the richest being oats with a dominant of polyunsaturated fatty acids. Raw, unrefined cereals are important sources of fibre (8%). These fibers are multiple: lignin (at the origin of lignans), cellulose, hemicelluloses such as beta-glucans from oats and barley, xylo-oligosaccharides (XOS) of variable molecular weight. Some of these fibers (XOS) have characterized prebiotic effects, other fibers (beta-glucans) have important metabolic effects (cholesterol-lowering effect, low glycemic index...). These values on dry weight are of course different after cooking since the weight is multiplied by two or three due to hydration. Cereals are also an interesting source of minerals, especially magnesium and to a lesser extent iron, calcium and vitamins B (except B12) and E.

#### *5.2.2. Changes related to the process*

Cooking cereals hydrates them without impoverishing them. On the other hand, preliminary refining, which leads for example from raw rice (paddy rice) to milled rice, depletes the grain in fiber and vitamin B1. The exclusive consumption of totally refined rice at the end of the 19th century led to an explosion of beriberi cases.

### 5.3. Bread

#### *5.3.1. Nutritional characteristics*

The composition of the bread, but also its metabolic effects (glycemic index), are eminently variable according to the stages of manufacture. Type 55 white bread (extraction rate = 75%) (Classic) contains 55% carbohydrates, 8% protein, 1% fat and 2g of fibre, few minerals except salt. Bread thus represents the first contribution to salt intake, with more than 25% of the salt supply in the French diet.

#### *5.3.2. Changes related to the process*

The first stage of bread making is the obtaining of the flour with the grinding leading through the bluing to a more or less refined flour. There are two ways of expressing the refining process: either it is expressed in terms of extraction rate, which is less than 100: an extraction rate of 98% (wholemeal flour) means that 98% of the wheat grain is extracted and retained, and 2% is removed (semolina rejection); or it is expressed in terms of residual ash content, corresponding to the mineral content obtained after combustion, which is strongly present in the husk: for the type thus defined, a figure of 100 may be exceeded, from the most refined 45 (totally white type 45 flour) to at least refined 150 (wholemeal type 150 flour). The fibre content increases as these values rise, but also the mineral and vitamin content with very large differences (× 2 to 5 in favour of wholemeal flour). The stages of bread making start with the mixing of the four ingredients of the bread (flour, water, yeast or leaven and salt) with the kneading, followed by the first fermentation or pointing followed by the baking (shaping), then the second fermentation or primer, followed by the incisions, finally the baking, and the sweating allowing the evaporation. Salt is not only a sapidity agent, it is also a technological agent. Kneading can be more or less fast (or slow). Rapid kneading leads to the incorporation of air into the dough which accelerates the oxidation processes and degrades some of the tocopherols and tocotrienols (vitamin E) and polyunsaturated fatty acids. It also plays a role in the degradation or conservation of aromatic molecules. Yeast fermentation leads to an alcoholic fermentation with the formation of evaporating alcohol, $H_2O$ and $CO_2$ with small cells; sourdough fermentation (from a previous dough), of the lactic type, leads to the production of lactic acid, which lowers the pH of the dough and bread with consequences on the bioavailability of minerals by allowing the activity of the phytase which degrades phytic acid: this prevents the formation of mineral phytates, insoluble complexes. The double baking surrounding a deep-freezing process (pre-baking - deep-freezing - baking) leads to an increase in the content of resistant starch, a major factor in lowering the glycemic index. Studies have shown that compliance with certain processes (double fermentation - slow kneading - long baking) such as in the so-called "French tradition" bread (decree in the Official Journal on 13-9-1993) improved the metabolic effects of the bread (glycemic index and insulin index). Conversely, bread, white bread and "American" bread have particularly high glycemic indexes. Some breads are subject to formal regulation: rye bread (with a maximum of 35% wheat flour), rye bread (at least 10% rye). Bran breads are less interesting than wholemeal breads (type 110 flour) which include the husk and germ, which is not the case with bran bread. A "good" bread can be identified by a thick uncracked crust, matt on percussion, an ochre crumb, elastic, well honeycombed, but irregularly, a little sour, not very salty.

## 5.4. Breakfast cereals

### *5.4.1. Nutritional characteristics*

Their nutritional interest is different with a wide variety of manufacturing and composition depending on the product. The less processed ones such as oat flakes, raw müeslis, whole wheat patties, corn and/or wheat petals differ little from the raw cereals. Most other breakfast cereals are fortified with sugars and become treats, or even food gadgets, whose only interest is to make children drink milk when they drink it.

#### 5.4.2. Changes related to the process

As far as breakfast cereals are concerned, raw müeslis require a lot of chewing and keep a high grain size and therefore a low glycemic index. Cereals obtained by cooking-extrusion lead to more cariogenic starches with higher Glycemic Indexes.

### 5.5. Rusks and related products

Biscottes and other "cracottes" have various recipes. Most often they are enriched with sugars, fats, proteins of lower biological value due to lysine degradation, higher contents of Maillard compounds (especially carboxymethyl-lysine [CML]). Their nutritional interest is modest.

## 6. Dried vegetables

Pulses are seeds of plants called legumes. They include lentils (blond, green, coral), beans whose varieties are very numerous (flageolets and goat, ingots, cocos, azukis, red, black, white, cowpeas ...), chickpeas, split peas, beans. Soya is also a legume but will be treated separately. Their cultivation is of particular interest because legumes enrich the soil with nitrogen through their nodules and therefore require very little nitrogen fertilizer: this is a net environmental asset.

### 6.1. Symbolic and historical

Pulses have been cultivated for a very long time. Their consumption is unfortunately declining due to their image (the meat of the poor), many prejudices, and the loss of culinary knowledge. They are associated with cereals in many traditional food modes in the great civilizations.

### 6.2. Nutritional characteristics

Pulses are primarily characterized by their high protein content (with the exception of fresh beans), which makes them one of the richest vegetable protein foods. They are also referred to as protein crops. However, pulses have a relative deficit in methionine, a sulphur amino acid, which is therefore the limiting factor, reducing the biological value and therefore the biological efficiency of legumes. Since cereals are deficient in lysine, and not in methionine, the combination of the two considerably improves their biological value (nitrogen absorbed/nitrogen retained), which is justified by the famous complementarity of cereals and legumes in a proportion that is usually 2/3-1/3. It is this complementarity which justifies the interest of the classical and empirical associations of traditional food modes : couscous, chickpeas, corn beans, rice-soja, rice-lentil, millet or rye and cowpeas (or cornips) ... In Europe the culinary associations are found in split pea soups with bread, pistou soups, lentil soups with bread, minestrones ... But many traditional dishes in France and Europe are associations of pulses, meat or delicatessen: lentils-sausage, cassoulet, leg of lamb flageolets, which should reduce the amount of meat. The protein content of pulses is about 20 to 25% of the dry weight, and about three times less after cooking given the necessary hydration of pulses. Pulses are also important sources of carbohydrates (about 60% of dry weight) and have very low glycemic indexes as one of their characteristics. As such, they are particularly interesting foods in the prevention or treatment of carbohydrate metabolism disorders. The carbohydrates of pulses also have the particularity of containing very specific carbohydrates which are fermentable galacto-oligosaccharides (GOS) such as stachyose, raffinose, verbascose ... This is the cause of digestive discomfort caused by their consumption, especially when the soaking is insufficient. However, lentils contain very little of it and are therefore particularly well tolerated. Pulses are also an important source of various fibres, which accounts for their moderately cholesterol-lowering effect, by reducing the reabsorption of bile acids: their content ranges from 11 to 25% of the dry weight. Nutritionally, pulses are also an interesting source of vitamins of the B group (except vitamin B12) including vitamin B9, minerals, magnesium, potassium, copper ... Finally, vegetables have a number of factors called anti-nutritional (antitryptic factor ...), alongside the GOS also considered as "anti-nutritional" factors. These anti-nutritional factors can alter the absorption of proteins, and in animals at high doses have negative effects on the pancreas.

### *6.2.1. Changes related to the process*

The culinary preparation of pulses requires, except for lentils, soaking. Coral lentils are shelled and cook faster. This soaking has the effect of eliminating galacto-oligosaccharides and therefore improves the tolerance of pulses. Cooking, on the other hand, eliminates other anti-nutritional factors, especially antitryptic factors. Many pulses are canned (canned), which increases their salt content. Some preparations also increase the lipid content, but, contrary to popular belief, dishes based on pulses are relatively low in lipids.

## 7. Soy

A distinction is made between green soybeans or Phaseolus mungo, a lentil-like bean whose germination results in the "soybean germ", which is related to a vegetable, and yellow soybeans, or Glycina max, a true soybean. The yellow soybean is an oil-protein legume, which, like legumes, enriches the soil with nitrogen. It is of course used in animal feed as a source of protein, just like corn meal, sunflower meal or alfalfa. But its use in human food is much older.

### 7.1. Symbolic and historical

This food has been consumed in China for several thousand years. In Asia, its consumption has taken various forms, especially after fermentation. Indeed, soy is not edible as is, and fermentation is the usual transformation in Asia leading to foods with tastes and flavors (spicy, salty) that Westerners are not used to: fermented tofu, miso, shoyu sauce and tamari. It is considered in China as the cow of the poor. Its consumption in the West is more recent, introduced more than a century ago in the United States and for about forty years in France and Europe.

### 7.2. Nutritional characteristics

Yellow soybeans are classified separately from other legumes. Firstly because it is not consumed as is, but always processed. Secondly, because its composition is significantly different. Its protein content is significantly higher, with 36% protein on total weight. Its lipid content is also much higher with 20% lipids, while its carbohydrate content is much lower. In terms of protein, its biological value is significantly higher, since there is no limiting amino acid. Nutritional factors (antitryptic factors) are destroyed during heating. Soya is characterized by a high content of polyunsaturated fatty acids, linoleic acid and alpha-linolenic acid, which are found in soya oil. Its specificity is its particularly high content of isoflavones, which are specific polyphenols and are the almost exclusive source of these. The isoflavones concerned are daidzein and genistein, which belong to the class of phytoestrogens. Their specific properties as ligands for estrogen receptors, especially beta receptors, give them estrogenic or anti-estrogenic effects depending on the endogenous hormonal environment. The AFSSA recommends not to exceed 1 mg/kg (of body weight) of phyto-oestrogens, a dose difficult to reach even with the consumption of three soya products per day. Phytoestrogens are not destroyed when heated. Part of the effects of soy isoflavones depends on the biotransformation by the colonic flora of daidzein into equol, which is its metabolite and most active compound.

#### *7.2.1. Changes related to the process*

Soya is used as a raw material in many industrial applications, from soya flour and from vegetable protein materials characterized by a very high protein content: isolates and concentrates. It is also processed into directly consumed foods. Mixed with water, crushed, it leads to soy juice, improperly called soy "milk" and whose commercial names are soy drink, soy juice, soy to drink, or according to the official name tonyu. The tonyu is then curdled to produce soy "cheese" or tofu, unfermented, which is presented as a firm white cheese. As for tonyu, it has the same uses as milk, but with a very different taste. Tofu has the same protein content as egg. Tonyu has a composition very similar to milk in terms of protein, carbohydrates and lipids. However, it is lactose-free, has a very different fatty acid composition and is cholesterol-free. Its mineral content is also very different: richer in iron, much lower in calcium (unless it is enriched) and sodium than milk. Of all the vegetable "milks" (the term milk is unsuitable) (oat, rice, almond juice...), it is the only one that can partially replace milk (except for calcium).

## 8. Vegetables and fruits

Vegetables and fruits are a group of foods characterized from a nutritional point of view by a high water content, a low fat and protein content and a richness in phytomicroconstituents. Botanically, it is a heterogeneous group of leaves, flowers, roots, tubers, fruits. Excluded are dried vegetables, rich in proteins; dried fruits rich in lipids. Floury fruits and dried fruits richer in carbohydrates can be compared to fresh fruits. From a culinary point of view, one distinguishes classically, according to the French gastronomic and culinary tradition, vegetables consumed as a starter, salty dishes, and fruits consumed as a dessert, sometimes as a starter, sweet dishes. As far as vegetables are concerned, we distinguish :

- ✓ Leafy vegetables: lettuce, spinach, sorrel, chard, parsley, celery, lamb's lettuce, purslane, red cabbage, green cabbage, white cabbage, kale;
- ✓ Flower and bud vegetables: cauliflower, Brussels sprouts, asparagus ;
- ✓ Bulbous vegetables: garlic, onion, shallots, leeks ;
- ✓ Root vegetables: carrots, turnips, radishes, celeriac, parsnips, beets, cassava, yams;
- ✓ Tuber vegetables: potatoes, Jerusalem artichokes ;
- ✓ Vegetables without chlorophyll: mushrooms ;
- ✓ Fruit vegetables: pumpkin, pumpkin, zucchini, tomatoes, eggplant, peppers, cucumbers, banana, plantain...
- ✓ Leguminous vegetables: green beans, peas, fresh broad beans ;
- ✓ Cereal vegetables: sweet corn.

### 8.1. Symbolic and historical

The symbolism of vegetables is marked by their low caloric intake, which has made them foods long considered as not very nutritious. However, their image has changed significantly with the emphasis on the nutrition-health concept, this time conveyed by their richness in micronutrients and phytomicroconstituents. The geographical origin of vegetables is most often marked by a cradle in the southern hemisphere, America, the East, Asia. Their acceptance by the Western population was sometimes long and difficult: this was the case for tomatoes and potatoes, which were long regarded with suspicion.

## 8.2. Nutritional characteristics

Their water content is high, between 75 and 95% of the total weight. They have been described as water eaters. They thus contribute a significant part of the water intake. In addition to this water intake, they also provide significant amounts of potassium with a particularly high potassium/sodium ratio, which is important for blood pressure regulation. They are also important sources of magnesium and, depending on the food, iron. In terms of macronutrients, vegetables and fruits are characterized by a low protein and lipid content, and a carbohydrate content of between 5% and 20%. Levels are 5% for leafy and flowering vegetables, 5-10% for root vegetables, 10-15% for leguminous vegetables, 15-20% for tuber vegetables. Fruits have close contents ranging from 5% for red fruits (currants, raspberries, strawberries, blackcurrants, blueberries), watermelon, melon; 5 to 10% for citrus fruits (oranges, grapefruits, lemons); 10 % for stone fruits (cherries, apricots, peaches, plums), kiwis; 12 % for fresh pineapple and pome fruits (apples, pears), 15 % for bananas and grapes, 18 % for fresh figs. Fruit carbohydrates are simple carbohydrates, except for the starchy banana: they include sucrose, free fructose and glucose. Some, such as plums (and prunes), cherry, contain sorbitol, which gives them special properties, especially laxative. The presence of fructose justifies a moderation of fruits in case of metabolic syndrome, this carbohydrate being more lipogenic and therefore hypertriglyceridemic. The carbohydrate content and their nature evolve with the maturity of the fruit. Thus during ripening, bananas become richer in saccharose to the detriment of starch. The presence of fructose and certain fibres gives the fruit a low glycemic index. Fruits and vegetables are also characterized by their high dietary fiber content. This is variable from 2 to 5% with especially a diversity in the nature of fibers: lignin, cellulose, hemicellulose, pectin ...

Fruits are an important source of pectin, especially certain red fruits (blueberries, currants ...) and pome fruits (apples, pears, quinces), which gives them special properties of viscosity ... suitable for jams! Salsify, artichokes, asparagus, Jerusalem artichokes are very rich in inulin, a fiber with characterized prebiotic properties. Of course, the nutritional characteristic, first in mind, of fruits and vegetables is their contribution in vitamins and phytomicroconstituents. Vitamin C is widely present in all fruits and vegetables, with cabbages, citrus fruits and kiwis as a priority. Sauerkraut is significantly enriched in vitamin C. Fruits and vegetables also make a significant contribution to vitamin E intakes with 20% to 30% of intakes; they are also major sources of vitamin B9 or folates, especially leafy vegetables such as spinach for example. Fruits and vegetables are high in carotenoids: beta-carotene, (carrots, pumpkin ...) lycopene (tomatoes), and carotenoids xanthophylls, lutein and zeaxanthin, (spinach, cabbage, salad ...). Carotenoids give vegetables their yellow, red, orange color, sometimes masked by green chlorophyll. Tomato concentrates and ketchup are major sources of lycopene. Vegetables and fruits are also important sources of polyphenols, of which there are more than 4000 different molecules. The best known are the quercitin of onions, naringenin and hesperitin of citrus fruits; but all fruits and vegetables contain polyphenols and in particular flavonoids: this is the case in priority for red fruits, blueberries, blackcurrants, raspberries...

Smoothies bring all the properties of fruit in a small volume and are therefore a good alternative to increase their consumption. All these molecules, vitamins, carotenoids and polyphenols, are responsible for the antioxidant power of fruits and vegetables, which can however be different from their in vivo effect, which is modulated by the sometimes very reduced bioavailability (absorption) of these microphytoconstituents, especially polyphenols. Fruits and vegetables are very limited sources of lipids. However, they are not free of them with about 1-2% lipids in most vegetables. Polyunsaturated fatty acids (linoleic acid and alpha-linolenic acid) are very widely represented: this is the case of alpha-linolenic acid in leafy vegetables such as lamb's lettuce, spinach, purslane. Their bioavailability is low but they are capable of enriching the flesh of animals that consume alpha-linolenic acid (snails, poultry). The potato has a special place because of its higher starch and therefore carbohydrate content, which often makes it classified among starchy foods, a vague term corresponding to a heterogeneous and poorly identified group. Although potatoes contain 19% carbohydrates, this is not very different from the carbohydrate content of certain roots, peas or bananas. Moreover, unlike cereals and pulses, potatoes contain very little protein and, unlike cereals, can never replace meat as a source of protein. It is an important source of vitamin C and polyphenols: it is even with apples the first source of polyphenols in the French diet. It therefore belongs to the class of vegetables although it is not included in the 400 g of fruits and vegetables of the PNNS.

### *8.2.1. Changes related to the process*

Fruits in syrup provide 2.5 times more carbohydrates and fewer vitamins than fresh fruit, as well as compotes when they are resweetened. All stages from harvest to plate are likely to alter the vitamin content of fruits and vegetables, especially those most sensitive to air, light and heat (storage, soaking, cooking), i.e., vitamin C and vitamin B9 in proportions that can be large but rarely lead to their total destruction except for prolonged cooking at high temperatures and repeated cooking. Hence the interest in preferring steamed vegetables, al dente. As far as minerals are concerned, losses are mainly due to cooking with water because of a high transfer, by osmotic exchange, of minerals, calcium and especially potassium and magnesium from the vegetables to the water. This transfer is limited by cooking with a little salt at the beginning of cooking. Cooking with steam, in soft steam, avoids these transfers. Another way to recover minerals is to drink the cooking water, i.e. to eat soup. Freezing vegetables preserves vitamins well. Canning is less favourable for heat-sensitive vitamins (vit B9 and C). But modern harvesting (short time between harvesting and storage) and processing techniques increasingly respect them. Cooking has little effect on the fibers. Mineral salts are better preserved. The addition of salt in canned or frozen ready meals is a negative element that varies according to recipes, as is the addition of fat. Freeze-drying depletes the vitamin content of vegetables, which is the case with bagged soups, except when they are supplemented. The starch in cooked potatoes, when cooled, changes and undergoes retrogradation, i.e. a structural change resulting in resistant starch. Resistant starch is poorly accessible to pancreatic enzymes, so that it reaches the colon undigested, where it behaves like a fibre with its effects on the colonic flora, but also on the glycemic index. The cold potato (in salad...) has therefore a low glycemic index, reinforced by the effect of vinegar which slows down gastric emptying and therefore also reduces the glycemic index. This phenomenon of retrogradation and appearance of resistant starch is observed for other foods (bread, cereals, rice ...) with other technological conditions.

### 8.3. Floury fruits

Floury fruits are in a class of their own, with few representatives since it is the chestnut. It is the chestnut, the fruit of the chestnut tree, which is edible although it is sometimes called "chestnut". It could be a major source of carbohydrates, starch, and therefore energy in a somewhat exceptional situation of scarcity and therefore poverty. It was then used as a "starchy food", a substitute for potatoes, and as a basic ingredient for flour and therefore for bread. Cooked chestnuts contain a majority of carbohydrates (28% cooked in water, 36% cooked over a fire) and a little protein (2%). It contains a little potassium, vitamins C and B9. Its calorie content is 132 kcal/100 g. Its composition is therefore close to that of the potato.

### 8.4. Dried fruit

Dried fruits include dried apricots, raisins, dried bananas which are the partially dehydrated fresh fruit, and figs and dates. They are very rich in carbohydrates (60 to 75%) due to this loss of water. Minerals (potassium, iron, calcium, magnesium...) are also in higher concentration. 100 g of dried apricots provide six times more calories than 100 g of fresh apricots but a dried apricot provides as many calories as a fresh apricot. These foods have the advantage of providing a lot of energy in a small volume. They are the food of long runs on foot or bicycle. Dates have a very high glycemic index.

## 9. Oil fruits and oil seeds

Fruits and oil seeds are plant foods characterized by a high lipid content. The oleaginous fruits are olive, pistachio, almond, nuts and cashews, hazelnuts, peanut, coconut, avocado. We add oil seeds: linseed, soybean, sesame, pine nuts, sunflower, squash, poppy, cotton ... The place of these foods is marginal because of their type of consumption (appetizer, snack, garnish ...). But in some cases, in some subjects, it can be important. These foods are therefore characterized by a high lipid content (11% for green olives, 20% for avocado, 60% for nuts). Several studies have shown that the accessibility of pancreatic enzymes to these lipids is limited, which greatly reduces their bioavailability. This results in an increase in steatorrhea. A majority of fatty acids is represented by unsaturated, monounsaturated (olive, almond, peanut, sesame) or polyunsaturated (walnut, sunflower ...). Walnuts are an important source of alpha-linoleic acid (12% of its lipids), flax is a major source with more than 50%, which is excessive. The coconut is very rich in saturated fatty acids and medium-chain fatty acids, which gives it a high digestibility. Oilseeds are also important but variable sources of protein (almond, peanut, soybean ...) which gives them additional interest. Oilseeds are an interesting source of fiber, phytosterols with a cholesterol-lowering effect, some isoflavones (soy), but also magnesium, calcium, vitamin B9 ... These effects explain why oilseeds are integrated into a diet called Portofolio which is likely to reduce weight, cholesterol and cardiovascular risk. However, certain transformations, consumption patterns can cause problems: this is the case of excessive consumption of peanuts (peanuts), often too salty, it is also the case of olives, green or black, very salty, the black ones being both fattier and saltier; note also the use in the United States of peanut paste as a substitute for butter and the presence in high quantities of hazelnut paste in some spreads, all known to be very caloric.

## 10. Fat or grease

Fat is a world. They are characterized by their generally high lipid content and are the exclusive representative of so-called visible fats, as opposed to hidden fats. Visible fats are both those that are added and those that can be separated and therefore easily measured.

### 10.1 Symbolism and history

Fats are therefore the foods used to add fat to the diet. They are therefore foods that are naturally sought after because they give taste and calories; they were naturally appreciated at a time when the important thing was to eat copiously. Their consumption has long been, in terms of their nature, highly dependent on local resources, firstly animal resources, derived from the fat of the animal: bacon and lard (pork), beef fat and horse fat, tallow (sheep), goose and duck fat, before being replaced more recently by fats of vegetable origin. Butter, a major fat, competed in the mid-19th century with margarine invented at the time of Napoleon III to provide the army with a cheap and available fat. This image of cheap ersatz has long been "stuck" to margarine. It must be said that all fats were used in the manufacturing process, regardless of their quality. For the past 25 years, fats other than butter, and still assimilated to "margarines", have been the object of very sophisticated technological and nutritional adaptations, which has enhanced their image; some have become true "functional" foods at the risk of being suspected of losing their neutrality.

### 10.2. Nutritional characteristics

Fats are foods added to other foods, for spreading, seasoning, cooking, frying or after cooking. They confer melting, softness and smoothness to food, although today the technology of ingredients and additives has made it possible to obtain these organoleptic characteristics without fat. They make food taste good, even better because of the lipids but also because of the aromas, this is the case of butter in particular. They are vectors of taste but also of fat-soluble vitamins. A classic distinction is made between fats of animal and vegetable origin, but a certain number of margarines and similar products can be mixed. One also distinguishes emulsions which are a water/oil mixture - and oils as well as concrete fats which do not contain water.

### 10.3 Fats of animal origin

#### *10.3.1. Fat from animal flesh*

Fat from animal flesh, lard, tallow, beef fat, horse fat, duck and goose fat (or even seal or whale fat in the past for the Eskimos) are derived from the subcutaneous fat (or under the skin), or perimuscular fat of land or sea animals. Their lipid content is variable, around 70 to 95%. It is the nature of the lipids and therefore of the fatty acids that is important to consider. Tallow, beef fat, horse fat, lard are rich in saturated fatty acids (49%). Duck and goose fats are rich in monounsaturated fatty acids (54%), which makes them famous, and the benefit is probably overestimated, even legendary. Fish fat is rich in omega 3 polyunsaturated fatty acids. These fats are rich in cholesterol. They support heating well.

### *10.3.2. Fat from milkfat*

It is fresh cream and butter. Fresh cream is obtained after skimming whole milk. Its lipid content is naturally 33%. Today, there is a range of fresh cream with a fat content of 15% - 12% - 10% or even 8%. The aroma depends partly on the fat phase and the proteins, even in small quantities: the lightening reduces it. In addition, fresh cream contains a little lactose, protein, calcium and vitamins A and D. Butter is also an emulsion, which from a regulatory point of view contains at least 82% lipids, the non-fat phase consisting of water and very little lactose, protein, vitamins A and D. Butter is obtained after churning the cream, followed by maturation and cooling. Milkfat contains 60 % saturated fatty acids, 35 % monounsaturated fatty acids, 5 % polyunsaturated fatty acids with a very favourable omega 6/omega 3 ratio of 2.4 even if the absolute values are modest. In total, milk fat contains more than 400 different fatty acids, including the butyric acid characteristic of milk fat, myristic, lauric, palmitic and stearic (saturated), but also oleic (monounsaturated), linoleic and alpha-linolenic (essential), trans- and rumenic (CLA) acids. The fatty acid composition depends to a small extent on the animal feed and the season. Of course, butter is a source of dietary cholesterol, (250mg) but also of vitamin A, of which it is the majority source and of vitamin D in small quantities. Its melting point is 35°C which gives it an ideal melting point in the mouth. Its raw and melted use does not pose any problem. When it is heated and therefore cooked oxystérols appear.

## 10.4 Vegetable fats

They include oils (and concrete fats) and emulsions (margarines and compound fats). Both oils and fats contain 100% lipids. Oils are fluid fats because their melting point is low, indicating a predominance of unsaturated fatty acids, while concrete fats have a high melting point and are therefore solid at room temperature, indicating a high content of saturated fatty acids. The higher the unsaturated fatty acid content and the higher the number of double bonds, the lower the melting point. Conversely, the higher the number of saturated bonds, the higher the melting point.

### *10.4.1. Concrete fats*

Tropical concrete fats (palm, palm kernel, copra [coconut] oil) only become fluid when the temperature rises significantly. Palm and palm kernel oil comes from the oil palm, while copra oil comes from the coconut. Their saturated fatty acid content is 92% (copra oil), 81% (palm kernel oil), 52% (palm oil) with a predominance of palmitic and stearic acid. Copra and palm kernel oils are also important sources of medium-chain triglycerides (and therefore fatty acids), which gives them a high digestibility (which is the case of Vegetaline), but also important uses in enteral and parenteral nutrition and for the dietetics of certain pathologies. Palm oil, particularly red palm oil, is an important source of beta-carotene. Palm oil is an important raw material for compound fats. Process modifications Concrete fats that are "low" in unsaturated fatty acids are very stable when heated. This is the advantage of their use.

### *10.4.2. Fluid oils*

Like concrete fats, they contain 100% lipids. Their fatty acid composition estimated by a fatty acid profile by gas chromatography allows them to be characterized very precisely. Fluid, they are all low in saturated fatty acids, the richest being olive oil with 15%, which precisely makes it less fluid. Oils can be distinguished according to fatty acids: polyunsaturated fatty acids (linoleic acid), such as sunflower, corn, soybean, grape seed, argan, safflower; monounsaturated fatty acids such as olive, peanut, rapeseed, sesame; alpha-linolenic acid such as rapeseed, walnut, wheat germ, soybean and flax. The latter, which is too rich in alpha-linolenic acid, is only allowed in "prefabricated" mixtures. Oils are also a source of vitamin E, all the more so as the content of polyunsaturated fatty acids is high, which makes them play an antioxidant role with regard to double bonds. The proportions of tocopherols and tocotrienols (eight isomers) are different from one oil to another. When they are virgin oils, they contain a significant unsaponifiable matter, i.e. a part that cannot generate soaps (fatty acids + base) and therefore does not contain lipids but contains a multitude of extremely diverse molecules: polyphenols, phytosterols, carotenoids, terpenes, hydroxytyrosol... Olive oil, argan oil, virgin walnut oil are particularly rich in this. This unsaponifiable matter gives oils particularly interesting properties in terms of health, cardiovascular protection and perhaps against cancer. This has been widely studied for olive oil and argan oil. The proper use of oils consists in favouring oils rich in alpha-linolenic acid, such as rapeseed and walnut oil for seasoning, especially as they are more sensitive to cooking (although today their heating in flat frying is allowed), and in choosing oils rich in monounsaturated fatty acids such as olive for flat frying, peanut for deep frying (even very stable vegetal).

**a. Process modifications :**

The manufacture of oils is different depending on whether they are obtained by pressure or by extraction. Oils obtained by pressing such as olive oil are cold pressed in mills (artisanal or industrial). The oil from the first pressing is the most interesting, while the pomace (the residual solid part after the first pressing) can be pressed again. The pomace from the second pressing can be subjected to solvent extraction leading to so-called lampante oils of lower quality. The flavors of olive oils are multiple and depend both on the unsaponifiable, the quality of the pressure, and the acidity index. A high acidity index, indicative of a high content of free fatty acids is not a quality criterion. The iodine index is a chemical criterion that indirectly indicates the number of double bonds. For oils obtained by extraction from seeds, the steps are quite different: after obtaining a paste from grinding + heating, the pressed crude oil is refined by solvent extraction and deodorized. This has the consequence on the one hand of eliminating nutritionally interesting substances and molecules, particularly those of unsaponifiable matter, while the decrease in tocopherols and tocotrienols is moderate, and on the other hand of eliminating undesirable compounds that are sources of trouble (waxes...) and acidity (free fatty acids), contaminants (mycotoxins, heavy metals, dioxins and other pollutants), but also of reducing the flavors. Refining induces the appearance of modified fatty acids, trans fatty acids in small quantities. The oils can then be restored to vitamin E and, more recently, fortified with vitamin D. The use of oils for cooking can also change the chemical quality of the oils with, on the one hand, the loss of normal compounds and, on the other hand, the appearance of new compounds. The loss of normal compounds concerns vitamin E whose loss is minimal in flat frying, more important in deep frying, but also polyunsaturated fatty acids whose content will decrease a little due to the appearance of products of thermo-oxidative alteration (PATO) secondary to heating; similarly, the percentage of saturated fatty acids can, in a negligible way, increase in relative value. The new chemical species that appear are oxidized fatty acids, trans isomers, cyclic monomers and polymers. This appearance depends both on the firing temperature (flat frying ≈ 140 °C), (deep frying = 180 to 240 °C), the firing time, the number of deep-frying baths, the renewal and filtering of these oils and, of course, the number of double bonds. The most resistant oils are those containing the fewest double bonds in descending order: Vegetal, palm, palm kernel, olive, peanut, rapeseed, corn, soybean, sunflower, grape seed, safflower oils...

However, these alterations occur especially under very severe conditions which, experimentally, are sometimes extreme.

### 10.5. Margarines and reduced fat and compound fats

They are emulsions. They contain margarines that traditionally do not contain butter or animal fat (less than 10% butyric acid) and have a lipid content of 82%. This type of product is disappearing, as most of them have a reduced lipid content. Many are mixed, i.e. fats composed with fats of vegetable and animal origin. The number of categories in this family of products, as well as the denominations, are incomprehensible and useless for both the consumer and the nutritionist. It is important to know that, to make a "solid" fat (in trays) from oils (fluids), several techniques exist. They all consist in increasing the percentage of saturated fatty acids around 25 to 30% to raise the melting point to make the product solid at room temperature, a necessary condition for their culinary uses:

- ✓ The first method is physicochemical, by interesterification, and consists of modifying the distribution of fatty acids on glycerol (dietary triglycerides) to increase the percentage of saturated fatty acids;
- ✓ The second method is chemical and consists in carrying out a hydrogenation aiming, by adding hydrogen, to saturate the double bonds either totally or partially, which leads in the latter case to the appearance of trans fatty acids on the remaining double bonds;
- ✓ The third method is of a physical nature, through fractionation: it consists of the physical separation of higher melting point oil fractions (e.g. palm stearin) mixed with other fluid oils. This mixture can also incorporate animal fats.

Low-end margarines, which are very solid, can still be made from mixtures with animal fats (tallow, or fully hydrogenated fish fats). Modern margarines can be very sophisticated with a very fine determination of fatty acid profiles for specific uses. Many have interesting omega 3 fatty acid contents, giving them a low omega 6/- omega 3 ratio. The addition of certain compounds, and in particular phytosterols, is proposed in certain margarines in order to obtain a marked hypocholesterolemic effect: but doubts exist as to the cardiovascular benefit of these constituents at supranutritional doses. The heating, in flat frying, of margarines rich in omega 3 does not involve an important degradation of these fatty acids.

## 11. Sweet foods

Sweet foods are an extremely heterogeneous group of foods as diverse as pastries, cakes, pies and other cookies, pastries, croissants; but also sweets, sweets, chocolates and other marshmallows, sweets; honey and jam, chestnut cream; desserts, milk desserts, custards, rice pudding and semolina cakes; ice cream and sorbets, not forgetting sugar.

## 11.1. Symbolic and historical

Sweetness is the flavor-tip, the flavor of instant pleasure, the reward. It has always been present in the history of food with honey, maple syrup and other natural sweet foods. Sugar cane and beet have made their appearance more recently. It is certain that sugar consumption has risen considerably since then, but it has remained stable for 40 years in metropolitan France, at around 30 to 35 kg/year/person. What has increased is the share of hidden sugar to the detriment of visible sugar, added, chunks or powder, itself competing with sweeteners, aspartame in mind. The symbolism of sugar is all sweetness. Isn't sugar in English sweetness, i.e. sweetness. What is sweetness if not something comforting. But the ambiguity is strong because if sugar is considered good because it is pleasant and tranquilizing at the same time, it is also considered dangerous because it is caloric. For more than a century, diatribes and righteous attacks have been poured out about the dangers of sugar, the source of all the evils of the earth. Sugar and sweetness do not leave one indifferent, yet it is indeed the innate flavor preferred by newborns.

## 11.2 Nutritional characteristics

### *11.2.1. Sugar, sweets, sweets, candies, fruit pastes*

These foods contain only simple carbohydrates, sucrose (especially), fructose, glucose. Poor or free of micronutrients, their subjects are talked about empty calories (carbohydrates). Only totally raw integral sugar is rich in minerals, trace elements and does not seem to be cariogenic. It is dark, "fat", thick, with a pleasant aroma. Its use is not very widespread.

### *11.2.2. Honey, jam, maple syrup, chestnut cream*

These foods contain mainly carbohydrates and a small amount of micronutrients and micro constituents; this is particularly the case with honey. The fructose content of honey is important, that of jams depends on the type of fruit. When the fructose intake is moderate, which is the case with ordinary consumption of honey, its metabolic effects are not deleterious. Some micro-constituents of honey seem to have specific effects.

### *11.2.3. Chocolate and chocolate spreads*

Chocolate is a fetish food that leaves no one indifferent. Brought back by the conquistadores of Mexico, where it was consumed by the Mayas, then by the Aztecs, cocoa is surrounded by legends. It was introduced to the courts of Europe in the 16th and 17th centuries. Before being a food, chocolate was considered a remedy. Its particular organoleptic qualities, sweet, fatty, melting, even unctuous, combined with the bitter aromas of cocoa have always made it a pleasant food, a symbol of food pleasure. It has positive psychotropic effects on the mood. It has not been established whether these beneficial effects are linked to the taste and the effects of the fat-sweet mixture or to psychostimulant molecules: anandamide (cannabinoid), phenylethylamine, theobromine, magnesium.... Its other microconstituents are polyphenols, whose numerous effects have been demonstrated in vivo, particularly on endothelial vasorelaxation. Chocolate is made from roasted cocoa beans, crushed, made into a paste, mixed with sugar (and milk for milk chocolate) and cocoa butter, kneaded, flavoured, refined; then it is the conching and tempering stage. Today the replacement of cocoa butter by other vegetable fats is unfortunately allowed but this must be mentioned on the product. Dark and milk chocolate provide about the same amount of calories 520 to 550kcal/100g. Their fat content is close (dark chocolate 35%, milk chocolate 32%), their carbohydrate content is 57% and 56% respectively. Only bitter chocolate with more than 85% is less rich in carbohydrates and richer in lipids. It is said to be an important source of magnesium: this is true in mg/100g but in relation to 100 kcalories its nutritional density in magnesium is modest due to its very high calorie content. The polyphenol content depends on the percentage of cocoa. Chocolate and/or hazelnut spreads have very high and very similar contents in calories, carbohydrates, lipids. They can be richer in certain micronutrients either because of enrichment or because of ingredients (milk, hazelnut...). Chocolate bars, cereal or not, caramelized or not, contain less chocolate but just as many carbohydrate and fat calories.

### *11.2.4. Ice cream and sorbets*

Ice creams are made with milk, fresh cream, sugar and fruit or fruit flavors. Sorbets contain sugar, fruit aromas and/or fruit flavours from coulis or even a little alcohol. This explains why ice creams are much richer in calories and fat, but they also contain some calcium. It is, however, a calcium source of modest interest. Their consumption has increased considerably over the past 40 years.

### *11.2.5. Cakes, pastries, pies, cookies, pastries, croissants, cakes, pastries, pies, cookies, pastries, croissants*

It is a range of foods whose consumption has exploded thanks to the ingenuity of agri-food technologies and a diversity of very many ingredients and technological agents (thickeners, emulsifiers, sugars, colorants, flavors, flavor enhancers, preservatives, antioxidants, salt ...). Like ice cream and chocolates (and related products), this category of food is carbohydrate-lipidic in variable proportions with energy intakes of 300 to 450 kcalories per 100g, up to twice the caloric content of bread, due to a high lipid intake. This lipid intake is of a very different nature in terms of quality depending on whether butter, vegetable oils, shortenings or partially hydrogenated fats are used in their manufacture: the latter two are likely to contain trans fatty acids in fairly large quantities. This is the case with low quality and low priced croissants. Several cookie manufacturers have removed trans fatty acids from their recipes. Fibre and micronutrient intakes are usually low, with poor nutrient density (mg/100kcal).

#### *11.2.6. Dairy desserts, desserts, dessert creams*

They could also be mentioned in dairy and soy products. Unlike yoghurt and related foods, heating these foods removes probiotics, and thus part of the interest. They are moderately caloric foods, rich in carbohydrates, sweet-tasting, low in fat and variable amounts of calcium.

#### *11.2.7. Rice pudding, semolina or semolina cakes*

These are foods that fall between cereal foods and dairy products. Their sweetness may make them part of the sweet foods group, but with special characteristics: complex carbohydrates (starch), low fat, variable sugar content, some calcium and fibre. They are also simple and interesting homemade recipes.

## 12. Beverages

Beverages are another heterogeneous group that should not be considered as a whole, as their composition and effects are so different. Alcoholic and alcoholic beverages in particular must be considered separately. They include water, herbal teas, tea, coffee, sweet drinks (fruit drinks, lemonades, colas, fruit juices...) soups, milk, wine, beer, cider, other alcoholic and alcoholic beverages... Their common characteristic is of course their water content, which is added to the water of the food. However, some drinks can be considered as real food; this is the case of milk, drinking yoghurts, soups, vegetable juices, fruit juices and smoothies.

### 12.1. Water

Drinking water meets well-defined criteria that give it a quality and safety that can be considered very high in France, both microbiologically and in terms of contaminants. Tap water is therefore free of pathogenic germs; it can be "hard" if its limestone content is high, which is not negative for health. The nitrate content is less than 50mg/L, the fluorine content is determined and information is disseminated to consumers. It often contains chlorine, which gives it an unpleasant taste (attenuated by leaving it in an open container in the fridge with a few mint leaves) but not dangerous. Table water is bottled ground water, like tap water. Spring water is water that has a prefectoral authorization. It comes from a spring and is not processed. Mineral water benefits from a ministerial authorization because it is considered to have properties beneficial to health. Its mineralization is extremely variable, sometimes very weakly mineralized (Volvic), moderately mineralized (Evian, Perrier), highly mineralized (Vittel, Contrex, Badoit), very highly mineralized (Vichy St Yorre, Célestin). The latter are rich in fluorine (too rich), and sodium bicarbonate, which does not have the same effects on blood pressure as sodium chloride, it seems.

## 12.2. Herbal teas, tea, coffee, chicory

### *12.2.1. Herbal teas and teas*

Herbal teas and teas are infusions. Herbal teas like tea are adorned with beneficial virtues. Apart from the added sugar and honey, their energy intake is nil, although their heat warms them up. Verbena, linden, mint, camomile, sage, lemon balm, orange blossom have sometimes soothing, sometimes digestive properties, always beneficial, according to tradition and classical herbal medicine. There are surely many molecules involved in these effects. They are both rehydrating and thirst-quenching drinks. Tea has been known since 3000 BC in China. Black tea undergoes fermentation while green tea is dried after plucking. Tea is considered exciting because of its content of caffeine, called theine, (25 to 45 mg/100mL) and other stimulating alkaloids (theobromine...). Tea is also rich in many polyphenols (catechins and tannins). This gives it antioxidant effects that would be more important for green tea. A too high consumption of tea (> 1 L/d) decreases the absorption of iron.

### *12.2.2. Café*

Coffee is the drink obtained by pouring boiling water over roasted and ground coffee beans. Arabica coffee contains 60-100 mg caffeine/100mL and Robusta coffee 120-150 mg. It provides very small amounts of minerals. It also contains molecules with an antioxidant effect.

### *12.2.3. Chicory*

Chicory is a drink still used in the North, alone or with coffee, possibly in milk. It is obtained from chicory roots, roasted, crushed. It enhances the taste of coffee. It has the reputation of a healthy food, a treasure trove of benefits. Its richness in fructo-oligosaccharides (FOS) derived from the inulin of chicory gives it laxative effects, and prebiotic properties that are the subject of multiple works.

## 12.3. Wine, beer and other alcoholic beverages

### *12.3.1. Symbolic and historical*

Most civilizations in all eras have made and consumed beverages containing alcohol. The Mediterranean culture and the Judeo-Christian civilization have always been associated with wine, which is very present in the Bible and the Scriptures. Wine is the blood of Christ after consecration. In France and in Europe, a distinction was traditionally made between the culture of wine (South - South-West - South-East) and that of beer (North and East). The styles of consumption are different, so are the consumers. But the differences are fading today. There is a gastronomy of wine and a gastronomy of beer. Alcoholic drinks are obtained either by natural fermentation of the sugars in the grape or various other vegetables (barley, apple, etc.) or by mixing alcoholic and non-alcoholic drinks (shandy, sangria, etc.). Alcoholic and non-alcoholic drinks therefore contain alcohol, which must be considered separately because of the non-essential nature of alcohol on the one hand and its potential psychotropic and toxic effects on the other.

### *12.3.2. Nutritional characteristics*

**a. Process modifications :**

The alcoholic degree of a drink is calculated by multiplying the degree by 0.8, the density of the alcohol. It is usually estimated that one glass of any alcoholic beverage provides 10g of alcohol, since the volume served is inversely proportional to the alcoholic degree. Wine is a drink obtained by alcoholic fermentation of whole or unripe grapes. For white wines, the pulp, the skin of the grapes and the woody part of the bunch are separated from the juice before fermentation. Red wine no longer contains sugar, while white wine may still contain sugar. It provides 700 kcalories per liter. It contains organic acids, some minerals and vitamins from the B group. Wine is rich in polyphenols, red wine is richer than white wine. These partially account for the effects considered beneficial for the cardiovascular system, especially when consumed in moderation. Beer is a drink obtained by fermenting a wort made from hops and barley malt (or sometimes other cereals). Beer also contains carbohydrates (35 g/L), vitamins and minerals and various molecules of the polyphenol family such as xanthohumol with complex, apparently positive properties. It provides about 440 kcalories per liter. Cider is obtained by alcoholic fermentation of apples. It contains varying amounts of residual sugar. Poor quality cider can be a source of patulin, a mycotoxin. Other alcoholic drinks are innumerable in their diversity and their recipes according to regions, countries and traditions: strong alcohols, sweet wines, liqueurs, digestives, made from all kinds of fermented plants, or with added alcohol.

### *12.3.3. Sweetened beverages*

These are also very numerous and very different. One must distinguish :

✓ Fruit juices are divided into three categories:

- Pure fruit juices without any additives or sugar additions, 100% fresh fruit juice without pasteurization and pure pasteurized juice after extraction,
- Fruit juices made from concentrated juice (100% fruit content) to which the same amount of water is added as that evaporated during concentration. The regulations authorize the addition of sugar within the limit of 1.5%, which must be mentioned, but this addition is rare,
- Nectars, fruit pulp or juice or puree (20-50%) with water and not more than 20% sugar. Preservatives and dyes are prohibited ;

✓ Fruit drinks containing a little fruit or fruit extracts and sugar ;

✓ Lemonades containing only sugar ;

✓ Colas and other drinks with plant extracts. They contain sugar and plant extracts with more or less stimulating properties;

✓ Syrups are mainly made of sugar and water is added (× 8 before use).

All these sweetened drinks contain about 100 to 150g of sugar per liter, or 20 to 30 sugar cubes. Fruit juices (pure juice and from concentrate) are important sources of vitamins C and B9 especially, and other antioxidants (carotenoids, polyphenols depending on the fruit).

### *12.3.4. Light drinks and flavoured waters*

The former contain a sweetener (aspartame), the latter fruit aromas and extracts, and do not contain sugar.

## 13. Food from organic farming

Organic agriculture is a method of cultivation or breeding subject to specific obligations requiring the absence of recourse, except in exceptional cases, to synthetic chemical products: for plants, no synthetic chemical pesticides and fertilizers, for animals, no use of antibiotics and other adjuvants, as well as to food from organic farming and without GMOs. It requires specific cultural practices: crop rotation, crop rotation, hedgerowing, plant associations, manual weeding, integrated biological control... It has been developing for about forty years with an increase in recent years under the pressure of ecological and political issues. It is a quality label with a logo corresponding to a specification requiring certification that corresponds to an obligation of means but not an obligation of results. The transition to organic farming requires a regulatory transition of three years of cultivated land. On the nutritional level, foods of plant origin from organic farming have been the subject of numerous analyses, studies and reports. These have all come to the same conclusions: a lack of consistent and relevant difference in nutritional composition. There is sometimes a discrete superiority for some micronutrients, but this difference is minimal and is not able to be significant in terms of nutrient intakes. A few studies suggest a higher polyphenol content, especially for apples, but this is not constant and seems to be mainly related to climatic conditions. The real or supposed benefits are the absence of pesticides and pesticide residues and lower nitrate levels in vegetables and cereals. However, the most objective studies do not show an excessive presence of pesticides in conventionally grown vegetables and fruits: a few exceedances of the maximum residue limit (MRL) are observed, but these almost never result in an exceedance of the acceptable daily intake (ADI), which is calculated to be at least one hundred times lower than the no-effect level. As for nitrates, they are natural compounds of all plants and their content is mainly influenced by the variety of the plant, the sunshine and the season. The main source of nitrates is water. Big consumers of nitrates are the big vegetable eaters. All studies showing a health benefit from a high consumption of vegetables and fruit have been carried out among consumers of vegetables and fruit from conventional agriculture. As far as animal products are concerned, they do not have a superior nutritional quality. The use of antibiotics is highly regulated in conventional animal husbandry. It thus appears that the essential interest of organic agriculture lies in the protection of the health of farmers directly exposed, by cutaneous and respiratory systemic way, to pesticides and in the safeguarding of the environment thanks to the agricultural practices that it requires: this in itself largely justifies its development.

## 14. Dietary supplements

Food supplements meet food regulations in terms of safety requirements. By definition they complement food, and from this point of view they correspond to nutritional supplements (vitamins, minerals ...). But food supplements also include compounds, ingredients, of different kinds: plant extracts, probiotics, functional peptides, plant "active", all naturally present in edible foodstuffs but isolated, concentrated ... Some are the subject of serious studies and are similar in their effects to drugs, which has led to talk about alicaments. This is the case of soy isoflavones to correct the effects of menopause. Others can contribute to prevention, such as dietary supplements based on omega-3 fatty acids for cardiovascular prevention, or based on xanthophyll carotenoids and omega-3 fatty acids for the prevention of AMD. Micronutrient based dietary supplements may be useful in nutritional doses, if the diet cannot meet nutritional requirements. The form of presentation of food supplements (capsules, packets, powders, liquids ...) takes them away from foods that are edible but also nutritious, palatable and customary to compare the functions of the food act which are to nourish, delight and bring together.

## 15. Functional foods

Functional foods are both a fashion and a reality. Indeed, all foods can be considered functional, that is to say that they exert effects on the body. Sweet foods provide fast glucose, fatty foods provide energy for reserves and for long-lasting physical effort of medium intensity, vegetables provide fiber for their metabolic and intestinal effects, fermented dairy products provide probiotics, meat and eggs provide protein for tissue renewal, Cholesterol for the membranes... A modern and hygienic vision of things only considers as functional those foods with effects considered today as positive, lowering cholesterol for example, or having reinforced effects compared to standard foods (new probiotics for example) or proven effects. On this last point, it is not because there is no demonstrated effect that there is no effect: tomorrow, using new markers, we will demonstrate effects that are "invisible" today due to the lack of relevant markers at nutritional doses. If a food has no positive effect, it is not a food but a poison. Certainly some foods have more effects, a richer nutritional content and therefore a promoter, but this vision of functional foods suggests the notion of good and bad foods. However, according to knowledge, what used to be good is no longer good, depending on the dose, what used to be bad becomes good. Is chocolate bad because it is fatty and sweet or functional thanks to these polyphenols? The identification of active compounds and mechanisms allows a better understanding of the functional effects of foods. This is one of the roles of nutritional science. The industry contributes positively to the improvement of foods when it better respects their intrinsic properties and compositions. On the basis of arbitrary nutrient profiles, considering that certain foods deserve functional claims raises questions.

# Chapter 3. Nutritional requirements and recommended intakes (adults, pregnant women, elderly people, athletes)

## 1. Introduction

Nutritional needs concern individuals and result from a physiological and medical approach. The recommended intakes are intended for a population and are part of a public health approach. Originally designed to solve states of deficit, they are valuable guides for formulating the nutrient content of the diet and provide benchmarks for optimal nutrition. They alone cannot describe what people eat since they ignore the symbolic value and hedonic dimension of food and food style.

Nutritional requirements express the amount of nutrient, micronutrient and energy to cover the net needs, taking into account the amount actually absorbed. Among the micronutrients, some are indispensable because, although they are necessary for the structure or proper functioning of the body, they cannot be synthesized. The endogenous synthesis of other micronutrients may not meet the increased specific needs in a particular physiological situation (pregnancy). These micronutrients are said to be "constitutionally indispensable". Minimum nutritional requirements express the amount needed to maintain major functions, possibly at the expense of reserves or other functions considered to be of low priority. They make it possible to avoid the installation of a deficiency.

### 1.1. Recommended Nutrient Intakes

The Recommended Nutrient Intakes or RNIs represent the amount of macro- and micronutrients needed to cover all physiological needs. They correspond to average nutritional requirements. They are estimated on the basis of scientific data and comply with rules set by the French Food Safety Agency (AFSSA). They are calculated in such a way that they cover the needs of 97.5% of the individuals in a population. The proposed values are not synonymous with binding standards but are benchmarks or references for the individuals in a population with the primary aim of avoiding deficiencies. NCAs are to be considered as optimal intakes for a given population. Doses higher than NCAs obtained through supplementation, supplementation or fortification of common foods lead to a zone close to pharmacological doses and may be questionable as to the actual benefits. Safety limits have been defined for some minerals and vitamins whose excess intake may not be without health consequences (Table 3-1). It is recognized that each individual in a population should approach the Nutrient Intake Levels for all macro and micronutrients, taking into account their individual characteristics, including when spontaneous intakes are above the Nutrient Intake Levels.

| **Recommended nutritional intake** | | | |
|---|---|---|---|
| A. Recommended energy intake for the population for an average level of activity | | | |
| | **Age (years)** | **Weight** | **Energy (kcal)** |
| **Men** | 20–4041–60 | 7070 | 2 7002<br>500 |
| **Women** | 20–4041–60 | 6060 | 2<br>2002 000 |
| **Seniors** | 60–75 | 36/kg body weight | |

| B. Vitamins | | | | | | | | | | |
|---|---|---|---|---|---|---|---|---|---|---|
| | **B1** | **B2** | **PP** | **B6** | **B9** | **B12** | **C** | **A** | **D** | **E** |
| | **mg** | **mg** | **mg** | **mg** | **µg** | **µg** | **mg** | **µg** | **µg** | **mg** |
| **Adult Men** | 1,3 | 1,6 | 14 | 1,8 | 330 | 3,4 | 110 | 800 | 5 | 12 |
| **Adult Women** | 1,1 | 1,5 | 11 | 1,5 | 300 | 2,4 | 110 | 600 | 5, | 12 |
| **Pregnant Women** | 1,8 | 1,6 | 16 | 2 | 400 | 2,6 | 120 | 700 | 10 | 12 |
| **Seniors** | 1,2 | 1,6 | 14 | 2,2 | 350 | 3,0 | 120 | 700 | 10–15 | 20–50 |

| C. Minerals and trace elements | | | | | | | |
|---|---|---|---|---|---|---|---|
| | **Ca** | **P** | **Mg** | **Fe** | **Zu** | **I** | **Go to** |
| | **mg** | **mg** | **mg** | **mg** | **mg** | **µg** | **µg** |
| **Adult Men** | 900 | 750 | 420 | 9 | 12 | 150 | 60 |
| **Adult Women** | 900 | 750 | 360 | 16 | 10 | 150 | 50 |
| **Pregnant Women** | 1 000 | 800 | 400 | 30 | 14 | 200 | 60 |
| **Seniors** | 1 200 | 800 | 400 | 10 | 12 | 150 | 80 |

**Table 5:** Recommended Nutrient Intakes.

Recommended Daily Allowance (RDA) :

They represent the sufficient quantity of the various nutrients needed to cover physiological needs. Evaluated on the basis of scientific data, they comply with rules set by AFSSA, and are calculated according to the average nutritional needs measured by group of individuals (e.g. children, pregnant women, elderly, etc.). However, it is unrealistic to want to apply these recommendations strictly speaking every day.

## 1.2 Definitions: food densities

### *1.2.1. Energy density*

It reflects the amount of energy provided by 100 g of food. The drier a food is (for example rusks compared to bread) or rich in lipids, the denser it is in energy. A meal providing the same amount of energy will have a variable volume depending on the energy density of the foods that make it up. Fruits and vegetables have a low energy density.

### *1.2.2. Nutritional density*

It reflects the micronutrient content per 1000 kcal. Saturated fats and simple carbohydrates (sugar) have a low nutritional density but a high energy density. Fruits and vegetables have a high nutritional density (intake of minerals, vitamins and micro components) and a low energy density. An optimal diet for health must have the highest possible nutritional density in relation to a low energy density while covering both energy and quality needs. This can be achieved by increasing the proportion of fruits and vegetables and complex, low-refined (high-fiber) carbohydrates.

## 1.3. Allegations

Functional foods either exert an effect on a body function in a health-promoting way, or they exert the same effect but beyond what a well-conducted diet could achieve. The health discourse carried by functional foods is a claim that is subject to specific regulation. A claim is defined as "any message or representation, not required by Community or national legislation, including a representation in the form of pictures, graphics or symbols, which states, suggests or implies that a food has particular characteristics. ». Nutrition claims refer to an energy intake or a significant content of a nutrient in relation to the recommended daily allowance. Health claims refer to a product's particular health benefits. Claims may be generic, listed in a consolidated list by the European Food Safety Authority, innovative, or refer to the development and growth of children or the reduction of a risk factor for disease. The latter claims will now be subject to mandatory a priori evaluation and have a scientific justification duly established by evidence-based studies conducted in humans. The aim of the regulation is to avoid consumer deception, to preserve nutritional relevance and to hunt down inaccurate claims. Finally, claims should only be attributed if the product has a satisfactory nutritional profile.

### 1.4. Nutritional profile

This new concept introduced in 2006 in the European regulation (for application in 2012) aims to temper the scope of nutritional or health claims. It implies a value judgment on foods and their place in relation to the overall nutritional balance. It is based on scientific knowledge about food and food related to health. In practice, a product can only carry a claim if it conforms to a defined nutritional profile, which implies a composition in accordance with certain rules. The definition of a nutrient profile for a food or for a food category is not easy since, beyond strictly scientific data, there are considerations of feasibility and pragmatism to be taken into account. As it stands, there are no precise criteria yet. Nevertheless, the favorable profile should be in line with the dietary balance: a product should not contain too much fat, SFAs, added sugars, trans fat and salt and will be all the more interesting as it will have a high nutritional density (fruits and vegetables) and will ensure a good coverage of NCAs (dairy products for calcium for example). Thus the attribution of a nutritional or health claim to a product will not only depend on the content of one component but on its overall composition. It will be proposed as a priority to products that are low in FFA, salt or sugars and rich in fiber and various micronutrients that confer a good nutritional density. At a later stage, the nutrient profile could be used for products without claims and appear on product labels to guide the consumer towards "healthy" foods.

## 2. Dietary balance

A concept widely used by the general public, where it is rightly associated with "eating well", balanced eating is also defended by nutritionists who, however, struggle to define it precisely. The principle is that a "balanced" distribution of nutrients not only ensures that needs are met, but also optimizes growth, overall health and physiological aging. Nutrient balance is seen as a means of preventing nutritionally deterministic chronic diseases. Implicit in its implementation is that free food choice should be guided and supported by nutrition education. In the current conditions of food availability, consumption incentives, socioeconomic constraints and psychosensory requirements, spontaneous feeding is quite far from the objectives of the nutritional balance proposed from the interpretation of epidemiological data. Today, the consumption profile too often results in an excess of energy intake with an over-representation of lipids compared to carbohydrates and an insufficient consumption of fruit and vegetables and dietary fibre.

## 2.1. Bases

The food balance advocated on scientific bases has great difficulty in integrating the diversity of behaviors and the chemical composition of food. It is formulated on the basis of a distribution of macronutrients expressed as a percentage of total energy intake.

## 2.2. For carbohydrates: 50 to 55

Low-processed cereal products and legumes are to be preferred because of their metabolic fate, their moderate energy density and their intake of protein, fiber and micronutrients. The share of so-called "simple" carbohydrates should be limited to about 10% of the energy intake.

## 2.3. For lipids: 35 to 40 % (AFSSA 2010)

As the main determinant of the energy density of food, overall lipid consumption should be limited. There is a desirable distribution according to the different types of FA with the primary objective of an adequate intake of linolenic acid and the desire to obtain a ratio of SFA n-6/n-3 close to 5. SFA, and in particular oleic acid, should be the main intake because of their relative neutrality on the incidence of cardiovascular disease while SFA should be limited. This means giving preference to vegetable oils and their variety over fats of animal origin and consuming foods with low energy density.

## 2.4. For proteins: 11 to 15%.

Although meeting protein requirements remains a major concern in human nutrition, limiting intake to 15% of the ration is justified by the limited capacity to increase body protein mass, the high metabolic cost of excess protein, potentially deleterious renal consequences and possible interactions with other nutrients that promote, for example, urinary calcium loss. The proportion of animal proteins (currently two thirds) would benefit from being reduced in favour of vegetable proteins because they are often associated with lipids and have a high energy cost with a rather negative impact on the environment. An intake of 1/3 of animal proteins would be sufficient to ensure the needs in essential amino acids, to satisfy the needs in vitamin B and to improve the bioavailability of certain micronutrients (Ca, Fe, Zn).

Dietary balance in practice :

The implementation of a balanced diet should ideally take place at each meal. In practice, the unit of time to remember is more likely to be the week, except in collective catering where the principle of a balanced meal seems desirable. Macronutrient balance should be achieved over a shorter period of time than micronutrient balance because of the difficulty of eliminating lipid stocks and the lack of storage of essential amino acids. Dietary diversification is a prerequisite for achieving balance as it facilitates the daily consumption of each of the major food classes. The adjustment of the frequency of consumption of certain foods may be necessary depending on the circumstances (festive food, fast food and snack food), insisting on the fact that no food jeopardizes the balance by itself but by the use that is made of it. The frequency of consumption of certain foods rich in fats and/or sugars can have harmful effects on the corpulence and the metabolisms by favouring, for example, the installation of insulin resistance and predisposing conditions for atherogenesis. On the contrary, the frequency of consumption of so-called "recharge foods" can be increased in physiological situations where the needs are increased. This is the case with meat for iron, zinc and proteins, marine products for iodine and selenium or dairy products for calcium and proteins. The food balance has been schematized for educational purposes in the form of a "boat" or food pyramid. The use of low-fat foods or foods intended for a particular diet or vitamin supplementation is not a priori in the design of the balanced diet. Dietary supplements should only be ingested to make up for established or assumed insufficiency of daily intakes. The practical formulation of dietary balance is not easy as it must take into account dietary habits and life constraints. It is facilitated by the structuring of meals (schedule and composition) that advocates dietary diversification and by nutritional information.

## 3. Special Nutritional Needs

### 3.1 Pregnancy

Pregnancy is a period of intense metabolic adaptation aimed at maintaining maternal homeostasis while providing for the quantitative and qualitative needs of the fetus. The nutritional status of the mother has a considerable impact on the evolution of the pregnancy and on fetal and neonatal development. Defining optimal nutritional intakes remains a difficult exercise that various expert committees have attempted to resolve through recommendations. The debate remains about the need to systematically provide nutritional supplements to a healthy pregnant woman from pre-conception to breastfeeding. Nutritional counseling provided early and throughout pregnancy increases the chances of a healthy pregnancy and fetal growth.

### 3.2. Metabolic adaptations during pregnancy

Pregnancy has a variable quantitative and qualitative cost depending on the stage. The first half of pregnancy is marked by the constitution of maternal reserves to be used during the second half during rapid fetal growth. The progressive development of hyperinsulinemia favors lipogenesis and fat storage during the first trimester. Thereafter, a state of peripheral insulin resistance induces lipolysis which, associated with an increase in hepatic glucose production, increases the availability of metabolic fuels to the fetus. Other mechanisms facilitate maternal protein storage. The net nitrogen balance becomes positive as early as the second trimester of gestation. The increased intestinal absorption capacity for various micronutrients (iron, calcium) protects the mother from excessive loss during the active transfer of amino acids, minerals and vitamins to the fetoplacental unit. These adaptations involving placental hormone secretions (placental lactogen hormone, estrogen-progestogens, etc.) require optimal availability in order to avoid inappropriate maternal weight changes, intrauterine growth retardation and increased perinatal morbidity and mortality. Fetomaternal transfers are flexible and are monitored. The iron, calcium and vitamin A status of newborns is to some extent independent of that of their mothers.

### 3.3. Energetics of Pregnancy

The energy cost of pregnancy is estimated at 75,000 kcal of which 15,000 are related to fetal growth. It is therefore necessary to adapt the energy intake to meet this additional load. The increase linked to the mother's own needs is due to the development of the fat mass, the uterus and the mammary glands and to the increase in maternal energy expenditure, which results in a significant increase in resting metabolism after the 24th week to reach 20% during the last month. This increase in energy expenditure alone represents 45% of the total theoretical cost of pregnancy. Longitudinal food consumption surveys indicate that the amount of energy consumed is much less than the theoretical cost of pregnancy. This difference reflects a physiological adaptation during pregnancy that allows for a better use of available energy. In addition, there is a wide range of variation in energy cost from one woman to another which depends in part on the level of fat mass and weight gain. In practice, the additional energy ingesta are of the order of half of the total theoretical cost. Weight gain is an essential parameter in pregnancy monitoring. Excessive weight gain is associated with the risk of macrosomia, which is responsible for obstetrical complications and neonatal morbidity. It also promotes the maintenance of excess weight during the postpartum period and may contribute to the development of obesity later on. Underweight gain is associated - especially in the third trimester - with a risk of intrauterine growth retardation, prematurity, neonatal morbidity and mortality, and a higher prevalence of metabolic syndrome and obesity in children later in life. Desirable weight gain during pregnancy depends to a large extent on the initial weight.

Preconceptional weight is another validated predictor of pregnancy outcome and child birth weight. A BMI < 20 is associated with a risk of prematurity and fetal hypoptrophy that can be corrected by sufficient weight gain during pregnancy. Conversely, moderate weight gain can reduce the consequences of maternal obesity. The critical threshold is difficult to define. Nevertheless, several recommendations, including recent North American recommendations, propose weight gain ranges that take into account preconceptional weight. These weight gain ranges should be increased in the case of twin pregnancies to take into account the additional mass of functional tissues. In short women, the targets are in the low end of the range, while in adolescents, the target is in the high end of the range. Energy intakes are provided by macronutrients in recommended proportions. Although there is no clear evidence on the effect of the fat content of the diet, and studies are more focused on fatty acid composition, the fat content of the energy intake should be around 30-35%, while the fat content of carbohydrates should be at least 55%. Protein requirements, whose purpose is less energetic than plastic and trophic, are largely covered by the usual diet and are of the order of 1 g/kg/d, i.e. 15% of the energy ration, even if the estimated requirements are conditioned by a sharp increase in protein deposition throughout pregnancy (0.7 g/d during the 1st trimester and more than 5 g/d in the 3rd trimester). The increase in energy intake is ensured by a spontaneous increase in energy nutrient intakes. It does not require dietary manipulation as long as the weight gain is in line with the objectives. Ultimately it is the carbohydrates that must cover most of the additional energy needs. Subject to an adaptation based on the initial BMI and a correct dietary balance, appetite is sufficient to guide the coverage of quantitative needs.

| Weight gain during pregnancy according to the pregestational BMI (BMI kg/m2): French and North American recommendations | | |
|---|---|---|
| IMC | Recommended weight gain in France (kg) | North American Recommendations |
| < 18,518,<br>5–2525–30><br>30 | 12,<br>5–1811,<br>5–167–11,<br>56–7 | 12–1811,<br>5–167–11,<br>55–9 |

**Table 6:** Weight gain during pregnancy according to pregtational BMI (BMI kg/m2 ): French and North American recommendations.

### 3.4. Qualitative needs

The deleterious effects of maternal nutritional deficiencies from the preconception phase are well documented. These are usually severe deficiencies observed in characterized situations: unfavorable social context, pathological context with undernutrition (digestive disorders, deficient diets), risk behaviors (alcoholism, heavy smoking, drug addiction). The minor deficiencies observed in a Western diet have little deleterious effect on fetal growth and maturation.

#### *3.4.1. Iron*

The iron content of a newborn baby is about 300 mg while the iron reserves are 500 mg in adults. The total requirement for pregnancy is estimated at 850 mg. These few figures are the key to the iron problem during pregnancy. The additional intake is 400 mg overall, which increases the NCA to 20 mg/d during pregnancy, compared to 10 mg normally, with a peak at 30 mg/d in the third trimester. In fact the fetal impact of maternal deficiency is limited because the fetus constitutes a martial stock independently of the mother's reserves. The increase in requirements is offset by the significant increase in intestinal iron absorption capacity during pregnancy, both for heme iron and mineral iron. This physiological adaptation makes it possible to meet the needs which reach 5 to 6 mg/d at the end of pregnancy. The needs are covered by a balanced diet provided that the food ration reaches at least 2,000 kcal/day. The consequences of a martial deficiency in the mother are dominated by fatigability and susceptibility to infections. The fetal consequences are minimal, and the iron status of the fetus is hardly improved by martial supplementation in the mother. Therefore, there is no justification for routine iron supplementation during pregnancy. The only indication is evidence of iron deficiency anemia or frank hypoferritinemia in the mother. A Cochrane systematic review indicates that martial supplementation justified by the actual presence of anemia was associated with a significantly lower rate of in utero or neonatal death and seizures than with routine supplementation.

#### *3.4.2. Folates*

Folic acid or vitamin B9 is involved in the metabolism of amino acids and nucleic acids. It contributes to the cellular multiplication of the embryo, the hematopoiesis of the mother and the overall growth of the fetus. The decrease in serum folate is almost constant during pregnancy. Partly due to hemodilution, it is also secondary to metabolic changes related to pregnancy. It is favoured by insufficient consumption, estimated at 300 µg/d for needs set at 400 µg/d during pregnancy. Folate deficiency has well-established repercussions on fetal development. It increases the risk of neural tube defects (encephalocele, anencephaly and spina bifida), cleft lip and extremity anomalies. Folate supplementation leads to an overall decrease in neural tube defects (about 50%). The consequences on pregnancy outcome are still being discussed, but it seems that folate deficiency could explain a number of premature births. The incidence of prematurity is reduced after high-dose supplementation (5g/d) in women whose diet is low in folate. Folate supplementation from the preconceptional phase is recommended since 1/3 of women of childbearing age have lowered red blood cell folate levels. In fact, simple dietary advice aimed at increasing fruit and vegetable consumption is insufficient to increase the intake of all women. By default, a systematic supplementation of 100 to 200 μg/d is recommended during the periconceptional period. In France, it is recommended to prescribe 5 mg of folates/d in women with an obstetrical history of neural tube closure anomalies or who are taking an anti-comial treatment. Individual supplementation of 4 mg/d should be initiated at least 1 month before conception in high-risk women and during the first trimester of pregnancy. Thereafter, supplementation of 2mg/d may be offered.

### *3.4.3 Iodine*

The importance of iodine, an unavoidable substrate for the synthesis of thyroid hormones, is due to the primordial action of thyroid hormones throughout fetal life on growth and development, particularly neurological development. Iodine requirements are increased due to a physiological increase in the renal clearance of iodine in the mother, transplacental transfer of iodine and iodine sequestration by increasing the thyroid hormone pool (increased synthesis of thyroid hormone carrier proteins and hormone catabolism by placental deiodase). Needs are estimated at 200 μg/d during pregnancy, while average intakes in France are around 80 to 100 μg/d. The result for the mother is thyroid hyperplasia partially reservable after pregnancy and relative hypothyroxinemia which can have harmful effects on the fetus which is totally dependent on the mother's thyroid hormonal intake until the 12th week. In the fetus whose thyroid is very sensitive to iodine deficiency, a goiter can occur as early as 16 weeks. Iodine deficiency with rough hypothyroidism during the first half of pregnancy would be responsible for a decrease in intellectual capacities in children at the age of 4 to 7 years because early maternal hypothyroxinemia seems to condition abnormalities in the psychomotor development of children. Neurocognitive alterations are prevented by correcting the mother's thyroid function. Marked iodine deficiency is associated with an increase in spontaneous abortions, perinatal mortality, birth hypotrophy and natal hypothyroidism. The deep iodine deficiency eradicated in Europe for a century is responsible for a severe impairment of psychomotor development leading to endemic cretinism in children. In the mother, iodine deficiency favours the appearance of goitre and hyperthyroidism. The consumption of naturally iodized foods is not sufficient to meet the needs during pregnancy. Beyond the need to correct severe deficiencies, it seems desirable to many to propose systematic iodine supplementation, as increasing the consumption of iodized salt is obviously not a realistic solution. The administration of potassium iodide tablets (100 to 200 µg/d) would make it possible to mitigate the disadvantages of an iodine deficiency. However, some placebo-controlled studies are not totally conclusive and suggest that, for this micronutrient as for others, a strong adaptability of the pregnant woman, in this case of the maternal thyroid, to the availability of substrate. In the immediate future, the application of systematic iodine supplementation in France remains subject to its approval by a consensus of experts, although it is already recommended by many professionals.

### *3.4.4 Calcium and Vitamin D*

Mineralization of the fetal skeleton theoretically increases maternal calcium requirements. This increase is globally covered by the increase in intestinal calcium absorption and its impact is limited by an adaptation of bone metabolism. Maternal bone accretion is increased during the first trimester, which makes it possible to subsequently provide the fetus with a significant calcium flow from the mobilization of these reserves. Nevertheless, repeated pregnancies and prolonged breastfeeding can lead to bone demineralization, which is only reversible with a significant vitamin and calcium intake. Calcium contributes to a healthy pregnancy regardless of its role in bone, as suggested by epidemiological studies and randomized supplementation studies carried out in at-risk populations by reducing the incidence of preeclampsia. There also appears to be a decrease in the incidence of preeclampsia but no causal link has been demonstrated. Concretely, three to four portions of milk and dairy products should be consumed per day to ensure a calcium intake of more than 1 g/d. It is interesting to note the existence of a correlation between milk calcium content and the amount of calcium intake during pregnancy. There is no need to recommend systematic calcium supplementation. A decrease in the plasma concentration of 25-hydroxy-vitamin D3 is not uncommon in pregnant women, especially when the end of pregnancy coincides with winter. The low level of sunshine and the low vitamin D content of food justify systematic supplementation measures in France. There is a relationship between maternal vitamin D deficiency and the occurrence of neonatal hypocalcemia or rickets.

The recommended intakes are 400 IU/d at the beginning of pregnancy, 1,000 IU/d during the second half of pregnancy or 200,000 IU as a single dose at the beginning of the 7th month. This last supplementation modality is essential in cases of low sun exposure.

#### *3.4.5. Fatty acids rich in omega-3*

N-3 FAs may play a role in the prevention of preeclampsia and prematurity, but a meta-analysis of supplementation studies during pregnancy showed no benefit on the risk of intrauterine growth retardation or preeclampsia with a reduced incidence of prematurity. The consumption of fish and the use of rapeseed or nut oil are recommended.

#### *3.4.6. Vitamins*

An adequate and diversified diet should meet all needs (with the exception of vitamin D). Vitamin A deserves special mention as it is involved in division and differentiation and in the regulation and expression of genes. In France, there is no need to consider supplementation, especially since excessive vitamin A intake is likely to be teratogenic. There is also no room for polyvitamin solutions.

### 3.5. Nutritional risk situations

A nutritional evaluation in search of dietary errors or deficiencies is desirable in principle in a woman expressing a desire to become pregnant or at the time of a pregnancy diagnosis because, at this age, many women consume few vegetables and dairy products and voluntarily limit their energy intake.

#### *3.5.1. Risk situations are to be known*

**a. Smoking :**

It induces a decrease in energy intake and a low consumption of fruits and vegetables.

**b. Consumption of alcohol :**

Alcohol has a direct toxic effect that causes foetal alcohol syndrome and promotes deficits in folate, vitamin A and protein. All consumption must be prohibited.

**c. Substance Abuse :**

It is often associated with a precarious socio-economic situation with a reduction in food intake, the consequences of which are all the more dreadful as smoking and alcoholism are frequently present.

**d. Special Dietary Modes :**

Vegetarianism globally satisfies all the needs on condition that it is carried out by respecting the principles of complementarity between cereals and legumes. On the other hand, the vegan diet, which excludes all products of animal origin, necessarily requires vitamin B12 supplementation.

**e. Underweight :**

As a consequence of risky behaviors, eating disorders or socio-economic problems, underweight is associated with perinatal morbimortality and fetal hypotrophy that can be prevented by nutritional management and vitamin and iron supplementation.

**f. Obesity :**

It exposes her to risks - pregnant MTA, preeclampsia, fetal macrosomia - controlled by a limitation of weight gain obtained through a moderately hypocaloric diet (no less than 1500kcal/d).

**g. Digestive disorders :**

Frequent nausea and vomiting at the end of the first trimester has little effect on final weight gain. Frequent meals with light carbohydrate snacks are fairly well tolerated. Constipation can be improved by consuming whole grains, fruits and vegetables, sufficient hydration and physical activity.

### *3.5.2 Foodborne infections*

**a. Toxoplasmosis :**

This parasitic disease is transmitted by Toxoplasma gondi and occurs after eating undercooked meat, poorly washed raw vegetables or through direct contact with cats. In non-immune pregnant women, toxoplasmosis can lead to abortion, fetal death or neurological and retinal damage. In France, serological tests for primary infection are mandatory when pregnancy is declared. In case of seronegativity, it is requested not to eat raw meat, to wash with plenty of water and to peel vegetables and fruits eaten raw and to avoid contact with cats.

**b. Listeriosis :**

Listeria monocytogenes is a cold-resistant but heat-sensitive bacterium that is transmitted through the consumption of contaminated food. Listeriosis is relatively benign for the mother and can cause spontaneous abortion or premature delivery. Preventive measures include avoiding raw milk or raw-milk cheeses, cold cuts (rillettes, pâtés, jelly products), removing cheese rinds, cooking foods of animal origin and properly reheating leftovers and cooked dishes, cleaning the refrigerator thoroughly and washing hands.

#### *3.5.3. Breastfeeding*

There are few accurate data regarding feeding during lactation. Overall the recommendations for pregnancy apply to this period and the nutritional requirements are those of pregnancy. Milk secretion is relatively unaffected by ambient nutritional conditions. Lactation is only compromised in extreme situations (famine). Part of the precursors of milk, mainly lipids, come from maternal reserves but the overall energy cost for the mother is rather modest, her resting metabolism being the same as that of a woman who was not breastfeeding. The weight gain acquired during this period and the non-return of the fat mass to its initial value are risk factors for later obesity. As the composition of the milk reflects the mother's diet, it is important that the mother provides the fatty acids essential for the development of the infant. The content of n-3 and n-6 fatty acids is particularly important in this respect. Calcium supplementation has little effect on bone density or the calcium content of milk. The calcium requirements of lactation are covered by the reversible mobilization of bone calcium and the reduction of urine loss. Lactation is therefore a situation which, like pregnancy, does not require specific supplementation but a sufficient, diversified and balanced diet.

### 3.6. Conclusion

A physiological adventure with specific nutritional needs, pregnancy is prepared from the preconception phase in order to obtain optimal nutritional status from the beginning of pregnancy. Apart from characterized deficiencies, only a few supplements can be discussed. Systematic vitamin D supplementation is recommended in areas with low levels of sunlight. Folate supplementation is desirable in the pre-conceptional phase and during the first trimester in case of risk of neural tube defects. Iron supplementation should be supported by anemia or hypoferritinemia. Iodine supplementation is recommended in areas of iodine subcarency (including France).

### 3.7. Nutrition of the elderly

Aging, an inescapable physiological phenomenon, is accompanied by significant nutritional changes that justify an adaptation of intakes independently of any pathology. Body compartments are modified with a lean mass that decreases and perivisceral adipose tissue that tends to increase at the expense of subcutaneous fat. Energy requirements decrease concomitantly with the energy ration, reflecting the reduction in the number of cells and structures involved in metabolism. The challenges are to maintain intakes adapted to needs, to maintain functionality and reserves to prevent the occurrence of disease from precipitating into malnutrition and sarcopenia.

### *3.7.1. Nutritional aging*

Eating behavior depends partly on vital momentum, socio-cultural status and food availability and partly on sensory status. The decrease in sensory perception with age

- odor perception and discrimination, taste perception threshold and ability to analyze flavors.

- results in a decrease in appetite which can be circumvented by adjusting the recipes. In addition, there is a dysregulation of appetite which prevents the adaptation of the contributions to the variations of the needs. The capacity for food adaptation progressively diminishes with age. In the short term, there is a delay in the establishment of early satiation mechanisms. In the long term, elderly subjects respond inadequately to a period of caloric restriction or excessive feeding due to an alteration in the self-regulation of food intake.

The process of digestion is weakened by a decrease in gastric emptying and by a decrease in digestive secretions with a lesser performance of absorption which is delayed.

Metabolisms are altered. A state of insulin resistance, mainly related to changes in body patterns, gradually develops. Protein anabolism is reduced due to less stimulation in the muscles as a consequence of sarcopenia. The decrease in energy expenditure is correlated with the decrease in muscle mass. Paradoxically, with comparable effort, the energy needs of the elderly subject are higher than those of the young adult (+ 20%). Contrary to a widespread opinion, energy requirements decrease little in the elderly subject but their net increase in the case of physical activity or metabolic aggression is not always sufficiently compensated by spontaneous intake, which exposes them to a risk of malnutrition.

### *3.7.2. Nutritional fragility*

Apart from the harmful effects of chronic or acute intercurrent pathologies, the nutritional status of the elderly is threatened by numerous circumstances that modern life sometimes accentuates. Isolation, insufficient income, insufficient household equipment, supply difficulties, physical handicap, polymedication, and reduced chewing capacity are classic causes, the consequences of which associations, home meal delivery and improved care provision are striving to mitigate. At this age, unjustified beliefs and food taboos develop and a propensity to follow diets, one of the most inadequate of which is the "salt-free" diet. The reduced sensation of thirst contributes to nutritional fragility and promotes anorexia.

### 3.7.3. Nutritional requirements

The energy needs of healthy seniors are comparable to those of adults. Just because a person is getting older does not mean that he or she has to eat less. The recommended nutritional intakes are estimated at 36 kcal/kg/day and 1g/kg/day of protein in the healthy elderly. This is the price to be paid to safeguard immune competence and muscle mass. Protein and micronutrient supplementation may be useful in the frailest individuals, but there are no conclusive intervention studies. As it stands, care should be taken to ensure that these frail people have intakes equivalent to healthy older people.

### 3.7.4. Nutritional target

Beyond the quantitative and qualitative aspect of nutrients, food is a social, convivial and pleasurable act. Environmental conditions are therefore essential. The hunt for taboos, taboos and food monotony comes from enlightened nutritional information. The fight against sarcopenia combines a sufficient energy and protein intake and, above all, the maintenance of regular and adapted physical activity. The prevention of osteopenia and osteoporosis depends on an energy, protein, calcium and vitamin D intake (if necessary through supplementation) and, above all, on significant physical activity. An annual dietary and nutritional check-up should be carried out by the attending physician in order to detect dietary errors and to remedy any impediments to a satisfactory diet (dental and body hygiene, the fight against sedentary life and against possible excesses... alcohol, tobacco, self-medication). A prevention of undernutrition must be instituted in case of acute incidental illness or surgical intervention to counter the deleterious effects of hypercatabolism which accompanies any aggression and exposes to an increased risk of undernutrition in the hospital environment. Increased energy and protein intake may require the prescription of oral nutritional supplements in these subjects who are willingly anorexic and adynamic during the acute phase of an illness.

### 3.7.5. Nutrition and longevity

Early experimental data showed that an energy reduction, not leading to deficiency, started immediately after weaning improved life expectancy in all animal species. The higher prevalence of centenarians in the Okinawa archipelago was explained by the consumption of a frugal and diversified diet providing less than 1,500kcal/day. The Mediterranean diet advocated for the prevention of cardiovascular and overload diseases is probably the one in our regions most likely to increase longevity when associated with significant physical activity.

### *3.7.6. Nutritional Factors and Cognitive Impairments*

Some epidemiological studies have suggested links between nutrition in the elderly and the development of cognitive disorders. High energy intake, overweight, hypercholesterolemia and hyperhomocysteinemia have been associated with an increased risk of cognitive impairment or Alzheimer's disease. On the other hand, moderate and regular consumption of alcohol, consumption of fish, polyunsaturated fatty acids and antioxidants are associated with a lower incidence. Intervention studies have not been able to confirm these data based on statistical links, but it is accepted that an optimal lifestyle such as the Mediterranean diet is protective. It is associated with less rapid cognitive decline and a lower risk of conversion of moderate cognitive impairment to Alzheimer's disease but there is no evidence of causality and there are many confounding factors.

### *3.7.7. Summary*

The recommendations given to the general population regarding diet and physical activity remain valid for the elderly with some adaptations. There is no specific diet for the elderly, but rather a nutritional aging exposing to the risk of undernutrition and requiring adapted and personalized measures taking into account the psychosocial and environmental decrease.

## 3.8. Diet of the sportsman

Often perceived by the athlete as a performance factor, nutrition is above all an important component of a healthy lifestyle, ensuring the best possible form. The knowledge of simple dietary rules aimed at avoiding errors and poor performance does not mean that the athlete's diet is specific except in exceptional situations where it must satisfy particular physiological needs. The usual practice of sport does not impose any fundamental modification to the principles of balanced diet. Globally, energy intakes adapted to the level of expenditure are provided by a balanced and diversified diet in the form of meals and snacks with an increase in carbohydrate intake. Protein, vitamin and mineral requirements are met by increasing energy intake. The use of food supplements or specific foods is not justified other than by beliefs, the search for psychological effects (magic potion) or because they contribute to an alternative to doping. The aim is to obtain an optimal level of fitness and to promote performance through optimal weight, to avoid any deficiency, to ensure substrate reserves as well as an anticipatory remineralization and hydration during long-duration events and to optimize the restoration of reserves after the event.

### *3.8.1. Energy requirements*

Physical activity, of which sport is a sub-category, leads to an increase in energy expenditure related to muscle contractions. Energy comes from the transformation of chemical energy into mechanical energy and heat by hydrolysis of ATP, each molecule providing 7.3 kcal. The muscle contains 3 g of ATP/kg which corresponds to 50 kcal. This immediately available source does not require oxygen but only allows a very brief effort of the order of three seconds at maximum power. Continued exercise depends on the resynthesis capacity of ATP, which is low in muscle concentrations. The anaerobic metabolism involved during intense and very brief efforts is insignificant from a nutritional point of view. It is first lactic thanks to a transfer of creatine phosphate from muscle phosphate to form ATP, then lactic by the transformation of muscle glycogen into lactates whose accumulation leads to a decrease in pH and muscle cramps. The latter starts as soon as the effort exceeds 10 to 15 seconds and produces 120 to 130 kcal/min. Its capacity depends on the stocks of muscle glycogen and the accumulation of lactates and acidosis which blocks the reaction after the use of about 60 g of glycogen. The anaerobic lactic system is improved by training and allows to expect a production of 350 to 400kcal at maximum power. Beyond a few minutes, the aerobic metabolism provides the necessary energy through the oxidation of glucose and fatty acids in the mitochondrial respiratory chain and, to a lesser degree, through the oxidation of amino acids (5 to 15%). It represents the main source of ATP and ensures the supply of energy under basal conditions and as soon as the effort is prolonged. The maximum oxygen consumption (VO2 max), easy to measure during exercise, reflects the maximum power of the aerobic metabolism (directly dependent on food intake). The capacity of the aerobic system depends on VO2 max and therefore on training and energy stores.

### *3.8.2. Nutrients*

By consensus, it is accepted that physically active people do not need additional nutritional supplements beyond those provided by a balanced diet. A few nuances complete this assertion, especially for sports with nutritional risk.

#### a. Carbohydrates :

Carbohydrates are the energy substrate of the effort. They help replace muscle glycogen and keep blood sugar levels within norms, stored as glycogen. They constitute at least 55% of the energy ration and up to 70% in endurance sports. Muscular glycogen is used locally without the possibility of resynthesis during the effort. Its exhaustion induces local fatigue. The hepatic glycogen, totally usable, aims to maintain the glycemia. Its exhaustion is at the origin of hypoglycemia. The totality of glycogen stocks ensures three hours of effort at 70% of VO2 max. It is customary to increase the proportion of carbohydrates during the three days preceding a competition in order to obtain a maximum level of muscle glycogen during the event. These are low or intermediate glycemic index carbohydrates, except during long-duration events where high index carbohydrates are used (glucose, sucrose, maltodextrin, polyglucose or fructose solution).

**b. Lipids :**

Recommended intakes are in the order of 20 to 30% of the ration in endurance sports. Their consumption is not recommended just before, during and after a competition. They contribute significantly to energy intake during the training phase and provide essential fatty acids and fat-soluble vitamins. Lipid oxidation covers two thirds of basic needs. At the beginning of exercise, it helps to save glycogen reserves, then it takes an increasingly important part and gradually replaces carbohydrate substrates but does not allow such intense effort.

**c. Protein requirements :**

Sustained physical activity causes protein loss due to muscle micro-injuries and oxidation during long-term training or competitions. It is therefore necessary to promote muscle protein synthesis (accretion) to obtain a positive protein balance after exercise. If the recommended dietary protein intake meets the needs of a leisure or occasional physical activity, preference should be given to proteins of high biological value and NCAs should be increased (× 1.5 to 2) in elite athletes or bodybuilders by using dietary proteins or protein powders. The latter can be administered in spaced doses for "slow" proteins (by analogy with slow and fast carbohydrates) such as casein or in close doses for "fast" proteins such as whey.

**d. Hydroelectrolytic needs :**

The early replacement of liquid losses during the effort helps to preserve performance. Thirst is a poor warning criterion for dehydration and occurs too late. Correct hydration helps to maintain an almost constant weight during a sporting event. It prevents the occurrence of heat stroke and reduces the risk of muscle cramps. Before an event, it is advisable to ingest a drink in portions (300-500 mL in 2 hours), especially if the weather conditions predispose to sweating. During the test, the amount of water is adjusted to the expected water loss and can be as much as 1.5 L/h when activity exceeds 1 hour. Beyond a duration of 3 hours, the needs are of the order of 0.5 L to 1 L/h. After exercise, it is advisable to quickly restore the water deficit by providing a quantity of drinks to compensate for 150% of the weight loss observed during the test. The addition of NaCl to the drink limits the decrease in plasma volume during exercise and prevents the occurrence of hyponatremia in long-term tests. Sodium is the most important electrolyte due to its excretion in sweat (NaCl 20 to 60mmol/L). A supplement of 1 to 1.5g of NaCl per liter of drink is advised during this type of test and during the recovery phase by avoiding the salt intake in the form of dragees or capsules.

#### e. Micronutrients :

Many vitamins and trace elements are involved in energy metabolism. Their needs increase in proportion to the effort, especially for B vitamins. Micronutrients are also involved in cell protection and the fight against oxidative stress. Their coverage requires a varied diet, or even supplementation in vitamin B, iron, calcium and magnesium depending on the type of diet. A healthy lifestyle partly prevents vitamin deficiencies: smoking cessation for vitamin C, low consumption of coffee and tea (absorption of vitamins A, B, and B12 and iron) and abstraction of alcoholic beverages (vitamins B and C). The consumption of sufficient quantities of dairy products, pulses and green vegetables, cereals, meat and possibly exercise drinks avoids any deficiency.

### 3.8.3. Sports drinks

Commercially available exercise energy drinks (energy cocktails, vitamins and electrolyte-enriched drinks) can be used during the waiting phase, during the event and during recovery. They can also be used at half-time during team sports. A distinction must be made between energy and energy drinks. The first are effort drinks and have a regulated composition and meet a formulation corresponding to the specific needs of the sportsman whose energy expenditure is intense. They must not be acidic, carbonated or too sweet. Energy drinks are the result of a marketing concept and contain stimulating, even exciting substances (caffeine, guarana, taurine, arginine, ginseng, vitamins, glucuronate, etc.). These drinks can cause a delay in the perception of the fatigue threshold and promote rhythm disorders.

### 3.8.4. Dietary supplements

The sports world is directly concerned by the problem of food supplements. The use of products added to the recommended diet is strongly supported by advertising and by individual experiences of undocumented consumption.

Decrease in fatigue, feeling of well-being, increase in performance are very often only the translation of a placebo effect. The objective analysis of the potential effects of supplements is most often disappointing. Creatine supplementation, widely practiced in all sports environments, causes only small physiological and performance changes, at the limit of significance, under specific and standardized evaluation conditions. In any case, the effects attributed to the massive use of creatinine on muscle mass and strength, on the decrease in the production of lactates and the increase in the energy available for short and intense efforts, are largely overrated. There is no assurance that the consumption of creatine in massive doses is free of health risks and several gastrointestinal, cardiovascular or muscular problems have been attributed to heavy consumption. It is likely that dietary and lifestyle changes are as effective as dietary supplements. For the AFSSA, the consumption of food supplements is justified only by the need to supplement intakes deemed insufficient by a doctor or dietician. The main supplements used are :

✓ Creatinine: The anaerobic ATP system, creatinine phosphate, which provides the phosphorus needed to replenish ATP, depends mostly on the muscle mass developed through training and little on diet. The goal of creatinine administration is to increase performance by increasing muscle energy reserves. It increases by 20% the reserve of muscular CRP that can be used in intense and short sports (sprint) but is forbidden in France. The tropical role on protein synthesis is not found in vivo;

✓ Carnitine: it participates in the transport of long-chain GA in the mitochondria and is believed to contribute to the saving of muscle glycogen by facilitating lipid oxidation. It would also increase the efficiency of ATP production through the Krebs cycle by raising the level of coenzyme A. In fact, the use of fatty acids during exercise is not determined by the availability of carnitine ;

✓ Polyunsaturated fatty acids: they improve tolerance to hypoxia by increasing the deformation of red blood cells;

✓ Caffeine: banned by the Olympic Committee, this sympathic stimulant can increase lipid oxidation and save muscle glycogen. It can increase endurance;

✓ Bicarbonates: supposed to buffer the accumulation of lactates in the muscle, this substance can increase performance during intense efforts of limited duration and promotes recovery after short and repeated intense efforts;

✓ Amino acids (AA) or derivatives such as beta-hydroxybetamethylbutyrate are not formally established. Branched chain AA supplementation has controversial effects, increases ammonia and probably does not improve performance or recovery.

### *3.8.5. Main benchmarks of the sportsman's diet*

**a. Training period :**

Three meals a day and one or two snacks if the time between meals is more than 4 hours. Each main meal contains four portions of carbohydrates (starchy foods, vegetables, raw or cooked fruit, simple sugar) and is based on the principles of a balanced and diversified diet taking into account individual preferences and intolerances. Energy intakes are evaluated by the portion size of the starchy foods (cereals, legumes, floury and starchy foods), bearing in mind that the needs are hardly modified in the occasional or leisure user (up to three sessions per week). Physical activity levels are shown in Table 7.

| Physical activity levels: resting metabolism increase factors (ROM) and "walking" equivalents (km/d) by weight | | | | |
|---|---|---|---|---|
| | **MdR Index** | **Walking distance** | | |
| | | 70kg | 120kg | 44kg |
| **Sedentary** | 1,25 | 0 | 0 | 0 |
| **Not very active** | 1,5 | 5 | 3,5 | 2,5 |
| **Active** | 1,75 | 15 | 12 | 8 |
| **Very active** | 2,20 | 30 | 25 | 20 |

**Table 7:** Physical Activity Levels: Resting Metabolism Increasing Factors (ROM) and "walking" equivalents (km/d) by weight

Starch These are to be consumed at each meal or snack in quantities proportional to the energy expenditure. The adaptation of intakes according to the expenditure which depends on the level of activity (rest, intensive training, competition) conditions weight management. Pasta and rice in a variety of culinary presentations are the starchy foods of reference at each meal.

**b. Dairy products :**

They contribute to bone building in the youngest age group and are a major source of easily assimilated calcium and protein. Semi-skimmed milk, yoghurt and cottage cheese should be preferred for three to four portions/day.

**c. Meat, fish and eggs :**

With dairy products, they ensure the coverage of the protein needs increased by the muscular activity (of the order of 1 to 2 g of protein/kg/day according to the intensity and the nature of the activity). It is advisable to limit those with a high lipid content.

**d. Fruits and vegetables :**

They provide essential micronutrients and participate in the fight against oxidative stress. They can be consumed in abundance due to their low energy density (five portions per day).

**e. Specific aspects for top athletes :**

Intense sport practice and preparation for competitions are the subject of nutritional strategies whose application must be personalized. They aim to correct the main mistakes made by athletes (destructuring, nibbling, imbalance, insufficient diversification and poor hydration), to match energy intake with measured expenditure, to maintain a healthy weight and to obtain an optimal muscular glycogenic reserve on the day of the competition. The preparation for a long and intense test type "bottom" is largely anticipatory. The week before the event, the dissociated diet modified in the "Scandinavian" way seems to have proved its effectiveness. After a conventional diet for three days (15% protein, 35% fat, 50% carbohydrates) follows a period of energy enrichment (+ 500 to 1000kcal) with an increase in "slow" carbohydrate portions up to 70% of the energy ration and a decrease in fat intake. Foods likely to cause digestive problems are discarded: cooked fats, irritating fibres from pulses, fermented, spicy or smoked foods, very carbonated or very sweet drinks, alcohol. The pre-competition meal (8 to 12 hours before an event such as a cycle race or marathon) includes a carbohydrate-reinforced dish (al dente cooked pasta, up to 400 g), a dairy product, a starter and a carbohydrate dessert. The breakfast or meal is at least 3 hours before the event. Very rich in carbohydrates, it provides 500 to 800 kcal in the form of cereals, müeslis, cookies or rice with possibly a portion of minced meat. The waiting ration consumed between 1 hour 30 and 15 minutes before the start of the event is more or less important depending on the breakfast or meal that precedes it and includes gingerbread, fruit paste or a cereal bar. A drink is recommended 10 to 15 minutes before: 250mL of diluted fruit juice or an exercise drink with carbohydrate intake.

**f. Per-competition power supply :**

For intermediate duration tests, it is essential to hydrate on a regular and anticipatory basis while providing electrolytes, vitamins and carbohydrates in the form of glucose, fructose, saccharose or maltodextrins (40 to 100g/L). For very long events (> 3 hours 30 minutes), solid carbohydrate foods should be ingested every hour with small meals every 2 hours providing protein (lean meat, ham, lean cheese).

**g. Post-competition feeding :**

Rehydration, sugaring and remineralization are to be undertaken at the end of the test by the regular intake (three times per hour for 3 to 4 hours) of energy drinks from the effort, fruit juice or bicarbonate of water, by the ingestion of a solid carbohydrate snack 1 hour later (rice or semolina cake, cereal bar) and / or a meal reinforced in carbohydrate intake (pasta, rice, potato, fruit). Thereafter, the diet will be that of the training period.

### *3.8.6. Regulatory provisions and recommendations*

It is recommended to perform an annual dietary check-up for high-level athletes in order to preserve their health and promote sporting success. This approach makes it possible to detect eating disorders, correct errors and beliefs, and provide advice adapted to the nature of the sport. The sport nutrition consultation was instituted by decree (April 11, 2004). It is the responsibility of the sports physician to implement it. The AFSSA considers that the consumption of food supplements should only be motivated by the need to supplement nutritional intakes judged insufficient by a doctor or a dietician. The AFSSA considers that the "micronutrition" is not a concept sufficiently supported scientifically and that the nutritional care of the sportsmen must be done in accordance with the consensual recommendations and the regulation.

# Chapter 4: Food Models

## 1. Introduction

A food model is a system developed from a set of rules, choices and practices of a technical, social and symbolic nature, shared by a group of individuals whose identity it contributes to. Food models vary across space and time. Their progressive elaboration depends partly on the control of food production and processing and partly on the cultural traditions of social organization. At the service of the vital act of eating to meet energy and nutrient requirements for the maintenance of good health, they are also a strong but evolving identity trait. Their diversity testifies to the great malleability of the biological and physiological mechanisms of omnivorous man. They also remind us that food is a social act supported by the symbolism of food and by the pleasure of eating associated with a desire for aestheticism underpinned by cultural values. These dimensions largely escape the nutritional rational which seeks to define the model most favourable to health. In France, the model of three structured meals presented as traditional, when it is less than a century old, and healthy, when the formal demonstration of its interest for health has not been established, prevails. Nevertheless, this model constitutes a guide for prescription and nutritional education in that it opposes certain practices whose deleterious effect on health is admitted, such as snacking or cognitive restriction. In fact, any dietary model is bound to evolve with food availability, the evolution of techniques leading to "new" foods, better knowledge of the effects of nutrients on health and socio-cultural values and constraints. Urbanization, new working conditions, the development of distant tourism, globalization, immigration, changes in societies and dietary background noise all take their toll on existing food models to reshape them in small ways and bring them into line with the changes. The evolution of "kitchens" is a good illustration of the possible changes in food models in whose wake it is taking place. Beyond the ingredients and techniques required for food processing and preparation, the kitchen is the emerging part of a complex system of more or less implicit rules and norms that structure representations and behaviours.

## 2. Food patterns and health

The basic question is whether there is one food model that is preferable to all others in terms of health, and whether the evolution of a so-called traditional model according to new social, cultural and technological conditions can be worrying for health. The answer is obviously not unequivocal. Epidemiology provides some elements of an answer by establishing interesting relationships between dietary style and health indicators. The main data, rarely supported by intervention studies but backed up by meta-analyses, allow us to identify foods or nutrients that are favourable or dangerous for health in general or in relation to certain so-called social diseases. Arguments highlighting a specific food style are less solid, but it is possible to construct a food model based on a traditional model with a partially reworked nutrient or food composition.

## 2.1. Caloric restriction

Reflection on the impact of dietary patterns on health takes on its full importance at a time when life expectancy continues to increase due to the decrease in mortality from infectious diseases and deficiencies and the rise in the incidence of degenerative diseases, of which nutrition is one of the etiopathogenic determinants. Longevity is a particularly demonstrative parameter. Experimentally, energy restriction without deficiency is associated with a significant increase in lifespan in several animal species. In humans, the dietary model of Okinawa, an archipelago located in the south of Japan with a high number of valid centenarians, supports the idea that frugality is one of the keys to longevity and the prevention of chronic diseases. Caloric restriction could preserve the integrity of the genome by decreasing DNA alterations and the frequency of age-related mutations and could improve the performance of DNA repair processes. Extrapolation to humans must be done with caution, although it is well known that excessive energy intake and physical inactivity in our current living conditions are at the origin of obesity and chronic cardiometabolic diseases falsely attributed to aging. Caloric restriction and frugality can be considered after the growth period in subjects with little activity provided that any risk of mal- or undernutrition is avoided and that the pleasure of eating is maintained. The medicalization of food leads to the promotion of models that seem to have proved their worth in terms of prevention by imposing them as new criteria for food choices that replace or add to traditional gastronomic criteria, symbolic at the risk of eroding identity models in favour of a standardization that is undesirable from a sociological point of view.

## 2.2. Mediterranean diet

The Mediterranean model highlighted for its benefits on cardiometabolic prevention is a guide for food choices approved by the WHO for its exemplary nature. Paradoxically, it is being advocated in northern regions at the same time as it is being abandoned in regions where it was atavistic in favor of the much decried Western-style diet. Indeed, its fundamental values - dietary diversity, richness in fruits, vegetables and fibers, preponderance of mono- and polyunsaturated fatty acids - are being beaten in Crete or Greece, which explains in part the increasing prevalence of cerebrovascular diseases, metabolic diseases and cancers in these countries. Voluntarily represented on the shape of a food pyramid, the Mediterranean-type diet remains a reference value for nutritional education and prevention in public health which has inspired, among other things, the National Plan for Nutrition and Health.

## 2.3 Meals and balanced diet

A dietary model is mainly expressed by the conditions and composition of food intake. The meal is the easiest marker to analyze. Defined by its shape, structure and schedule, it contributes to the food day in very variable ways. It involves a combination of elements cooked with techniques whose aim is both to make them more digestible and to bring them into line with social and cultural values in an implicit logic of social identity. It is by acting at the level of the meal that it is possible to reshape the food day. The traditional French meal consisting of a starter, a garnished dish, cheese or a dessert makes it possible to consume all the food necessary for a balanced diet in a variety of forms ensuring the dietary diversity that is one of the foundations of a balanced diet. The meal structured in this way offers considerable freedom of choice in terms of food that it is however desirable to influence by simple recommendations, not very prescriptive, in order to avoid unstructured meals defined by single dishes or monotonous fast food type food, and excessive energy intake. The promotion of a desirable food model relies more on the organization of a diet globally with low energy density and high nutritional density. Simple and validated messages need to be tailored to individual dietary behaviours and the social and cultural context. The knowledge of consumer typologies makes it possible to better identify dietary patterns and to assert advice if it proves necessary by adjusting the frequency of consumption of certain foods. Frequency of consumption is an important parameter of dietary balance to fight against deficiencies. The rhythm of food intake determines the distribution of intakes during the day when intakes are traditionally three without there being scientific support setting an ideal number of meals. Breakfast, which provides about 20-30% of total energy intake, is recommended but cannot be considered mandatory for adults if a snack in the morning shortens the prolonged fast that separates dinner from lunch. The 4 p.m. snack or 4th meal recommended for children is not mandatory either. There are no scientific arguments to define an ideal number of meals. A controlled fractioning of the diet is necessary to prevent anarchic extra-prandial food intake that disrupts the control of energy intake through snacking or compulsive intake confining to genuine eating disorders. The structure appears more important than the frequency of meals.

### 3. Micronutrition

This dietary model, which is close to a diet, is intended to limit the harmful effects of aging and to precede the onset of chronic diseases. Micronutrition highlights the impact on health of micronutrient deficiencies and excess intakes of saturated fat and free radicals. It recommends, on the basis of a balanced diet, micronutrient supplementation.

### 4. Food models with exclusion

Food exclusions that focus primarily on foods of animal origin but with little nutritional risk have been recommended by some religions. Some communities advocate the exclusion of some or all foods of animal origin and impose a vegetarian (exclusion of meat and fish) or even vegan (exclusion of dairy products and eggs) diet. In practice, a distinction should be made between vegetarianism, which does not seem to have any nutritional disadvantages as long as it is applied in accordance with dietary rules of diversity - and veganism, whose nutritional dangers are established. Vegetarianism exposes to potentially severe deficits in vitamin B12, iodine, taurine, omega 3 fatty acids and calcium, which can have serious consequences, especially in young children, where growth disorders and megaloblastic anaemia have been described.

#### 4.1. Vegetarianism

The practice of a vegetarian diet that conforms to the basic principles of a balanced diet advocating the consumption of the three main plant families (cereals, legumes and oilseeds) and the consumption of dairy products could even have beneficial effects on health, all the more so since the practitioners of this diet usually adopt a rather favorable lifestyle. The basic problem is related to the complementarity of the various proteins of plant origin whose biological value is lower than that of proteins of animal origin in that they all contain a different limiting essential amino acid. The followers of this type of diet point out the economic advantages (meat consumption appears to be a waste as it takes 8 kg of vegetable proteins to produce 1.5 kg of animal proteins), and nutritional advantages due to the reduction in the consumption of saturated fats, the increase in fibre intake and the limitation of salt intake. Nutritional risks are mainly related to poor practice. Protein deficiency exists only in vegan areas. The same is true of vitamin deficiencies apart from vitamin B12, whose intake by plants is insignificant.

#### 4.2. Vegetarianism and adolescents

Vegetarianism particularly appeals to teenagers, whose fragility with regard to food is known and is the subject of a questioning in connection with the problem of empowerment. Nutritional experiences are part of a process of individualization in relation to the ambient model, particularly in the family, and are an opportunity for transgression. In some cases, it is the nibbling behaviour that is favoured, in others, it is exclusion that is preferred. The choice of vegetarianism by teenagers is often accompanied by a desire to restrict their consumption, with the concern for better health in order to control their weight with, incidentally, ethical and religious preoccupations. The long-term nutritional consequences are difficult to evaluate. The differences do not concern energy intakes or micronutrient intakes, which are comparable in vegetarians and omnivores, but rather protein intakes, which are lower, and fiber intakes, which are higher. Nevertheless, the desire for better weight control and a greater frequency of eating disorders justify some medical and environmental vigilance even if the choice of vegetarian food has no documented adverse effects on growth and health. The objective is to ensure that the exclusion only applies to meat and fish and to prevent any drift towards veganism.

### 4.3. Other dietary patterns of exclusion

There are various variants of the vegetarian diet depending on the combination of foods: lacto-vegetarianism excluding eggs, veganism excluding any product or by-product of animal origin. These practices can be accompanied by particularities: crudivorism considers that foods consumed as part of a lacto-ovo vegetarian diet must be consumed in the state where nature provides them, hygienism considers that the combination of foods and their order of ingestion have a great importance on digestion and absorption. Macrobiotic food is part of a lifestyle and a philosophical system. Foods are classified according to the antagonistic principles of Yin and Yang. All these variations around a given theme - eating - have no scientific basis, as man is obviously omnivorous.

## 5. Conclusion

No food model with a precise food composition has a scientific rationale strong enough to impose itself on all the others. As a bearer of socio-cultural and hedonic values, but also of identity because it is rooted in a tradition with multiple determinants, a dietary model can nevertheless be partially reshaped in the light of what has been achieved in the field of chronic disease prevention, for example by taking the Mediterranean diet pyramid as a reference point. There are, in fact, both harmful and protective dietary factors, knowledge of which, if properly understood, will help to influence practices towards a balanced diet adapted to particular characteristics.

# Chapter 5. Food and covid-19

## 1. Introduction

The Covid-19 pandemic is a pandemic of an emerging infectious disease, coronavirus 2019 (Covid-19), caused by the coronavirus SARS-CoV-2. It appears on November 17, 2019 in the city of Wuhan, central China, and then spreads worldwide. To date, all countries in the world are already affected by this pandemic. Added to this is the emergence of chronic non-communicable diseases (NCDs) including hypertension, cardiovascular disease, diabetes and cancer, which are on the rise. These are linked to hereditary factors but above all to unhealthy eating habits. In the face of this overwhelming health emergency, recommended essential protective measures to prevent saturation of intensive care units and to reinforce preventive hygiene. These measures include the elimination of physical contact, handshaking, hugging, gatherings and large demonstrations, as well as unnecessary travel and movement; promotion of handwashing, quarantine of suspicious cases, etc. This global pandemic is causing a series of cancellations of scientific, sporting and cultural events around the world in general and in Morocco in particular. In addition, the implementation by many countries of containment measures and the closing of borders to curb the formation of new contagion foci is causing a sharp drop in stock markets due to the uncertainties and fears it is causing to the regional economy and especially food and nutritional security. In the response against Covid-19, the governments of the affected countries and the scientific community have shown a good reactivity observable through the various clinical studies in progress and the proposals for therapeutic measures derived from traditional medicine and proven in vitro studies. However, the implications of the pandemic for food and nutritional security are rarely mentioned, even though early consideration of this aspect is essential in the fight against COVID-19. The importance of nutrition in the body's response to bacterial and viral infections is widely documented.

## 2. Concept of food and nutritional security

At the 1996 World Food Summit, food security was defined as follows: "Food security exists when all people, at all times, have physical and economic access to sufficient, safe and nutritious food to meet their dietary needs and food preferences for an active and healthy life". The analysis of food security is based on three pillars:

- ✓ Food availability.
- ✓ Access to food.

- ✓ The use of food.

Food availability is the food available in the study area from domestic production in all its forms, commercial imports and food aid. It is determined by: production, trade, stocks (food stored by traders and in national warehouses); transfers (government-supplied food), aid organizations, or both. Access to food refers to a household's ability to regularly obtain food through production, stockpiling, purchase, barter, donation, borrowing, food aid, or a combination of these sources.

Examples: family production (crops, livestock, etc.); hunting, fishing, and wild food gathering; purchases at markets, stores, etc.; barter (exchange of credit for food); gifts from friends or relatives, donations from the community, government, aid organizations, etc.; and other activities. Food may be available but inaccessible to some households if they cannot obtain sufficient food or properly diversify their diet from these different sources.

Food utilization refers to how households prepare and distribute the food they have access to, and how individuals are able to assimilate and metabolize food (efficiency of food processing by the body). It includes: how food is stored, processed and prepared, including water and fuel used for cooking and hygienic conditions; feeding practices, especially for people with special nutritional needs, such as infants, young children, the elderly, the sick and pregnant or lactating women; the sharing of food within the household and the extent to which this sharing corresponds to the nutritional needs of individual members (growth, pregnancy, breastfeeding, etc.); and the health status of each member of the household. Food may be available and accessible, but some household members may not take full advantage of it if the share they receive is not sufficiently large or diverse, or if their bodies are unable to assimilate it due to poor preparation or illness.

Food insecurity is one of the three underlying causes of malnutrition, so wherever it exists, there is a risk of malnutrition, including micronutrient deficiencies.

### 3. Healthy and balanced diet

Strengthening the immune system to prevent infections such as COVID-19 and maintaining health requires a healthy and balanced diet. Food is made up of micronutrients and macronutrients. Micronutrients such as vitamins and minerals are needed only in small amounts, while macronutrients such as carbohydrates, proteins and fats are needed in larger amounts. The body cannot function properly if it lacks one or more nutrients. A healthy and balanced diet provides a good quantity and variety of food, which is healthy and free of disease and harmful substances. Food is essential for our body to develop, replace and repair cells and tissues; to produce the energy used to stay warm, move and work; to carry out chemical processes such as food digestion; and to strengthen the immune system, which is essential for the body's protection, resistance and fight against infections and to heal disease processes. To cope effectively with COVID-19, people need to eat a diet rich in protein, carbohydrates, fats, vitamins and minerals with an optimal intake of water. These food components provide calories for activity, growth and all body functions, such as breathing, digestion and temperature maintenance; and elements for growth, maintenance of the body, and maintenance of a good immune system. Dietary proteins, especially essential amino acids, are important in adult diets to help the body repair and regenerate cells and are important in the diets of children and adolescents for their growth and development. Carbohydrates provide the body's share of energy. Lipids provide energy that can be stored in fat tissue or used at the same time. Vitamins and minerals support the body's healthy development and strengthen the immune system. Water intake is very important for the hydration of the body and its harmonious functioning, as the body is largely made up of water. The basis for a balanced diet is also the number of meals per day. In relation to the frequency of food intake, it is recommended to take 4 meals a day: breakfast, lunch, snack and dinner. For a healthy and balanced nutrition, the following guidelines should be adopted :

### 3.1. Varying foods

#### *3.1.1. Eat starchy foods at every meal*

Starchy foods include cereals (such as fonio, rice, maize, millet, sorghum, wheat and barley), roots and tubers (such as potatoes, sweet potatoes, cassava and yams) and starchy fruits such as plantains.

#### *3.1.2. Eat legumes (or pulses) if possible every day.*

Legumes include dried beans, peas, lentils, peanuts (including peanut butter) and soybeans.

#### *3.1.3. Eating animal foods and dairy products regularly*

These foods include all forms of meat, poultry, fish, eggs and dairy products such as milk, sour milk, buttermilk, yogurt, cheese, eggs, native chicken and rabbit meat, the latter are excellent sources of protein, phosphorus, zinc and iron, of animal origin of excellent nutritional quality.

### *3.1.4. Eating fruits and vegetables every day*

Fruits and vegetables are essential components of a healthy and balanced meal. They provide the vitamins and minerals that keep the body functioning and strengthen the immune system. The nutritional quality provided by the consumption of vegetables and legumes has been the subject of several in-depth studies. Legumes and vegetables are rich sources of protein, fat, carbohydrates, minerals, antioxidants, fibre and water, as well as being excellent sources of β-carotene (provitamin A), thiamin (B1), riboflavin (B2), niacin, pyridoxine ( B6), pantothenic acid, folic acid (folacin), ascorbic acid and vitamins E and K. Fruits such as cabbage, oranges, tangerines, grapefruit, lemons, guavas, mangoes, passion fruit, ripe pineapple, baobab fruit, tomatoes are very rich in vitamin C. Green leafy vegetables such as spinach, pumpkin, green bell pepper, squash, carrot, papaya and mango are good sources of vitamin A. These food resources are particularly important for people already affected by VIDOC-19 to fight infection.

### *3.1.5. Consume oils, fats, sugar and sweet products in moderation.*

They are an important source of energy, even in small quantities. They are found in butter, bacon, margarine, cooking oils (vegetable, coconut and palm), cream, mayonnaise and coconut cream. They can also be found in avocados, oil seeds (sunflower, peanut and sesame), the fatty parts of meat and fish, curdled milk and cheese. Sugar and sweet products are honey, jam, table sugar, cakes and cookies. Although fats and sugars are good sources of energy, they do not contain other nutrients. Therefore, they should be eaten as a supplement to other foods, not as a replacement.

### *3.1.6. Drink a lot of drinking water*

Water is important for life and must be drunk daily. To this drinking water consumption can be added all the liquids from juices, soups, vegetables and fruits as well as dishes with sauces. However, you should avoid drinking tea or coffee with meals because they can decrease the absorption of iron from food. As well as failing to avoid alcoholic beverages that absorb water from the body and they can also interact with the action of medications especially for people being treated for Coronavirus disease.

### *3.1.7. Increasing vitamin and mineral intake*

Vitamins and minerals are essential for good health. They protect you from opportunistic infections by keeping your skin, lungs and intestines healthy and keeping your immune system functioning properly. Vitamins A, C, E, some of the B group and minerals such as selenium, zinc and iron are of great importance.

Vitamin A is important for the maintenance of healthy skin and mucous membranes (lungs and intestines). Dark green, yellow, orange and red fruits and vegetables are good sources of vitamin A. These include spinach, pumpkin, green bell pepper, squash, carrots, yellow peach, apricot, papaya and mango. This vitamin is also found in palm oil, corn, sweet potatoes, egg yolks and liver. Vitamin C protects the body against infections and promotes healing. It is particularly found in citrus fruits such as oranges, grapefruit, lemons and mandarins. Mangoes, tomatoes and potatoes are also a good source of vitamin C. Vitamin E protects cells and promotes resistance to infection. Foods containing vitamin E include green leafy vegetables, vegetable oils, peanuts and egg yolks. Group B vitamins are necessary for the maintenance of a healthy immune and nervous system. White beans, potatoes, meat, fish, chicken, watermelon, corn, seeds, nuts, avocado, broccoli and green leafy vegetables contain a lot of them. Iron. Iron is found in leafy greens, oilseeds, whole grain products, dried fruit, sorghum, millet, dried beans, alfalfa, red meat, chicken, liver, fish, seafood and eggs. Selenium is an important mineral because it has a stimulating effect on immunity. It is found in wholegrain cereals such as wholegrain bread, corn and millet, and dairy products such as milk, yogurt and cheese. Meat, fish, poultry, eggs and all protein-rich foods such as peanut butter, dried beans and nuts are rich in selenium. Zinc is also important for the immune system. Zinc deficiency reduces appetite. It is found in meat, fish, poultry, shellfish, whole grains, corn, dried beans, peanuts, milk and dairy products.

The macronutrient and micronutrient composition of some foods is recorded in the appendices.

### 4. Physical activities

During the VIDOC-19 pandemic, the movement of many people is restricted; where restrictions permit an activity, physical distancing and hand hygiene should be practiced routinely. Indoor exercise and online physical activity classes are encouraged in preference to outdoor activities.

Obviously, for the maintenance of good health and optimal weight, especially during this period of confinement, daily physical activity (30 to 60 minutes each day) goes hand in hand with a healthy diet. Physical activity is recommended both to prevent cardiovascular diseases, their occurrence and to limit their consequences when they occur. The main conditions concerned are coronary artery disease, chronic heart failure and arteriopathy of the lower limbs. The regular practice of physical activity can delay or slow down certain deleterious processes related to aging. There is a close relationship between physical activity and respiratory disease. Physical activity is the most effective therapeutic tool in the treatment of dyspnea and chronic obstructive pulmonary disease (COPD) and the respiratory disease of smokers whose prevalence is galloping.

## 5. Fasting

Fasting is considered fasting when there is no food intake for at least 16 hours. If there is no caloric intake and only water is allowed, it is called complete fasting. If the individual does not eat or drink water, it is called a dry fast. Fasting can also be partial: in this case, the diet is based on fruit or vegetable juices, which limits caloric intake to about 300 kilocalories per day (for an adult, the recommended intake is about 2,000 kilocalories). Fasting can be continuous or intermittent (one or two days a week). Deprivation of food for several days leads to a disruption of the metabolism. The body must make every effort to compensate for this loss of energy intake by alternately using its various fuel reserves. On the first day of fasting, nothing changes. The body uses the glucose circulating in the blood or the glucose stored in the liver. But by the second day, these stocks are exhausted. The body and the brain then draw from the fat and protein reserves to make glucose. After five days, the body stops pumping out its protein stores. And with good reason: when half of these reserves have been consumed, the individual is no longer viable. Energy must then be produced from fat reserves - more precisely fatty acids - which can be directly used or transformed into ketone bodies, substitutes for glucose.

### 5.1. The benefits of fasting

Fasting is a way to achieve good health where the body can rest and purify itself. The human body is then cleansed of old cells, fats, waste and toxins that have accumulated during the year through diet and activity.

The benefits of fasting have been proven by several physiological studies:

✓ It fights against metabolic pathologies (excess cholesterol, triglycerides, fatty diabetes...),

✓ Improves physical and intellectual fitness,

Fasting also rejuvenates the skin, strengthens the teeth, reinforces the hair, soothes the senses, increases concentration, etc.

Fasting remains accessible to everyone (except in a few pathological cases), because the human body has the necessary reserves to last for several hours without problems:

✓ The liver can supply large quantities of glycogen, an excellent internal fuel.

✓ Blood and lymph carry many nutrients.

- ✓ Bone marrow contains nutrients.
- ✓ Several kilos of fat (even in slim people) are stored: the body draws primarily on its fat reserves.

No studies have been conducted on youth and the risk of infection by the COVID-19 virus. Healthy people should be able to fast during Ramadan, as in previous years, while COVID-19 patients should consider not fasting, according to religious deviations, in consultation with their physician, as with any other illness.

## 6. Conclusion

The immune system defends our body, especially against external aggressions. When it is healthy, no infection can resist it, including infections such as the flu or viruses. In the face of the COVID-19 epidemic, it is necessary to plan and follow an appropriate diet capable of boosting the immune system and therefore effectively fighting the virus.

# Appendices

# Appendix 1

| viandes, oeuf ,poisson et assimilé | Energie, Règlement UE N° 1169/2011 (kJ/100 g) | Energie, Règlement UE N° 1169/2011 (kcal/100 g) | Energie, N x facteur Jones, avec fibres (kJ/100 g) | Energie, N x facteur Jones, avec fibres (kcal/100 g) | Eau (g/100 g) | Protéines, N x facteur de Jones (g/100 g) | Protéines, N x 6.25 (g/100 g) | Glucides (g/100 g) | Lipides (g/100 g) | Fibres alimentaires (g/100 g) | Sel chlorure de sodium (g/100 g) | Calcium (mg/100 g) | Chlorure (mg/100 g) | Cuivre (mg/100 g) | Fer (mg/100 g) | Iode (µg/100 g) | Magnésium (mg/100 g) | Manganèse (mg/100 g) | Phosphore (mg/100 g) | Potassium (mg/100 g) | Sélénium (µg/100 g) | Sodium (mg/100 g) | Zinc (mg/100 g) | Vitamine D (µg/100 g) | Vitamine E (mg/100 g) | Vitamine K1 (µg/100 g) | Vitamine K2 (µg/100 g) | Vitamine C (mg/100 g) | Vitamine B1 ou Thiamine (mg/100 g) | Vitamine B2 ou Riboflavine (mg/100 g) | Vitamine B3 ou PP ou Niacine (mg/100 g) | Vitamine B5 ou Acide pantothénique (mg/100 g) | Vitamine B6 (mg/100 g) | Vitamine B9 ou Folates totaux (µg/100 g) | Vitamine B12 (µg/100 g) |
|---|---|---|---|---|---|---|---|---|---|---|---|---|---|---|---|---|---|---|---|---|---|---|---|---|---|---|---|---|---|---|---|---|---|---|---|
| Abat, cuit (aliment moyen) | 682 | 162 | 682 | 162 | 64,8 | 25,1 | 25,1 | 1,36 | 6,24 | 0,12 | 0,6 | 9,73 | | 4,93 | 6,54 | 11,6 | 17,8 | 0,18 | 265 | 252 | 24,5 | 244 | 4,23 | 0,77 | 0,39 | 1,4 | | 10,3 | 0,2 | 1,34 | 9,44 | 2,89 | 0,52 | 238 | 21,4 |
| Accra de poisson | 1140 | 275 | 1140 | 275 | 49,5 | 12,4 | 12,4 | 15,1 | 17,6 | 3 | 1,24 | 55,3 | 743 | < 0,1 | 1,1 | 20 | 18 | 0,3 | 110 | 216 | 18 | 583 | 0,8 | < 0,5 | 4,84 | - | - | 0,5 | 0,064 | 0,1 | 1,1 | 0,35 | 0,19 | 33 | 0,43 |
| Agneau, collier, braisé ou bouilli | 1520 | 365 | 1520 | 365 | 40,6 | 33,6 | 33,6 | traces | 25,6 | 0 | - | - | - | - | 0,66 | - | - | - | - | - | 13,9 | - | 7,14 | - | 0,36 | - | - | - | 0,1 | - | 4,26 | 0,69 | 0,095 | - | 2,06 |
| Agneau, collier, cru | 812 | 195 | 812 | 195 | 68,2 | 18 | 18 | traces | 13,7 | 0 | 0,2 | - | - | - | 1,14 | - | - | - | - | - | 7,41 | 80,1 | 3,82 | - | - | - | - | - | - | - | 4,27 | - | 0,16 | - | 2,18 |
| Agneau, côte découverte, crue | 854 | 205 | 854 | 205 | 66,5 | 16,3 | 16,3 | 1,38 | 15 | 0 | 0,21 | - | - | - | 1,58 | - | - | - | - | - | 7,5 | 82 | 3,77 | - | - | - | - | - | - | - | - | - | - | - | 1,77 |
| Agneau, côte filet, crue | 1000 | 242 | 1000 | 242 | 64 | 17,6 | 17,6 | 2,3 | 18 | 0 | 0,18 | 9 | - | - | 1,15 | - | - | - | - | - | 5,7 | 69,6 | 2,08 | - | - | - | - | - | - | - | 5,51 | - | 0,23 | - | 1,47 |
| Agneau, côte filet, grillée/poêlée | 660 | 157 | 660 | 157 | 67,7 | 26,3 | 26,3 | traces | 5,78 | 0 | - | - | - | - | 1,27 | - | - | - | - | - | 10,1 | - | 3,55 | - | 0,16 | - | - | - | 0,1 | - | 6,72 | 0,62 | 0,3 | - | 1,22 |
| Agneau, côte ou côtelette, crue (aliment moyen) | 1130 | 273 | 1130 | 273 | 60 | 16,3 | 16,3 | 0,92 | 22,6 | 0 | 0,19 | 9,97 | | | 1,34 | | | | | | 4,85 | 74,7 | 2,78 | | | | | | | | 4,98 | | 0,21 | | 1,52 |
| Agneau, côte ou côtelette, cuite (aliment moyen) | 660 | 157 | 660 | 157 | 67,7 | 26,3 | 26,3 | 0 | 5,78 | 0 | | | | | 1,27 | | | | | | 10,1 | | 3,55 | | 0,16 | | | | 0,1 | | 6,72 | 0,62 | 0,3 | | 1,22 |
| Agneau, côte ou côtelette, grillée/poêlée (aliment moyen) | 660 | 157 | 660 | 157 | 67,7 | 26,3 | 26,3 | 0 | 5,78 | 0 | | | | | 1,27 | | | | | | 10,1 | | 3,55 | | 0,16 | | | | 0,1 | | 6,72 | 0,62 | 0,3 | | 1,22 |
| Agneau, côte première, crue | 1140 | 276 | 1140 | 276 | 59,3 | 16,9 | 16,9 | 0 | 23,1 | 0 | 0,2 | 8 | - | - | 1,12 | - | - | - | - | - | 4,81 | 81,2 | 1,84 | - | - | - | - | - | - | - | 5,14 | - | 0,21 | - | 1,64 |
| Agneau, côte première, grillée/poêlée | 777 | 185 | 777 | 185 | 64,1 | 26,1 | 26,1 | 0 | 9,01 | 0 | 0,18 | 22 | - | 0,12 | 1,27 | - | 20 | 0,019 | 166 | 271 | 9,36 | 73 | 3,43 | - | 0,1 | - | - | 0 | 0,1 | 0,21 | 6,64 | 0,7 | 0,28 | 15 | 1,38 |
| Agneau, côtelette, crue | 1520 | 368 | 1520 | 368 | 50,2 | 14,4 | 14,4 | 0 | 34,5 | 0 | 0,17 | 12,9 | - | 0,12 | 1,5 | 0,7 | 17 | 0,01 | 136 | 218 | 1,4 | 66 | 3,42 | 0,4 | 0,7 | 0 | - | 1 | 0,18 | 0,3 | 4,3 | 0,55 | 0,2 | 2,6 | 1,2 |
| Agneau, côtelette, grillée | 999 | 240 | 999 | 240 | 58 | 25,7 | 25,7 | traces | 15,2 | 0 | 0,31 | 19,7 | - | 0,077 | - | 5 | 20,1 | 0,0079 | 184 | - | 9,97 | - | 3,5 | - | - | - | - | - | - | - | 7,45 | - | - | - | 2,8 |
| Agneau, épaule, crue | 711 | 170 | 711 | 170 | 70,2 | 18,3 | 18,3 | 0,45 | 10,6 | 0 | 0,2 | 12,7 | - | 0,11 | 1,63 | 0,7 | 19 | 0,015 | 170 | 224 | 7,53 | 78,8 | 3,43 | 0,4 | 0,46 | 3,4 | - | 1 | 0,15 | 0,26 | 4,98 | 0,62 | 0,17 | 10,8 | 1,65 |
| Agneau, épaule, maigre, crue | 582 | 139 | 582 | 139 | 73 | 19,6 | 19,6 | 0 | 6,76 | 0 | 0,18 | 15 | - | 0,11 | 1,66 | - | 24 | 0,024 | 184 | 274 | - | 70 | 4,77 | 0,1 | 0,22 | 3,4 | - | 0 | 0,12 | 0,22 | 5,45 | 0,71 | 0,16 | 23 | 2,78 |
| Agneau, épaule, maigre, rôtie/cuite au four | 822 | 197 | 822 | 197 | 63,3 | 24,9 | 24,9 | 0 | 10,8 | 0 | 0,17 | 19 | - | 0,11 | 2,13 | 5 | 25 | 0,026 | 200 | 265 | 6 | 68 | 6,04 | 0,1 | 0,18 | 4,5 | - | 0 | 0,09 | 0,26 | 5,76 | 0,73 | 0,15 | 25 | 2,7 |
| Agneau, épaule, rôtie/cuite au four | 1050 | 252 | 1050 | 252 | 53,7 | 30,5 | 30,5 | < 0,1 | 14,5 | 0 | 0,24 | 8,8 | 61 | 0,2 | 2,2 | 8,6 | 25,4 | < 0,1 | 231 | 376 | < 5 | 95,5 | 4,7 | < 0,5 | 0,14 | 4,6 | - | 6,9 | 0,064 | 0,3 | 7,2 | 0,3 | 0,21 | 24,2 | 2,72 |
| Agneau, gigot, braisé | 932 | 222 | 932 | 222 | 55,2 | 35,2 | 35,2 | traces | 9,03 | 0 | - | - | - | - | 0,97 | - | - | - | - | - | 15,6 | - | 5,28 | - | - | - | - | - | - | - | 6,52 | - | 0,16 | - | 1,88 |
| Agneau, gigot, cru | 538 | 128 | 538 | 128 | 74,5 | 20 | 20 | 0,48 | 5,13 | 0 | 0,12 | 10 | - | 0,12 | 1,51 | 0,7 | 21,9 | 0,015 | 177 | 255 | 7,73 | 48,7 | 3 | 0,4 | 0,46 | 0 | - | 1 | 0,15 | 0,27 | 6,52 | 0,63 | 0,26 | 9,8 | 2,17 |
| Agneau, gigot, grillé/poêlé | 731 | 174 | 731 | 174 | 64,8 | 27,6 | 27,6 | traces | 7,08 | 0 | - | - | - | - | 1,51 | - | - | - | - | - | 10,7 | - | 4,15 | - | - | - | - | - | - | - | 6,52 | - | 0,23 | - | 1,88 |
| Agneau, gigot, rôti/cuit au four | 708 | 169 | 708 | 169 | 65,9 | 26,8 | 26,8 | 0 | 6,85 | 0 | 0,19 | 21,8 | 61 | 0,12 | 1,4 | 5 | 30,7 | 0,015 | 183 | 301 | 10,3 | 75,5 | 4,01 | - | 0,13 | - | - | 0 | 0,1 | 0,28 | 6,52 | 0,83 | 0,24 | 17 | 1,88 |
| Agneau, selle, crue | 979 | 236 | 979 | 236 | 63,8 | 17,5 | 17,5 | 0 | 18,4 | 0 | 0,14 | 12,8 | - | 0,13 | 1,16 | 0,7 | 19 | 0,014 | 175 | 222 | 5,73 | 55 | 2,11 | 0,4 | 0,44 | 0 | - | 0 | 0,1 | 0,19 | 5,41 | 0,59 | 0,22 | 9,2 | 1,73 |
| Agneau, selle, partie maigre, grillée/poêlée | 645 | 153 | 645 | 153 | 68,3 | 26,1 | 26,1 | traces | 5,45 | 0 | - | - | - | - | 1,28 | - | - | - | - | - | 10,3 | - | 3,67 | - | - | - | - | - | - | - | 6,67 | - | 0,29 | - | 1,53 |
| Agneau, selle, partie maigre, rôtie/cuite au four | 645 | 153 | 645 | 153 | 68,3 | 26,1 | 26,1 | 0 | 5,45 | 0 | 0,16 | 18 | - | 0,12 | 1,21 | - | 23 | 0,02 | 180 | 246 | 10,3 | 64 | 3,67 | 0,1 | 0,11 | 4,7 | - | 0 | 0,1 | 0,24 | 6,67 | 0,83 | 0,27 | 19 | 1,53 |
| Agneau, viande, cuite (aliment moyen) | 868 | 208 | 868 | 208 | 60,5 | 28,1 | 28,1 | 0,011 | 10,6 | 0 | | | | | 1,74 | | | | | | 7,86 | | 4,94 | | 0,17 | | | | 0,083 | | 6,37 | 0,58 | 0,2 | | 2,27 |
| Anchois au sel (anchoité, semi-conserve) | 540 | 128 | 540 | 128 | - | 25 | 25 | 0 | 3,1 | 0 | 9,01 | 542 | - | - | 6,9 | - | - | - | 481 | 215 | - | 3600 | 1,7 | 4,6 | 0,29 | - | - | 0 | 0,06 | 0,26 | 14 | - | 0,14 | 8 | - |
| Anchois commun, cru | 541 | 129 | 541 | 129 | 71,8 | 18,6 | 18,6 | traces | 6,07 | 0 | 0,18 | 86,5 | - | 0,38 | 2,63 | 55 | 22,3 | 0,23 | 162 | 294 | 21,3 | 73 | 1,86 | 11 | 0,57 | 0,1 | - | 0 | 0,061 | 0,32 | 12,7 | 0,65 | 0,14 | 9 | 0,62 |
| Anchois commun, mariné, préemballé | 573 | 136 | 573 | 136 | 60,2 | 23,9 | 23,9 | 2,7 | 3,26 | 0 | 5,58 | 120 | 5310 | 0,17 | 2,8 | 8 | 102 | 0,14 | 148 | 159 | 46 | 2230 | 1,95 | 2,4 | 1,1 | - | - | - | < 0,04 | 0,2 | 2,75 | 0,41 | 0,18 | 37 | 15,3 |
| Anchois, filets à l'huile, semi-conserve, égoutté | 765 | 182 | 765 | 182 | 55,1 | 26,4 | 26,4 | < 0,2 | 8,45 | < 0,25 | 11,3 | 189 | 6090 | 0,25 | 3,22 | 30 | 43,7 | 0,1 | 231 | 355 | 8 | 4210 | 2,92 | 1,7 | 1,16 | 12,1 | - | < 0,5 | < 0,04 | 0,25 | 3,9 | 0,63 | 0,37 | 14,5 | 20,9 |
| Anchois, filets roulés aux câpres, semi-conserve, égoutté | 830 | 198 | 830 | 198 | - | 24,8 | 24,8 | 1,48 | 10,3 | < 1 | - | - | - | - | - | - | - | - | - | - | - | - | - | - | - | - | - | - | - | - | - | - | - | - | - |
| Andouille | 878 | 211 | 878 | 211 | 60,5 | 18,4 | 18,4 | 0,3 | 15,1 | traces | 2,08 | - | 1730 | - | - | 2 | - | - | - | - | - | 820 | - | - | - | - | - | - | - | - | - | - | - | - | - |
| Andouille de Guéméné | 972 | 234 | 972 | 234 | - | 18 | 18 | traces | 18 | traces | - | - | - | - | - | - | - | - | - | - | - | - | - | - | - | - | - | - | - | - | - | - | - | - | - |
| Andouille de Vire | 989 | 238 | 989 | 238 | - | 19 | 19 | traces | 18 | traces | - | - | - | - | - | - | - | - | - | - | - | - | - | - | - | - | - | - | - | - | - | - | - | - | - |
| Andouille, réchauffée à la poêle | 1120 | 271 | 1120 | 271 | 55,5 | 22,5 | 22,5 | 0 | 20 | < 1 | 2,82 | 18 | - | 0,2 | 2,4 | 24 | 12,2 | < 0,1 | 150 | 104 | 24 | 1130 | 3,4 | < 0,5 | 0,48 | - | - | < 1 | 0,056 | 0,28 | 1,55 | 0,48 | 0,12 | < 20 | 1,35 |
| Andouillette de Troyes, à cuire | 1050 | 254 | 1050 | 254 | 59,9 | 19,7 | 19,7 | 0,5 | 19 | 0,9 | 1,97 | 26,1 | - | 0,2 | 2,8 | - | 9,4 | < 0,2 | 110 | 58,4 | 22 | 830 | 2,6 | < 0,25 | 0,55 | - | - | - | 0,02 | - | 0,8 | - | < 0,1 | < 2 | 0,5 |
| Andouillette, à cuire | 946 | 227 | 946 | 227 | 63 | 18,8 | 18,8 | 1 | 16,3 | 0,71 | 1,46 | 34,1 | - | 0,28 | 1,67 | 2 | 12,5 | 0,16 | 112 | 55,6 | 18,3 | 584 | 2,6 | < 0,25 | 0,39 | - | - | - | 0,028 | 0,15 | 0,81 | - | < 0,1 | 1,2 | 0,49 |
| Andouillette, sautée/poêlée | 1260 | 303 | 1260 | 303 | 48,7 | 24,5 | 24,5 | 1 | 22,3 | < 1 | 2,73 | 35,6 | - | 0,2 | 2,3 | 15 | 9,6 | < 0,1 | 98,3 | 64 | 19 | 1090 | 4,3 | < 0,5 | 0,41 | - | - | < 1 | 0,035 | 0,15 | 0,96 | 0,24 | 0,09 | < 20 | 0,91 |
| Anguille, bouillie/cuite à l'eau | 820 | 197 | 820 | 197 | 62 | 20,5 | 20,5 | traces | 12,7 | 0 | - | - | - | 0,029 | - | 8 | - | 0,04 | - | - | 28 | - | 2,08 | - | 5,1 | - | - | - | - | - | - | - | - | - | - |
| Anguille, crue | 963 | 232 | 963 | 232 | 62,1 | 16,1 | 16,1 | traces | 18,6 | 0 | 0,16 | 23,6 | 57 | 0,041 | 0,71 | 36,3 | 18 | 0,078 | 285 | 260 | 28,7 | 65,5 | 2,27 | 16 | 6 | 1,25 | - | 1,6 | 0,17 | 0,24 | 3,16 | 0,17 | 0,21 | 13,8 | 2,8 |
| Anguille, cuite (aliment moyen) | 888 | 213 | 888 | 213 | 60,7 | 22,1 | 22,1 | 0 | 13,8 | 0 | | | | 0,029 | | 8 | | 0,04 | | | | | 2,08 | | | | | | | | | | | | |
| Anguille, rôtie/cuite au four | 955 | 229 | 955 | 229 | 59,3 | 23,6 | 23,6 | traces | 15 | 0 | 0,16 | 26 | - | 0,029 | 0,64 | 8 | 26 | 0,04 | 277 | 349 | - | 65 | 2,08 | - | - | - | - | 1,8 | 0,18 | 0,051 | 4,49 | 0,28 | 0,077 | 17 | 2,89 |
| Araignée de mer, cuite | - | - | - | - | 75,3 | 18,3 | 18,3 | - | 3,8 | 0 | 0,96 | 230 | 662 | 0,91 | 2 | 80 | 64 | 0,07 | 250 | 260 | 110 | 385 | 5,9 | < 0,25 | 6,42 | < 0,8 | - | < 0,5 | < 0,015 | 1,25 | 1,43 | 0,79 | 0,12 | 95,1 | 9,13 |
| Autruche, viande crue | 666 | 159 | 666 | 159 | 71,1 | 20,2 | 20,2 | 0 | 8,7 | 0 | 0,18 | 7 | - | 0,13 | 2,91 | - | 20 | 0,017 | 199 | 291 | - | 72 | 3,51 | - | 0,24 | - | - | 0 | 0,18 | 0,27 | 4,38 | 1,08 | 0,48 | 7 | 4,61 |
| Autruche, viande cuite | 706 | 168 | 706 | 168 | 67,1 | 26,2 | 26,2 | 0 | 7,07 | 0 | 0,2 | 8 | - | 0,14 | 3,43 | - | 23 | 0,017 | 224 | 323 | - | 80 | 4,33 | 0,1 | 0,24 | 3,5 | - | 0 | 0,21 | 0,27 | 6,56 | 1,21 | 0,5 | 14 | 5,74 |
| Bar commun ou loup (Méditerranée), cru, élevage | 521 | 124 | 521 | 124 | 72,8 | 21,4 | 21,4 | traces | 4,25 | 0 | 0,12 | 5,6 | - | < 0,1 | 0,39 | 9,36 | 32,3 | < 0,1 | 209 | 450 | 7,57 | 46,6 | 0,41 | 2,31 | 1,35 | - | - | - | 0,18 | 0,09 | 6,79 | 0,53 | 0,39 | - | 4,33 |
| Bar commun ou loup (Méditerranée), cru, sauvage | 414 | 97,9 | 414 | 97,9 | 76,8 | 20,1 | 20,1 | traces | 1,94 | 0 | 0,18 | 1,56 | - | < 0,1 | 0,4 | 23,3 | 28,3 | < 0,1 | 191 | 371 | 25,4 | 71,4 | 0,43 | 3,65 | 0,71 | - | - | - | 0,064 | 0,098 | 3,72 | 0,33 | 0,39 | - | 4,16 |
| Bar commun ou loup, cru, sans précision | 350 | 82,5 | 350 | 82,5 | 78,1 | 19,1 | 19,1 | traces | 0,7 | 0 | 0,23 | 10 | 119 | 0,04 | 0,26 | 21,9 | 27 | < 0,01 | 207 | 390 | 30 | 92,3 | 0,3 | 5,59 | 0,79 | < 0,8 | - | 0,66 | 0,025 | 0,058 | 3,67 | 0,55 | 0,34 | 22,2 | 3,12 |
| Bar commun ou loup, rôti/cuit au four | 617 | 147 | 617 | 147 | 69,1 | 22,2 | 22,2 | traces | 6,51 | 0 | 0,17 | 11 | 119 | 0,05 | 0,28 | 13,4 | 30,3 | 0,01 | 183 | 410 | < 50 | 68,8 | 0,42 | 4,15 | 0,97 | < 0,8 | - | - | 0,17 | 0,08 | 3,49 | 0,53 | 0,26 | 12,6 | 3,7 |
| Bar ou loup de l'Atlantique, cru | 379 | 89,9 | 379 | 89,9 | 79,6 | 16,6 | 16,6 | traces | 2,61 | 0 | 0,25 | 20,3 | - | 0,04 | 0,72 | 60 | 26,4 | 0,013 | 185 | 286 | 57,5 | 99 | 0,74 | 0,77 | 2,4 | 0 | - | traces | 0,19 | 0,064 | 2,33 | 0,57 | 0,37 | 3 | 2,08 |
| Bar rayé ou bar d'Amérique, cru | 477 | 113 | 477 | 113 | 74,6 | 20,3 | 20,3 | traces | 3,58 | 0 | 0,19 | 66 | - | 0,081 | 1,41 | 60 | 36 | 0,68 | 218 | 356 | - | 76,3 | 0,63 | - | - | - | - | 2,05 | 0,087 | 0,065 | 1,62 | 0,79 | 0,19 | 13,7 | 2,71 |
| Baudroie rousse ou Lotte, crue | 298 | 70,2 | 298 | 70,2 | 81,5 | 16,7 | 16,7 | traces | 0,37 | 0 | 0,32 | 1,02 | - | < 0,1 | 0,26 | 41,2 | 28,7 | < 0,1 | 149 | 339 | 414 | 127 | 0,44 | 0,33 | 0,33 | - | - | - | 0,025 | 0,026 | 2,43 | 0,17 | 0,14 | - | 0,46 |
| Beignet de crevette | 930 | 222 | 930 | 222 | 54,2 | 6,5 | 6,5 | 26 | 10 | 1 | 1,9 | 77 | 1000 | - | 1 | 22 | 18,1 | - | - | 102 | 41,7 | 770 | 0,5 | - | - | - | - | - | - | - | - | - | - | - | - |
| Bigorneau, cuit | - | - | - | - | 74,9 | 16,3 | 16,3 | - | 3,5 | 0 | 0,48 | < 0,2 | 318 | < 0,01 | < 0,01 | 570 | < 0,05 | < 0,01 | < 0,3 | < 0,3 | < 20 | 193 | < 0,05 | < 0,25 | 2,18 | 2,59 | - | < 0,5 | 0,034 | 0,061 | 1,3 | 0,28 | 0,014 | 32,7 | 60,7 |
| Boeuf, à bourguignon ou pot-au-feu, cru | 686 | 164 | 686 | 164 | - | 24 | 24 | 0 | 7,5 | 0 | 0,1 | 18 | - | - | 2,1 | - | - | - | - | - | - | 41,8 | - | - | - | - | - | - | - | - | - | - | - | - | - |
| Boeuf, à bourguignon ou pot-au-feu, cuit | 728 | 172 | 728 | 172 | 59,4 | 34 | 34 | traces | 4,05 | 0 | 0,15 | - | - | - | 4,3 | 6 | - | - | 171 | - | 4,1 | - | 9,7 | - | 0,51 | - | - | - | 0,04 | - | 2,6 | 0,5 | 0,24 | 4,5 | 1,2 |
| Boeuf, basse-côte, crue | 837 | 201 | 837 | 201 | 65,8 | 19 | 19 | 1,45 | 13,2 | 0 | 0,17 | - | - | - | 2,75 | - | - | - | - | - | 8,68 | 68 | 5,97 | - | - | - | - | - | - | - | - | - | - | - | 1,22 |
| Boeuf, bavette d'aloyau, crue | 557 | 133 | 557 | 133 | 73,7 | 20,4 | 20,4 | 0 | 5,67 | 0 | 0,085 | 14,2 | - | 0,072 | 3,31 | - | 22 | 0,012 | 195 | 328 | 11 | 34,4 | 6,8 | - | 0,31 | 1,2 | - | 0 | 0,066 | 0,1 | 4,17 | 0,61 | 0,26 | 11 | 3,12 |
| Boeuf, bavette d'aloyau, grillée/poêlée | 682 | 162 | 682 | 162 | 67,8 | 25 | 25 | traces | 6,94 | 0 | - | - | - | - | 3,38 | - | 24,8 | - | - | - | 13,5 | - | 8,33 | - | 0,54 | - | - | - | 0,06 | - | 4,17 | 0,38 | 0,26 | - | 3,12 |
| Boeuf, boule de macreuse, crue | 495 | 118 | 495 | 118 | 74,5 | 21,8 | 21,8 | 0 | 3,36 | 0 | 0,097 | 4,95 | - | 0,09 | 2,86 | - | 22 | - | 192 | 354 | 10,7 | 37,9 | 4,59 | - | 0,13 | - | - | 0 | 0,12 | 0,19 | 4,43 | - | 0,44 | 7 | 1,89 |
| Boeuf, boule de macreuse, grillée/poêlée | 607 | 144 | 607 | 144 | 68,8 | 26,7 | 26,7 | traces | 4,11 | 0 | - | - | - | - | 2,86 | - | - | - | - | - | 13 | - | 5,62 | - | - | - | - | - | - | - | 4,42 | - | 0,44 | - | 1,89 |
| Boeuf, boule de macreuse, rôtie/cuite au four | 619 | 147 | 619 | 147 | 68,2 | 27,3 | 27,3 | traces | 4,2 | 0 | - | - | - | - | 3,03 | - | - | - | - | - | 13,3 | - | 5,74 | - | - | - | - | - | - | - | 4,42 | - | 0,44 | - | 1,89 |
| Boeuf, boulettes cuites | 906 | 217 | 906 | 217 | 62 | 18,5 | 18,5 | 5,65 | 13,3 | 0,63 | 1,13 | 29 | - | 0,083 | 2,2 | 9,01 | 26,9 | 0,15 | 172 | 326 | 5,66 | 452 | 2,88 | 0,2 | 0,38 | - | - | 6,94 | 0,072 | 0,1 | 2,8 | 0,48 | 0,23 | 25,8 | 1,23 |
| Boeuf, braisé | 1000 | 240 | 1000 | 240 | 55,7 | 32,1 | 32,1 | 0 | 12,4 | 0 | 0,15 | - | - | - | 5,9 | 6 | 24,3 | - | - | - | - | - | 10,5 | - | - | - | - | - | - | - | - | - | - | - | - |
| Boeuf, collier, braisé | 776 | 184 | 776 | 184 | - | 33 | 33 | traces | 5,8 | 0 | - | - | - | - | 4,5 | - | - | - | - | - | 5,9 | - | 9,3 | - | 0,62 | - | - | - | 0,04 | - | 3,9 | 1,2 | 0,28 | - | 1,9 |
| Boeuf, collier, cru | 825 | 198 | 825 | 198 | 66,7 | 20,1 | 20,1 | 0,78 | 12,7 | 0 | 0,17 | - | - | - | 2,42 | - | - | - | - | - | 6,64 | 67,5 | 6,06 | - | - | - | - | - | - | - | - | - | - | - | 1,3 |
| Boeuf, côte, crue | 1040 | 251 | 1040 | 251 | - | 18,7 | 18,7 | traces | 19,6 | 0 | 0,16 | 11 | - | - | 2,8 | 2,4 | 18 | - | 188 | 355 | - | 65 | - | - | - | - | - | 0 | 0,08 | 0,17 | 4,5 | - | 0,42 | 6,9 | 1,4 |
| Boeuf, entrecôte, crue | 961 | 231 | 961 | 231 | 62,5 | 19,4 | 19,4 | 0 | 17,1 | 0 | 0,075 | 5,44 | - | 0,066 | 2,21 | 1,1 | 19,5 | 0,011 | 154 | 270 | 8,13 | 29,8 | 4,25 | 0,45 | 0,66 | 1,5 | - | 0 | 0,069 | 0,2 | 3,91 | 0,6 | 0,29 | 4 | 1,85 |

| | | | | | | | | | | | | | | | | | | | | | | | | | | | | | | | | | | | |
|---|---|---|---|---|---|---|---|---|---|---|---|---|---|---|---|---|---|---|---|---|---|---|---|---|---|---|---|---|---|---|---|---|---|---|---|
| Boeuf, entrecôte, partie maigre, grillée/poêlée | 829 | 198 | 829 | 198 | 62,3 | 25,5 | 25,5 | < 0,1 | 10,7 | 0 | 0,15 | 12 | - | 0,085 | 2,55 | 6 | 23 | - | 200 | 279 | 12,3 | 58 | 6,34 | 0,1 | 0,58 | 1,6 | - | 0 | 0,07 | 0,15 | 4,4 | 0,22 | 0,34 | 6 | 1,85 |
| Boeuf, épaule, crue | 703 | 168 | 703 | 168 | 69 | 20,4 | 20,4 | 0 | 9,64 | 0 | 0,17 | 7,38 | - | 0,073 | 2,29 | 0,95 | 22,5 | 0,012 | 192 | 342 | 7,25 | 67,5 | 5,46 | 0,55 | 0,32 | 1,5 | - | 0 | 0,057 | 0,19 | 5,38 | 0,77 | 0,5 | 5 | 1,84 |
| Boeuf, faux-filet, cru | 639 | 152 | 639 | 152 | 70,5 | 22,3 | 22,3 | 0,6 | 6,74 | 0 | 0,082 | 9,28 | - | 0,076 | 2,26 | 0,9 | 21,3 | 0,011 | 189 | 318 | 10,6 | 32,2 | 3,26 | 0,55 | 0,37 | 1,4 | - | 0 | 0,059 | 0,15 | 5,77 | 0,73 | 0,48 | 9,6 | 1,19 |
| Boeuf, faux-filet, grillé/poêlé | 762 | 182 | 762 | 182 | 64,3 | 27,1 | 27,1 | 0 | 8,17 | 0 | 0,14 | 20 | - | 0,081 | 2,26 | 6 | 22 | 0,009 | 209 | 336 | 12,9 | 56 | 3,95 | 0,2 | 0,44 | 1,6 | - | 0 | 0,04 | 0,13 | 5,77 | 0,3 | 0,48 | 8 | 1,19 |
| Boeuf, faux-filet, rôti/cuit au four | 812 | 194 | 812 | 194 | 62 | 28,8 | 28,8 | traces | 8,71 | 0 | - | - | - | - | 2,41 | - | - | - | - | - | 13,8 | - | 4,21 | - | - | - | - | - | - | - | 5,77 | - | 0,47 | - | 1,19 |
| Boeuf, filet, cru | 553 | 132 | 553 | 132 | 72,8 | 21,6 | 21,6 | 0,22 | 4,95 | 0 | 0,15 | - | - | - | 2,57 | - | - | - | - | - | 8,83 | 61,8 | 3,22 | - | - | - | - | - | - | - | - | - | - | - | 1,73 |
| Boeuf, gîte à la noix, cru | 622 | 148 | 622 | 148 | 68,9 | 21,3 | 21,3 | 0 | 7,05 | 0 | 0,14 | 10,3 | - | 0,086 | 1,85 | 0,9 | 22 | 0,012 | 185 | 334 | 6,5 | 54,3 | 4,16 | 0,4 | 0,29 | 1,4 | - | 0 | 0,063 | 0,13 | 6,09 | 0,75 | 0,56 | 9 | 1,45 |
| Boeuf, gîte à la noix, cuit | 431 | 102 | 431 | 102 | - | 21 | 21 | 0,2 | 1,9 | 0 | - | - | - | - | - | - | - | - | - | - | - | - | - | - | - | - | - | - | - | - | - | - | - | - | - |
| Boeuf, hampe, crue | 643 | 154 | 643 | 154 | 71,5 | 19 | 19 | 0 | 8,63 | 0 | 0,15 | 6 | - | 0,17 | 3,68 | - | 19 | 0,17 | 142 | 242 | 11,8 | 59,3 | 4,51 | 0,1 | 0,15 | 1,5 | - | 0 | 0,1 | 0,56 | 3,97 | 1,13 | 0,29 | 3 | 4,61 |
| Boeuf, hampe, grillée/poêlée | 728 | 174 | 728 | 174 | 67,6 | 21,6 | 21,6 | traces | 9,76 | 0 | - | - | - | - | 3,79 | - | - | - | - | - | 13,4 | - | 5,11 | - | - | - | - | - | - | - | 3,93 | - | 0,28 | - | 4,6 |
| Boeuf, jarret, bouilli/cuit à l'eau | 679 | 161 | 679 | 161 | - | 31 | 31 | traces | 4,1 | 0 | - | - | - | - | 3,9 | - | - | - | - | - | 2,8 | - | 11 | - | 0,53 | - | - | - | 0,03 | - | 3,2 | 0,4 | 0,21 | - | 1,2 |
| Boeuf, jarret, cru | 513 | 122 | 513 | 122 | 72,5 | 20,9 | 20,9 | 0 | 4,25 | 0 | 0,14 | 5 | - | 0,074 | 1,96 | - | 24 | 0,011 | 215 | 363 | - | 54 | 5,32 | - | 0,21 | 0 | - | 0 | 0,071 | 0,19 | 5,01 | 0,73 | 0,53 | 11 | 4,49 |
| Boeuf, joue, braisée ou bouillie | 992 | 236 | 992 | 236 | 52,9 | 39,2 | 39,2 | traces | 8,82 | 0 | - | - | - | - | 2,04 | - | - | - | - | - | 25,3 | - | 4,75 | - | - | - | - | - | - | - | 4,67 | - | 0,096 | - | 7,23 |
| Boeuf, joue, crue | 571 | 136 | 571 | 136 | 73,2 | 22,3 | 22,3 | 0,4 | 5,02 | 0 | 0,18 | - | - | - | 3,18 | - | - | - | - | - | 14,4 | 71 | 2,7 | - | - | - | - | - | - | - | 4,67 | - | 0,15 | - | 7,24 |
| Boeuf, onglet, cru | 577 | 138 | 577 | 138 | 72,4 | 19,9 | 19,9 | 0,62 | 6,18 | 0 | 0,18 | - | - | - | 3,63 | - | - | - | - | - | 8,12 | 70,9 | 4,67 | - | - | - | - | - | - | - | - | - | - | - | 3,6 |
| Boeuf, onglet, grillé | 717 | 171 | 717 | 171 | - | 23 | 23 | traces | 8,8 | 0 | - | - | - | - | 4 | - | - | - | - | - | 4,1 | - | 6 | - | 0,53 | - | - | - | 0,09 | - | 4,8 | 1,4 | 0,43 | - | 3,9 |
| Boeuf, paleron, braisé ou bouilli | 1050 | 251 | 1050 | 251 | 51,8 | 36,8 | 36,8 | 0,2 | 11,4 | 0 | 0,17 | 14 | - | 0,14 | 1,63 | - | 22 | 0,025 | 230 | 298 | 17,7 | 67 | 9,57 | 0,1 | 0,11 | 1,6 | - | 0 | 0,083 | 0,28 | 3,67 | 0,91 | 0,15 | 8 | 2,77 |
| Boeuf, paleron, cru | 602 | 144 | 602 | 144 | 72,3 | 21,2 | 21,2 | 0 | 6,54 | 0 | 0,12 | 7,45 | - | 0,087 | 2,5 | - | 25 | 0,012 | 223 | 343 | 10,2 | 49 | 5,51 | 0,1 | 0,2 | 1,5 | - | 0 | 0,08 | 0,21 | 3,67 | 0,86 | 0,27 | 3 | 2,77 |
| Boeuf, plat de côtes, braisé | 1130 | 269 | 1130 | 269 | 48,9 | 37,3 | 37,3 | 0 | 13,3 | 0 | 0,13 | 12 | - | 0,099 | 1,41 | - | 15 | 0,013 | 162 | 224 | 18,4 | 50 | 8,02 | 0,7 | 0,29 | 2,4 | - | 0 | 0,03 | 0,15 | 4,87 | 1 | 0,23 | 5 | 2,57 |
| Boeuf, plat de côtes, cru | 1190 | 287 | 1190 | 287 | 58,2 | 18,4 | 18,4 | 0 | 23,7 | 0 | 0,12 | 9 | - | 0,053 | 1,9 | - | 14 | 0,011 | 137 | 232 | 6,68 | 49 | 3,08 | - | - | - | - | 0 | 0,071 | 0,12 | 3,72 | 0,24 | 0,25 | 5 | 2,57 |
| Boeuf, queue, bouillie/cuite à l'eau | 820 | 196 | 820 | 196 | - | 28 | 28 | traces | 9,3 | 0 | - | - | - | - | 2,9 | - | - | - | - | - | - | - | 8 | - | 0,16 | - | - | - | 0,02 | - | 2,1 | 0,23 | 0,15 | - | 1,1 |
| Boeuf, queue, crue | 849 | 203 | 849 | 203 | 65 | 21,5 | 21,5 | 0,23 | 12,9 | 0 | 0,28 | - | - | - | 2,55 | - | - | - | - | - | 5,45 | 111 | 4,52 | - | - | - | - | - | - | - | - | - | - | - | 1,47 |
| Boeuf, rond de gîte, cru | 470 | 111 | 470 | 111 | 74 | 22,9 | 22,9 | 0,38 | 2,02 | 0 | 0,15 | - | - | - | 2,08 | - | - | - | - | - | 6,9 | 60,2 | 3,28 | - | - | - | - | - | - | - | - | - | - | - | 0,93 |
| Boeuf, rosbif, rôti/cuit au four | 494 | 117 | 494 | 117 | 72,5 | 21,9 | 21,9 | 0,3 | 3,16 | 0 | < 0,1 | 10,1 | - | 0,1 | 3,1 | < 5 | 32 | 0,012 | 203 | 369 | 16,1 | 70,4 | 5,43 | 0 | 0,2 | - | - | - | 0,1 | 0,12 | 6,1 | 1,8 | 0,66 | 9,1 | 1,3 |
| Boeuf, rumsteck, cru | 482 | 114 | 482 | 114 | 70,9 | 22,5 | 22,5 | 0,4 | 2,5 | 0 | 0,13 | 7,33 | - | 0,067 | 2,48 | 0,9 | 16,5 | 0,0075 | 195 | 326 | 9,3 | 50,5 | 3,54 | 0,5 | 0,29 | 1,5 | - | 0 | 0,062 | 0,21 | 6,17 | 0,58 | 0,57 | 5 | 1,52 |
| Boeuf, rumsteck, grillé | 518 | 123 | 518 | 123 | - | 25 | 25 | traces | 2,5 | 0 | - | - | - | - | 2,9 | - | - | - | - | - | 4,6 | - | 4,2 | - | 0,44 | - | - | - | 0,1 | - | 7,3 | 1,5 | 0,56 | - | 1,5 |
| Boeuf, steak haché 10% MG, cru | 713 | 171 | 713 | 171 | 69,5 | 20 | 20 | < 0,05 | 10,1 | 0 | 0,16 | 12 | - | 0,072 | 2,24 | 6,8 | 20 | 0,01 | 184 | 321 | 13,9 | 64,7 | 4,79 | 0,1 | 0,17 | 0,8 | 6,7 | 0 | 0,042 | 0,15 | 5,07 | 0,6 | 0,37 | 6 | 2,21 |
| Boeuf, steak haché 10% MG, cuit | 879 | 210 | 879 | 210 | 61,2 | 26,1 | 26,1 | 0 | 11,8 | 0 | 0,17 | 13 | - | 0,09 | 2,71 | 6,1 | 22 | 0,013 | 202 | 333 | - | 68 | 6,37 | 0 | 0,12 | 1,1 | - | 0 | 0,044 | 0,18 | 5,66 | 0,66 | 0,4 | 8 | 2,56 |
| Boeuf, steak haché 15% MG, cru | 873 | 209 | 873 | 209 | 67,2 | 20,2 | 20,2 | 0,47 | 14,1 | 0 | 0,1 | 9,33 | - | 0,069 | 2,64 | 1 | 20 | 0,011 | 171 | 302 | 6,78 | 41,7 | 4,8 | 0,35 | 0,26 | 1,3 | - | 1 | 0,048 | 0,17 | 4,1 | 0,51 | 0,2 | 5,75 | 1,9 |
| Boeuf, steak haché 15% MG, cuit | 996 | 239 | 996 | 239 | 59,1 | 23,6 | 23,6 | 0 | 16,1 | 0 | 0,21 | 15 | 61 | 0,078 | 2,6 | 6,1 | 29,5 | 0,06 | 198 | 318 | < 5 | 84,3 | 4,9 | 0 | 0,12 | 1,2 | - | 0 | 0,046 | 0,18 | 5,2 | 0,66 | 0,29 | 9 | 2,3 |
| Boeuf, steak haché 20% MG, cru | 1030 | 249 | 1030 | 249 | 61,9 | 17,3 | 17,3 | 0 | 20 | 0 | 0,16 | 12,7 | - | 0,061 | 1,94 | 6,8 | 17 | 0,01 | 158 | 270 | 12,7 | 64 | 4,18 | 0,1 | 0,17 | 1,8 | - | 0 | 0,043 | 0,15 | 4,23 | 0,5 | 0,32 | 7 | 2,14 |
| Boeuf, steak haché 20% MG, cuit | 1100 | 265 | 1100 | 265 | 56,7 | 23 | 23 | 0,14 | 19,2 | 0 | 0,19 | 24 | - | 0,08 | 2,48 | 6,1 | 20 | 0,011 | 115 | 304 | 19,4 | 75 | 4,5 | 0 | 0,12 | 1,6 | - | 0 | 0,047 | 0,18 | 5,18 | 0,66 | 0,27 | 27,4 | 3,31 |
| Boeuf, steak haché 5% MG, cru | 547 | 130 | 547 | 130 | 74 | 21,9 | 21,9 | 0,3 | 4,59 | 0 | 0,097 | 7 | - | 0,082 | 2,65 | 0,8 | 22 | 0,011 | 184 | 353 | 6,72 | 37,8 | 4,54 | 0,25 | 0,2 | 0,3 | - | 0 | 0,048 | 0,17 | 4,7 | 0,64 | 0,34 | 6 | 2,12 |
| Boeuf, steak haché 5% MG, cuit | 650 | 155 | 650 | 155 | 68,9 | 25,5 | 25,5 | 0 | 5,85 | 0 | 0,16 | 7 | - | 0,096 | 2,83 | 6,1 | 22 | 0,014 | 180 | 347 | - | 65 | 6,43 | 0 | 0,12 | 1,3 | - | 0 | 0,042 | 0,18 | 5,94 | 0,66 | 0,41 | 7 | 2,47 |
| Boeuf, steak haché, cuit (aliment moyen) | 963 | 231 | 963 | 231 | 60 | 23,8 | 23,8 | 0,0018 | 15,1 | 0 | 0,21 | 14,3 | 61 | 0,08 | 2,62 | 6,1 | 28,6 | 0,054 | 195 | 321 | 2,75 | 82,2 | 5,05 | 0 | 0,12 | 1,21 | | 0 | 0,046 | 0,18 | 5,27 | 0,66 | 0,31 | 9,04 | 2,33 |
| Boeuf, steak ou bifteck, cru | 926 | 223 | 926 | 223 | 63,1 | 19,2 | 19,2 | 0 | 16,2 | 0 | 0,15 | 6 | 52 | 0,071 | 1,93 | 1 | 19,8 | 0,012 | 174 | 307 | 6,5 | 58,7 | 4,03 | 0,6 | 0,48 | 0 | - | 0 | 0,075 | 0,16 | 4,44 | 0,52 | 0,42 | 5,77 | 2,18 |
| Boeuf, steak ou bifteck, grillé | 541 | 128 | 541 | 128 | 69,1 | 27,6 | 27,6 | traces | 1,95 | 0 | 0,16 | 7,07 | 61 | 0,076 | 2,21 | 6 | 28,9 | 0,012 | - | 371 | < 10 | 63,9 | 5,39 | - | - | - | - | - | - | - | 5,15 | - | 0,46 | - | 2,7 |
| Boeuf, tende de tranche, crue | 488 | 116 | 488 | 116 | 74,8 | 23,1 | 23,1 | 0,57 | 2,34 | 0 | 0,078 | 7,93 | - | 0,061 | 2,75 | 0,9 | 17 | 0,006 | 196 | 329 | 10,1 | 31,1 | 3,46 | 0,4 | 0,23 | 1,5 | - | 0 | 0,06 | 0,18 | 5,21 | 0,63 | 0,51 | 5,5 | 1,16 |
| Boeuf, tende de tranche, grillée/poêlée | 597 | 141 | 597 | 141 | 68,5 | 28,7 | 28,7 | traces | 2,92 | 0 | - | - | - | - | 2,75 | - | - | - | - | - | 12,6 | - | 4,32 | - | - | - | - | - | - | - | 5,21 | - | 0,51 | - | 1,16 |
| Boeuf, tende de tranche, rôtie/cuite au four | 618 | 146 | 618 | 146 | 67,4 | 29,8 | 29,8 | traces | 3,03 | 0 | - | - | - | - | 2,93 | - | - | - | - | - | 13 | - | 4,48 | - | - | - | - | - | - | - | 5,21 | - | 0,5 | - | 1,16 |
| Bogue, crue | 380 | 90 | 380 | 90 | 78,5 | 17,9 | 17,9 | traces | 2,03 | 0 | 0,28 | 52,5 | 120 | 0,33 | 2,3 | 60 | 22,5 | 0,12 | 230 | 309 | - | 110 | 0,96 | 0 | 0 | - | - | 0 | 0,08 | 0,1 | 5,4 | 0,21 | 0,46 | - | 2 |
| Bonite, crue | 675 | 161 | 675 | 161 | 70 | 22,3 | 22,3 | traces | 8,01 | 0 | 1,35 | 16 | - | 0,26 | 2,08 | - | 10 | 0,1 | - | 35 | - | 540 | 0,68 | - | - | - | - | - | - | - | - | - | - | - | - |
| Bouchées ou émincé au soja et blé (ne convient pas aux véganes ou végétaliens), préemballé | 775 | 185 | 747 | 178 | 58,9 | 17,6 | 19,3 | 8,19 | 7 | 5,24 | 1,24 | 74 | 683 | 0,2 | 1,8 | < 20 | 59 | 0,73 | 190 | 570 | < 20 | 494 | 0,94 | < 0,25 | 1,75 | 6,32 | - | 2,86 | 0,15 | 0,11 | 1,12 | 0,37 | 0,095 | 45,6 | 0,13 |
| Bouchées ou émincé végétal au soja et blé (convient aux véganes ou végétaliens), préemballé | 662 | 157 | 631 | 150 | 61,2 | 19,5 | 21,3 | 6,76 | 3,7 | 4,86 | 1,68 | 72 | 934 | 0,54 | 2,8 | < 20 | 84 | 1 | 220 | 680 | < 20 | 670 | 2 | < 0,25 | 0,36 | 6,74 | - | < 0,5 | 0,1 | 0,066 | 0,79 | 0,86 | 0,12 | 48,7 | - |
| Boudin antillais, à cuire | - | - | - | - | 63,9 | 11,5 | 11,5 | 9,2 | 16,5 | - | 1,22 | - | - | - | - | - | - | - | - | - | - | 544 | - | - | - | - | - | - | - | - | - | - | - | - | - |
| Boudin blanc truffé, à cuire | 944 | 227 | 944 | 227 | 58,8 | 11,3 | 11,3 | 4,8 | 17,8 | 1,4 | 1,48 | 54 | - | - | 1,9 | - | 15,1 | - | - | 142 | - | 711 | 1,3 | - | - | - | - | - | - | - | - | - | - | - | - |
| Boudin blanc, à cuire | - | - | - | - | 64,2 | 10 | 10 | - | 21 | - | 1,42 | - | - | - | - | - | - | - | - | - | - | 523 | - | - | - | - | - | - | - | - | - | - | - | - | - |
| Boudin blanc, sauté/poêlé | - | - | - | - | 60,1 | 9,89 | 9,89 | 4,6 | 17,1 | - | 1,64 | 43,8 | 85 | < 0,1 | 0,45 | 35,1 | 10,9 | < 0,1 | 134 | 160 | < 5 | 655 | 0,9 | 0,61 | 0,46 | - | - | < 0,5 | 0,16 | 0,14 | 1,87 | 0,38 | 0,11 | 20,7 | 0,54 |
| Boudin noir, à cuire | 1300 | 313 | 1300 | 313 | 56 | 12,2 | 12,2 | 4,5 | 27,1 | 1,42 | 1,29 | 22,4 | - | 0,3 | 17,4 | 5 | 6,84 | < 0,2 | 137 | 162 | 8 | 574 | 0,37 | < 0,25 | 0,12 | - | - | - | 0,034 | 0,13 | 0,41 | - | < 0,1 | 2,22 | 0,19 |
| Boudin noir, sauté/poêlé | 1020 | 246 | 1020 | 246 | 61,8 | 11,9 | 11,9 | 3,78 | 20,3 | 0 | 1,31 | 27,2 | 1150 | < 0,1 | 22,8 | 40 | 8,5 | < 0,1 | 27,2 | 155 | 11 | 525 | 0,7 | 0,75 | 0,13 | 0 | - | 0 | 0,052 | 0,062 | 0,53 | < 0,16 | 0,094 | 25,4 | 0,2 |
| Boudin, sauté/poêlé (aliment moyen) | 1020 | 246 | 1020 | 246 | 61,3 | 11,3 | 11,3 | 4,02 | 19,4 | 0 | 1,41 | 32,2 | 834 | 0,05 | 16,1 | 38,5 | 9,22 | 0,05 | 59 | 157 | 8,46 | 564 | 0,76 | 0,71 | 0,23 | 0 | | 0,075 | 0,084 | 0,085 | 0,93 | 0,17 | 0,097 | 24 | 0,3 |
| Boulettes au boeuf et à l'agneau (type kefta), préemballées, crues | 1030 | 247 | 1030 | 247 | 55,5 | 15,7 | 15,7 | 9,46 | 15,7 | 0,8 | 1,03 | 24 | 641 | 0,1 | 2,4 | < 20 | 26 | 0,33 | 150 | 330 | < 20 | 413 | 3,1 | < 0,25 | 1,04 | 13,5 | - | - | < 0,015 | < 0,01 | 3,12 | 0,43 | 0,13 | 56,6 | 1,75 |
| Brème, cru | 435 | 104 | 435 | 104 | 77 | 16,9 | 16,9 | traces | 4 | 0 | 0,11 | 63 | - | 0,028 | 0,44 | 7 | 26 | 0,087 | 210 | 350 | 24 | 43 | 0,42 | 5,5 | 1,3 | 0 | - | 0 | 0,07 | 0,07 | 4 | - | - | - | - |
| Bresaola | 697 | 165 | 697 | 165 | 57,1 | 31,6 | 31,6 | 0 | 43 | 0 | 3,46 | 35 | - | - | 7,1 | - | - | - | 200 | 323 | - | 1390 | 4,1 | 0,5 | 0,13 | - | - | 0 | 0,12 | 0,6 | 2,4 | - | 0,53 | 19 | - |
| Brochet, cru | 355 | 83,7 | 355 | 83,7 | 79,1 | 18,8 | 18,8 | traces | 0,94 | 0 | 0,13 | 33 | - | 0,09 | 0,9 | 15,3 | 21 | 0,15 | 176 | 228 | 12,7 | 51,9 | 0,97 | 4,33 | 0,45 | 0,1 | - | 3,6 | 0,076 | 0,058 | 1,83 | 0,75 | 0,13 | 15 | 2 |
| Brochet, rôti/cuit au four | 425 | 100 | 425 | 100 | 74,7 | 23,1 | 23,1 | traces | 0,88 | 0 | 0,12 | 73 | - | 0,058 | 0,71 | 12 | 40 | 0,28 | 282 | 331 | - | 49 | 0,77 | 2,7 | - | - | - | 3,8 | 0,067 | 0,077 | 2,8 | 0,81 | 0,14 | 16 | 2,15 |
| Brochette d'agneau | 467 | 112 | 467 | 112 | 72,8 | 12,6 | 12,6 | 2,54 | 5,1 | 2,6 | 0,48 | 15,8 | - | 0,061 | 1,34 | 2,59 | 19,5 | 0,042 | 109 | 240 | 1,37 | 190 | 0,87 | 0 | 0,5 | - | - | 11,7 | 0,079 | 0,15 | 3,71 | 0,43 | 0,18 | 8,61 | 0,93 |
| Brochette de boeuf | 793 | 191 | 793 | 191 | 62,6 | 17,1 | 17,1 | 1,02 | 12,9 | 1 | 0,13 | 7,8 | - | 0,062 | 1,8 | 4,49 | 21 | 0,021 | 172 | 307 | 7,02 | 50 | 1,17 | 0 | 0,66 | - | - | 9,46 | 0,062 | 0,21 | 3,02 | 0,28 | 0,32 | 14,2 | 1,39 |
| Brochette de volaille | 811 | 195 | 811 | 195 | 64,2 | 18,6 | 18,6 | 0,8 | 13 | 0 | 0,29 | 14 | - | 0,042 | 1 | 3,74 | 22,2 | 0,023 | 151 | 239 | 9,76 | 114 | 0,85 | 0,06 | 0,64 | - | - | 9,46 | 0,054 | 0,13 | 5,12 | 0,71 | 0,33 | 9,43 | 0,56 |
| Brochette mixte de viande | 866 | 208 | 866 | 208 | 63,4 | 19,7 | 19,7 | 1,02 | 13,8 | 0,33 | 0,61 | 10,9 | - | 0,052 | 1,4 | 4,32 | 21,4 | 0,022 | 162 | 273 | 8,36 | 247 | 1,01 | 0,03 | 0,65 | - | - | 9,46 | 0,058 | 0,17 | 4,09 | 0,49 | 0,32 | 11,5 | 0,98 |
| Bulot ou Buccin, cru | 555 | 131 | 555 | 131 | 66 | 23,8 | 23,8 | 7,61 | 0,57 | 0 | 0,52 | 57 | - | 1,03 | 5,02 | - | 86 | 0,45 | 141 | 347 | - | 206 | 1,62 | 0 | 0,13 | 0,1 | - | 4 | 0,028 | 0,11 | 1,08 | 0,21 | 0,34 | 6 | 9,07 |
| Bulot ou Buccin, cuit | 415 | 97,7 | 415 | 97,7 | 74,1 | 20,7 | 20,7 | 2,69 | 0,47 | 0 | 0,97 | 64,5 | 701 | 0,27 | 0,64 | 114 | 144 | < 0,1 | 108 | 173 | 31,4 | 387 | 1,58 | < 0,5 | 0,8 | - | - | 3,4 | < 0,04 | 0,096 | 1,44 | 0,41 | 0,083 | 7,67 | 4,61 |
| Cabillaud, cru | 329 | 77,6 | 329 | 77,6 | 79,8 | 18,1 | 18,1 | traces | 0,57 | < 0,2 | 0,23 | 4,43 | 110 | < 0,1 | 0,49 | 101 | 25,6 | < 0,1 | 163 | 357 | 136 | 65,2 | 0,39 | 1,41 | 0,51 | 0,1 | - | 1 | < 0,04 | 0,049 | 2,18 | 0,14 | 0,16 | 21,2 | 1,32 |
| Cabillaud, cuit à la vapeur | 451 | 106 | 451 | 106 | 73,4 | 24,5 | 24,5 | traces | 0,95 | 0 | 0,16 | 10 | 120 | 0,02 | 0,1 | 110 | 21 | 0,01 | 240 | 360 | 29,1 | 65 | 0,5 | 0 | 0,48 | 0 | - | 0 | 0,025 | 0,04 | 1,55 | 0,23 | 0,19 | 13 | 2 |
| Cabillaud, cuit, sans précision (aliment moyen) | 420 | 98,9 | 420 | 98,9 | 75 | 23,1 | 23,1 | 0 | 0,73 | 0 | 0,2 | 15,2 | 368 | 0,018 | 0,22 | 122 | 26,2 | 0,013 | 193 | 321 | 24,9 | 80,8 | 0,5 | 0,75 | 0,62 | 0,062 | | 0,62 | 0,046 | 0,055 | 2,08 | 0,22 | 0,22 | 11,1 | 1,7 |
| Cabillaud, rôti/cuit au four | 401 | 94,6 | 401 | 94,6 | 76 | 22,3 | 22,3 | traces | 0,6 | 0 | 0,23 | 18,4 | 520 | 0,017 | 0,3 | 130 | 29,4 | 0,016 | 164 | 297 | 22,4 | 90,5 | 0,5 | 1,2 | 0,7 | 0,1 | - | 1 | 0,059 | 0,065 | 2,41 | 0,22 | 0,24 | 10 | 1,53 |
| Caille, viande et peau, crue | 570 | 136 | 570 | 136 | 69,7 | 21,5 | 21,5 | 0 | 5,55 | 0 | 0,13 | 14 | - | 0,51 | 3,97 | 1,5 | 23 | 0,019 | 275 | 216 | - | 50 | 2,42 | 0,5 | 0,05 | - | - | 6,1 | 0,24 | 0,26 | 7,54 | 0,77 | 0,6 | 8 | 0,43 |
| Caille, viande et peau, cuite | 860 | 206 | 860 | 206 | 60,5 | 26,8 | 26,8 | 0 | 10,9 | 0 | 0,13 | 15 | - | 0,59 | 4,43 | 3 | 22 | - | 279 | 216 | - | 52 | 3,1 | 0,2 | 0,7 | 4,2 | - | 2,3 | 0,22 | 0,3 | 7,92 | - | 0,62 | 6 | 0,36 |
| Caille, viande, crue | 532 | 126 | 532 | 126 | 70 | 22,5 | 22,5 | 0,5 | 3,81 | 0 | 0,12 | 14 | - | 0,59 | 4,51 | 1,5 | 25 | 0,019 | 307 | 237 | - | 49 | 2,7 | - | - | - | - | 7,2 | 0,28 | 0,29 | 8,2 | 0,79 | 0,53 | 7 | 0,47 |
| Calmar ou Calamar ou encornet, à la romaine (beignet) | 1050 | 252 | 1050 | 252 | 51,5 | 8,69 | 8,69 | 21,2 | 14,4 | 1,66 | 2,13 | 32,1 | 845 | < 0,1 | < 1 | 18,8 | 20,4 | 0,11 | 236 | 88,8 | < 3 | 852 | < 1 | < 0,2 | 1,8 | - | - | 0 | < 0,05 | 1,1 | 0,56 | 0,3 | < 0,05 | 11,5 | 1,2 |
| Calmar ou calamar ou encornet, bouilli/cuit à l'eau | 632 | 149 | 632 | 149 | 61,1 | 32,5 | 32,5 | 1,64 | 1,4 | 0 | 1,86 | 180 | - | 1 | 5,77 | - | 60 | 0,21 | - | 637 | - | 744 | 3,46 | - | - | - | - | 8,5 | 0,017 | 1,73 | 2,19 | 0,9 | 0,27 | 24 | 5,4 |
| Calmar ou calamar ou encornet, cru | 326 | 77 | 326 | 77 | 80,9 | 14,4 | 14,4 | 2,17 | 1,19 | 0 | 0,61 | 17,5 | - | 0,15 | 0,2 | 12,8 | 44,2 | < 0,1 | 166 | 227 | 23,8 | 243 | 1,13 | 0,36 | 1,39 | 0 | - | 4,7 | < 0,04 | 0,032 | 1,57 | 0,24 | 0,096 | 11 | 3,15 |
| Calmar ou calamar ou encornet, frit ou poêlé avec matière grasse | 725 | 173 | 725 | 173 | 64,5 | 17,9 | 17,9 | 8,45 | 7,48 | 0 | 0,77 | 39 | - | 2,11 | 1,01 | - | 38 | 0,07 | 251 | 279 | - | 306 | 1,74 | - | - | - | - | 4,2 | 0,056 | 0,46 | 2,6 | 0,51 | 0,058 | 5 | 1,23 |
| Canard, cuisse avec peau, sans os, crue | 433 | 103 | 433 | 103 | 72,3 | 18,7 | 18,7 | traces | 3,1 | 0 | 0,26 | - | - | - | - | - | - | - | - | - | - | 103 | - | - | - | - | - | - | - | - | - | - | - | - | - |
| Canard, magret fumé | 1510 | 364 | 1510 | 364 | - | 21,9 | 21,9 | 0,55 | 30,5 | 0 | 3,28 | - | - | - | - | - | - | - | - | - | - | 1310 | - | - | - | - | - | - | - | - | - | - | - | - | - |
| Canard, magret, cru | 1410 | 340 | 1410 | 340 | 53,2 | 17,9 | 17,9 | 0,85 | 29,4 | 0 | 0,21 | 5 | - | - | - | - | - | - | - | - | - | 54,4 | - | - | - | - | - | - | - | - | - | - | - | - | - |
| Canard, magret, grillé/poêlé | 928 | 222 | 928 | 222 | 61,7 | 26,7 | 26,7 | traces | 12,8 | 0 | 1,09 | 5 | - | 0,5 | 4,8 | 15 | 31,2 | < 0,1 | 273 | 390 | 34 | 435 | 2,1 | 0,93 | 0,46 | - | - | 3,5 | 0,38 | 0,88 | 13,5 | 3,4 | 0,98 | < 20 | 3,3 |
| Canard, viande et peau, cru | 1180 | 284 | 1180 | 284 | 58,6 | 17,4 | 17,4 | 2,75 | 22,6 | 0 | 0,14 | 5,1 | 72,5 | 0,23 | 2,2 | < 20 | 20 | 0,05 | 180 | 310 | < 20 | 56,9 | 2,1 | 0,68 | 0,69 | < 0,8 | - | < 0,5 | 0,18 | 0,72 | 4,87 | 1,97 | 0,26 | 24,9 | 1,65 |
| Canard, viande et peau, rôti/cuit au four | 1370 | 331 | 1370 | 331 | 51,8 | 19 | 19 | 0 | 28,4 | 0 | 0,15 | 11 | - | 0,23 | 2,7 | - | 16 | 0,019 | 156 | 204 | - | 59 | 1,86 | 0,1 | 0,7 | 5,1 | - | 0 | 0,17 | 0,27 | 4,83 | 1,1 | 0,18 | 6 | 0,3 |
| Canard, viande, crue | 527 | 126 | 527 | 126 | 72,7 | 19,4 | 19,4 | 0 | 5,33 | 0 | 0,19 | 9 | - | 0,2 | 2,63 | 1,2 | 20,5 | 0,019 | 199 | 277 | 24 | 77,7 | 1,47 | 0,55 | 0,7 | 2,8 | 3,6 | 5,85 | 0,38 | 0,42 | 4,83 | 1,6 | 0,41 | 36,5 | 0,49 |
| Canard, viande, rôtie/cuite au four | 811 | 194 | 811 | 194 | 63,1 | 23,3 | 23,3 | 0 | 11,2 | 0 | 0,39 | 6,78 | 245 | 0,35 | 3,35 | 4 | 35,8 | 0,036 | 203 | 392 | 9,08 | 154 | 1,92 | 0,1 | 0,7 | 3,8 | - | 0 | 0,26 | 0,47 | 11,1 | 1,5 | 0,71 | 10 | 2,9 |

| | | | | | | | | | | | | | | | | | | |
|---|---|---|---|---|---|---|---|---|---|---|---|---|---|---|---|---|---|---|
| Capelan, cru | 380 | 90 | 380 | 90 | 78,6 | 18,7 | 18,7 | traces | 1,71 | 0 | 0,27 | - | - | - | - | - | - | - |
| Carangue, cru | 438 | 104 | 438 | 104 | - | 19,9 | 19,9 | traces | 2,7 | 0 | 0,22 | 47 | - | 0,027 | 0,7 | - | 30 | - |
| Cardine franche, crue | 395 | 93,4 | 395 | 93,4 | - | 19,9 | 19,9 | traces | 1,53 | 0 | - | - | - | - | - | 17 | - | - |
| Carpaccio de saumon avec marinade | 810 | 194 | 810 | 194 | 64,3 | 19,8 | 19,8 | 0,22 | 12,7 | 0 | 0,098 | 14,9 | - | 0,08 | 0,64 | 37,9 | 24,7 | 0,06 |
| Carpe, crue, élevage | 477 | 114 | 477 | 114 | 77,3 | 17,7 | 17,7 | traces | 4,36 | 0 | 0,077 | 29,2 | 37 | < 0,1 | 1,58 | 18,3 | 23,7 | < 0,1 |
| Carpe, rôtie/cuite au four | 633 | 151 | 633 | 151 | 69,6 | 21,6 | 21,6 | traces | 7,17 | 0 | 0,16 | 52 | - | 0,12 | 1,59 | 6 | 38 | 0,23 |
| Carrelet ou plie, cru | 375 | 88,6 | 375 | 88,6 | 77,7 | 20 | 20 | traces | 0,94 | 0 | 0,2 | 127 | - | < 0,1 | 0,21 | 191 | 28,5 | < 0,1 |
| Carrelet ou plie, cuit à la vapeur | 393 | 93 | 393 | 93 | 78 | 19 | 19 | traces | 1,89 | 0 | - | - | - | 0,05 | - | 30 | - | 0,04 |
| Carrelet ou plie, pané, frit | 1010 | 242 | 1010 | 242 | 58,5 | 13,4 | 13,4 | 8,86 | 16,8 | 1 | 0,66 | 27 | 240 | 0,052 | 0,54 | 14 | 22,4 | 0,29 |
| Caviar, semi-conserve | 1060 | 254 | 1060 | 254 | - | 25 | 25 | traces | 17,1 | 0 | - | - | - | 0,04 | - | 117 | - | 0,07 |
| Cerf, cru | 479 | 113 | 479 | 113 | 73,6 | 23,7 | 23,7 | 0 | 2,04 | 0 | 0,13 | 5,25 | - | 0,25 | 3,79 | - | 23 | 0,041 |
| Cerf, rôti/cuit au four | 632 | 150 | 632 | 150 | 65,2 | 30,2 | 30,2 | 0 | 3,19 | 0 | 0,14 | 7 | - | 0,3 | 4,47 | - | 24 | 0,046 |
| Cervelas | 1150 | 278 | 1150 | 278 | 61,1 | 11,9 | 11,9 | 1,56 | 24,8 | 0,4 | 1,95 | 32,6 | 1340 | 0,11 | 0,5 | 37,9 | 20,4 | < 0,1 |
| Cervelas obernois | 1260 | 305 | 1260 | 305 | - | 13 | 13 | 1 | 27,5 | 0,6 | 2 | - | - | - | 0,5 | - | - | - |
| Cervelle, agneau, crue | 508 | 122 | 508 | 122 | 79,2 | 10,4 | 10,4 | 0,8 | 8,58 | 0 | 0,28 | 9 | - | 0,24 | 1,75 | - | 12 | 0,044 |
| Cervelle, agneau, cuite | 603 | 145 | 603 | 145 | 75,7 | 12,6 | 12,6 | 0,8 | 10,2 | 0 | 0,34 | 12 | - | 0,21 | 1,68 | 1 | 14 | 0,059 |
| Cervelle, veau, crue | 488 | 117 | 488 | 117 | 79,8 | 10,3 | 10,3 | 0,5 | 8,21 | 0 | 0,32 | 10 | - | 0,22 | 2,13 | 0,62 | 14 | 0,037 |
| Cervelle, veau, cuite | 551 | 133 | 551 | 133 | 76,9 | 11,5 | 11,5 | 0 | 9,63 | 0 | 0,39 | 16 | - | 0,26 | 1,67 | 1 | 16 | 0,038 |
| Chair à saucisse, crue | 1340 | 323 | 1340 | 323 | 54,2 | 14,7 | 14,7 | 0,6 | 29,1 | 0 | 1,39 | 14,1 | - | - | 1,2 | 6 | 17,1 | - |
| Chapon, viande et peau, cru | 764 | 183 | 764 | 183 | 63,2 | 21,8 | 21,8 | 0 | 10,6 | 0 | 0,56 | 11 | - | 0,05 | 1,09 | - | 21 | 0,02 |
| Chapon, viande et peau, rôti/cuit au four | 923 | 221 | 923 | 221 | 58,7 | 29 | 29 | 0 | 11,7 | 0 | 0,12 | 14 | - | 0,069 | 1,49 | - | 24 | 0,021 |
| Charcuterie (aliment moyen) | 991 | 238 | 991 | 238 | 58,5 | 19,7 | 19,7 | 1,47 | 17,1 | 0,25 | 2,47 | 14,6 | 1530 | 0,19 | 1,61 | 9,77 | 21,8 | 0,075 |
| Cheval, entrecôte, crue | 499 | 118 | 499 | 118 | - | 22,1 | 22,1 | traces | 3,34 | 0 | - | - | - | - | 2,84 | - | - | - |
| Cheval, entrecôte, grillée/poêlée | 711 | 169 | 711 | 169 | 64,4 | 28,2 | 28,2 | traces | 6,24 | 0 | - | - | - | - | 3,38 | - | - | - |
| Cheval, faux-filet, cru | 510 | 121 | 510 | 121 | - | 22,2 | 22,2 | traces | 3,58 | 0 | - | - | - | - | 2,55 | - | - | - |
| Cheval, faux-filet, grillé/poêlé | 685 | 163 | 685 | 163 | 65,5 | 26,6 | 26,6 | traces | 6,27 | 0 | - | - | - | - | 3,3 | - | - | - |
| Cheval, faux-filet, rôti/cuit au four | 729 | 173 | 729 | 173 | 63,2 | 28,3 | 28,3 | traces | 6,68 | 0 | - | - | - | - | 3,3 | - | - | - |
| Cheval, steak, cru | 541 | 129 | 541 | 129 | 72,6 | 21,4 | 21,4 | 0,4 | 4,6 | 0 | 0,13 | 6 | 9 | 0,14 | 3,82 | 1,2 | 24 | 0,019 |
| Cheval, tende de tranche, crue | 448 | 106 | 448 | 106 | - | 22,7 | 22,7 | traces | 1,68 | 0 | - | - | - | - | 3,06 | - | - | - |
| Cheval, tende de tranche, grillée/poêlée | 570 | 135 | 570 | 135 | 68,7 | 27,9 | 27,9 | traces | 2,58 | 0 | - | - | - | - | 3,99 | - | - | - |
| Cheval, tende de tranche, rôtie/cuite au four | 607 | 144 | 607 | 144 | 66,6 | 29,7 | 29,7 | traces | 2,75 | 0 | - | - | - | - | 3,99 | - | - | - |
| Cheval, viande, crue | 699 | 167 | 699 | 167 | 70 | 18,8 | 18,8 | 0,6 | 10 | 0 | 0,16 | 10 | - | 0,14 | 3,5 | 1 | 20 | 0,02 |
| Cheval, viande, rôtie/cuite au four | 581 | 138 | 581 | 138 | 68,6 | 28 | 28 | 0 | 2,85 | 0 | 0,1 | 4,8 | - | 0,16 | 3,4 | 4 | 28,1 | < 0,1 |
| Chevreau, cru | 436 | 103 | 436 | 103 | 75,8 | 20,6 | 20,6 | 0 | 2,31 | 0 | 0,21 | 13 | - | 0,26 | 2,83 | - | - | 0,038 |
| Chevreau, cuit | 573 | 136 | 573 | 136 | 68,2 | 27,1 | 27,1 | 0 | 3,03 | 0 | 0,22 | 17 | - | 0,3 | 3,73 | - | - | 0,042 |
| Chevreuil, cru | 465 | 110 | 465 | 110 | - | 23 | 23 | 0 | 2 | 0 | 0,053 | 5 | - | - | 3,4 | - | - | - |
| Chevreuil, rôti/cuit au four | 632 | 150 | 632 | 150 | 65,2 | 30,2 | 30,2 | 0 | 3,19 | 0 | 0,14 | 7 | - | 0,3 | 4,47 | 4 | 24 | 0,046 |
| Chinchard gras, cru | 616 | 147 | 616 | 147 | 71,7 | 19,6 | 19,6 | traces | 7,66 | 0 | 0,14 | 5,72 | - | < 0,1 | 0,85 | 20,8 | 31,7 | < 0,1 |
| Chinchard maigre, cru | 408 | 96,7 | 408 | 96,7 | 77,4 | 18,7 | 18,7 | traces | 2,42 | 0 | 0,16 | 20,1 | - | < 0,1 | 0,96 | 37,3 | 31,7 | < 0,1 |
| Chinchard, cru | 487 | 116 | 487 | 116 | 78 | 19 | 19 | traces | 4,42 | 0 | - | 92 | - | 0,99 | - | - | 69 | 0,2 |
| Chipolata, crue | 1050 | 253 | 1050 | 253 | 62,2 | 15,8 | 15,8 | 1 | 20,6 | 0,4 | 1,58 | < 20 | 1510 | 0,8 | 0,48 | 7 | 16,6 | < 0,2 |
| Chipolata, cuite | 1170 | 282 | 1170 | 282 | 55,4 | 18,8 | 18,8 | 0,8 | 22,4 | 0 | 2,15 | 19 | 1150 | 0,07 | 0,79 | < 20 | 23 | 0,04 |
| Chorizo | 1790 | 433 | 1790 | 433 | 34 | 23,5 | 23,5 | 3,05 | 36,1 | 0,93 | 4,24 | - | - | - | - | 7 | - | - |
| Chorizo supérieur, doux ou fort, type charcuterie en tranches | 1340 | 322 | 1340 | 322 | 43,6 | 22,6 | 22,6 | 2,1 | 23,9 | 0 | 4,3 | 28 | 2380 | 0,1 | 1,4 | < 20 | 30 | 0,23 |
| Chorizo supérieur, doux ou fort, type saucisse sèche | 1970 | 476 | 1970 | 476 | 28 | 20,9 | 20,9 | 2,6 | 41,9 | 0 | 3,91 | 24 | 2480 | 0,11 | 1,8 | 30 | 27 | 0,27 |
| Clam, Praire ou Palourde, bouilli/cuit à l'eau | 420 | 99,2 | 420 | 99,2 | 74,1 | 16,2 | 16,2 | 5,33 | 1,48 | 0 | 1,56 | 72,5 | - | 0,5 | 9,67 | 80 | 12,9 | 0,75 |
| Clam, Praire ou Palourde, cru | 339 | 80,5 | 339 | 80,5 | 81,1 | 11,5 | 11,5 | 2,66 | 2,65 | 0 | 0,82 | 46 | - | 0,053 | 7,81 | 80 | 19 | 0,085 |
| Coeur, agneau, cru | 497 | 119 | 497 | 119 | 76,7 | 16,5 | 16,5 | 0,4 | 5,68 | 0 | 0,22 | 6 | - | 0,4 | 4,6 | - | 17 | 0,046 |
| Coeur, agneau, cuit | 750 | 179 | 750 | 179 | 64,2 | 25 | 25 | 1,93 | 7,91 | 0 | 0,16 | 14 | - | 0,61 | 5,52 | - | 24 | 0,055 |
| Coeur, boeuf, cru | 435 | 103 | 435 | 103 | 79 | 18,5 | 18,5 | 0,7 | 2,95 | 0 | 0,2 | 5,48 | - | 0,35 | 5,14 | 1,8 | 19 | 0,034 |
| Coeur, boeuf, cuit | 560 | 133 | 560 | 133 | - | 23 | 23 | 0,15 | 4,5 | 0 | 0,15 | 5 | - | 0,56 | 6,7 | 3 | 21 | 0,033 |
| Coeur, dinde, cru | 566 | 135 | 566 | 135 | 74,5 | 16,7 | 16,7 | 0,4 | 7,44 | 0 | 0,32 | 18 | - | 0,49 | 3,7 | - | 21 | 0,1 |
| Coeur, dinde, cuit | 701 | 167 | 701 | 167 | 67,2 | 24,9 | 24,9 | 0 | 7,52 | 0 | 0,35 | 21 | - | 0,83 | 6,96 | - | 28 | 0,12 |
| Coeur, poulet, cru | 622 | 149 | 622 | 149 | 73,6 | 15,6 | 15,6 | 0,7 | 9,32 | 0 | 0,17 | 12 | - | 0,35 | 5,96 | - | 16 | 0,089 |
| Coeur, poulet, cuit | 744 | 177 | 744 | 177 | 64,9 | 26,4 | 26,4 | 0,1 | 7,92 | 0 | 0,12 | 19 | - | 0,5 | 9,03 | - | 20 | 0,11 |
| Coeur, veau, cru | 446 | 106 | 446 | 106 | 77,3 | 16,1 | 16,1 | 1,8 | 3,84 | 0 | 0,23 | 4,71 | - | 0,32 | 3,92 | 1,9 | 20 | 0,034 |
| Confit de canard | 1140 | 273 | 1140 | 273 | 49,7 | 25,7 | 25,7 | 0,15 | 18,8 | 0,4 | 1,44 | 5,3 | - | < 0,1 | 2,7 | 17 | 18,8 | < 0,1 |
| Confit de canard, viande (cuisse), sans peau, réchauffé | 818 | 195 | 818 | 195 | 58 | 32 | 32 | traces | 7,4 | 0 | 1,89 | 13 | 1220 | 0,16 | 2,3 | < 20 | 17 | < 0,01 |
| Confit de foie de volaille | 1470 | 356 | 1470 | 356 | 46,7 | 14 | 14 | 7,42 | 30 | 0 | 1,93 | - | - | - | - | - | - | - |
| Congre, cru | 481 | 115 | 481 | 115 | 73,2 | 18,3 | 18,3 | traces | 4,6 | 0 | 0,13 | 71 | 100 | 0,2 | 1,3 | 30 | 20 | 0,01 |
| Coppa | 1180 | 284 | 1180 | 284 | - | 25 | 25 | 1 | 20 | 0 | 4,5 | - | - | - | - | 12 | - | - |
| Coquille Saint-Jacques, noix et corail, crue | 351 | 83 | 351 | 83 | 79,3 | 17 | 17 | 0,78 | 1,31 | 0 | 0,48 | 220 | - | < 0,09 | 1,16 | 42,6 | 51,4 | 2,4 |
| Coquille Saint-Jacques, noix et corail, cuite | 463 | 110 | 463 | 110 | 73 | 20,2 | 20,2 | 3,22 | 1,77 | 0 | 0,46 | 17,1 | 410 | 0,073 | 1,4 | 20 | 62,8 | 1,27 |
| Coquille Saint-Jacques, noix, crue | 354 | 83,6 | 354 | 83,6 | 78,7 | 17,9 | 17,9 | 1,15 | 0,84 | 0 | 0,34 | 11,9 | - | < 0,1 | 0,58 | 28 | 44 | 1,15 |
| Corb, cru | 370 | 87,2 | 370 | 87,2 | 78 | 20 | 20 | traces | 0,8 | 0 | - | 74 | - | 0,8 | - | - | 61 | 0,3 |
| Cordon bleu de volaille, préemballé | 944 | 226 | 944 | 226 | 54,1 | 14 | 14 | 14,6 | 12,2 | 0,78 | 1,73 | 60,6 | - | 0,071 | 0,68 | 8 | 37,6 | 0,18 |
| Corégone lavaret, cru | 486 | 115 | 486 | 115 | 74,7 | 20,5 | 20,5 | traces | 3,72 | 0 | 0,13 | 43 | - | 0,049 | 0,44 | 12 | 31,5 | 0,059 |
| Corned-beef, appertisé | 787 | 188 | 787 | 188 | 63,4 | 23,1 | 23,1 | < 0,1 | 10,5 | < 1,5 | 1,95 | 14,5 | 665 | 0,064 | 2,24 | - | 14 | 0,014 |
| Coulirou, cru | 301 | 70,9 | 301 | 70,9 | - | 17,5 | 17,5 | traces | 0,1 | 0 | 0,15 | 50 | - | - | 0,4 | - | 30 | - |
| Crabe ou Tourteau, bouilli/cuit à l'eau | 520 | 124 | 520 | 124 | 72,3 | 19,5 | 19,5 | 1,79 | 4,29 | 0 | 1,23 | 53 | 640 | 1,99 | 1,62 | 100 | 55 | 0,087 |
| Crabe, cru | 438 | 104 | 438 | 104 | 75,7 | 19,5 | 19,5 | 0,027 | 2,86 | 0 | 0,69 | 84 | - | 0,44 | 1,67 | 60 | 42,5 | - |
| Crabe, miettes et ou pattes décortiquées, appertisé, égoutté | 468 | 111 | 468 | 111 | 72,4 | 14,8 | 14,8 | 8,72 | 1,8 | < 0,33 | 1,59 | 120 | 830 | 0,44 | 3,5 | 60 | 32 | 0 |
| Crevette géante tigrée, cuite | - | - | - | - | 73,5 | 23,4 | 23,4 | - | 0,98 | 0 | 1,43 | 44,4 | 870 | 1,21 | 0,34 | 22,1 | 49 | < 0,1 |
| Crevette grise, cuite | 381 | 89,9 | 381 | 89,9 | 74,7 | 18,3 | 18,3 | 1,47 | 1,2 | 0 | 1,15 | 1000 | 671 | 1 | 0,79 | 260 | 53 | 0,14 |
| Crevette pattes blanches, cuite | - | - | - | - | 73,9 | 22,6 | 22,6 | - | 0,96 | 0 | 1,82 | 109 | 1070 | 0,73 | 1,64 | 4,12 | 43,5 | < 0,1 |
| Crevette rose bouquet, cuite | - | - | - | - | 74,6 | 22,8 | 22,8 | - | 1,27 | 0 | 0,47 | 35,4 | 369 | 0,94 | 0,2 | 14,1 | 29,7 | 0,064 |
| Crevette rose, crue | 392 | 92,3 | 392 | 92,3 | 76,3 | 21,9 | 21,9 | 0 | 0,53 | 0 | 0,3 | 64 | - | 0,39 | 0,52 | - | 35 | 0,033 |
| Crevette royale rose, cuite | - | - | - | - | 71,2 | 26,6 | 26,6 | - | 1,18 | 0 | 0,34 | 81,1 | 159 | 0,29 | 0,84 | 40,9 | 54,3 | < 0,1 |
| Crevette, crue | 420 | 99 | 420 | 99 | 74,3 | 19,7 | 19,7 | 3,21 | 0,84 | 0 | 0,45 | 84,8 | - | 0,72 | 3,21 | 120 | 49,8 | 0,17 |
| Crevette, cuite | 397 | 93,8 | 397 | 93,8 | 74,9 | 19 | 19 | 1,87 | 1,16 | 0 | 1,35 | 340 | 1510 | 0,68 | 1,98 | < 5 | 61 | 0,077 |
| Crevette, surgelée, crue | 446 | 105 | 446 | 105 | - | 23,4 | 23,4 | 0,74 | 0,9 | < 0,5 | 0,6 | - | - | - | - | - | - | - |
| Denté, cru | 418 | 99,3 | 418 | 99,3 | 78,1 | 17 | 17 | traces | 3,5 | 0 | 0,19 | 34 | - | 0,015 | 1 | - | - | - |

| | | | | | | | | | | | | | | | | | |
|---|---|---|---|---|---|---|---|---|---|---|---|---|---|---|---|---|---|
| Capelan, cru | - | - | - | - | - | - | - | - | - | - | - | - | - | - | - | - | - |
| Carangue, cru | 193 | 415 | - | 89 | 0,42 | - | - | - | - | 0 | 0,19 | 0,1 | 5,6 | - | 0,65 | 2,9 | 9,1 |
| Cardine franche, crue | - | - | - | - | - | - | - | - | - | - | - | - | - | - | - | - | - |
| Carpaccio de saumon avec marinade | 173 | 279 | 19,5 | 39 | 0,55 | 4,21 | 3,25 | - | - | 4,21 | 0,25 | 0,08 | 6,26 | 1,46 | 0,55 | 16,3 | 4,33 |
| Carpe, crue, élevage | 143 | 310 | 20 | 30,6 | 0,58 | 3,84 | 0,63 | 0,1 | - | 1,3 | < 0,04 | 0,072 | 2,81 | 1,25 | 0,13 | 16 | 3,16 |
| Carpe, rôtie/cuite au four | 531 | 427 | 20,2 | 63 | 1,58 | 3,25 | - | - | - | 1,6 | 0,14 | 0,07 | 2,1 | 0,87 | 0,22 | 17 | 1,47 |
| Carrelet ou plie, cru | 174 | 352 | 31,4 | 79,5 | 0,49 | 1,11 | 0,72 | 0,15 | 2,2 | 1,5 | 0,42 | 0,096 | 4,32 | 1,29 | 0,25 | 9,5 | 0,64 |
| Carrelet ou plie, cuit à la vapeur | - | - | 30,4 | - | 0,68 | - | - | - | - | - | - | - | - | - | - | - | - |
| Carrelet ou plie, pané, frit | 132 | 234 | 20 | 264 | 0,73 | 0 | 0 | 0 | - | 0 | 0,18 | 0,06 | 0,8 | 0,82 | 0,12 | 27 | 0 |
| Caviar, semi-conserve | - | - | 8 | - | 0,8 | - | 1,6 | - | - | - | 0,49 | 0,071 | 0,08 | 0,28 | 0,08 | 25 | 5,45 |
| Cerf, cru | 202 | 318 | - | 51,4 | 2,09 | - | 0,2 | 1,1 | - | 0 | 0,22 | 0,48 | 6,37 | - | 0,37 | 4 | 6,31 |
| Cerf, rôti/cuit au four | 226 | 335 | - | 54 | 2,75 | - | - | - | - | 0 | 0,18 | 0,6 | 6,71 | - | - | - | - |
| Cervelas | 197 | 187 | < 5 | 763 | 1 | < 0,5 | 0,56 | - | - | 18,1 | 0,21 | 0,11 | 3,43 | 0,43 | 0,17 | 32,9 | 0,61 |
| Cervelas obernois | - | - | - | 800 | - | - | - | - | - | - | - | - | - | - | - | - | - |
| Cervelle, agneau, crue | 270 | 296 | - | 112 | 1,17 | - | - | - | - | 16 | 0,13 | 0,3 | 3,9 | 0,92 | 0,29 | 3 | 11,3 |
| Cervelle, agneau, cuite | 337 | 205 | - | 134 | 1,36 | - | - | - | - | 12 | 0,11 | 0,24 | 2,47 | 0,99 | 0,11 | 5 | 9,25 |
| Cervelle, veau, crue | 274 | 315 | 11,6 | 127 | 1,11 | - | - | - | - | 14 | 0,13 | 0,26 | 4,3 | 2,72 | 0,28 | 3 | 12,2 |
| Cervelle, veau, cuite | 385 | 214 | - | 156 | 1,61 | - | - | - | - | 13 | 0,08 | 0,2 | 2,43 | 1 | 0,17 | 3 | 9,65 |
| Chair à saucisse, crue | - | 293 | - | 556 | 2,3 | - | - | - | - | - | - | - | - | - | - | - | - |
| Chapon, viande et peau, cru | 183 | 217 | - | 272 | 1,17 | - | 0,32 | 2,4 | - | 1,7 | 0,064 | 0,13 | 7,27 | 0,97 | 0,36 | 6 | 0,34 |
| Chapon, viande et peau, rôti/cuit au four | 246 | 255 | - | 49 | 1,74 | - | - | - | - | 0 | 0,07 | 0,17 | 8,95 | 1,1 | 0,43 | 6 | 0,33 |
| Charcuterie (aliment moyen) | 194 | 380 | 10,7 | 992 | 2,23 | 0,51 | 0,4 | | | | 0,46 | 0,27 | 5,63 | 0,7 | 0,29 | 14,3 | 0,96 |
| Cheval, entrecôte, crue | - | - | 5,37 | - | 2,51 | - | - | - | - | - | - | - | 6,15 | - | 0,66 | - | 1,44 |
| Cheval, entrecôte, grillée/poêlée | - | - | 7,92 | - | 3,18 | - | - | - | - | - | - | - | 5,55 | - | 0,66 | - | 1,44 |
| Cheval, faux-filet, cru | - | - | 5,35 | - | 1,91 | - | - | - | - | - | - | - | 5,6 | - | 0,66 | - | 1,51 |
| Cheval, faux-filet, grillé/poêlé | - | - | 7,54 | - | 2,25 | - | - | - | - | - | - | - | 5,22 | - | 0,69 | - | 1,51 |
| Cheval, faux-filet, rôti/cuit au four | - | - | 8,03 | - | 2,4 | - | - | - | - | - | - | - | 5,22 | - | 0,69 | - | 1,51 |
| Cheval, steak, cru | 221 | 360 | 12,2 | 53 | 2,9 | - | - | - | - | 1 | 0,13 | 0,1 | 4,6 | - | 0,38 | - | 3 |
| Cheval, tende de tranche, crue | - | - | 5,22 | - | 2,32 | - | - | - | - | - | - | - | 5,39 | - | 0,7 | - | 1,63 |
| Cheval, tende de tranche, grillée/poêlée | - | - | 7,47 | - | 3,32 | - | - | - | - | - | - | - | 5,28 | - | 0,69 | - | 1,63 |
| Cheval, tende de tranche, rôtie/cuite au four | - | - | 7,96 | - | 3,53 | - | - | - | - | - | - | - | 5,28 | - | 0,69 | - | 1,63 |
| Cheval, viande, crue | 185 | 291 | 6 | 62 | 4,61 | 0,3 | 0,23 | 0 | - | 0,6 | 0,14 | 0,27 | 4,5 | 0,6 | 0,5 | 8 | 3,1 |
| Cheval, viande, rôtie/cuite au four | 214 | 333 | 5 | 39,8 | 3,1 | < 0,5 | 1,15 | - | - | < 1 | 0,15 | 0,36 | 9,5 | 0,45 | 0,7 | < 20 | 4,05 |
| Chevreau, cru | 180 | 385 | - | 82 | 4 | - | - | - | - | 0 | 0,11 | 0,49 | 3,75 | - | - | 5 | 1,13 |
| Chevreau, cuit | 201 | 405 | - | 86 | 5,27 | 0 | 0,34 | 1,2 | - | 0 | 0,09 | 0,61 | 3,95 | - | 0 | 5 | 1,19 |
| Chevreuil, cru | - | - | - | 21 | - | - | - | - | - | - | - | - | - | - | - | - | - |
| Chevreuil, rôti/cuit au four | 226 | 335 | 8,9 | 54 | 2,75 | - | - | 2 | 0,7 | 0 | 0,18 | 0,6 | 6,71 | - | 0,65 | 6 | 0,8 |
| Chinchard gras, cru | 195 | 382 | 42,8 | 54,7 | 0,38 | 48,5 | 0,48 | - | - | - | 0,08 | 0,12 | 6,7 | 0,29 | 0,37 | - | 7,5 |
| Chinchard maigre, cru | 206 | 386 | 16,6 | 62 | 0,36 | 41,2 | 0,64 | - | - | - | 0,1 | 0,12 | 4,12 | 0,38 | 0,3 | - | 6,99 |
| Chinchard, cru | - | 986 | 26 | - | 1,59 | - | - | - | - | - | - | - | - | - | - | - | - |
| Chipolata, crue | 144 | 404 | 5,44 | 722 | 1,76 | 0,19 | 0,34 | - | - | - | 0,4 | - | 4,6 | - | 0,21 | 1,5 | 0,36 |
| Chipolata, cuite | 170 | 370 | < 20 | 859 | 2 | 0,46 | 0,36 | - | - | 11 | 0,38 | 0,041 | 4,66 | 0,66 | 0,13 | < 5 | 0,81 |
| Chorizo | - | - | - | 1690 | - | - | - | - | - | - | - | - | - | - | - | - | - |
| Chorizo supérieur, doux ou fort, type charcuterie en tranches | 300 | 650 | < 20 | 1720 | 2,7 | 0,73 | 1,6 | - | - | - | 0,35 | 0,21 | 4,92 | 1,19 | 0,32 | 49,1 | 0,82 |
| Chorizo supérieur, doux ou fort, type saucisse sèche | 220 | 490 | < 20 | 1560 | 2 | 0,83 | 2,25 | - | - | - | 0,39 | 0,22 | 6,14 | 1,24 | 0,27 | 51,4 | 1,11 |
| Clam, Praire ou Palourde, bouilli/cuit à l'eau | 231 | 432 | 51,8 | 625 | 2,05 | 0 | 1 | - | - | 22,1 | 0,11 | 0,29 | 2,03 | 0,47 | 0,13 | 20,1 | 39,5 |
| Clam, Praire ou Palourde, cru | 168 | 180 | - | 329 | 0,96 | 0 | 0,68 | 0,2 | - | 13 | 0,048 | 0,13 | 1,08 | 0,15 | 0,035 | 10,5 | 11,3 |
| Coeur, agneau, cru | 175 | 316 | - | 89 | 1,87 | 0,51 | 0,48 | - | - | 5 | 0,37 | 0,99 | 6,14 | 2,63 | 0,39 | 2 | 10,3 |
| Coeur, agneau, cuit | 254 | 188 | - | 63 | 3,68 | - | - | - | - | 7 | 0,17 | 1,19 | 4,36 | 1,37 | 0,3 | 2 | 11,2 |
| Coeur, boeuf, cru | 213 | 264 | 23,5 | 81,7 | 1,49 | 1 | 0,21 | 0 | - | 2 | 0,29 | 0,83 | 6,78 | 2,15 | 0,11 | 11,5 | 8,44 |
| Coeur, boeuf, cuit | 254 | 219 | - | 59 | 2,3 | 0,1 | 0,29 | 0,5 | - | - | 0,36 | 1,21 | 6,9 | 1,06 | 0,33 | 5 | 11,5 |
| Coeur, dinde, cru | 183 | 179 | - | 129 | 3,21 | 0,4 | 0,31 | 0 | - | 3 | 0,17 | 1,13 | 6,44 | 3,12 | 0,48 | 6 | 13,3 |
| Coeur, dinde, cuit | 241 | 203 | - | 140 | 4,6 | 0 | 0,13 | 0 | - | - | 0,24 | 1,54 | 7,76 | 1,58 | 0,61 | 8 | 13,9 |
| Coeur, poulet, cru | 177 | 177 | - | 69 | 6,59 | - | - | - | - | 3,2 | 0,23 | 0,73 | 4,88 | 2,56 | 0,3 | 58,5 | 7,29 |
| Coeur, poulet, cuit | 199 | 132 | - | 48 | 7,3 | - | - | - | - | 1,8 | 0,07 | 0,74 | 2,8 | 2,65 | 0,32 | 80 | 7,29 |
| Coeur, veau, cru | 213 | 261 | 14 | 90,5 | 1,44 | 1 | 0,8 | 0 | - | 6,5 | 0,43 | 0,84 | 6,2 | 2,79 | 0,37 | 11 | 12,9 |
| Confit de canard | 179 | 241 | 28 | 568 | 5,7 | 1,8 | 0,46 | - | - | < 1 | 0,056 | 0,34 | 7,1 | 0,74 | 0,3 | < 20 | 2,95 |
| Confit de canard, viande (cuisse), sans peau, réchauffé | 170 | 300 | 40 | 757 | 5,5 | 1,14 | 0,59 | < 0,8 | - | < 0,5 | 0,043 | 0,23 | 5,16 | 0,92 | 0,18 | 23,8 | 4,24 |
| Confit de foie de volaille | - | - | - | 770 | - | - | - | - | - | - | - | - | - | - | - | - | - |
| Congre, cru | 270 | 240 | - | 50 | 0,9 | 0 | 0 | - | - | traces | 0,06 | 0,04 | 4,3 | 0,24 | 0 | 0 | 3 |
| Coppa | - | - | - | 1800 | - | - | - | - | - | - | - | - | - | - | - | - | - |
| Coquille Saint-Jacques, noix et corail, crue | 225 | 388 | 21,4 | 191 | 2,91 | < 0,5 | 0 | 0 | - | 0 | < 0,04 | 0,14 | 1,9 | 0,4 | 0,1 | 17 | 4,99 |
| Coquille Saint-Jacques, noix et corail, cuite | 357 | 348 | 61,4 | 183 | 2,89 | 0 | 0 | 0 | - | 0 | 0,046 | 0,037 | 0,99 | 0,26 | 0,11 | 19 | 8,39 |
| Coquille Saint-Jacques, noix, crue | 224 | 392 | 7,7 | 137 | 1,56 | < 0,5 | 1,04 | - | - | - | < 0,04 | 0,046 | 1,97 | 0,092 | 0,13 | - | 2 |
| Corb, cru | - | 527 | - | - | 2,1 | - | - | - | - | - | - | - | - | - | - | - | - |
| Cordon bleu de volaille, préemballé | 172 | 336 | < 10 | 683 | 1,16 | < 0,5 | 2,4 | - | - | 0 | 0,13 | 0,12 | 4,8 | 0,6 | 0,28 | < 16 | 0,58 |
| Corégone lavaret, cru | 280 | 349 | 16 | 50 | 0,77 | 7,5 | 1,45 | 0,1 | - | 0 | 0,11 | 0,095 | 3,5 | 0,75 | 0,3 | 15 | 1 |
| Corned-beef, appertisé | 111 | 244 | - | 781 | 3,57 | 0,2 | 0,15 | 1,6 | - | 0 | 0,055 | 0,19 | 3,32 | 0,63 | 0,13 | 9 | 1,62 |
| Coulirou, cru | 115 | 614 | - | 61 | - | - | - | - | - | 0 | 0,03 | 0,08 | 3,2 | - | - | - | - |
| Crabe ou Tourteau, bouilli/cuit à l'eau | 302 | 248 | 39,6 | 478 | 4,42 | traces | 2,4 | 0 | - | 3,6 | 0,084 | 0,53 | 2,1 | 0,73 | 0,2 | 25,5 | 6,79 |
| Crabe, cru | 288 | 268 | 82 | 275 | 7,7 | 0 | 2,4 | 0 | - | 2,5 | 0,097 | 0,17 | 2,4 | 0,71 | 0,21 | 25,9 | 9,55 |
| Crabe, miettes et ou pattes décortiquées, appertisé, égoutté | 220 | 110 | 31 | 638 | 11,9 | 0 | 2,4 | 0 | - | 0 | 0,05 | 0,4 | 1,7 | 0,71 | 0,3 | 20 | 13,5 |
| Crevette géante tigrée, cuite | 183 | 291 | 43,8 | 570 | 2 | 0,36 | 1,92 | - | - | - | 0,024 | 0,029 | 3,77 | 0,44 | 0,18 | - | 1,85 |
| Crevette grise, cuite | 280 | 97 | 40 | 458 | 2,4 | < 0,25 | 4,32 | < 0,8 | - | < 0,5 | < 0,015 | 0,28 | 0,49 | 0,11 | 0,094 | 33,5 | 12,6 |
| Crevette pattes blanches, cuite | 177 | 288 | 31,8 | 730 | 1,38 | < 0,5 | 2,15 | - | - | - | < 0,04 | 0,041 | 3,16 | 0,26 | 0,19 | - | 6,62 |
| Crevette rose bouquet, cuite | 171 | 318 | 20,4 | 188 | 1,32 | < 0,5 | 5,98 | - | - | - | 0,057 | 0,046 | 1,91 | 0,09 | 0,12 | - | 1,77 |
| Crevette rose, crue | 214 | 264 | - | 119 | 1,34 | - | - | - | - | - | - | - | - | - | - | - | - |
| Crevette royale rose, cuite | 203 | 207 | 52,2 | 138 | 1,99 | < 0,5 | 1,91 | - | - | - | < 0,04 | 0,042 | 2,91 | 0,13 | 0,18 | - | 2,46 |
| Crevette, crue | 195 | 278 | 34 | 182 | 2,81 | < 0,2 | 1,7 | - | - | 0 | 0,035 | 0,03 | 3,2 | 0,37 | 0,15 | 19 | 1 |
| Crevette, cuite | 57 | 254 | 21,5 | 541 | 1,69 | < 0,5 | 2,48 | 0,2 | - | 0,83 | 0,019 | 0,04 | 1,77 | 0,31 | 0,1 | 14,6 | 1,64 |
| Crevette, surgelée, crue | - | - | - | 240 | 0,9 | - | 2,85 | - | - | 0 | - | 0,02 | - | - | 0,07 | 13 | - |
| Denté, cru | 226 | 350 | - | 77 | 0,41 | - | 0,4 | - | - | 0 | 0,08 | 0,13 | 2,7 | - | 0,19 | 5 | - |

| | | | | | | | | | | | | | | | | | | | | | | | | | | | | | | | | | | | |
|---|---|---|---|---|---|---|---|---|---|---|---|---|---|---|---|---|---|---|---|---|---|---|---|---|---|---|---|---|---|---|---|---|---|---|---|
| Dés, allumettes, râpé ou haché de jambon | 508 | 121 | 508 | 121 | - | 17,9 | 17,9 | 2,08 | 4,46 | 0,36 | 2,06 | - | - | - | - | - | - | - | - | - | - | 811 | - | - | - | - | - | - | - | - | - | - | - | - | - |
| Dés, allumettes, râpé ou haché de jambon de volaille | 554 | 132 | 554 | 132 | - | 18 | 18 | 1,58 | 5,88 | 0,5 | 2,2 | - | - | - | - | - | - | - | - | - | - | 850 | - | - | - | - | - | - | - | - | - | - | - | - | - |
| Dinde, aile, crue | 800 | 192 | 800 | 192 | 66,5 | 20,2 | 20,2 | 0 | 12,3 | 0 | 0,14 | 14 | - | 0,077 | 1,26 | - | 21 | 0,017 | 165 | 240 | - | 55 | 1,54 | - | - | - | - | 0 | 0,05 | 0,11 | 4,43 | 0,55 | 0,41 | 7 | 0,39 |
| Dinde, cuisse, viande et peau, crue | 627 | 150 | 627 | 150 | 73,4 | 19,8 | 19,8 | 0 | 7,85 | 0 | 0,24 | 14 | - | 0,11 | 1,36 | - | 21,5 | 0,018 | 172 | 241 | - | 95,3 | 2,71 | 0,5 | 0,11 | 0 | - | 0 | 0,066 | 0,22 | 4,16 | 0,97 | 0,37 | 8 | 1,12 |
| Dinde, cuisse, viande sans peau, crue | 461 | 109 | 461 | 109 | 76 | 21,3 | 21,3 | 0,4 | 2,5 | 0 | 0,31 | 11 | - | 0,092 | 1,04 | - | 25 | 0,014 | 176 | 226 | 28 | 124 | 2,59 | 0,3 | 0,12 | 0 | - | 0 | 0,062 | 0,26 | 5,7 | 0,94 | 0,44 | 7 | 2,05 |
| Dinde, escalope viennoise ou milanaise ou escalope panée | 882 | 211 | 882 | 211 | - | 14,1 | 14,1 | 13,5 | 10,9 | 1,3 | 1,05 | - | - | - | - | - | - | - | - | - | - | 356 | - | - | - | - | - | - | - | - | - | - | - | - | - |
| Dinde, escalope, crue | 463 | 109 | 463 | 109 | 74,3 | 24,1 | 24,1 | 0,51 | 1,22 | 0 | 0,2 | 16,4 | - | 0,07 | 1,02 | - | 28 | 0,011 | 201 | 242 | 10 | 82,8 | 1,28 | 0,1 | 0,06 | 0 | - | 0 | 0,042 | 0,15 | 9,92 | 0,78 | 0,81 | 7 | 0,63 |
| Dinde, escalope, rôtie/cuite au four | 539 | 128 | 539 | 128 | 67,9 | 24,6 | 24,6 | 0,5 | 3,04 | 0 | 0,25 | 9 | - | 0,063 | 0,57 | - | 32 | 0,011 | 230 | 249 | 30,8 | 99 | 1,72 | 0,3 | 0,06 | 0 | - | 0 | 0,035 | 0,21 | 11,8 | 0,9 | 0,81 | 9 | 0,39 |
| Dinde, escalope, sautée/poêlée | 524 | 124 | 524 | 124 | 66,3 | 28,5 | 28,5 | 0 | 1,09 | 0 | 0,16 | 5,42 | 124 | 0,054 | 0,49 | 4 | 44,4 | 0,012 | 293 | 462 | 6,48 | 65,2 | 1,22 | < 0,5 | 0,08 | - | - | 0,75 | 0,049 | 0,083 | 13,5 | 0,84 | 0,59 | - | 0,61 |
| Dinde, viande et peau, crue | 557 | 132 | 557 | 132 | 73,1 | 23,4 | 23,4 | 0 | 4,31 | 0 | 0,25 | 10,5 | - | 0,11 | 0,83 | 1,5 | 23 | 0,012 | 167 | 227 | 24 | 103 | 1,57 | 0,3 | 0,3 | 0 | - | 0 | 0,056 | 0,17 | 5,86 | 0,78 | 0,5 | 8,5 | 1,11 |
| Dinde, viande, crue | 464 | 110 | 464 | 110 | 75,4 | 22,4 | 22,4 | 0,8 | 1,88 | 0 | 0,25 | 10,5 | - | 0,067 | 0,82 | 1,5 | 24 | 0,012 | 170 | 233 | 9,9 | 106 | 2,22 | 0,2 | 0,15 | 0 | - | 0 | 0,07 | 0,18 | 8 | 0,82 | 0,56 | 11 | 1,62 |
| Dinde, viande, rôtie/cuite au four | 636 | 151 | 636 | 151 | 66,7 | 29,1 | 29,1 | 0 | 3,84 | 0 | 0,25 | 9,66 | - | 0,094 | 1,14 | 6 | 44,3 | 0,02 | 222 | 483 | < 10 | 101 | 3,75 | 0,3 | 0,06 | 0 | - | 0 | 0,047 | 0,28 | 9,5 | 0,95 | 0,64 | 9 | 0,94 |
| Diot, cru | - | - | - | - | - | 18 | 18 | - | 30 | 0 | 2,25 | - | - | - | - | - | - | - | - | - | - | 900 | - | - | - | - | - | - | - | - | - | - | - | - | - |
| Dorade (Daurade) royale, cuite au four | 620 | 148 | 620 | 148 | 68,8 | 23,6 | 23,6 | traces | 5,9 | 0 | 0,18 | 15 | 132 | 0,02 | 0,2 | < 20 | 29 | < 0,01 | 230 | 420 | < 20 | 71 | 0,48 | 5,71 | 0,65 | 1,18 | - | 0,59 | < 0,015 | < 0,01 | 6,41 | 0,4 | 0,26 | 17,2 | 2,37 |
| Dorade grise, ou daurade grise, ou griset, crue | 547 | 130 | 547 | 130 | 72,9 | 20,5 | 20,5 | traces | 5,27 | < 1 | 0,15 | 100 | - | < 0,1 | 0,39 | 30,9 | 31,2 | < 0,1 | 250 | 400 | 20,2 | 60,6 | 0,8 | 0,57 | 0,69 | - | - | - | 0,025 | 0,062 | 4,83 | 0,28 | 0,36 | - | 2,65 |
| Dorade grise, ou daurade grise, ou griset, rôtie/cuite au four | 512 | 121 | 512 | 121 | 71 | 22,9 | 22,9 | traces | 3,3 | 0 | - | - | - | - | - | - | - | - | - | - | 50 | - | - | - | - | - | - | - | - | - | - | - | - | - | - |
| Dorade rose, ou daurade rose, crue | 431 | 102 | 431 | 102 | - | 21 | 21 | traces | 2 | 0 | - | 34 | - | 0,4 | 4,3 | - | 22 | 0,1 | - | 690 | - | - | 1,6 | - | - | - | - | - | - | - | - | - | - | - | - |
| Dorade royale ou daurade ou vraie daurade, crue, élevage | 543 | 129 | 543 | 129 | 72,5 | 20,9 | 20,9 | traces | 5,1 | 0 | 0,13 | 2,34 | - | < 0,1 | 0,39 | 5 | 30,7 | < 0,1 | 248 | 458 | 3,7 | 51,5 | 0,41 | 3,11 | 0,95 | - | - | 0 | 0,077 | 0,065 | 7,4 | 0,28 | 0,43 | 5 | 3,16 |
| Dorade royale, ou daurade ou vraie daurade, crue, sauvage | 393 | 93,2 | 393 | 93,2 | 78,6 | 18,1 | 18,1 | traces | 2,31 | 0 | 0,12 | 33,3 | - | 0,11 | 1,6 | 40 | 19 | 0,12 | 178 | 459 | 20 | 46,5 | 0,76 | 7,15 | 0,85 | 0 | - | 1 | 0,15 | 0,095 | 1,2 | - | 0,24 | 7 | - |
| Écrevisse, crue | 290 | 68,3 | 290 | 68,3 | 82,6 | 14,8 | 14,8 | 0,81 | 0,64 | 0 | 0,39 | 53,5 | 280 | 0,37 | 1,71 | 68 | 25,5 | 0,15 | 245 | 300 | 70 | 157 | 1,18 | 0,25 | 1,5 | 0 | - | 1,7 | 0,056 | 0,056 | 2,1 | 0,57 | 0,48 | 26 | 2,2 |
| Écrevisse, cuite | 312 | 73,5 | 312 | 73,5 | 81,6 | 16,3 | 16,3 | 0,56 | 0,7 | 0 | 0,21 | 73 | 96,8 | 0,89 | 0,65 | < 20 | 26 | 0,32 | 150 | 140 | < 50 | 85,7 | 1,5 | < 0,5 | - | < 0,8 | - | 0,5 | 0,01 | 0,05 | 0,36 | 0,21 | 0,028 | 17,7 | 4,87 |
| Églefin, cru | 307 | 72,3 | 307 | 72,3 | 80,4 | 17,3 | 17,3 | traces | 0,35 | 0 | 0,16 | 9,23 | 86 | < 0,1 | 0,28 | 126 | 26,4 | < 0,1 | 144 | 292 | 36 | 65,9 | 0,32 | < 0,5 | 0,42 | 0,1 | - | 0 | < 0,04 | < 0,04 | 2,64 | 0,11 | 0,32 | 11,3 | 1,07 |
| Églefin, cuit à la vapeur | 377 | 89 | 377 | 89 | 78,3 | 20,9 | 20,9 | traces | 0,6 | 0 | 0,18 | 26 | - | 0,02 | 0,1 | 260 | 24 | 0,01 | 200 | 370 | 34,3 | 73 | 0,5 | traces | 0,41 | - | - | traces | 0,04 | 0,11 | 4,1 | 0,25 | 0,41 | 9 | 2 |
| Églefin, grillé/poêlé | 360 | 84,9 | 360 | 84,9 | 79,7 | 20 | 20 | traces | 0,55 | 0 | 0,65 | 14 | - | 0,026 | 0,21 | - | 26 | 0,013 | 278 | 351 | - | 261 | 0,4 | 0,6 | 0,55 | 0,1 | - | 0 | 0,023 | 0,069 | 4,12 | 0,49 | 0,33 | 13 | 2,13 |
| Empereur, filet, sans peau, cru | 311 | 73,5 | 311 | 73,5 | - | 15 | 15 | traces | 1,5 | 0 | 0,21 | - | - | - | - | - | - | - | - | - | - | 83 | - | - | - | - | - | - | - | - | - | - | - | - | - |
| Éperlan, cru | 361 | 85,4 | 361 | 85,4 | 79,5 | 17,5 | 17,5 | traces | 1,7 | 0 | 0,39 | - | - | - | - | 67 | 24 | - | 245 | 357 | - | 156 | - | - | - | - | - | 0 | - | 0,12 | 1,45 | 0,64 | - | 37 | 3,44 |
| Escargot en sauce au beurre persillé, préemballé, cuit | 1150 | 277 | 1150 | 277 | 57,2 | 12,9 | 12,9 | 2,26 | 23,7 | 1,69 | 1,66 | - | 992 | 2,3 | 2,5 | < 20 | 35 | 0,34 | 150 | 89 | < 50 | 665 | 1,2 | < 0,5 | - | 10,4 | - | - | 0,065 | 0,04 | 0,2 | 0,25 | 0,069 | 14,1 | 0,26 |
| Escargot, cru | 330 | 77,9 | 330 | 77,9 | 79,1 | 16,1 | 16,1 | 1,29 | 0,92 | 0 | 0,35 | 90 | - | 2,8 | 2,6 | 6 | 48,3 | 1,2 | 157 | 35 | - | 138 | 1,17 | < 0,5 | 5 | 0,1 | - | 15 | 0,02 | 0,12 | 1,7 | - | 0,028 | 8,5 | 0,92 |
| Escargot, sans matière grasse ajoutée, cuit | 338 | 80 | 338 | 80 | 78,8 | 16,6 | 16,6 | 0,23 | 1,4 | 0 | 0,27 | - | - | - | - | - | - | - | - | - | - | 109 | - | - | - | - | - | - | - | - | - | - | - | - | - |
| Espadon, cru | 543 | 130 | 543 | 130 | - | 18,9 | 18,9 | traces | 5,97 | 0 | 0,28 | 12,1 | 130 | 0,035 | 0,65 | 40 | 33,6 | 0,017 | 289 | 362 | 52,7 | 112 | 0,79 | 12,5 | 0,5 | 0,1 | - | 0,5 | 0,089 | 0,11 | 7,38 | 0,39 | 0,47 | 2 | 2,46 |
| Espadon, rôti/cuit au four | 1110 | 266 | 1110 | 266 | 52,9 | 28,7 | 28,7 | traces | 16,9 | 0 | 0,18 | 5,5 | 170 | 0,046 | 0,53 | 9,3 | 34,1 | 0,021 | 200 | 475 | 122 | 73,4 | 0,78 | 16,6 | 7,2 | 0,1 | - | 0 | 0,047 | 0,13 | 9,8 | 0,46 | 0,62 | 2 | 2,91 |
| Esturgeon, cru | 517 | 123 | 517 | 123 | 74,5 | 17,7 | 17,7 | traces | 5,84 | 0 | 0,12 | 15,6 | - | 0,041 | 0,62 | - | 40,9 | 0,025 | 225 | 307 | - | 49,2 | 0,63 | 10,3 | 0,5 | 0,1 | - | 0 | 0,07 | 0,07 | 6,96 | 0,76 | 0,32 | 15 | 1,74 |
| Faisan, viande et peau, cru | 688 | 164 | 688 | 164 | 67,8 | 23,2 | 23,2 | 0 | 7,95 | 0 | 0,09 | 15 | - | 0,065 | 1,58 | 8 | 20 | 0,017 | 214 | 285 | - | 36 | 0,96 | - | - | 0 | - | 5,3 | 0,072 | 0,14 | 6,43 | 0,93 | 0,66 | 6 | 0,77 |
| Faisan, viande, cru | 535 | 127 | 535 | 127 | 72,2 | 23,3 | 23,3 | 0 | 3,78 | 0 | 0,094 | 14,8 | - | 0,071 | 1,22 | 8 | 20,1 | 0,017 | 231 | 261 | 14 | 37,6 | 1 | - | - | 0 | - | 5,91 | 0,076 | 0,15 | 6,56 | 0,96 | 0,73 | 6,25 | 0,83 |
| Faisan, viande, rôtie/cuite au four | 999 | 239 | 999 | 239 | 54,2 | 32,4 | 32,4 | 0 | 12,1 | 0 | 0,11 | 16 | - | 0,084 | 1,43 | 4 | 22 | - | 242 | 271 | - | 45 | 1,37 | 0,2 | 0,27 | 4,9 | - | 2,3 | 0,07 | 0,18 | 7,53 | - | 0,75 | 5 | 0,72 |
| Filet de bacon | 500 | 118 | 500 | 118 | 70,1 | 23,1 | 23,1 | 0,7 | 2,6 | 0 | 3,18 | 3,8 | 1820 | 0,03 | 0,4 | < 20 | 23 | < 0,01 | 190 | 470 | < 20 | 1270 | 1,4 | < 0,25 | 0,45 | < 0,8 | - | 6,55 | 0,6 | 0,082 | 6,53 | 0,93 | 0,34 | 49,1 | 0,38 |
| Flétan de l'Atlantique ou flétan blanc, cru | 409 | 96,5 | 409 | 96,5 | 76,8 | 21,2 | 21,2 | traces | 1,31 | 0 | 0,15 | 15 | - | 0,2 | 0,65 | 12,6 | 23 | 0,01 | 221 | 424 | - | 59 | 0,45 | - | 0,64 | - | - | 0 | 0,05 | 0,063 | 4,9 | 0,28 | 0,38 | 12 | 0,96 |
| Flétan du Groënland ou flétan noir ou flétan commun, cru | 700 | 168 | 700 | 168 | 72,3 | 13,5 | 13,5 | traces | 12,7 | 0 | 0,43 | 17,7 | - | 0,03 | 0,4 | 17,3 | 24 | 0,012 | 161 | 287 | 30,5 | 202 | 0,4 | 21,2 | 0,79 | 0,1 | - | 1 | 0,06 | 0,08 | 1,5 | 0,25 | 0,43 | 6,5 | 1 |
| Flétan du Groënland ou flétan noir ou flétan commun, cuit à la vapeur | 722 | 173 | 722 | 173 | 68,8 | 17,3 | 17,3 | traces | 11,6 | 0 | 0,36 | 8,6 | 133 | 0,02 | 0,07 | < 20 | 21 | < 0,01 | 160 | 240 | 60 | 143 | 0,36 | 4,36 | - | < 0,8 | - | - | 0,03 | 0,02 | 0,66 | 0,16 | 0,041 | < 5 | 0,69 |
| Foie de morue, appertisé, égoutté | 1790 | 433 | 1790 | 433 | 45,7 | 7,25 | 7,25 | 1,1 | 44,4 | 0 | 1,12 | 10 | - | 0,4 | 2,1 | 105 | 10,3 | < 0,1 | 145 | 130 | 107 | 446 | 2 | 54,3 | 4,96 | - | - | 4 | < 0,04 | 0,6 | 3,28 | 0,54 | 0,15 | 188 | 14,4 |
| Foie de morue, cru | 2550 | 619 | 2550 | 619 | 26,8 | 5 | 5 | traces | 66,6 | 0 | 1,47 | 10 | - | 0,66 | 4 | 500 | 8 | - | 100 | 130 | 63,5 | 589 | 1,94 | 100 | 20 | - | - | 4 | 0,1 | 0,65 | 2,5 | 0,64 | 0,15 | 300 | 10 |
| Foie gras de canard, cru | 2470 | 600 | 2470 | 600 | - | 5,94 | 5,94 | 1,41 | 63,4 | 0 | 0,52 | 10 | - | - | 6,4 | - | - | - | - | - | - | 385 | - | - | - | - | - | - | - | - | - | - | - | - | - |
| Foie gras, canard, bloc (aliment moyen) | 2010 | 489 | 2010 | 489 | 39,1 | 6,88 | 6,88 | 2,43 | 50,1 | 0 | 1,29 | | | | 2,1 | | | | | | | 514 | | | | | | | | | | | | | |
| Foie gras, canard, bloc, 30% de morceaux | 2010 | 489 | 2010 | 489 | 39,1 | 6,88 | 6,88 | 2,43 | 50,1 | 0 | 1,29 | - | - | - | 2,1 | - | - | - | - | - | - | 514 | - | - | - | - | - | - | - | - | - | - | - | - | - |
| Foie gras, canard, bloc, 50% de morceaux | 1970 | 479 | 1970 | 479 | 38,5 | 6,39 | 6,39 | 2 | 49,5 | 0 | 1,29 | - | - | - | 2,1 | - | - | - | - | - | - | 516 | - | - | - | - | - | - | - | - | - | - | - | - | - |
| Foie gras, canard, bloc, sans morceaux | 1980 | 482 | 1980 | 482 | 40,8 | 5,75 | 5,75 | 2,1 | 50 | 0,35 | 1,14 | 1,7 | 798 | 0,8 | 5,2 | 2 | 8,4 | 0,1 | 109 | 130 | - | 457 | 1,2 | 2,75 | 2,4 | - | - | 14,7 | 0,1 | 0,57 | 5,8 | 2,15 | 0,95 | 142 | 4,55 |
| Foie gras, canard, entier, cuit | 2200 | 535 | 2200 | 535 | 35,1 | 8,41 | 8,41 | 2,32 | 54,6 | 0,35 | 1,11 | 5,68 | - | 0,91 | 4,13 | - | 9,33 | 0,11 | - | 135 | 12,3 | 440 | 1,19 | 0,2 | - | - | - | 31 | - | - | - | - | 0,3 | - | 15 |
| Foie, agneau, cru | 622 | 148 | 622 | 148 | 71,2 | 21,8 | 21,8 | 2,92 | 5,45 | 0 | 0,14 | 7 | - | 6,98 | 6,43 | 4 | 19 | 0,18 | 364 | 313 | 48,7 | 57,4 | 4 | 0,45 | 0,69 | 1 | - | 4 | 0,34 | 3,63 | 18 | 6,13 | 0,37 | 230 | 95,8 |
| Foie, agneau, cuit | 659 | 157 | 659 | 157 | 65 | 23 | 23 | 3,16 | 5,8 | 0 | 0,23 | 8,5 | - | 8,45 | 4,1 | 5 | 22,5 | 0,56 | 424 | 287 | - | 90 | 4,7 | 0,51 | 0,38 | 1,6 | - | 17,6 | 0,46 | 4,31 | 18,5 | 12,6 | 0,54 | 457 | 60 |
| Foie, canard, cru | 549 | 131 | 549 | 131 | 71,8 | 18,7 | 18,7 | 3,5 | 4,62 | 0 | 0,27 | 11 | - | 5,96 | 30,5 | - | 22 | 0,26 | 269 | 235 | 124 | 109 | 3,07 | - | - | - | - | 4,5 | 0,56 | 0,89 | 6,5 | 6,18 | 0,76 | 738 | 54 |
| Foie, dinde, cru | 514 | 123 | 514 | 123 | 75,5 | 18,3 | 18,3 | 0 | 5,5 | 0 | 0,33 | 20 | - | 0,86 | 8,94 | - | 24 | 0,3 | 279 | 214 | - | 131 | 3,37 | 1,3 | 0,24 | 0 | - | 24,5 | 0,21 | 2,25 | 11,2 | 6,28 | 1,04 | 677 | 19,7 |
| Foie, dinde, cuit | 762 | 182 | 762 | 182 | 64 | 27 | 27 | 0 | 8,18 | 0 | 0,25 | 19 | - | 1,05 | 1,79 | - | 26 | 0,33 | 312 | 153 | - | 98 | 4,53 | - | 0,15 | 0 | - | 22,6 | 0,26 | 2,69 | 11,1 | 4,35 | 0,88 | 691 | 28,2 |
| Foie, génisse, cru | 579 | 138 | 579 | 138 | 70 | 21 | 21 | 3,67 | 4,3 | 0 | 0,15 | 5,95 | - | 8,08 | 5,92 | 4,3 | 19,5 | 0,35 | 368 | 307 | 39,3 | 59,9 | 3,57 | 0,67 | 0,4 | 53,6 | 7,94 | 23 | 0,18 | 2,88 | 15,4 | 7,44 | 0,52 | 1300 | 95,6 |
| Foie, génisse, cuit | 695 | 165 | 695 | 165 | 64,1 | 26,3 | 26,3 | 2,2 | 5,71 | 0 | 0,19 | 6,14 | 0,1 | 3,46 | 5,8 | < 5 | 17,8 | 0,32 | 485 | 351 | < 2,2 | 77 | 3,82 | 0,51 | 0,46 | 3,9 | - | 20 | 0,33 | 3,43 | 18 | 8,5 | 1,4 | 536 | 1,9 |
| Foie, lapin, cru | 556 | 132 | 556 | 132 | 70,8 | 19 | 19 | 5 | 4 | 0 | 0,17 | 12,3 | - | 0,81 | 6,61 | 2,69 | 24,2 | 0,14 | 241 | 335 | 75,1 | 69,5 | 2,04 | - | 0,13 | - | - | 9 | 0,31 | 0,5 | 5,7 | - | 0,5 | 5 | - |
| Foie, oie, cru | 543 | 129 | 543 | 129 | 71,8 | 16,3 | 16,3 | 6,3 | 4,29 | 0 | 0,27 | 43 | - | 7,52 | 30,5 | 2,8 | 22 | 0 | 261 | 235 | 61 | 109 | 3,07 | - | - | 10,9 | 369 | 4,5 | 0,56 | 2,05 | 6,5 | 6,18 | 0,85 | 738 | 54 |
| Foie, poulet, cru | 509 | 121 | 509 | 121 | 74,5 | 18,8 | 18,8 | 1,1 | 4,62 | 0 | 0,17 | 10,5 | 130 | 0,5 | 8,68 | 2,4 | 19 | 0,24 | 309 | 233 | 49,8 | 69,3 | 2,79 | 0,21 | 0,51 | 80 | - | 19,7 | 0,35 | 2,36 | 9,54 | 6,41 | 0,83 | 1640 | 19,3 |
| Foie, poulet, cuit | 670 | 160 | 670 | 160 | 66,8 | 24,5 | 24,5 | 0,8 | 6,51 | 0 | 0,19 | 11 | - | 0,5 | 11,6 | 5 | 25 | 0,36 | 405 | 263 | - | 76 | 3,98 | 0 | 0,82 | 0 | - | 27,9 | 0,29 | 1,99 | 11 | 6,67 | 0,76 | 578 | 16,9 |
| Foie, veau, cru | 505 | 120 | 505 | 120 | 71,6 | 15,5 | 15,5 | 5,64 | 3,4 | 0 | 0,16 | 4,9 | 121 | 22 | 4,6 | < 20 | 18 | 0,13 | 280 | 320 | 40 | 62,7 | 5,3 | 0,62 | 0,79 | < 0,8 | - | 9,75 | 0,14 | 0,67 | 17,1 | 3,13 | 0,5 | 1180 | 46,5 |
| Foie, veau, cuit | 511 | 121 | 511 | 121 | 67,7 | 19 | 19 | 4,47 | 3,02 | 0 | 0,2 | 6,5 | - | 20,1 | 4,5 | 5 | 21,5 | 0,29 | 302 | 341 | 40,9 | 81,5 | 4,6 | 2,52 | 0,64 | 1,5 | - | 13,2 | 0,15 | 1,72 | 18,8 | 1,81 | 1,03 | 592 | 52,6 |
| Foie, volaille, cru | 561 | 133 | 561 | 133 | - | 21,3 | 21,3 | 1,13 | 4,88 | 0 | 0,21 | 12 | - | - | 9 | - | - | - | - | - | - | 83 | - | - | - | - | - | 25 | - | - | - | - | - | - | - |
| Foie, volaille, cuit | 687 | 164 | 687 | 164 | 65,4 | 25,8 | 25,8 | 0 | 6,4 | 0 | 0,25 | 6,6 | 196 | 0,51 | 12 | < 20 | 24 | 0,37 | 380 | 340 | 90 | 99 | 3,9 | 1,29 | 1,51 | 3,2 | - | < 0,5 | 0,28 | 1,4 | 32,6 | 8,77 | 0,59 | 1440 | 50,5 |
| Fromage de tête | 720 | 173 | 720 | 173 | 67,7 | 14,1 | 14,1 | 0,2 | 12,9 | 0 | 1,91 | 13,1 | 0,26 | 0,11 | 1,23 | 2,1 | 9 | 0,02 | 80 | 69 | 7,4 | 754 | 1,24 | 0,9 | 0,17 | 3,4 | - | 0 | 0,26 | 0,11 | 0,77 | 0,9 | 0,14 | 2,5 | 0,78 |
| Fruits de mer (aliment moyen) | 438 | 104 | 438 | 104 | 74 | 18,9 | 18,9 | 3,06 | 1,76 | 0,00092 | 1,1 | 130 | 912 | 0,64 | 2,67 | 68,3 | 62,2 | 0,3 | 157 | 271 | 37,4 | 439 | 4 | 0,18 | 1,91 | 0,13 | | 3,28 | 0,032 | 0,23 | 1,71 | 0,41 | 0,12 | 22,1 | 9,94 |
| Fruits de mer, cuits, surgelés | 336 | 79,5 | 336 | 79,5 | - | 13,5 | 13,5 | 2,83 | 1,57 | 0 | 0,99 | 70 | - | - | 8,48 | - | - | - | - | - | - | 402 | - | - | - | - | - | - | - | - | - | - | - | - | - |
| Galantine (aliment moyen) | 1150 | 279 | 1150 | 279 | 57 | 15,2 | 15,2 | 1,57 | 23,4 | 0,67 | 2,04 | 9 | 1240 | | 2,3 | 7 | 9 | | 174 | 230 | | 783 | | | | | | 5 | 0,16 | 0,22 | 2,6 | | | | |
| Gésier, canard, confit, appertisé | 632 | 149 | 632 | 149 | 63,4 | 32,4 | 32,4 | 0 | 2,1 | < 1 | 1,92 | 8,5 | - | < 0,1 | 9,4 | 32 | 18,3 | < 0,1 | 126 | 214 | 37 | 766 | 6 | < 0,5 | 0,12 | - | - | < 1 | 0,14 | 0,2 | 3,6 | 0,3 | 0,12 | < 20 | 6,8 |
| Gésier, poulet, cru | 443 | 105 | 443 | 105 | 77,8 | 17,3 | 17,3 | 0,6 | 3,75 | 0 | 0,18 | 11 | - | 0,12 | 2,49 | 0,4 | 14 | 0,055 | 148 | 213 | - | 73,3 | 2,72 | 0,6 | 0,27 | 0 | - | 3,7 | 0,059 | 0,14 | 4,09 | 0,63 | 0,12 | 21,5 | 0,91 |
| Gibier à plumes, viande, cuit (aliment moyen) | 908 | 217 | 908 | 217 | 61,6 | 25,1 | 25,1 | 0 | 13 | 0 | 0,73 | 6,81 | | 0,45 | 4,2 | 10,1 | 31,3 | 0,043 | 246 | 369 | 24,4 | 290 | 2,12 | 0,55 | 0,54 | | | 2,08 | 0,32 | 0,67 | 11,8 | 2,58 | 0,82 | 9,46 | 2,78 |
| Gibier à poil, cuit (aliment moyen) | 641 | 152 | 641 | 152 | 64,4 | 29,6 | 29,6 | 0 | 3,74 | 0 | 0,14 | 11,8 | | 0,18 | 3,03 | | 25,8 | | 187 | 362 | | 56 | 2,84 | | | 1,63 | | 0 | 0,23 | 0,36 | 5,59 | | 0,5 | 6,18 | 1,26 |
| Grenadier (de roche), cru | 292 | 68,9 | 292 | 68,9 | 82,7 | 15,4 | 15,4 | traces | 0,81 | 0 | 0,24 | 1,64 | - | < 0,1 | 0,12 | 13,4 | 22,1 | < 0,1 | 103 | 281 | 356 | 94 | 0,22 | 0,52 | 1,36 | - | - | - | < 0,04 | < 0,04 | 1,29 | 0,077 | 0,12 | - | 0,33 |
| Grenadier bleu ou hoki de Nouvelle-Zélande, cru | 319 | 75,5 | 319 | 75,5 | 82,8 | 15,6 | 15,6 | traces | 1,46 | 0 | 0,14 | 8 | 73 | 0,02 | 0,84 | - | 20 | 0,016 | 154 | 306 | 47 | 54,4 | 0,23 | 0,004 | 0,24 | - | - | 0,3 | 0,02 | 0,002 | 1,84 | 0,44 | 0,09 | 20 | 0,7 |
| Grenouille, cuisse, crue | 297 | 70 | 297 | 70 | 82 | 16,2 | 16,2 | 0,66 | 0,28 | 0 | 0,14 | 18,5 | - | 0,25 | 2,58 | 6 | 21,5 | - | 241 | 302 | - | 54,5 | 1,5 | 0,35 | 1 | 0,1 | - | 5 | 0,15 | 0,19 | 1,2 | - | 0,84 | 12,5 | 0,4 |
| Grenouille, cuisse, grillée/poêlée | - | - | - | - | 76,7 | 21,9 | 21,9 | - | 0,9 | 0 | 0,22 | 39 | 86,6 | 0,03 | 0,52 | < 20 | 22 | 0,06 | 92 | 110 | < 50 | 87,8 | 0,56 | 1,12 | - | < 0,8 | - | - | 0,035 | 0,07 | 1,47 | 0,23 | 0,15 | < 5 | 2,29 |
| Grondin perlon, cru | 412 | 97,4 | 412 | 97,4 | 76,3 | 20,3 | 20,3 | traces | 1,79 | 0 | 0,14 | - | - | - | - | - | - | - | - | - | 37,2 | - | - | - | - | - | - | - | - | - | - | - | - | - | - |
| Grondin, cru | 397 | 94 | 397 | 94 | 75 | 19 | 19 | traces | 2 | 0 | - | - | - | - | - | - | - | - | - | - | - | - | - | - | - | - | - | - | - | - | - | - | - | - | - |

| | | | | | | | | | | | | | | | | | | | | | | | | | | | | | | | | | | | |
|---|---|---|---|---|---|---|---|---|---|---|---|---|---|---|---|---|---|---|---|---|---|---|---|---|---|---|---|---|---|---|---|---|---|---|---|
| Haché à base de boeuf ou Préparation de viande hachée de boeuf, 15% MG, cru, préemballé | - | - | - | - | 62,5 | 15,8 | 15,8 | 2,48 | 13,6 | 1,55 | 0,87 | - | - | - | 2,5 | - | - | - | - | - | - | 313 | - | - | - | - | - | - | - | - | - | - | - | - | - |
| Haché de volaille | 575 | 137 | 575 | 137 | - | 17,7 | 17,7 | 3,5 | 5,68 | 0,63 | 1,3 | - | - | - | - | - | - | - | - | - | - | 550 | - | - | - | - | - | - | - | - | - | - | - | - | - |
| Haddock (fumé) ou églefin fumé | 411 | 97 | 411 | 97 | 76,8 | 22,6 | 22,6 | 0 | 0,73 | 0 | 1,91 | 49 | 1200 | 0,042 | 1,4 | 255 | 54 | 0,03 | 251 | 415 | 28 | 763 | 0,5 | 0,8 | 0,55 | 0,1 | - | 0 | 0,047 | 0,049 | 5,07 | 0,17 | 0,4 | 15 | 1,6 |
| Hareng fumé, à l'huile | 699 | 168 | 699 | 168 | 65,3 | 15,1 | 15,1 | 2 | 11 | 0 | 4,08 | 59,2 | 3820 | 0,1 | 0,9 | 20 | 27,4 | < 0,1 | 125 | 173 | 17 | 1630 | 0,5 | 14,5 | 0,96 | - | - | - | < 0,04 | 0,2 | 2,25 | 0,41 | 0,28 | 10 | 9,4 |
| Hareng fumé, au naturel | 721 | 173 | 721 | 173 | 68,3 | 16,5 | 16,5 | 0,5 | 11,7 | 0 | 3,98 | 62,4 | 2910 | < 0,1 | 1 | 40 | 34,8 | < 0,1 | 155 | 226 | 28 | 1590 | 0,6 | 22 | 1,35 | 0 | - | 0 | < 0,04 | 0,23 | 3,45 | 0,55 | 0,37 | 10 | 11,8 |
| Hareng fumé, filet, doux | 676 | 162 | 676 | 162 | 47,7 | 18,1 | 18,1 | 0 | 9,9 | < 0,5 | 3,78 | 62 | 2310 | 0,08 | 0,7 | < 20 | 39 | 0,03 | 250 | 380 | 30 | 1510 | 0,51 | 17,7 | 1,72 | < 0,8 | - | < 0,5 | < 0,015 | 0,2 | 4,53 | 1,22 | 0,28 | 10,8 | 13,4 |
| Hareng gras, cru | 719 | 172 | 719 | 172 | 68,3 | 18,7 | 18,7 | traces | 10,8 | 0 | 0,13 | 47,7 | 650 | < 0,1 | 0,88 | 8,94 | 32,7 | < 0,1 | 247 | 421 | 25,4 | 52,4 | 0,44 | 8,36 | 1,8 | 0 | - | 0,5 | < 0,04 | 0,15 | 6,14 | 0,72 | 0,4 | 10 | 8,28 |
| Hareng maigre, cru | 466 | 111 | 466 | 111 | 76,2 | 18,3 | 18,3 | traces | 4,22 | 0 | 0,19 | 23,8 | 255 | < 0,1 | 0,87 | 5,5 | 36,2 | < 0,1 | 261 | 450 | 18,6 | 75 | 0,49 | 9,59 | 0,57 | - | - | - | < 0,04 | 0,13 | 4,11 | 0,57 | 0,42 | - | 8,47 |
| Hareng mariné ou rollmops | 880 | 211 | 880 | 211 | 64,7 | 13,7 | 13,7 | 8 | 13,8 | 0 | 2,56 | 53,5 | - | 0,11 | 1,07 | 47 | 7,2 | 0,035 | 70,5 | 56,5 | 46 | 1030 | 0,59 | 13,2 | 1,48 | 0,2 | - | 0 | 0,033 | 0,13 | 2,9 | 0,081 | 0,19 | 1,9 | 5,1 |
| Hareng, cru | 734 | 176 | 734 | 176 | 71,5 | 17,7 | 17,7 | traces | 11,7 | 0 | 0,22 | 61,6 | 331 | 0,15 | 1,21 | 54,1 | 27 | 0,027 | 235 | 342 | 31,2 | 89,2 | 0,79 | 10,7 | 1,21 | 0,45 | 0,21 | 0,75 | 0,035 | 0,24 | 3,95 | 0,84 | 0,38 | 9,4 | 11,5 |
| Hareng, frit | 840 | 201 | 840 | 201 | - | 23 | 23 | traces | 12,1 | 0 | - | - | - | 0,14 | - | 38 | - | 0,02 | - | - | 31,3 | - | 0,75 | - | - | - | - | - | - | - | - | - | - | - | - |
| Hareng, grillé/poêlé | 788 | 189 | 788 | 189 | 64,2 | 21,6 | 21,6 | traces | 11,4 | 0 | 0,34 | 76,5 | 220 | 0,15 | 1,51 | 47,7 | 41,5 | 0,045 | 307 | 425 | 46 | 138 | 1,24 | 10,8 | 1,37 | 0,1 | - | 0,7 | 0,11 | 0,28 | 4,06 | 0,76 | 0,35 | 11 | 14,1 |
| Hoki, tout lieu de pêche, cru | 389 | 92,3 | 389 | 92,3 | 78,7 | 17,8 | 17,8 | traces | 2,36 | 0 | 0,18 | 6,46 | - | < 0,1 | 0,23 | 13,5 | 28,2 | < 0,1 | 167 | 345 | 27,8 | 70,8 | 0,26 | 2,49 | 0,47 | - | - | - | 0,046 | 0,045 | 1,67 | 0,52 | 0,12 | - | 1,11 |
| Homard, bouilli/cuit à l'eau | 385 | 90,9 | 385 | 90,9 | 76,5 | 19,6 | 19,6 | 0,11 | 1,32 | 0 | 0,9 | 68,8 | 530 | 1,3 | 0,63 | 120 | 34,2 | 0,19 | 212 | 208 | < 2,2 | 361 | 2,96 | 0 | 1,6 | 0 | - | 0 | 0,051 | 0,029 | 1,61 | 1,43 | 0,14 | 12,3 | 1,64 |
| Homard, cru | 362 | 85,5 | 362 | 85,5 | 77,9 | 17,9 | 17,9 | 0,94 | 1,15 | 0 | 0,9 | 68,8 | - | 1,3 | 0,45 | 240 | 34,2 | 0,19 | 150 | 204 | < 2,2 | 361 | 2,96 | 0 | 1,19 | 0 | - | 2,5 | 0,038 | 0,038 | 1,7 | 1,47 | 0,12 | 12 | 3,78 |
| Huître creuse, crue | 283 | 67,2 | 283 | 67,2 | 83,3 | 8,64 | 8,64 | 3,86 | 1,91 | 0 | 1,41 | 81 | - | 1,45 | 2,07 | 101 | 78,1 | 0,58 | 94,6 | 215 | 14,9 | 566 | 21,8 | < 0,5 | 1,21 | - | - | 6,5 | 0,037 | 0,15 | 1,95 | 0,65 | 0,095 | 6,5 | 28,6 |
| Huître plate, crue | 226 | 53,5 | 226 | 53,5 | 85,7 | 10,2 | 10,2 | 1,16 | 0,9 | 0 | 1,28 | 186 | - | - | 6 | - | - | - | 267 | 260 | - | 510 | 45 | 0,54 | 0,46 | - | - | 0 | 0,1 | 0,2 | 1,5 | - | 0,16 | 10 | - |
| Huître, sans précision, crue | 283 | 67,2 | 283 | 67,2 | 83,3 | 8,64 | 8,64 | 3,86 | 1,91 | 0 | 1,45 | 92,7 | 820 | 1,29 | 2,18 | 101 | 90,2 | 0,55 | 94,6 | 175 | 23,5 | 578 | 22,5 | < 0,5 | 1,21 | 0 | - | 5,13 | 0,037 | 0,15 | 1,95 | 0,65 | 0,095 | 12,5 | 28,6 |
| Jambon à l'os braisé | 992 | 238 | 992 | 238 | 58,4 | 21,6 | 21,6 | 0,3 | 16,8 | 0 | 2,97 | 7 | - | 0,083 | 0,87 | - | 19 | 0,014 | 214 | 286 | - | 1190 | 2,32 | 0,7 | 0,36 | 0 | - | - | 0,6 | 0,22 | 4,46 | 0,46 | 0,38 | 3 | 0,64 |
| Jambon cru | 939 | 225 | 939 | 225 | 50,1 | 25,9 | 25,9 | 0,76 | 13,2 | traces | 5,67 | 10,4 | 3210 | 0,079 | 0,97 | 1,2 | 36,2 | 0,012 | 228 | 591 | 5,66 | 2250 | 2,37 | 0,6 | 0,21 | - | - | 13 | 0,89 | 0,21 | 8,1 | - | 0,56 | 1,24 | 0,33 |
| Jambon cru, fumé | 1020 | 246 | 1020 | 246 | 55,8 | 24,2 | 24,2 | 0,98 | 16,1 | 0 | 5,17 | 5,16 | - | < 0,1 | 0,9 | < 5 | 20 | < 0,1 | 275 | 256 | 24,2 | 2050 | 0,28 | < 0,5 | 0,03 | 0 | - | 0 | 0,41 | 0,19 | 3,4 | 0,6 | 0,4 | 1,5 | 1 |
| Jambon cru, fumé, allégé en matière grasse | 598 | 142 | 598 | 142 | 63,9 | 27,7 | 27,7 | 0,1 | 3,4 | 0 | 4,24 | - | - | - | - | - | - | - | - | - | - | 1700 | - | - | - | - | - | - | - | - | - | - | - | - | - |
| Jambon cuit, choix | 528 | 125 | 528 | 125 | 72,3 | 19,5 | 19,5 | 1,7 | 4,5 | 0,08 | 2,1 | - | 1330 | - | - | - | - | - | 146 | - | - | 945 | - | - | - | - | - | - | - | - | - | - | - | - | - |
| Jambon cuit, choix, avec couenne | 572 | 137 | 572 | 137 | 71,5 | 19,4 | 19,4 | 0,55 | 6,3 | 0 | 3,06 | - | - | - | - | - | - | - | 136 | - | - | 1200 | - | - | - | - | - | - | - | - | - | - | - | - | - |
| Jambon cuit, choix, découenné dégraissé | 481 | 114 | 481 | 114 | 72,1 | 19,6 | 19,6 | 1,61 | 3,24 | 0,15 | 1,96 | 10 | 1040 | - | - | - | - | - | 145 | 358 | - | 774 | - | - | - | - | - | - | - | - | - | - | - | - | - |
| Jambon cuit, de Paris, découenné dégraissé | 485 | 115 | 485 | 115 | - | 20 | 20 | 1,08 | 3,42 | 0 | - | - | - | - | - | - | - | - | 426 | - | - | - | - | - | - | - | - | - | - | - | - | - | - | - | - |
| Jambon cuit, fumé | 569 | 135 | 569 | 135 | 73,2 | 20 | 20 | 0,79 | 5,8 | 0,1 | 1,93 | 4,3 | 1,3 | 0,077 | 1,77 | < 5 | 23,5 | < 0,1 | 188 | 261 | 14 | 797 | 1,8 | 0,63 | 0,1 | 0 | - | - | 0,65 | 0,17 | 6,95 | 0,42 | 0,46 | 17,4 | 0,28 |
| Jambon cuit, supérieur | 527 | 125 | 527 | 125 | 73,3 | 20,8 | 20,8 | 0,81 | 4,28 | < 0,15 | 1,79 | < 20 | 935 | 0,2 | 0,54 | < 5 | 22,5 | < 0,2 | 193 | 362 | 10,4 | 749 | 2 | < 0,5 | 0,25 | 0 | - | 18,1 | 0,54 | 0,2 | 5,73 | 0,4 | 0,3 | 16 | 0,31 |
| Jambon cuit, supérieur, à teneur réduite en sel | 509 | 121 | 509 | 121 | - | 20,8 | 20,8 | 0,74 | 3,83 | < 0,42 | 1,48 | - | - | - | - | - | - | - | - | - | - | 575 | - | - | - | - | - | - | - | - | - | - | - | - | - |
| Jambon cuit, supérieur, avec couenne | 576 | 137 | 576 | 137 | 72,2 | 20,2 | 20,2 | 0,91 | 5,82 | 0,33 | 1,87 | - | 1480 | - | - | - | - | - | 265 | - | - | 744 | - | - | - | - | - | - | - | - | - | - | - | - | - |
| Jambon cuit, supérieur, découenné | 492 | 117 | 492 | 117 | 73 | 20,5 | 20,5 | 0,77 | 3,52 | 0,05 | 1,87 | 14 | 1090 | - | 1,5 | - | - | - | 155 | 440 | - | 742 | 6,5 | - | 0,05 | - | - | - | 0,52 | 0,25 | 2,6 | - | 0,22 | 0 | - |
| Jambon cuit, supérieur, découenné dégraissé | 501 | 119 | 501 | 119 | 72,8 | 20,3 | 20,3 | 1,03 | 3,66 | 0,32 | 1,82 | < 20 | 1130 | 0,2 | 0,8 | 6 | 22,6 | < 0,2 | 210 | 313 | 11 | 722 | 2,2 | 0,7 | 0,17 | 0 | - | - | 0,64 | 0,27 | 5,6 | 0,7 | 0,3 | 2,7 | 0,32 |
| Jambon de Bayonne | 953 | 228 | 953 | 228 | 48,6 | 28 | 28 | 0,63 | 12,6 | traces | 5,75 | - | 2720 | - | - | - | - | - | - | 554 | - | 2270 | - | - | - | - | - | 13 | - | - | - | - | - | - | - |
| Jambon de dinde ou Blanc de dinde en tranche | 441 | 104 | 441 | 104 | 74 | 20,9 | 20,9 | 1,29 | 1,67 | 0,39 | 1,9 | 7,1 | 1140 | 0,04 | 0,35 | < 20 | 26 | 0,02 | 230 | 370 | < 50 | 763 | 0,92 | < 0,5 | 0,11 | < 0,8 | - | 26,3 | 0,058 | 0,065 | 8,86 | 0,61 | 0,35 | 17,3 | 0,6 |
| Jambon de poulet ou Blanc de poulet en tranche | 446 | 106 | 446 | 106 | 74,3 | 20,7 | 20,7 | 1,39 | 1,79 | 0,53 | 1,86 | 6,3 | 1150 | 0,04 | 0,34 | < 20 | 29 | 0,01 | 240 | 380 | < 50 | 766 | 0,58 | < 0,5 | 0,51 | < 0,8 | - | 17,9 | 0,086 | 0,08 | 9,62 | 1,16 | 0,3 | 5,64 | 0,29 |
| Jambon en croûte | 810 | 193 | 810 | 193 | 52,2 | 14,1 | 14,1 | 17,8 | 7 | 1,1 | 2,11 | 21,3 | - | 0,063 | 0,96 | 6,12 | 19,5 | 0,052 | 148 | 280 | 6,72 | 727 | 1,28 | 0,46 | 1,03 | - | - | 7,21 | 0,5 | 0,16 | 3,34 | 0,38 | 0,23 | 4,14 | 0,52 |
| Jambon persillé en gelée | 637 | 152 | 637 | 152 | 69,6 | 19 | 19 | 0,48 | 7,9 | 0 | 1,67 | 12 | 985 | 0,08 | 0,72 | < 20 | 15 | 0,04 | 130 | 280 | < 20 | 669 | 1,7 | 0,32 | 0,33 | 5,89 | - | 10,4 | 0,22 | 0,057 | 2,11 | 0,41 | 0,15 | 11,8 | 0,38 |
| Jambon sec | 961 | 230 | 961 | 230 | 53,2 | 28,7 | 28,7 | 0,48 | 12,6 | traces | 5,7 | - | - | - | - | - | - | - | - | - | - | 2250 | - | - | - | - | - | - | - | - | - | - | - | - | - |
| Jambon sec de Parme | 1040 | 248 | 1040 | 248 | 52,2 | 27,2 | 27,2 | 0,3 | 15,4 | 0 | 5,53 | 16 | - | - | 0,7 | - | - | - | - | - | - | 2180 | - | - | - | - | - | 13 | - | - | - | - | - | - | - |
| Jambon sec Serrano | 985 | 235 | 985 | 235 | 54,1 | 30,4 | 30,4 | 0,76 | 12,3 | traces | 5,02 | - | - | - | - | - | - | - | - | - | - | 1990 | - | - | - | - | - | - | - | - | - | - | - | - | - |
| Jambon sec, découenné, dégraissé | 804 | 192 | 804 | 192 | 56,2 | 26,3 | 26,3 | 0,3 | 9,5 | 0 | 6,09 | - | 3280 | - | 1,4 | 1,2 | - | - | 230 | - | - | 2410 | - | - | - | - | - | - | 1,2 | 0,3 | 8,7 | - | 0,6 | - | 0,5 |
| Jambonneau, cuit | 679 | 162 | 679 | 162 | 65,5 | 23,2 | 23,2 | 0,4 | 7,52 | 0 | 2,12 | - | - | - | - | - | - | - | 123 | - | - | 910 | - | - | - | - | - | - | 0,4 | 1,81 | 4,2 | - | 0,29 | - | 0,32 |
| Joëls (petits poissons entiers) pour friture, crus | 375 | 88,9 | 375 | 88,9 | - | 17,5 | 17,5 | traces | 2,1 | 0 | 0,27 | 60 | - | - | 0,9 | - | - | - | - | - | - | 108 | - | - | - | - | - | - | - | - | - | - | - | - | - |
| Julienne ou Lingue, crue | 343 | 80,8 | 343 | 80,8 | 79,3 | 19,2 | 19,2 | traces | 0,44 | 0 | 0,26 | 27,6 | 0 | 0,065 | 0,68 | 43,5 | 62,5 | 0,03 | 204 | 365 | 37,5 | 128 | 0,59 | traces | 0,3 | 0 | - | traces | 0,09 | 0,11 | 2,08 | 0,32 | 0,3 | 7 | 0,78 |
| Julienne ou Lingue, cuite | 447 | 105 | 447 | 105 | 73,9 | 24,5 | 24,5 | traces | 0,82 | 0 | 0,21 | 13 | 117 | 0,02 | 0,14 | 49 | 31 | < 0,01 | 180 | 330 | 60 | 84,3 | 0,62 | < 0,5 | - | < 0,8 | - | 0 | 0,024 | 0,01 | 2,26 | 0,22 | 0,22 | 19,5 | 0,78 |
| Langouste, bouillie/cuite à l'eau | 449 | 106 | 449 | 106 | - | 21,8 | 21,8 | 1,3 | 1,52 | 0 | 0,46 | 65,5 | - | 0,42 | 1,36 | 65 | 51 | 0,018 | 229 | 208 | - | 182 | 7,27 | 0,5 | 1,5 | - | - | 2,1 | 0,009 | 0,056 | 4,9 | 0,4 | 0,17 | 1 | 4,04 |
| Langouste, crue | 368 | 86,9 | 368 | 86,9 | 78,6 | 17,7 | 17,7 | 1,07 | 1,34 | 0 | 0,45 | 60,6 | - | 0,38 | 1,12 | 65 | 30,3 | 0,015 | 241 | 340 | - | 180 | 5,69 | - | 1,47 | - | - | 2 | 0,044 | 0,097 | 3,06 | 0,35 | 0,17 | 1 | 3,5 |
| Langoustine, bouillie/cuite à l'eau | - | - | - | - | 76,3 | 20,9 | 20,9 | - | 0,98 | 0 | 0,67 | 84 | 413 | 0,83 | 1,28 | 394 | 45 | 0,06 | 208 | 283 | 55,6 | 270 | 1,7 | < 0,5 | 3,01 | - | - | - | 0,049 | 0,033 | 2,32 | 0,22 | 0,16 | - | 3,35 |
| Langoustine, crue | 381 | 89,8 | 381 | 89,8 | 76,5 | 19,1 | 19,1 | 1,53 | 0,79 | 0 | 0,72 | 41,5 | - | 0,55 | 2,92 | 141 | 43,3 | 0,07 | 144 | 505 | 55 | 287 | 3,4 | 0 | 1,5 | 0 | - | 3 | 0,13 | 0,01 | 2 | 1,5 | 0,21 | 17 | 0,9 |
| Langoustine, panée, frite | 1180 | 282 | 1180 | 282 | 44,6 | 10,9 | 10,9 | 29 | 13,6 | traces | 1,65 | - | 610 | 0,16 | 1,7 | 41 | 24 | 0,35 | 310 | 130 | 17 | 660 | 0,6 | 0 | 0 | - | - | 0 | 0,11 | 0,04 | 1,2 | 0,26 | 0,09 | 0 | 1 |
| Langue, agneau, crue | 695 | 167 | 695 | 167 | 71 | 15,4 | 15,4 | 1,75 | 10,9 | 0 | 0,25 | - | - | - | 2,3 | - | - | - | - | - | 13,8 | 102 | 2,08 | - | - | - | - | - | - | - | - | - | - | - | 3,9 |
| Langue, boeuf, crue | 804 | 193 | 804 | 193 | 67,7 | 16,8 | 16,8 | 0,4 | 13,8 | 0 | 0,17 | 6,5 | - | 0,15 | 2,27 | 2,3 | 17 | 0,028 | 147 | 303 | 11,9 | 68,6 | 3,31 | 0 | 0,1 | 0 | - | 2,9 | 0,1 | 0,34 | 4,2 | 1,33 | 0,17 | 13 | 5,1 |
| Langue, boeuf, cuite | 981 | 235 | 981 | 235 | 59,6 | 23,7 | 23,7 | 0 | 15,6 | 0 | 0,21 | 14,5 | - | 0,13 | 3,5 | 5 | 9,44 | 0,044 | 144 | 184 | < 2,2 | 82,5 | 2,2 | 0,4 | 0,3 | 1,2 | - | 1,3 | 0,05 | 0,3 | 3,9 | 1,07 | 0,12 | 7 | 6,62 |
| Langue, veau, crue | 632 | 151 | 632 | 151 | 71,9 | 17 | 17 | 0,9 | 8,84 | 0 | 0,22 | 5,7 | - | 0,18 | 2,51 | 2,3 | 17 | 0,026 | 166 | 256 | 2 | 87,5 | 3,02 | 0 | 0,1 | 0 | - | 3,85 | 0,13 | 0,35 | 2,61 | 1,6 | 0,16 | 12 | 5,05 |
| Langue, veau, cuite | 986 | 237 | 986 | 237 | 64,1 | 21 | 21 | 0 | 17 | 0 | 0,16 | 9 | - | 0,21 | 2,6 | - | 18 | 0,047 | 166 | 162 | - | 64 | 3,8 | - | 1,1 | - | - | 6 | 0,11 | 0,35 | 2,6 | 1,1 | 0,16 | 9 | 6,8 |
| Lapin de garenne, viande, crue | 456 | 108 | 456 | 108 | 74,5 | 21,8 | 21,8 | 0 | 2,32 | 0 | 0,13 | 12 | - | - | 3,2 | - | 29 | - | 226 | 378 | - | 50 | - | - | - | - | - | 0 | 0,03 | 0,06 | 6,5 | - | - | - | - |
| Lapin de garenne, viande, cuite | 691 | 164 | 691 | 164 | 61,4 | 33 | 33 | 0 | 3,51 | 0 | 0,11 | 18 | - | 0,18 | 4,85 | - | 31 | - | 240 | 343 | - | 45 | 2,38 | 0 | 0,41 | 1,5 | - | 0 | 0,02 | 0,07 | 6,4 | - | 0,34 | 8 | 6,51 |
| Lapin, viande braisée | 828 | 197 | 828 | 197 | 58,8 | 30,4 | 30,4 | 0 | 8,41 | 0 | 0,093 | 20 | - | 0,18 | 2,37 | 3 | 20 | 0,032 | 226 | 300 | - | 37 | 2,37 | 0 | 0,44 | 1,6 | - | 0 | 0,06 | 0,17 | 7,16 | 0,67 | 0,34 | 9 | 6,51 |
| Lapin, viande crue | 787 | 189 | 787 | 189 | 68,1 | 20,4 | 20,4 | 0,66 | 11,6 | 0 | 0,14 | 10,3 | 165 | 0,15 | 1,11 | 0,55 | 23,5 | 0,032 | 212 | 357 | 10 | 56,3 | 1,64 | 0 | 0,4 | 0 | - | 3 | 0,04 | 0,18 | 10,4 | 1,1 | 0,18 | < 20 | 2,9 |
| Lapin, viande cuite | 697 | 167 | 697 | 167 | 68,2 | 20,5 | 20,5 | 0,5 | 9,2 | 0 | 0,16 | 13,5 | 185 | 0,063 | 2,27 | - | 21,8 | 0,015 | 200 | 340 | 73 | 63,2 | 1,79 | - | 1,22 | - | - | 0 | 0,1 | 0,21 | 9 | 0,5 | 0,13 | < 2,1 | 2,2 |
| Lardon fumé, cru | 1140 | 275 | 1140 | 275 | 55,7 | 16,7 | 16,7 | 0,96 | 22,6 | 0,34 | 2,63 | 18,2 | 1760 | < 0,1 | 0,6 | 11,1 | 23,8 | < 0,1 | 122 | 596 | < 5 | 1100 | 1,2 | 0,48 | 0,19 | 0 | - | 0,9 | 0,38 | 0,11 | 4,9 | 0,44 | 0,25 | 21,1 | 0,42 |
| Lardon fumé, cuit | 1080 | 261 | 1080 | 261 | 54,6 | 14,7 | 14,7 | 0,8 | 22,1 | 0 | 2,59 | 10 | - | 0,054 | 0,82 | - | 21 | 0,027 | 296 | 428 | - | 1040 | 1,7 | 1,1 | 0,34 | 0 | - | 0 | 0,43 | 0,2 | 6,92 | 0,52 | 0,45 | 4 | 0,78 |
| Lardon nature, cru | 1120 | 270 | 1120 | 270 | 56,9 | 16,6 | 16,6 | 1,01 | 22,1 | 0,38 | 2,72 | < 20 | - | 0,35 | 0,42 | 0,7 | 15,6 | < 0,2 | 147 | 540 | 5,77 | 1130 | 1,67 | 0,75 | 0,3 | 0 | - | 0,8 | 0,42 | 0,16 | 4,87 | 0,4 | 0,21 | < 2 | 0,38 |
| Lardon nature, cuit | 1340 | 324 | 1340 | 324 | 48,2 | 23,8 | 23,8 | 2 | 24,5 | 0 | 3,8 | 7,5 | - | < 0,1 | 0,5 | 12 | 22 | < 0,1 | 156 | 572 | 12 | 1650 | 2,1 | < 0,5 | 0,24 | - | - | - | 0,59 | 0,16 | 6,9 | 0,8 | 0,42 | < 16 | 0,67 |
| Lieu jaune ou colin, cru | 320 | 75,5 | 320 | 75,5 | 80,8 | 17,7 | 17,7 | traces | 0,51 | 0 | 0,29 | 13,7 | 0 | 0,04 | 0,39 | 80 | 38,5 | 0,01 | 235 | 331 | 35 | 129 | 0,4 | 0 | 0,78 | 0 | - | 0 | 0,085 | 0,08 | 1,65 | 0,14 | 0 | 3 | 3 |
| Lieu jaune ou colin, cuit | 425 | 100 | 425 | 100 | 74,1 | 24,4 | 24,4 | traces | < 0,5 | 0 | 0,23 | 18,9 | 127 | 0,042 | 0,2 | 150 | 34,3 | 0,0065 | 230 | 388 | 22,1 | 91,4 | 0,46 | 0,73 | 0,84 | < 0,8 | - | 1,09 | < 0,015 | 0,033 | 1,09 | 0,25 | 0,072 | 33,7 | 0,94 |
| Lieu noir, cru | 350 | 82,5 | 350 | 82,5 | 79,4 | 18,8 | 18,8 | traces | 0,8 | 0 | 0,2 | 14,7 | - | < 0,1 | 1,3 | 143 | 30,7 | < 0,1 | 207 | 385 | 120 | 80,5 | 0,57 | 1,77 | 0,36 | 0 | - | 0 | 0,13 | 0,15 | 2,26 | 0,4 | 0,25 | 10 | 4,38 |
| Lieu noir, cuit | 423 | 99,7 | 423 | 99,7 | 75,5 | 23,1 | 23,1 | traces | 0,81 | 0 | 0,25 | 37,5 | 0,2 | 0,064 | 0,4 | 12 | 36 | 0,011 | 152 | 224 | 23,8 | 102 | 0,63 | < 0,5 | 1,7 | - | - | < 1 | 0,058 | 0,08 | 0,8 | 0,14 | 0,081 | 15,2 | 2,1 |
| Lieu noir, surgelé, cru | 350 | 82,4 | 350 | 82,4 | 78,3 | 19,8 | 19,8 | traces | 0,37 | 0 | 0,2 | 19,4 | - | 0,03 | 0,48 | 73 | 21,4 | 0,037 | 107 | 255 | 19,8 | 80,9 | 0,35 | 0,89 | 0,36 | 0 | - | 0 | 0,05 | 0,1 | 1,6 | - | 0,12 | 10 | 2,79 |
| Lieu ou colin d'Alaska, cru | 300 | 70,8 | 300 | 70,8 | 81,2 | 16,3 | 16,3 | traces | 0,61 | < 0,17 | 0,48 | 12,8 | 297 | 0,039 | 0,4 | 78,2 | 36,9 | 0,014 | 160 | 296 | 15,7 | 190 | 0,41 | 1,1 | 0,36 | 0,1 | - | traces | 0,07 | 0,07 | 1,1 | < 0,16 | 0,14 | 3 | 2,3 |
| Lieu ou colin d'Alaska, fumé | 344 | 81,3 | 344 | 81,3 | - | 18,3 | 18,3 | 0 | 0,9 | 0 | - | - | - | - | - | - | - | - | - | - | - | - | - | - | - | - | - | - | - | - | - | - | - | - | - |
| Lièvre, viande crue | 466 | 111 | 466 | 111 | 73,3 | 21,6 | 21,6 | 0 | 2,67 | 0 | 0,15 | 11,6 | - | 0,24 | 1,8 | 0,55 | 26 | 0,038 | 210 | 318 | 10 | 60,5 | 1,7 | 0 | 0,1 | 0 | 0,1 | 0 | 0,09 | 0,06 | 8,07 | 0,8 | 0,3 | 10 | 1 |
| Limande, crue | 342 | 80,8 | 342 | 80,8 | 79,7 | 18,3 | 18,3 | traces | 0,85 | 0 | 0,28 | 47,5 | 0 | 0,06 | 0,29 | 30 | 24,7 | 0,04 | 176 | 262 | 13,8 | 111 | 0,46 | < 0,5 | 0,4 | 0 | - | 0,5 | 0,1 | 0,081 | 2,64 | 0,54 | 0,23 | 5 | 1,67 |
| Limande-sole, crue | 341 | 80,6 | 341 | 80,6 | 80 | 17 | 17 | traces | 1,4 | 0 | 0,24 | 17 | - | 0,1 | 0,5 | 94 | 17 | - | 200 | 230 | 75,3 | 95 | 0,45 | 0 | 0 | 0 | - | 0 | 0,09 | 0,08 | 3,5 | 0,3 | 0,3 | 11 | 1 |
| Limande-sole, cuite à la vapeur | 380 | 90,1 | 380 | 90,1 | 76,9 | 17,4 | 17,4 | traces | 2,27 | 0 | 0,3 | 21 | 120 | 0,01 | 0,6 | 45,5 | 20 | 0 | 250 | 280 | 73 | 120 | 0,5 | 0 | 0,3 | - | - | 0 | 0,1 | 0,1 | 3,8 | 0,29 | 0,28 | 13 | 1 |
| Limande-sole, panée, frite | 863 | 206 | 863 | 206 | 60,5 | 16,7 | 16,7 | 11,9 | 10,2 | 0 | 0,69 | - | - | 0,01 | - | 25 | - | traces | - | - | 73 | 270 | 0,5 | - | 0,1 | - | - | - | 0,1 | 0,1 | 3,8 | 0,29 | 0,14 | 13 | 1 |
| Lingue bleue ou Lingue, crue | 338 | 79,6 | 338 | 79,6 | 79,6 | 18,9 | 18,9 | traces | 0,44 | 0 | 0,23 | 3,96 | - | < 0,1 | 0,26 | 9,82 | 26,8 | < 0,1 | 151 | 302 | 23,2 | 92,8 | 0,4 | < 0,5 | 0,17 | - | - | - | < 0,04 | < 0,04 | 2,03 | 0,11 | 0,18 | - | 0,64 |
| Lompe, crue | 695 | 168 | 695 | 168 | 74,6 | 10 | 10 | traces | 14,2 | 0 | 0,17 | 20 | - | - | 0,5 | 50 | - | - | 203 | 485 | 16,5 | 69 | - | 3,9 | - | 0 | - | 0 | 0,07 | 0,08 | 2 | - | - | - | - |
| Lotte de rivière, crue | 326 | 77 | 326 | 77 | 80,6 | 17,8 | 17,8 | traces | 0,66 | 0 | 0,22 | 44 | - | 0,11 | 0,65 | 40 | 26,5 | 0,37 | 195 | 337 | 20 | 86 | 0,65 | 0,5 | 0,85 | - | - | 0 | 0,28 | 0,14 | 1,56 | 0,15 | 0,3 | 1 | 0,8 |
| Lotte ou baudroie, crue | 285 | 67,2 | 285 | 67,2 | 82,6 | 15,1 | 15,1 | traces | 0,74 | < 0,2 | 0,19 | 18,2 | 370 | 0,027 | 0,3 | 32,2 | 20,7 | 0,022 | 241 | 296 | 34,5 | 77,8 | 0,55 | 1,43 | 0 | - | - | 0,67 | 0,027 | 0,06 | 2,1 | 0,15 | 0,24 | 7 | 0,9 |
| Lotte ou baudroie, grillée/poêlée | 417 | 98,2 | 417 | 98,2 | 75 | 23 | 23 | traces | 0,68 | 0 | 0,061 | 11 | 0,4 | 0,3 | 0,46 | 40,6 | 28,5 | < 0,1 | 368 | 472 | 42,5 | 24,5 | 0,64 | traces | 0,6 | - | - | 0,5 | 0,035 | 0,077 | 3,09 | 0,19 | 0,21 | 20,5 | 0,93 |

| | | | | | | | | | | | | | | | | | | | | | | | | | | | | | | | | | | | |
|---|---|---|---|---|---|---|---|---|---|---|---|---|---|---|---|---|---|---|---|---|---|---|---|---|---|---|---|---|---|---|---|---|---|---|---|
| Loup tacheté, cru | 454 | 108 | 454 | 108 | 78,2 | 16,3 | 16,3 | traces | 4,8 | 0 | 0,26 | 35 | - | - | 1 | 60 | 25 | - | 180 | 282 | - | 105 | - | 1,3 | 2,4 | 0 | - | 0 | 0,18 | 0,045 | 2,5 | 0,57 | 0,35 | 1 | 2,2 |
| Maquereau espagnol ou maquereau blanc ou billard, cru | 567 | 135 | 567 | 135 | 71,8 | 22,1 | 22,1 | traces | 5,15 | 0 | 0,15 | 11 | - | 0,055 | 0,44 | - | 33 | 0,014 | 205 | 446 | - | 59 | 0,49 | 7,3 | 0,69 | 0,1 | - | 1,6 | 0,13 | 0,17 | 2,3 | 0,75 | 0,4 | 1 | 2,4 |
| Maquereau, au naturel, appertisé, égoutté | 787 | 189 | 787 | 189 | 65,7 | 21,6 | 21,6 | 0 | 11,4 | 0 | 1,21 | 128 | 610 | 0,13 | 1,57 | 21,9 | 30,5 | 0,04 | 261 | 225 | 31,9 | 485 | 1,31 | 5,01 | 0,97 | 0,1 | - | 0 | 0,05 | 0,19 | 6,39 | 0,58 | 0,23 | 7,5 | 7,32 |
| Maquereau, cru | 808 | 194 | 808 | 194 | 66,4 | 18,1 | 18,1 | traces | 13,5 | 0 | 0,16 | 4,92 | 82 | < 0,1 | 0,48 | 87,2 | 28,4 | < 0,1 | 190 | 340 | 37,6 | 64 | 0,6 | 6,44 | 1,08 | 3,6 | 0,4 | 0,2 | 0,086 | 0,19 | 9,13 | 0,32 | 0,53 | 8,16 | 4,9 |
| Maquereau, filet au vin blanc, appertisé, égoutté | 862 | 207 | 862 | 207 | 64,4 | 17,8 | 17,8 | 2,25 | 14,1 | 0 | 0,97 | 184 | 524 | < 0,1 | 0,84 | 14,4 | 17 | < 0,1 | 123 | 165 | 28,8 | 387 | 1,02 | 5,44 | 0,88 | - | - | traces | 0,025 | 0,24 | 4,42 | 0,27 | 0,2 | - | 9,14 |
| Maquereau, filet grillé, appertisé, nature, égoutté | 1020 | 246 | 1020 | 246 | 59,9 | 19,7 | 19,7 | 0,55 | 18,3 | 0 | 0,84 | 38 | 571 | 0,1 | 1,1 | < 20 | 23 | 0,03 | 170 | 200 | 70 | 337 | 0,8 | 4,56 | 1,23 | < 0,8 | - | < 0,5 | 0,027 | 0,14 | 5,32 | 0,47 | 0,062 | 8,37 | 10,5 |
| Maquereau, filet sauce moutarde, appertisé, égoutté | 902 | 217 | 902 | 217 | 61,7 | 20,8 | 20,8 | 1,09 | 14,2 | 0,58 | 1,01 | 22,9 | 745 | < 0,1 | 1,3 | 20 | 26,7 | < 0,1 | 127 | 180 | < 5 | 405 | 0,4 | 7,3 | 1,63 | - | - | - | 0,051 | 0,15 | 5,1 | 0,23 | 0,25 | - | 8,2 |
| Maquereau, filet sauce tomate, appertisé, égoutté | 946 | 227 | 946 | 227 | 61,9 | 17,7 | 17,7 | 1,94 | 16,3 | 1,02 | 0,64 | 20 | 200 | < 0,1 | 1,1 | 63,8 | 25 | 0,06 | 135 | 326 | 37,4 | 257 | 1 | 2,38 | 0,9 | 0 | - | 0 | 0,06 | 0,16 | 4,55 | 0,6 | 0,3 | 19 | 7,5 |
| Maquereau, frit | 777 | 186 | 777 | 186 | 64,6 | 23,6 | 23,6 | traces | 10,2 | 0 | 0,21 | - | - | 0,16 | - | 107 | - | 0,06 | 273 | 323 | 94,7 | 83,3 | 1 | 12,3 | - | - | - | - | - | 0,29 | 10,5 | 0,49 | 0,58 | - | 12 |
| Maquereau, fumé | 1220 | 294 | 1220 | 294 | 55,7 | 18,8 | 18,8 | 0 | 24,3 | 0 | 0,96 | 20 | 1130 | 0,09 | 1,1 | 46,5 | 29 | 0,02 | 267 | 312 | 31,9 | 384 | 0,1 | 5,2 | 1,6 | 0 | - | 0 | 0,14 | 0,35 | 10 | 0,52 | 0,5 | 8 | 9,91 |
| Maquereau, mariné | 768 | 185 | 768 | 185 | 66,6 | 16,3 | 16,3 | traces | 13,3 | 0 | 2,29 | 14,6 | - | 0,09 | 0,6 | - | 13,6 | < 0,1 | 92,9 | 90,8 | 16 | 915 | 0,25 | 7,3 | - | - | - | - | 0,037 | 0,13 | 3,18 | 0,22 | 0,25 | 12,4 | - |
| Maquereau, rôti/cuit au four | 949 | 228 | 949 | 228 | 61,5 | 21,5 | 21,5 | traces | 15,8 | 0 | 0,21 | 12,8 | 1,5 | < 0,1 | 1,57 | 45,8 | 28,5 | < 0,1 | 207 | 401 | 51,6 | 83 | 0,94 | 7,72 | 0,71 | - | - | 0,4 | 0,16 | 0,41 | 6,85 | 0,37 | 0,62 | 22,6 | 19 |
| Merguez, boeuf et mouton, crue | 1300 | 315 | 1300 | 315 | - | 14 | 14 | 2,8 | 27,6 | 0 | 1,58 | - | - | - | - | - | - | - | - | - | - | 630 | - | - | - | - | - | - | - | - | - | - | - | - | - |
| Merguez, boeuf et mouton, cuite | 1170 | 283 | 1170 | 283 | 54,8 | 19,8 | 19,8 | 1,8 | 21,8 | 0 | 2,33 | 38 | - | 0,081 | 2,04 | 25 | 22,8 | 0,04 | 128 | 513 | < 10 | 950 | 3,3 | < 0,5 | 1,7 | - | - | < 1 | 0,059 | 0,2 | 4,7 | 0,26 | 0,24 | < 20 | 1,45 |
| Merguez, crue | 1310 | 316 | 1310 | 316 | 56,2 | 13,5 | 13,5 | 2,5 | 28 | 0 | 1,48 | 22,1 | - | 0,083 | - | 7 | 23,3 | 0,055 | 60 | - | 2,38 | 900 | 2,78 | - | - | - | - | - | - | - | - | - | - | - | - |
| Merguez, pur boeuf, crue | - | - | - | - | 39,8 | 11 | 11 | - | 43,5 | 0 | - | - | 1190 | - | - | - | - | - | - | - | - | - | - | - | - | - | - | - | - | - | - | - | - | - | - |
| Merlan, cru | 336 | 79,3 | 336 | 79,3 | 79,3 | 18,8 | 18,8 | traces | 0,47 | 0 | 0,16 | 18 | 109 | < 0,1 | 0,31 | 94,9 | 31,7 | < 0,1 | 196 | 417 | 30,2 | 63,6 | 0,39 | 4,11 | 0,27 | 0,1 | - | 0 | 0,025 | 0,024 | 2,26 | 0,12 | 0,18 | 13 | 1,33 |
| Merlan, cuit à la vapeur | 417 | 98,3 | 417 | 98,3 | 75,2 | 23 | 23 | traces | 0,7 | 0 | 0,28 | 21 | 140 | 0,01 | 0,1 | 75 | 28 | 0,01 | 190 | 400 | 25 | 110 | 0,4 | 0 | 0,44 | - | - | 0 | 0,05 | 0,31 | 1,8 | 0,24 | 0,19 | 0 | 0 |
| Merlan, frit | 526 | 125 | 526 | 125 | 71,4 | 23,1 | 23,1 | traces | 3,61 | 0 | 0,43 | 26,5 | 0,2 | 0,024 | 0,5 | 213 | 34,1 | 0,014 | 165 | 402 | 25,2 | 171 | 0,47 | - | 1,25 | - | - | - | 0,047 | 0,087 | 2,58 | 0,39 | 0,26 | 28,9 | 1,89 |
| Merlan, pané | 797 | 190 | 797 | 190 | 62,6 | 12,6 | 12,6 | 13 | 9,57 | 0,9 | 0,64 | 48 | 190 | 0 | 0,7 | - | 33 | 0 | 260 | 320 | - | 250 | 0 | 0 | 2,48 | - | - | 0 | 0 | 0 | 0 | 0 | 0 | 0 | 0 |
| Merlu blanc du Cap, surgelé, cru | 330 | 78 | 330 | 78 | - | 16,1 | 16,1 | traces | 1,52 | 0 | 0,28 | 32 | - | - | 0,4 | - | - | - | - | - | 29 | 87,5 | - | - | - | - | - | - | - | - | - | - | - | - | - |
| Merlu, cru | 349 | 82,6 | 349 | 82,6 | 80 | 17,6 | 17,6 | traces | 1,35 | 0 | 0,22 | 25,3 | 83 | 0,082 | 1 | 37,6 | 22,9 | 0,035 | 160 | 332 | 31,2 | 98,6 | 0,64 | 2,15 | 0,25 | 0 | - | traces | 0,08 | 0,17 | 1,7 | 0 | 0,18 | 9 | 1 |
| Merlu, cuit à l'étouffée | 472 | 112 | 472 | 112 | 74,4 | 21,2 | 21,2 | traces | 3 | 0 | 0,25 | 41,8 | 0,2 | < 0,1 | 0,3 | 42 | 29,2 | < 0,1 | 190 | 290 | 22 | 99,2 | 0,55 | 3,2 | 0,6 | - | - | 0 | 0,052 | 0,086 | 2,65 | 0,24 | 0,23 | 26,4 | 1,6 |
| Merlu, filet, surgelé, cru | 346 | 82 | 346 | 82 | 80,1 | 15,1 | 15,1 | traces | 2,4 | 0 | 0,53 | 12,5 | - | 0,03 | 0,16 | 20,3 | 36,8 | - | 39,2 | 57,9 | 32,7 | 262 | 0,13 | - | - | - | - | - | - | - | - | - | - | - | - |
| Mérou, cru | 349 | 82,3 | 349 | 82,3 | 78,6 | 18,6 | 18,6 | traces | 0,86 | 0 | 0,13 | 19 | - | 0,02 | 0,6 | - | 31 | 0,014 | 145 | 483 | - | 53 | 0,49 | - | 0,5 | - | - | 0 | 0,055 | 0,063 | 0,36 | 0,75 | 0,3 | 9 | 0,6 |
| Mortadelle | 1260 | 305 | 1260 | 305 | 55,1 | 15 | 15 | 1,4 | 26,7 | 0 | 2,13 | 36,5 | 1340 | < 0,1 | 0,8 | 27,8 | 26,9 | < 0,1 | 127 | 270 | < 5 | 851 | 1,5 | 0,78 | 0,5 | - | - | - | 0,33 | 0,16 | 4,09 | 0,52 | 0,25 | 31,5 | 0,68 |
| Morue, salée, bouillie/cuite à l'eau | 480 | 113 | 480 | 113 | 67,4 | 26 | 26 | traces | 1,01 | 0 | 5,8 | 64,9 | 5 | 0,089 | 0,35 | 74 | 13,7 | 0,019 | 42 | 110 | 20,7 | 2320 | 0,84 | 1,25 | 0,55 | - | - | - | < 0,04 | 0,052 | 0,47 | < 0,16 | 0,075 | 23,5 | 2,2 |
| Morue, salée, sèche | 871 | 206 | 871 | 206 | 35,2 | 47,6 | 47,6 | traces | 1,67 | 0 | 12,3 | 124 | - | 0,098 | 2,09 | 230 | 80,7 | 0,05 | 551 | 1300 | - | 4900 | 1,07 | 4 | 1,67 | 0,4 | - | 0 | 0,13 | 0,2 | 5,26 | 1,01 | 0,58 | 22,3 | 5,43 |
| Moule commune, crue | 303 | 71,8 | 303 | 71,8 | 82,7 | 11,2 | 11,2 | 2,69 | 1,82 | 0 | 0,85 | 51,5 | 462 | 0,18 | 6,78 | 138 | 31 | 1 | 213 | 276 | 48,3 | 339 | 2,36 | 0 | 1,6 | 0,1 | - | 6,4 | 0,11 | 0,23 | 1,6 | 0,44 | 0,077 | 39,5 | 12,3 |
| Moule de Méditerranée, crue | 290 | 68,7 | 290 | 68,7 | 83,7 | 10,7 | 10,7 | 2,77 | 1,67 | 0 | - | 88 | - | 1,1 | 10,9 | - | 44 | 0,8 | - | 382 | - | - | 4,6 | - | - | - | - | - | - | - | - | - | - | - | - |
| Moule, appertisée, égouttée | 431 | 102 | 431 | 102 | 75,1 | 14,6 | 14,6 | 5,93 | 2,2 | 0 | 1,6 | 57 | - | 0,16 | 8,44 | 197 | 42 | 0,7 | 189 | 98 | 57,2 | 639 | 3,56 | 0 | 3,5 | 0 | - | 0 | 0,01 | 0,11 | 1 | - | 0,08 | 27 | 10,2 |
| Moule, bouillie/cuite à l'eau | 457 | 108 | 457 | 108 | 74,1 | 17,2 | 17,2 | 5,12 | 2,09 | 0 | 0,8 | 59,4 | 411 | 0,11 | 3,99 | 106 | 66,6 | 0,19 | 179 | 166 | 55,7 | 322 | 2,85 | < 0,34 | 2,27 | 0 | - | 5,87 | < 0,04 | 0,29 | 1,16 | 0,28 | 0,044 | 48 | 17,6 |
| Moules à la sauce catalane ou escabèche (tomate), appertisée, égouttée | 605 | 145 | 605 | 145 | - | 12,5 | 12,5 | 5 | 7,9 | 2 | 1,73 | - | - | - | - | - | - | - | - | - | - | 691 | - | - | - | - | - | - | | - | - | - | - | - | - |
| Mousse de canard | 1630 | 395 | 1630 | 395 | 44,4 | 10,1 | 10,1 | 7,07 | 36,1 | 0,45 | 1,58 | - | - | - | - | - | - | - | - | - | - | 627 | - | - | - | - | - | - | - | - | - | - | - | - | - |
| Mouton, épaule, crue | 848 | 204 | 848 | 204 | 62,7 | 18,3 | 18,3 | 0,1 | 14,5 | 0 | 0,23 | 8,25 | - | 0,095 | 2,15 | 3 | 16 | 0,01 | 130 | 301 | 0,4 | 90,5 | 3,5 | 0,4 | 0,43 | 0 | - | 1 | 0,17 | 0,27 | 5,4 | 0,95 | 0,2 | 2 | 5 |
| Mouton, gigot, cru | 935 | 225 | 935 | 225 | 64 | 18 | 18 | traces | 17 | 0 | 0,2 | 10 | - | 0,1 | 2,5 | 1,8 | 23 | 0,02 | 213 | 380 | 2 | 78 | 3,7 | - | - | - | - | - | - | - | - | - | - | - | - |
| Mouton, pied, cru | 848 | 204 | 848 | 204 | - | 18,1 | 18,1 | traces | 14,6 | 0 | 0,19 | 10 | - | - | 1,5 | - | 16 | - | 165 | 295 | - | 75 | - | - | - | - | - | - | 0,16 | 0,22 | 5,2 | - | 0,33 | - | 2,2 |
| Mouton, tête, crue | 848 | 204 | 848 | 204 | - | 18,1 | 18,1 | traces | 14,6 | 0 | 0,19 | 10 | - | - | 1,5 | - | 16 | - | 165 | 295 | - | 75 | - | - | - | - | - | - | 0,16 | 0,22 | 5,2 | - | 0,33 | - | 2,2 |
| Mouton, viande, crue | 576 | 137 | 576 | 137 | 74,2 | 20,6 | 20,6 | traces | 6,11 | 0 | 0,22 | 9,5 | - | 0,12 | 2,29 | 4 | 24,6 | 0,015 | 114 | 279 | 2,97 | 86,4 | 3,6 | - | - | - | - | - | - | - | - | - | - | - | - |
| Mulet, cru | 475 | 113 | 475 | 113 | 75,5 | 20,1 | 20,1 | traces | 3,62 | 0 | 0,18 | 13,6 | 130 | < 0,1 | 0,7 | 14 | 26,7 | < 0,1 | 171 | 305 | 168 | 70,4 | 0,58 | 2,12 | 0,64 | 0,1 | - | 1,2 | 0,16 | 0,24 | 6,31 | 0,76 | 0,37 | 9 | 5,56 |
| Mulet, rôti/cuit au four | 602 | 143 | 602 | 143 | 70,5 | 24,8 | 24,8 | traces | 4,86 | 0 | 0,18 | 31 | - | 0,11 | 1,41 | 190 | 33 | 0,051 | 244 | 458 | - | 71 | 1,21 | 0 | 0 | - | - | 1,2 | 0,08 | 0,13 | 5,05 | 0,88 | 0,49 | 10 | 0,25 |
| Museau de boeuf | 637 | 151 | 637 | 151 | 66,1 | 26,2 | 26,2 | 0 | 5,17 | traces | - | - | - | - | - | - | - | - | - | - | - | - | - | - | - | - | - | - | - | - | - | - | - | - | - |
| Museau de boeuf en vinaigrette | 694 | 167 | 694 | 167 | - | 11,3 | 11,3 | 1 | 13,1 | traces | 1,59 | - | - | - | - | - | - | - | - | - | - | 634 | - | - | - | - | - | - | - | - | - | - | - | - | - |
| Oeuf au jambon en gelée | - | - | - | - | 81,9 | 11,5 | 11,5 | - | 4,7 | - | 1,35 | 26,2 | - | - | 1 | - | 7,1 | - | - | 113 | - | 541 | 0,8 | - | - | - | - | - | - | - | - | - | - | - | - |
| Oeuf de caille, cru | 639 | 154 | 639 | 154 | 74,4 | 13,1 | 13,1 | 0,41 | 11,1 | 0 | 0,35 | 64 | - | 0,062 | 3,65 | - | 13 | 0,038 | 226 | 132 | - | 141 | 1,47 | 1,4 | 1,08 | 0,3 | - | 0 | 0,13 | 0,79 | 0,15 | 1,76 | 0,15 | 66 | 1,58 |
| Oeuf de cane, cru | 753 | 181 | 753 | 181 | 70,8 | 13 | 13 | 1,31 | 13,8 | 0 | 0,37 | 64 | - | 0,062 | 3,85 | - | 17 | 0,038 | 220 | 222 | - | 146 | 1,41 | 1,7 | 1,34 | 0,4 | - | 0 | 0,16 | 0,4 | 0,2 | 1,86 | 0,25 | 80 | 5,4 |
| Oeuf de dinde, cru | 691 | 166 | 691 | 166 | 72,5 | 13,7 | 13,7 | 1,13 | 11,9 | 0 | 0,38 | 99 | - | 0,062 | 4,1 | - | 13 | 0,038 | 170 | 142 | - | 151 | 1,58 | 1,75 | 1,11 | - | - | 0 | 0,11 | 0,47 | 0,024 | 1,89 | 0,13 | 71 | 1,69 |
| Oeuf d'oie, cru | 750 | 180 | 750 | 180 | 70,4 | 13,8 | 13,8 | 1,4 | 13,3 | 0 | 0,35 | 60 | - | 0,062 | 3,64 | - | 16 | 0,038 | 208 | 210 | - | 138 | 1,33 | 1,7 | 1,29 | 0,4 | - | 0 | 0,15 | 0,38 | 0,19 | 1,76 | 0,24 | 76 | 5,1 |
| Oeuf, à la coque | 590 | 142 | 590 | 142 | 76 | 12,2 | 12,2 | 1,08 | 9,82 | 0 | 0,2 | 150 | - | < 0,1 | 1,9 | 50,4 | 11,7 | < 0,1 | 155 | 164 | 23,8 | 150 | 0,93 | 1,28 | 2,17 | - | - | < 0,5 | 0,081 | 0,41 | 0,06 | 1,34 | 0,06 | 57 | 0,96 |
| Oeuf, au plat, frit, salé | 847 | 204 | 847 | 204 | 67,9 | 14,3 | 14,3 | 0,74 | 16 | 0 | 0,36 | 57 | 0,3 | 0,051 | 2,2 | 58,4 | 10,4 | 0,026 | 134 | 141 | 8,95 | 144 | 0,87 | 0,55 | 3,06 | 5,6 | - | 0 | 0,044 | 0,41 | 0,082 | 1,27 | 0,18 | 62,6 | 1,08 |
| Oeuf, au plat, sans matière grasse | 611 | 147 | 611 | 147 | 74,6 | 13,8 | 13,8 | 1,01 | 9,72 | 0 | 0,4 | - | 243 | - | 2,6 | 54,5 | - | - | - | - | - | 159 | - | 0,74 | - | - | - | - | - | - | - | - | - | - | - |
| Oeuf, blanc (blanc d'oeuf), cru | 204 | 48,1 | 204 | 48,1 | 87,6 | 10,8 | 10,8 | 0,85 | 0,19 | 0 | 0,42 | 6 | - | 0,023 | 0,05 | 1,9 | 11 | 0,011 | 15 | 163 | 6 | 166 | 0,03 | 0 | 0 | 0 | 0,9 | 0 | 0,004 | 0,44 | 0,093 | 0,2 | 0,0035 | 5,5 | 0,09 |
| Oeuf, blanc (blanc d'oeuf), cuit | 201 | 47,3 | 201 | 47,3 | 87,8 | 10,3 | 10,3 | 1,12 | 0,17 | 0 | 0,39 | 6,67 | - | 0,023 | 0,067 | - | 9,67 | 0,011 | 14,7 | 147 | 6,1 | 155 | 0,025 | 0 | 0 | - | - | 0 | 0,006 | 0,35 | 0,078 | 0,18 | 0,007 | 7,9 | 0,056 |
| Oeuf, blanc (blanc d'oeuf), en poudre | 1510 | 356 | 1510 | 356 | 5,8 | 81,2 | 81,2 | 7,73 | 0 | 0 | 3,2 | 62 | - | 0,11 | 0,15 | - | 88 | 0,007 | 111 | 1130 | - | 1280 | 0,1 | 0 | 0 | 0 | - | 0 | 0,005 | 2,53 | 0,87 | 0,78 | 0,036 | 18 | 0,18 |
| Oeuf, brouillé, avec matière grasse | 604 | 145 | 604 | 145 | 76,4 | 9,99 | 9,99 | 1,62 | 11 | 0 | 0,36 | 66 | - | 0,059 | 1,31 | 52 | 11 | 0,022 | 165 | 132 | 22,5 | 145 | 1,04 | 1,8 | 0,98 | 4 | - | 0 | 0,04 | 0,38 | 0,076 | 1,22 | 0,13 | 36 | 0,18 |
| Oeuf, cru | 584 | 140 | 584 | 140 | 76,3 | 12,7 | 12,7 | 0,27 | 9,83 | 0 | 0,31 | 76,8 | - | 0,055 | 1,88 | 21 | 11 | 0,027 | 204 | 134 | < 2,58 | 124 | 1,01 | 1,88 | 1,43 | 0,3 | - | 0 | 0,055 | 0,45 | 0,063 | 1,57 | 0,15 | 34 | 1,45 |
| Oeuf, dur | 557 | 134 | 557 | 134 | 76,5 | 13,5 | 13,5 | 0,52 | 8,62 | 0 | 0,31 | 41 | 182 | 0,068 | 1,72 | 49,7 | 14 | 0,031 | 172 | 120 | 7,01 | 123 | 1,27 | 1,12 | 1,03 | 0,3 | - | 0 | 0,066 | 0,51 | 0,064 | 1,4 | 0,12 | 44 | 1,11 |
| Oeuf, en poudre | 2420 | 582 | 2420 | 582 | 3,39 | 47,9 | 47,9 | 3,11 | 42 | 0 | 1,25 | 234 | - | 0,2 | 5,56 | 21 | 38 | 0,058 | 730 | 517 | 87,4 | 502 | 4,22 | 3,1 | 1,11 | 1,2 | - | 0 | 0,19 | 1,76 | 0,32 | 5,73 | 0,44 | 145 | 3,67 |
| Oeuf, jaune (jaune d'oeuf), cru | 1270 | 307 | 1270 | 307 | 55 | 15,5 | 15,5 | 1,09 | 26,7 | 0 | 0,048 | 50,2 | 0,3 | < 0,1 | 0,7 | 175 | 1,5 | < 0,1 | 141 | 43 | 83,5 | 19,1 | 1,1 | 2 | 3,89 | 0,7 | 32,1 | 0 | 0,17 | 0,38 | < 0,16 | 2,87 | 0,29 | 159 | 3,03 |
| Oeuf, jaune (jaune d'oeuf), cuit | 1410 | 340 | 1410 | 340 | 50,3 | 16 | 16 | 1,31 | 30,1 | 0 | 0,035 | 76,4 | 0,2 | 0,15 | 0,95 | 192 | 4,5 | < 0,1 | 121 | 58,6 | 76,4 | 14 | 1,4 | 2,11 | 5 | - | - | 0 | 0,17 | 0,39 | < 0,16 | 4,54 | 0,3 | 166 | 2,43 |
| Oeuf, jaune (jaune d'oeuf), en poudre | 2730 | 660 | 2730 | 660 | 2,87 | 34,1 | 34,1 | 2,77 | 57 | 0 | 0,37 | 296 | - | 0,012 | 7,46 | - | 19,5 | 0,19 | 980 | 254 | - | 149 | 6,33 | 11,4 | 4,81 | 1,5 | - | 0 | 0,34 | 1,57 | 0,089 | 8,42 | 0,75 | 227 | 5,68 |
| Oeuf, poché | 575 | 138 | 575 | 138 | 75,9 | 12,5 | 12,5 | 0,71 | 9,47 | 0 | 0,2 | 68,2 | - | 0,069 | 1,75 | 52 | 10,8 | 0,03 | 197 | 138 | 7,68 | 79,3 | 1,1 | 2 | 1,04 | 0,3 | - | 0 | 0,032 | 0,39 | 0,063 | 1,53 | 0,14 | 35 | 0,71 |
| Oeufs de cabillaud, fumés, semi-conserve | 614 | 146 | 614 | 146 | 67 | 26,9 | 26,9 | 0,1 | 42 | 0 | 0,73 | 8,2 | - | 0,082 | 1,11 | 230 | 12,8 | 0,13 | 453 | 280 | - | 290 | 4,3 | 27,2 | - | - | - | - | - | - | - | - | - | - | - |
| Oeufs de lompe, semi-conserve | 462 | 111 | 462 | 111 | 76 | 10,4 | 10,4 | 1 | 7,27 | 0 | 4,87 | 11,6 | 3000 | < 0,1 | 0,4 | 60 | 3 | < 0,1 | 83 | 68,4 | 8 | 1950 | 0,8 | 2,65 | 1,89 | 0,6 | - | < 0,5 | < 0,04 | 0,071 | < 0,16 | 0,28 | 0,08 | 25 | 5,45 |
| Oeufs de saumon, semi-conserve | 938 | 224 | 938 | 224 | 49 | 30,8 | 30,8 | 1,5 | 10,5 | 0 | 2,34 | 5,6 | 1710 | < 0,01 | 0,08 | 330 | 8,1 | 0,09 | 2,6 | 94 | < 20 | 936 | < 0,05 | 27 | 7,52 | < 0,8 | - | < 0,5 | 0,35 | 0,5 | 0,33 | 1,68 | 0,11 | 124 | 48,6 |
| Oeufs de truite, semi-conserve | - | - | - | - | 58,7 | 26,3 | 26,3 | - | 9,5 | 0 | 2,68 | 58 | 1860 | 0,17 | 1,1 | 70 | 46 | 0,28 | 400 | 180 | 80 | 1070 | 3,1 | 7,17 | 9,74 | 1,58 | - | < 0,5 | 0,5 | 0,88 | < 0,1 | 3 | 0,1 | 165 | 10,4 |
| Oie, viande crue | 648 | 155 | 648 | 155 | 68,3 | 22,6 | 22,6 | 0 | 7,12 | 0 | 0,2 | 13 | - | 0,31 | 2,57 | - | 24 | 0,024 | 312 | 368 | 24 | 78 | 2,34 | 1 | 0 | 0 | 31 | 7,2 | 0,13 | 0,38 | 4,28 | 1,97 | 0,64 | 21 | 0,49 |
| Oie, viande et peau, crue | 1510 | 365 | 1510 | 365 | 49,7 | 15,7 | 15,7 | 0 | 33,6 | 0 | 0,18 | 12 | - | 0,2 | 2,5 | 1,2 | 21 | 0,02 | 234 | 312 | 24 | 71 | 1,54 | 1 | 0 | 0 | - | 4,2 | 0,11 | 0,3 | 5,8 | 1,29 | 0,49 | 7,5 | 0,34 |
| Oie, viande et peau, rôtie/cuite au four | 1240 | 298 | 1240 | 298 | 52 | 25,2 | 25,2 | 0 | 21,9 | 0 | 0,18 | 13 | - | 0,26 | 2,83 | - | 22 | 0,023 | 270 | 329 | - | 70 | 2,62 | 0,1 | 1,74 | 5,1 | - | 0 | 0,077 | 0,32 | 4,17 | 1,53 | 0,37 | 2 | 0,41 |
| Oie, viande, rôtie/cuite au four | 961 | 230 | 961 | 230 | 57,2 | 29 | 29 | 0 | 12,7 | 0 | 0,19 | 14 | - | 0,28 | 2,87 | 4 | 25 | 0,024 | 309 | 388 | - | 76 | 3,17 | 0,1 | - | - | - | 0 | 0,092 | 0,39 | 4,08 | 1,83 | 0,47 | 12 | 0,49 |
| Omble chevalier, cru | 436 | 103 | 436 | 103 | 77,6 | 19,8 | 19,8 | traces | 27 | 0 | 0,14 | 8,8 | - | - | 0,49 | - | 30 | - | 260 | 370 | 16 | 57,5 | 0,67 | 9 | - | 0 | - | 0 | 0,16 | 0,09 | - | - | 0,37 | - | - |
| Omelette au fromage | 1060 | 255 | 1060 | 255 | 58,5 | 17,7 | 17,7 | 2,56 | 19,3 | 0 | 0,58 | - | - | 0,051 | 1,16 | 42,8 | 23,6 | 0,036 | 388 | 112 | 10,8 | 231 | 2,48 | 1,39 | 0,82 | - | - | 0 | 0,059 | 0,34 | 0,08 | 0,85 | 0,09 | 25,8 | 1,55 |
| Omelette aux champignons | 520 | 125 | 520 | 125 | 79 | 8,09 | 8,09 | 2,28 | 9,2 | 0,45 | 0,35 | 48 | - | 0,16 | 1,4 | 35,6 | 11,4 | 0,04 | 152 | 193 | 11,5 | 141 | 0,86 | 1,11 | 0,82 | - | - | - | 0,075 | 0,37 | 1,25 | 1,39 | 0,12 | 34,6 | 0,81 |
| Omelette aux fines herbes | 612 | 147 | 612 | 147 | 75,7 | 9,5 | 9,5 | 1,98 | 11,2 | 0,55 | 0,43 | 86 | - | 0,044 | 1,83 | 43 | 14,5 | 0,19 | 157 | 188 | 11,6 | 173 | 0,92 | 1,38 | 1,3 | - | - | - | 0,071 | 0,33 | 0,2 | 1,06 | 0,12 | 56,1 | 0,97 |
| Omelette aux lardons | 1110 | 267 | 1110 | 267 | 61,1 | 12,5 | 12,5 | 1,46 | 23,5 | 0 | 0,98 | 48,2 | - | 0,034 | 1,39 | 56,3 | 12 | 0,022 | 157 | 184 | 11,3 | 392 | 0,99 | 1,11 | 0,99 | - | - | 0 | 0,21 | 0,29 | 1,13 | 0,95 | 0,14 | 25,6 | 0,9 |
| Omelette, garnitures diverses : légumes, fromages, viandes... (aliment moyen) | 764 | 184 | 764 | 184 | 68,9 | 9,82 | 9,82 | 5,07 | 13,5 | 1 | 0,71 | 43,4 | | 0,1 | 1,15 | 29,8 | 17,8 | 0,072 | 177 | 255 | 10,9 | 284 | 1,08 | 0,88 | 1,99 | | | 0,95 | 0,084 | 0,27 | 0,64 | 1,07 | 0,11 | 32,2 | 0,83 |
| Ormeau, cru | 422 | 99,6 | 422 | 99,6 | 74,6 | 14,4 | 14,4 | 8,69 | 0,83 | 0 | 0,47 | 31 | - | 0,2 | 5,9 | - | 48 | 0,04 | 154 | 215 | - | 189 | 0,82 | 0 | 4 | - | - | 2 | 0,15 | 0,19 | 1,5 | 3 | 0,15 | 5 | 0,73 |
| Orphie commune, crue | 558 | 133 | 558 | 133 | 70,1 | 20 | 20 | traces | 5,91 | 0 | 0,21 | 73 | - | 0,2 | 0,73 | 20 | 27,5 | 0,02 | 196 | 377 | 30,9 | 83,5 | 0,77 | 5 | 1,7 | 0 | - | 1,5 | 0,012 | 0,063 | 5,75 | 0,8 | 0,9 | 1 | 2 |
| Pancetta ou Poitrine roulée sèche | 1420 | 342 | 1420 | 342 | 43,9 | 20,2 | 20,2 | 0,75 | 28,8 | 0 | 4,14 | 6,2 | 2190 | 0,06 | 0,71 | < 20 | 18 | 0,05 | 160 | 370 | < 20 | 1650 | 2,5 | 0,66 | 0,94 | < 0,8 | - | 14,3 | 0,51 | 0,12 | 4,54 | 0,84 | 0,27 | 18,3 | 0,93 |
| Panga, Pangasius, ou poisson-chat du Mékong, filet, cuit | 335 | 79 | 335 | 79 | 80,6 | 17,5 | 17,5 | traces | 1 | 0 | 0,58 | 7,5 | 176 | 0,02 | 0,09 | < 20 | 18 | < 0,01 | 130 | 200 | < 20 | 233 | 0,3 | < 0,25 | 0,28 | < 0,8 | - | < 0,5 | < 0,015 | < 0,01 | 1,44 | 2,35 | 0,13 | 11,1 | 0,89 |

| | | | | | | | | | | | | | | | | | | | | | | | | | | | | | | | | | | | |
|---|---|---|---|---|---|---|---|---|---|---|---|---|---|---|---|---|---|---|---|---|---|---|---|---|---|---|---|---|---|---|---|---|---|---|---|
| Pangasius ou Poisson-chat, cru | 276 | 65,2 | 276 | 65,2 | 83,8 | 13,5 | 13,5 | traces | 1,27 | 0 | 0,68 | 36,7 | - | < 0,1 | 0,21 | 7,95 | 17,6 | < 0,1 | 107 | 200 | 9,6 | 274 | 0,24 | < 0,5 | 0,16 | 2,1 | - | 0 | < 0,04 | 0,022 | 1,35 | 1,96 | 0,1 | 10 | 0,3 |
| Pâté (aliment moyen) | 1350 | 325 | 1350 | 325 | 51,2 | 13,6 | 13,6 | 4,61 | 27,9 | 0,29 | 1,58 | 14 | | 0,68 | 5,58 | 15,1 | 15,3 | 0,15 | 173 | 193 | 18,6 | 627 | 2,74 | 0,17 | 0,58 | | | 9,08 | 0,15 | 0,81 | 6,07 | | 0,19 | 93,8 | 5,11 |
| Pâté au jambon | - | - | - | - | 56,6 | 15,2 | 15,2 | - | 28,6 | traces | 0,18 | 18 | - | - | - | - | 26 | - | - | - | - | 72 | - | - | - | - | - | - | - | - | - | - | - | - | - |
| Pâté au poivre vert | 1480 | 358 | 1480 | 358 | 48,1 | 10,1 | 10,1 | 1,45 | 34,6 | traces | - | - | - | - | - | - | - | - | - | - | - | - | - | - | - | - | - | - | - | - | - | - | - | - | - |
| Pâté breton | 1500 | 363 | 1500 | 363 | 50 | 11 | 11 | 2,88 | 34 | 0,5 | 1,67 | - | - | - | - | - | - | - | - | - | - | 669 | - | - | - | - | - | - | - | - | - | - | - | - | - |
| Pâté de foie de volaille | 1120 | 269 | 1120 | 269 | 59,8 | 13,8 | 13,8 | 1,96 | 22,9 | 0 | 1,22 | 10 | - | 0,18 | 9,19 | 10 | 13 | 0,16 | 175 | 95 | - | 467 | 2,14 | 0 | 0,98 | 0 | - | 10 | 0,052 | 1,4 | 7,52 | 2,62 | 0,26 | 321 | 8,07 |
| Pâté de foie d'oie | 1900 | 459 | 1900 | 459 | 37 | 11,4 | 11,4 | 4,67 | 43,8 | 0 | 1,74 | 70 | - | 0,4 | 5,5 | - | 13 | 0,12 | 200 | 138 | - | 697 | 0,92 | - | - | - | - | 2 | 0,088 | 0,3 | 2,51 | 1,2 | 0,06 | 60 | 9,4 |
| Pâté de gibier | 1180 | 284 | 1180 | 284 | - | 15 | 15 | 4 | 23 | 0,5 | 1,53 | - | - | - | - | - | - | - | - | - | - | 610 | - | - | - | - | - | - | - | - | - | - | - | - | - |
| Pâté de lapin | 1020 | 247 | 1020 | 247 | 61,5 | 16,3 | 16,3 | 0,00027 | 20,2 | 0 | 1,53 | 12,5 | - | - | 3,05 | 4 | 26 | - | 212 | 311 | - | 341 | - | 0,47 | 0,4 | - | - | 0,74 | 0,23 | 0,14 | 5,03 | - | 0,27 | - | 1 |
| Pâté en croûte | 1220 | 292 | 1220 | 292 | 45,4 | 11 | 11 | 22,9 | 17,1 | 1,5 | 1,83 | 19,7 | 947 | 0,1 | 1,1 | 6 | 15,8 | 0,2 | 130 | 151 | - | 723 | 1,3 | < 0,5 | 0,41 | - | - | 4,5 | 0,17 | 0,26 | 2,7 | 0,41 | 0,091 | 30 | 0,72 |
| Pâté ou terrine aux champignons (forestier) | 1360 | 329 | 1360 | 329 | 50,8 | 12,6 | 12,6 | 5,93 | 28,1 | 0,58 | 1,73 | - | 1220 | - | - | - | - | - | - | - | - | 690 | - | - | - | - | - | - | - | - | - | - | - | - | - |
| Pâté ou terrine de campagne | 1340 | 323 | 1340 | 323 | 52,2 | 15,5 | 15,5 | 1,5 | 28,2 | 0 | 1,54 | 12,3 | - | 0,92 | 6,1 | 10 | 15,4 | 0,15 | 185 | 212 | 17,2 | 617 | 3,13 | 0,2 | 0,38 | - | - | 5,9 | 0,14 | 0,8 | 6,8 | - | 0,19 | 24 | 3,94 |
| Pecten d'Amérique ou Peigne du canada, noix, crue | 345 | 81,3 | 345 | 81,3 | 78,7 | 17,3 | 17,3 | 1,78 | 0,54 | 0 | - | 22 | - | - | - | - | - | - | 234 | - | - | - | - | - | - | - | - | - | - | - | - | - | - | - | - |
| Perche du Nil, crue | 349 | 82,4 | 349 | 82,4 | 79 | 19,1 | 19,1 | traces | 0,68 | 0 | 0,14 | 4,96 | - | < 0,1 | 0,26 | 6,54 | 26,8 | < 0,1 | 156 | 308 | 30,6 | 57,5 | 0,35 | 0,46 | 0,49 | - | - | - | 0,071 | < 0,04 | 2,01 | 0,34 | 0,11 | - | 1,17 |
| Perche, crue | 348 | 82,2 | 348 | 82,2 | 79,7 | 17,9 | 17,9 | traces | 1,19 | 0 | 0,14 | 70,8 | - | 0,037 | 0,58 | 13 | 28,5 | 0,15 | 208 | 290 | 24 | 55,2 | 0,66 | 5,75 | 0,7 | 0,1 | - | 2,9 | 0,061 | 0,081 | 2,65 | 0,75 | 0,16 | 11 | 2 |
| Perche, rôtie/cuite au four | 503 | 119 | 503 | 119 | 73,7 | 22,1 | 22,1 | traces | 3,42 | 0 | 0,096 | 73 | - | 0,058 | 0,71 | < 5 | 26 | 0,28 | 105 | 331 | 26 | 38,3 | 0,77 | 9 | 1,32 | - | - | 3,8 | 0,09 | 0,077 | 2,98 | 0,81 | 0,13 | 16 | 0,83 |
| Pétoncle ou Peigne du Pérou, noix, crue | 376 | 88,7 | 376 | 88,7 | - | 17,5 | 17,5 | 3,1 | 0,7 | 0 | 0,38 | - | - | - | - | - | - | - | - | - | - | 150 | - | - | - | - | - | - | - | - | - | - | - | - | - |
| Pigeon, cru | 924 | 222 | 924 | 222 | 64,7 | 18,9 | 18,9 | 0,7 | 16 | 0 | 0,17 | 12,5 | - | 0,52 | 4,03 | - | 22,8 | 0,019 | 278 | 244 | - | 67,8 | 2,45 | 0,4 | 0,1 | - | - | 6,2 | 0,25 | 0,25 | 6,45 | 0,77 | 0,47 | 9,75 | 0,44 |
| Pigeon, viande, rôtie/cuite au four | 887 | 213 | 887 | 213 | 62 | 23,9 | 23,9 | 0 | 13 | 0 | 0,14 | 17 | - | 0,76 | 5,91 | 4 | 26 | - | 332 | 256 | - | 57 | 3,83 | 0,2 | 0,06 | 4 | - | 2,9 | 0,28 | 0,35 | 7,6 | - | 0,57 | 6 | 0,41 |
| Pilchard, sauce tomate, appertisé, égoutté | 696 | 167 | 696 | 167 | 70,1 | 13 | 13 | 3,99 | 11 | 0 | 1 | 250 | 520 | 0,16 | 2,5 | 64 | 29 | 0,11 | 280 | 310 | 30 | 400 | 1,3 | 14 | 2,5 | - | - | traces | 0,01 | 0,33 | 5,9 | 0,85 | 0,27 | 0 | 13 |
| Pintade, crue | 601 | 143 | 601 | 143 | 71,7 | 22,8 | 22,8 | 0 | 5,75 | 0 | 0,18 | 11,5 | - | 0,044 | 0,8 | - | 23 | 0,018 | 161 | 207 | - | 70,5 | 1,17 | 0,4 | 0,3 | - | - | 1,5 | 0,063 | 0,11 | 8,22 | 0,91 | 0,43 | 5,5 | 0,36 |
| Pintade, cuisse, crue | 510 | 121 | 510 | 121 | 73,6 | 21,3 | 21,3 | 0 | 4 | 0 | 0,2 | 13 | - | - | 2,5 | - | - | - | 220 | 340 | - | 78 | 1,9 | 0,3 | 0,01 | - | - | 0 | 0,29 | 0,35 | 5,1 | - | 0,61 | 17 | - |
| Pintade, poitrine, crue | 453 | 107 | 453 | 107 | 75,3 | 25,1 | 25,1 | 0 | 0,7 | 0 | 0,13 | 4 | - | - | 0,3 | - | - | - | 230 | 360 | - | 50 | 1 | 0,3 | traces | - | - | 0 | 0,14 | 0,15 | 9,1 | - | 0,81 | 9 | - |
| Poisson pané, frit | 809 | 193 | 809 | 193 | 62,2 | 11,9 | 11,9 | 13,4 | 10 | 0,9 | 0,84 | 15,1 | 501 | 0,055 | 0,44 | 5 | 32,4 | 0,14 | 164 | 286 | 6,24 | 338 | 0,39 | traces | 3,8 | - | - | < 0,5 | 0,062 | 0,073 | 1,6 | 0,25 | 0,12 | 29 | 1,95 |
| Poisson pané, surgelé, cru | 679 | 162 | 679 | 162 | 64,8 | 11 | 11 | 16,1 | 5,73 | 0,9 | 0,69 | 20 | 800 | 0,07 | 0,51 | 110 | 29 | 0,25 | 108 | 322 | - | 270 | 0,53 | traces | - | - | - | traces | 0,09 | 0,14 | 1,7 | 0,33 | 0,28 | - | 1 |
| Poisson, croquette ou beignet ou nuggets, frit | 1060 | 255 | 1060 | 255 | 50,5 | 13,2 | 13,2 | 18,4 | 13,7 | 2,5 | 1,04 | 31,1 | 0,8 | 0,045 | 0,7 | 48,7 | 22,3 | 0,12 | 130 | 259 | 2,13 | 418 | 0,41 | traces | 0 | - | - | traces | 0,11 | 0,07 | 1,24 | 0,32 | 0,15 | 32,2 | 1,54 |
| Poule, cuisse, crue | 551 | 131 | 551 | 131 | 74,4 | 19,6 | 19,6 | 0 | 5,92 | 0 | 0,27 | 9,5 | - | 0,087 | 0,99 | - | 21,5 | 0,018 | 182 | 234 | - | 108 | 2,07 | 0 | 0,21 | 2,65 | - | 4,6 | 0,13 | 0,25 | 5,3 | 1,18 | 0,36 | 7 | 0,44 |
| Poule, viande ,crue | 523 | 124 | 523 | 124 | 73,7 | 20,9 | 20,9 | 0 | 4,51 | 0 | 0,16 | 8,5 | - | 0,11 | 0,95 | 0,9 | 23,5 | 0,021 | 170 | 249 | 12,4 | 65,5 | 1,38 | 0,2 | 0,17 | 2 | - | 3,3 | 0,12 | 0,19 | 6,99 | 1,06 | 0,52 | 8 | 0,45 |
| Poule, viande et peau, crue | 894 | 215 | 894 | 215 | 63,7 | 17,8 | 17,8 | 0,4 | 15,8 | 0 | 0,18 | 32,7 | - | 0,11 | 1,25 | 0,9 | 18,7 | 0,019 | 169 | 196 | 8 | 72,7 | 1,27 | - | 0,32 | 2,4 | - | 0 | 0,086 | 0,14 | 6,42 | 0,9 | 0,33 | 7,5 | 0,32 |
| Poulet (var. blanc), viande et peau, cru | 519 | 123 | 519 | 123 | 72,1 | 21,2 | 21,2 | 0 | 4,3 | 0 | 0,2 | 9,73 | - | 0,11 | 0,65 | 4,1 | 25,1 | 0,02 | 138 | 250 | 7,26 | 78,3 | 1,07 | - | - | - | - | - | - | - | - | - | - | - | - |
| Poulet éviscéré sans abats, cru | 875 | 210 | 875 | 210 | - | 18,6 | 18,6 | traces | 15,1 | 0 | 0,18 | 11 | - | - | 0,9 | - | 20 | - | 147 | 189 | - | 70 | - | - | - | - | - | 1,6 | 0,06 | 0,12 | 6,8 | - | 0,35 | 6 | 0,31 |
| Poulet fermier, viande et peau, cru | 569 | 135 | 569 | 135 | 67,5 | 21,5 | 21,5 | 0,2 | 5,4 | 0 | 0,21 | 10,2 | - | 0,13 | 0,78 | 4,1 | 24,9 | 0,02 | 142 | 267 | 8,95 | 100 | 1,27 | - | - | - | - | - | - | - | - | - | - | - | - |
| Poulet, aile, viande et peau, cru | 754 | 181 | 754 | 181 | 66,9 | 20,4 | 20,4 | 0 | 11 | 0 | 0,22 | 16 | 121 | 0,03 | 0,54 | < 20 | 17 | 0,01 | 140 | 230 | < 20 | 86,1 | 1,1 | < 0,25 | 0,78 | < 0,8 | - | < 0,5 | 0,017 | 0,054 | 9,62 | 1,01 | 0,36 | 39,7 | 1,73 |
| Poulet, aile, viande et peau, rôti/cuit au four | 1030 | 247 | 1030 | 247 | 59,4 | 23,8 | 23,8 | 0 | 16,9 | 0 | 0,25 | 18 | - | 0,041 | 0,84 | - | 19 | 0,011 | 147 | 212 | - | 98 | 1,64 | 0,2 | 0,69 | 0,3 | - | 0 | 0,065 | 0,15 | 6,32 | 0,86 | 0,56 | 9 | 0,35 |
| Poulet, croquette panée ou nuggets | 1050 | 250 | 1050 | 250 | 48,6 | 15,6 | 15,6 | 19,4 | 11,6 | 2,4 | 1,37 | 25 | 792 | 0,07 | 1 | 60 | 25 | 0,23 | 190 | 270 | < 20 | 547 | 0,67 | < 0,25 | 2,41 | 0,99 | - | < 0,5 | 0,083 | 0,025 | 7,15 | 0,77 | 0,098 | 10,9 | 0,23 |
| Poulet, cuisse, viande et peau, bouilli/cuit à l'eau | 790 | 188 | 790 | 188 | 64,1 | 26,1 | 26,1 | < 0,55 | 9,1 | 0 | 0,18 | 15 | 93,6 | 0,09 | 0,96 | < 20 | 23 | 0,02 | 180 | 250 | < 20 | 73 | 2,3 | < 0,25 | 0,54 | < 0,8 | - | < 0,5 | 0,059 | 0,092 | 4,36 | 1,22 | 0,17 | 16,3 | 0,39 |
| Poulet, cuisse, viande et peau, cru | 799 | 192 | 799 | 192 | 69,2 | 17,3 | 17,3 | 0 | 13,5 | 0 | 0,22 | 6,7 | 109 | 0,05 | 0,68 | < 20 | 21 | 0,01 | 170 | 280 | < 20 | 86 | 1,4 | < 0,25 | 0,74 | 2,27 | - | 3,26 | 0,09 | 0,058 | 4,35 | 1,29 | 0,18 | 12,9 | 0,4 |
| Poulet, cuisse, viande et peau, cru, bio | 805 | 193 | 805 | 193 | 68,3 | 18,1 | 18,1 | 0 | 13,3 | 0 | 0,19 | 6,6 | 102 | 0,06 | 0,92 | < 20 | 21 | 0,01 | 180 | 300 | 50 | 77 | 1,5 | 0,46 | < 0,08 | - | - | 1,39 | 0,067 | 0,12 | 4,55 | 1,76 | 0,19 | 5,17 | 0,56 |
| Poulet, cuisse, viande et peau, cru, label rouge | 723 | 173 | 723 | 173 | 69,9 | 19,1 | 19,1 | 0 | 10,6 | 0 | 0,21 | 6,1 | 95 | 0,07 | 1 | < 20 | 23 | 0,02 | 200 | 330 | 20 | 84 | 1,8 | 0,73 | 0,89 | - | - | 1,09 | 0,08 | 0,095 | 5,1 | 1,62 | 0,19 | 6,04 | 0,48 |
| Poulet, cuisse, viande et peau, rôtie/cuite au four | 893 | 213 | 893 | 213 | 62,2 | 25,9 | 25,9 | 2 | 11,3 | 0 | 0,23 | 14,6 | 137 | 0,063 | 1,2 | 5 | 25,1 | 0,015 | 200 | 324 | 17,3 | 91 | 2,16 | < 0,5 | 0,19 | 3,55 | - | 3 | 0,059 | 0,18 | 6,35 | 0,88 | 0,28 | 5,33 | 0,42 |
| Poulet, cuisse, viande, bouilli/cuit à l'eau | 785 | 188 | 785 | 188 | 64,3 | 24,8 | 24,8 | 0,7 | 9,52 | 0 | 0,23 | 5,8 | 100 | < 0,1 | 1,2 | 3 | 26,2 | < 0,1 | 164 | 262 | - | 90 | 2,4 | < 0,5 | 0,27 | 3,55 | - | 2,2 | 0,059 | 0,15 | 4,9 | 0,64 | 0,2 | 7,5 | 0,49 |
| Poulet, cuisse, viande, cru | 478 | 114 | 478 | 114 | 76 | 19,3 | 19,3 | 0 | 4,05 | 0 | 0,23 | 10,3 | - | 0,061 | 0,99 | 2 | 22,3 | 0,019 | 173 | 229 | - | 92 | 1,8 | 0 | 0,2 | 2,53 | 34,3 | 3,1 | 0,079 | 0,18 | 5,9 | 1,2 | 0,35 | 7,67 | 0,42 |
| Poulet, cuisse, viande, rôti/cuit au four | 718 | 171 | 718 | 171 | 66,9 | 24,8 | 24,8 | 0 | 8,03 | 0 | 0,26 | 11,6 | - | 0,07 | 1,2 | - | 22,6 | 0,018 | 197 | 252 | 29 | 104 | 2,3 | 0,13 | 0,2 | 3,73 | - | 0 | 0,084 | 0,21 | 6,03 | 1,17 | 0,39 | 5,8 | 0,36 |
| Poulet, escalope panée | 1410 | 339 | 1410 | 339 | 41,8 | 7,25 | 7,25 | 21,7 | 24 | 3,4 | 1,26 | 22 | 786 | 0,05 | 1,6 | < 20 | 26 | 0,33 | 160 | 230 | < 50 | 504 | 0,8 | < 0,5 | 4,55 | < 0,8 | - | - | 0,11 | 0,02 | 3,71 | 0,71 | 0,14 | 9,85 | 0,45 |
| Poulet, filet, sans peau, cru | 464 | 110 | 464 | 110 | 75,1 | 23,4 | 23,4 | 0 | 1,5 | 0 | 0,11 | 3,3 | 64,1 | 0,03 | 0,33 | < 20 | 32 | < 0,01 | 240 | 390 | < 20 | 43 | 0,52 | < 0,25 | 0,36 | 1,3 | 8,9 | 4,89 | 0,12 | 0,052 | 10,7 | 1,69 | 0,41 | 14,6 | 0,17 |
| Poulet, filet, sans peau, cru, bio | 497 | 118 | 497 | 118 | 74,1 | 24,6 | 24,6 | 0 | 1,8 | 0 | 0,095 | 3,6 | 133 | 0,03 | 0,32 | < 20 | 29 | < 0,01 | 220 | 350 | < 20 | 38 | 0,51 | 0,32 | 0,31 | - | - | 0,99 | 0,032 | 0,02 | 12,6 | 1,36 | 0,39 | 14 | 0,25 |
| Poulet, filet, sans peau, cru, label rouge | 478 | 113 | 478 | 113 | 73,8 | 25,2 | 25,2 | 0 | 1 | 0 | 0,1 | 3,6 | 76,4 | 0,02 | 0,3 | < 20 | 30 | < 0,01 | 230 | 370 | < 20 | 40 | 0,54 | < 0,25 | 0,35 | - | - | 1,12 | 0,058 | 0,022 | 12,7 | 1,12 | 0,37 | 12 | 0,21 |
| Poulet, filet, sans peau, sauté/poêlé | 598 | 141 | 598 | 141 | 67,5 | 30,1 | 30,1 | 0 | 2 | 0 | 0,14 | 4,7 | 107 | 0,03 | 0,39 | < 20 | 37 | < 0,01 | 270 | 440 | < 20 | 56 | 0,72 | < 0,25 | 0,11 | < 0,8 | - | 1,43 | 0,087 | 0,049 | 11,8 | 1,74 | 0,19 | 9,55 | 0,18 |
| Poulet, filet, sans peau, sauté/poêlé, bio | 611 | 144 | 611 | 144 | 67,5 | 31,1 | 31,1 | 0 | 1,8 | 0 | 0,12 | 4,5 | 106 | 0,03 | 0,36 | < 20 | 34 | < 0,01 | 270 | 420 | 20 | 46 | 0,64 | 0,43 | 0,082 | - | - | < 0,5 | 0,035 | 0,051 | 14 | 1,51 | 0,2 | 8,97 | 0,19 |
| Poulet, haut de cuisse, viande, cru | 487 | 116 | 487 | 116 | 76,2 | 19,7 | 19,7 | 0 | 4,12 | 0 | 0,24 | 7 | - | 0,062 | 0,81 | - | 23 | 0,013 | 185 | 242 | - | 95 | 1,58 | 0 | 0,18 | 2,9 | - | 0 | 0,088 | 0,2 | 5,56 | 1,23 | 0,45 | 4 | 0,61 |
| Poulet, manchons marinés, rôtis/cuits au four | 890 | 213 | 890 | 213 | 60,2 | 22,4 | 22,4 | 2,1 | 12,6 | 0,9 | 1,69 | 34,1 | 985 | < 1 | 1,1 | 41 | 18,4 | < 1 | 168 | - | 5,7 | 674 | 1,5 | 0,4 | 0,7 | - | - | 1,4 | < 0,05 | < 0,05 | 6,3 | 0,77 | 0,15 | < 5 | 1,8 |
| Poulet, pilon, cru | 648 | 155 | 648 | 155 | 71,7 | 18,4 | 18,4 | 0 | 9,05 | 0 | 0,23 | 8,5 | - | 0,062 | 1,26 | - | 19 | 0,014 | 168 | 202 | - | 91 | 1,73 | 0,8 | 0,35 | 2,5 | - | 2 | 0,072 | 0,17 | 5,12 | 1,03 | 0,3 | 5 | 0,42 |
| Poulet, poitrine, viande et peau, cru | 657 | 157 | 657 | 157 | 69,7 | 21,1 | 21,1 | 0 | 8,08 | 0 | 0,16 | 11 | - | 0,039 | 0,82 | - | 25 | 0,018 | 186 | 220 | - | 63 | 0,8 | 0,95 | 0,39 | 0 | - | 1 | 0,062 | 0,088 | 9,9 | 0,8 | 0,53 | 4 | 0,34 |
| Poulet, poitrine, viande et peau, rôti/cuit au four | 794 | 189 | 794 | 189 | 62,4 | 29,8 | 29,8 | 0 | 7,78 | 0 | 0,18 | 14 | - | 0,05 | 1,07 | - | 27 | 0,018 | 214 | 245 | - | 71 | 1,02 | 0,1 | 0,27 | 0,3 | - | 0 | 0,066 | 0,12 | 12,7 | 0,94 | 0,56 | 4 | 0,32 |
| Poulet, viande et peau, cru | 723 | 173 | 723 | 173 | 68,1 | 20,2 | 20,2 | 0,3 | 10,1 | 0 | 0,16 | 9,69 | - | 0,14 | 1,63 | 0,4 | 19,8 | 0,026 | 174 | 227 | 12,4 | 62,3 | 1,3 | 0,53 | 0,32 | 1,5 | - | 1,6 | 0,12 | 0,48 | 7,04 | 0,98 | 0,54 | 10,4 | 1,29 |
| Poulet, viande et peau, rôti/cuit au four | 892 | 213 | 892 | 213 | 59,7 | 28,9 | 28,9 | 2,1 | 9,88 | 0 | 0,4 | 19,4 | 243 | 0,05 | 0,68 | 5 | 39,6 | 0,013 | 200 | 414 | 5,66 | 160 | 1,27 | 1,5 | 0,39 | 2,4 | - | 0 | 0,06 | 0,16 | 8,55 | 0,97 | 0,32 | 5 | 0,52 |
| Poulet, viande, crue | 477 | 113 | 477 | 113 | 74,6 | 20 | 20 | 0 | 3,73 | 0 | 0,2 | 11 | - | 0,044 | 0,77 | 0,4 | 24,1 | 0,016 | 179 | 257 | 12,4 | 81 | 1,17 | 0,15 | 0,25 | 2,1 | - | 2,3 | 0,084 | 0,15 | 8,15 | 1,05 | 0,41 | 14,1 | 0,39 |
| Poulpe, cru | 255 | 60,1 | 255 | 60,1 | 85,4 | 12,9 | 12,9 | 0,97 | 0,5 | 0 | 0,68 | 50,1 | 0 | 0,61 | 2,32 | 20,3 | 33 | 0,11 | 166 | 265 | 38 | 271 | 2,36 | 0,5 | 1,65 | 0,1 | - | 5 | 0,048 | 0,059 | 3,2 | 0,5 | 0,36 | 14 | 15 |
| Poulpe, cuit | 659 | 156 | 659 | 156 | 60,5 | 29,8 | 29,8 | 4,4 | 2,08 | 0 | 1,15 | 106 | - | 0,74 | 9,54 | - | 60 | 0,047 | 279 | 630 | - | 460 | 3,36 | 0 | 1,2 | 0,1 | - | 8 | 0,057 | 0,076 | 3,78 | 0,9 | 0,65 | 24 | 36 |
| Protéine de soja texturée, réhydratée | 630 | 150 | 600 | 143 | 62,8 | 18,6 | 20,4 | 7,03 | 2,9 | 5,61 | < 0,013 | 90 | < 20 | 0,69 | 3,3 | < 20 | 100 | 1,3 | 260 | 820 | < 20 | < 5 | 2,1 | < 0,25 | 0,3 | 5,7 | - | < 0,5 | 0,062 | 0,018 | 0,77 | 0,73 | 0,21 | 52,7 | - |
| Quenelle au tofu (ne convient pas aux véganes ou végétaliens), préemballée | 570 | 136 | 552 | 132 | 73,3 | 11 | 12,1 | 4,97 | 7,2 | < 3 | 1,01 | 120 | 604 | 0,22 | 1,4 | < 20 | 41 | 0,58 | 130 | 130 | < 20 | 405 | 1 | < 0,25 | 1,28 | 20,4 | - | < 0,5 | 0,027 | < 0,1 | 0,15 | 0,2 | 0,038 | 28,3 | < 0,01 |
| Quenelle de poisson, crue | 1030 | 245 | 1030 | 245 | 52,3 | 8,1 | 8,1 | 23,6 | 12,7 | 2,15 | 0,95 | 27,7 | - | 0,09 | 0,84 | - | 11,9 | 0,13 | 71,1 | 98,8 | - | 370 | 0,82 | - | - | - | - | - | - | - | - | - | - | - | - |
| Quenelle de poisson, cuite | - | - | - | - | 65,4 | 5,72 | 5,72 | - | 13,1 | - | 0,8 | 27,7 | - | 0,08 | 0,71 | 12,6 | 11,8 | 0,12 | 215 | 145 | 7,54 | 734 | 0,57 | 0,69 | 1,26 | - | - | - | 0,047 | 0,076 | 0,45 | 0,37 | 0,045 | 30,5 | 0,29 |
| Quenelle de poisson, en sauce | 564 | 135 | 564 | 135 | - | 3,79 | 3,79 | 10,5 | 8,51 | 0,72 | 1,11 | - | - | - | - | - | - | - | - | - | - | 453 | - | - | - | - | - | - | - | - | - | - | - | - | - |
| Quenelle de veau, en sauce | 688 | 165 | 688 | 165 | 68,8 | 3,52 | 3,52 | 15,6 | 9,5 | 1,39 | 1,17 | - | - | - | - | - | - | - | - | - | - | 457 | - | - | - | - | - | - | - | - | - | - | - | - | - |
| Quenelle de volaille, crue | 780 | 187 | 780 | 187 | - | 9,3 | 9,3 | 14,1 | 10,1 | 1,1 | 0,74 | - | - | - | - | - | - | - | - | - | - | 295 | - | - | - | - | - | - | - | - | - | - | - | - | - |
| Quenelle de volaille, en sauce | 574 | 138 | 574 | 138 | - | 3,55 | 3,55 | 10,3 | 8,92 | 1,2 | 1,1 | - | - | - | - | - | - | - | - | - | - | 432 | - | - | - | - | - | - | - | - | - | - | - | - | - |
| Quenelle nature, crue | 904 | 217 | 904 | 217 | - | 7,68 | 7,68 | 14,8 | 13,7 | 2,15 | 1,04 | - | - | - | - | - | - | - | - | - | - | 410 | - | - | - | - | - | - | - | - | - | - | - | - | - |
| Raie, crue | 382 | 89,9 | 382 | 89,9 | 76,8 | 21,4 | 21,4 | traces | 0,47 | 0 | 0,35 | 11,8 | 210 | < 0,1 | 0,23 | 23,6 | 23,1 | < 0,1 | 136 | 279 | 27,8 | 139 | 0,41 | 0,36 | 0,31 | - | - | 0 | < 0,04 | 0,05 | 3,44 | 0,15 | 0,19 | 5 | 1,13 |
| Raie, cuite au court-bouillon | 415 | 97,8 | 415 | 97,8 | 73,7 | 23,2 | 23,2 | traces | 0,57 | 0 | 0,4 | - | - | 0,03 | - | 8,9 | 24,3 | 0 | 163 | - | - | 161 | 0,6 | 0 | 0 | - | - | - | < 0,04 | - | - | 0 | 0,23 | 0 | 2,7 |
| Raie, rôtie/cuite au four | 409 | 96,3 | 409 | 96,3 | 75,9 | 23 | 23 | traces | 0,5 | 0 | 0,38 | 50 | 260 | 0,03 | 0,6 | 22,5 | 37 | 0 | 220 | 320 | - | 150 | 0,6 | 0 | 0 | - | - | traces | 0,15 | 0,25 | 2,5 | 0 | 0,46 | 0 | 8 |
| Rascasse, crue | 371 | 87,6 | 371 | 87,6 | 78,9 | 18,9 | 18,9 | traces | 1,33 | 0 | - | 61 | - | 0,5 | - | 11 | 21 | 0,3 | - | 465 | - | - | 1,8 | 4,3 | - | - | - | - | - | - | - | - | - | - | - |
| Rascasse, cuite à la vapeur | 533 | 126 | 533 | 126 | 71,7 | 24,6 | 24,6 | traces | 3,1 | 0 | 0,22 | 14 | 115 | 0,02 | 0,18 | < 20 | 33 | 0,01 | 200 | 380 | 80 | 88,8 | 0,37 | 1,09 | - | < 0,8 | - | - | 0,025 | 0,04 | 1,32 | 0,084 | 0,14 | < 5 | 1,65 |
| Requin, cru | 461 | 109 | 461 | 109 | 73,6 | 20,8 | 20,8 | traces | 2,91 | 0 | 0,2 | 33 | - | 0,033 | 1,12 | - | 37 | 0,015 | 201 | 355 | - | 79 | 0,43 | 0,6 | 1 | 0,1 | - | 0 | 0,031 | 0,046 | 3,67 | 0,7 | 0,4 | 3 | 1,49 |
| Rillettes de canard | 1560 | 377 | 1560 | 377 | 47,3 | 15,3 | 15,3 | 1,03 | 34,6 | traces | 1,25 | 140 | 742 | 0,1 | 1,6 | < 20 | 19 | 0,08 | 200 | 210 | < 20 | 500 | 2,5 | < 0,25 | 0,6 | 85,4 | - | 106 | < 0,015 | 0,069 | 3,05 | 0,69 | 0,086 | 15 | 1,76 |
| Rillettes de crabe, préemballées | 798 | 192 | 798 | 192 | - | 12,2 | 12,2 | 3,5 | 13,6 | 3,5 | 1,11 | - | - | - | - | - | - | - | - | - | - | 440 | - | - | - | - | - | - | - | - | - | - | - | - | - |
| Rillettes de maquereau, préemballées | 1020 | 246 | 1020 | 246 | - | 11 | 11 | 1 | 22 | 0 | - | - | - | - | - | - | - | - | - | - | - | - | - | - | - | - | - | - | - | - | - | - | - | - | - |
| Rillettes de poisson, préemballées | 1010 | 243 | 1010 | 243 | 59,9 | 13,9 | 13,9 | 0,3 | 20,7 | 0 | 0,96 | 26 | 626 | 0,04 | 0,55 | < 20 | 22 | 0,02 | 130 | 200 | 30 | 385 | 0,39 | 2,58 | 4,05 | 10,1 | - | < 0,5 | 0,083 | 0,062 | 5,42 | 0,4 | 0,14 | 11,7 | 3,57 |

| | | | | | | | | | | | | | | | | | | | | | | | | | | | | | | | | | | | |
|---|---|---|---|---|---|---|---|---|---|---|---|---|---|---|---|---|---|---|---|---|---|---|---|---|---|---|---|---|---|---|---|---|---|---|---|
| Rillettes de poulet | 1410 | 340 | 1410 | 340 | 52,2 | 16,4 | 16,4 | 0,49 | 30,3 | 0,25 | 1,19 | - | - | - | - | - | - | - | - | - | - | 470 | - | - | - | - | - | - | - | - | - | - | - | - | - |
| Rillettes de saumon, préemballées | 985 | 237 | 985 | 237 | 62,8 | 15,1 | 15,1 | 2 | 18,4 | 1,6 | 1,3 | 14 | 737 | < 0,1 | 0,4 | 10 | 21 | < 0,1 | 130 | 222 | 20 | 510 | 0,4 | 2,5 | 4,65 | - | - | - | 0,056 | 0,11 | 4,25 | 0,42 | 0,24 | 19 | 2,15 |
| Rillettes de thon, préemballées | 826 | 199 | 826 | 199 | 63,5 | 16,4 | 16,4 | 1,65 | 13,8 | 1,4 | 1,24 | 18,4 | 707 | < 0,1 | 0,7 | 6 | 25,5 | < 0,1 | 121 | 202 | 8 | 490 | 0,5 | 1,2 | 7,09 | - | - | - | 0,22 | 0,07 | 9,1 | < 0,16 | 0,36 | 27 | 2,6 |
| Rillettes de Tours | 2160 | 522 | 2160 | 522 | - | 18 | 18 | traces | 50 | traces | - | - | - | - | - | - | - | - | - | - | - | - | - | - | - | - | - | - | - | - | - | - | - | - | - |
| Rillettes d'oie | 1510 | 364 | 1510 | 364 | 48,9 | 14,9 | 14,9 | 0,99 | 33,4 | traces | 1,29 | 10 | 751 | 0,11 | 1 | < 20 | 13 | 0,07 | 110 | 200 | < 20 | 516 | 1,6 | 2,53 | < 0,08 | < 0,8 | - | 19,3 | 0,025 | 0,058 | 2,97 | 0,4 | 0,082 | 11,6 | 0,99 |
| Rillettes du Mans | 1740 | 420 | 1740 | 420 | 40 | 15,6 | 15,6 | 4,7 | 37,6 | 0,21 | 1,28 | - | 810 | - | - | - | - | - | - | - | - | 504 | - | - | - | - | - | - | - | - | - | - | - | - | - |
| Rillettes pur oie | 1720 | 417 | 1720 | 417 | - | 14,7 | 14,7 | traces | 39,8 | traces | 1,3 | - | - | - | - | - | - | - | - | - | - | - | - | - | - | - | - | - | - | - | - | - | - | - | - |
| Ris, agneau, cru | 508 | 122 | 508 | 122 | 74,6 | 14,6 | 14,6 | 0 | 7,04 | 0 | 0,18 | 5,5 | - | 0,13 | 1,64 | 1,8 | 13,6 | 0,061 | 400 | 380 | 6 | 70,3 | 1,92 | 1 | 0,04 | - | - | 38 | 0,063 | 0,21 | 3,95 | 1 | 0,05 | 13 | 4,6 |
| Ris, agneau, cuit | 948 | 227 | 948 | 227 | 59,7 | 22,8 | 22,8 | 0 | 15,1 | 0 | 0,13 | 12 | - | 0,083 | 2,12 | - | 19 | 0,043 | 431 | 291 | - | 32 | 2,68 | - | - | - | - | 20 | 0,02 | 0,21 | 2,56 | 0,85 | 0,05 | 13 | 5,54 |
| Ris, veau, braisé ou sauté/poêlé | 572 | 136 | 572 | 136 | 73,3 | 21 | 21 | 0 | 5,8 | 0 | 0,15 | 4 | - | 0,072 | 0,5 | 4 | 24 | 0,021 | 627 | 435 | - | 59 | 1 | 0 | 0,09 | 0 | - | 39,4 | 0,09 | 0,1 | 4,4 | 1,77 | 0,08 | 20 | 4,9 |
| Ris, veau, cru | 407 | 96,6 | 407 | 96,6 | 76,6 | 17,4 | 17,4 | 0 | 3,02 | 0 | 0,19 | 2,67 | - | 0,08 | 1,52 | 3 | 14,6 | 0,019 | 477 | 414 | 6 | 72 | 1,43 | 0,25 | 0,09 | 0 | - | 49,2 | 0,13 | 0,19 | 3,89 | 1,23 | 0,035 | 22 | 3,33 |
| Rognon, agneau, braisé | 536 | 127 | 536 | 127 | 70,5 | 23,7 | 23,7 | 0 | 3,62 | 0 | 0,38 | 18 | - | 0,37 | 12,4 | 10 | 20 | 0,14 | 290 | 178 | - | 151 | 3,8 | 0 | 0,41 | - | - | 12 | 0,35 | 2,07 | 5,99 | 2,04 | 0,12 | 81 | 78,9 |
| Rognon, agneau, cru | 396 | 94,1 | 396 | 94,1 | 79,7 | 14,7 | 14,7 | 1,13 | 3,45 | 0 | 0,52 | 13 | - | 0,45 | 3,97 | 7 | 17 | 0,12 | 246 | 277 | 95,5 | 209 | 2,37 | - | 0,43 | - | - | 11 | 0,62 | 2,24 | 7,51 | 4,22 | 0,22 | 28 | 31,5 |
| Rognon, boeuf, cru | 404 | 95,9 | 404 | 95,9 | 79,8 | 17,1 | 17,1 | 0,9 | 2,65 | 0 | 0,42 | 11,2 | - | 0,43 | 7,04 | 4,2 | 16 | 0,11 | 243 | 236 | 118 | 169 | 1,52 | 1,05 | 0,21 | 0 | - | 12,2 | 0,4 | 2,37 | 7,91 | 3,91 | 0,36 | 89 | 21,1 |
| Rognon, boeuf, cuit | 718 | 171 | 718 | 171 | 66,9 | 27 | 27 | 0 | 7 | 0 | 0,24 | 19 | - | 0,56 | 9,5 | 10 | 12 | 0,19 | 304 | 135 | - | 94 | 3,1 | 1,1 | 0,08 | 0 | - | - | 0,43 | 2,97 | 6,5 | 2,34 | 0,66 | 83 | 26 |
| Rognon, cuit (aliment moyen) | 678 | 161 | 678 | 161 | 67,8 | 26,1 | 26,1 | 0 | 6,31 | 0 | 0,25 | 17,1 | | 0,58 | 7,35 | 10 | 14,6 | 0,17 | 283 | 142 | | 95,5 | 3,96 | 0,44 | 0,57 | | | | 0,44 | 2,46 | 6,26 | 3,51 | 0,55 | 70,2 | 31,2 |
| Rognon, veau, braisé ou sauté/poêlé | 701 | 167 | 701 | 167 | 67,7 | 26 | 26 | 0 | 7 | 0 | 0,28 | - | - | - | 5 | 10 | - | - | - | - | - | - | 4,9 | 0 | 1,2 | - | - | - | 0,5 | - | 6,3 | 5,6 | 0,55 | - | 42 |
| Rognon, veau, cru | 363 | 86,1 | 363 | 86,1 | 78 | 16 | 16 | 0,2 | 2,36 | 0 | 0,46 | 10,2 | - | 0,44 | 5,08 | 4,2 | 15,5 | 0,1 | 243 | 248 | 120 | 185 | 1,84 | 1 | 0,2 | - | - | 9,5 | 0,35 | 1,9 | 6,5 | 3,65 | 0,44 | 50,5 | 25,6 |
| Rond de jambon cuit | 761 | 182 | 761 | 182 | 66 | 20,3 | 20,3 | 1,4 | 10,6 | 0 | - | - | - | - | - | - | - | - | 90 | - | - | - | - | - | - | - | - | - | - | - | - | - | - | - | - |
| Rosette ou Fuseau | 1580 | 380 | 1580 | 380 | 34,9 | 24,2 | 24,2 | 1,79 | 30,7 | traces | 4,56 | 44 | 2850 | 0,3 | 1,2 | 36,2 | 40,3 | < 0,1 | 205 | 488 | 7 | 1790 | 2,9 | 1,11 | 0,35 | - | - | 2,3 | 0,68 | 0,22 | 6,87 | 1,21 | 0,48 | 25,3 | 0,82 |
| Rôti de volaille en salaison, cuit | 464 | 110 | 464 | 110 | - | 22 | 22 | 1,28 | 1,73 | 0,44 | 2 | - | - | - | - | - | - | - | - | - | - | 806 | - | - | - | - | - | - | - | - | - | - | - | - | - |
| Rouget-barbet de roche, cru | 660 | 158 | 660 | 158 | 70,6 | 18,4 | 18,4 | traces | 9,37 | 0 | 0,17 | 22,8 | 140 | 0,08 | 0,49 | 11,5 | 30,7 | < 0,1 | 218 | 343 | 36,2 | 66,3 | 0,37 | 1,12 | 0,51 | - | - | traces | 0,06 | 0,065 | 4,46 | 0,34 | 0,37 | 11 | 4,52 |
| Rouget-barbet de roche, vapeur | 602 | 143 | 602 | 143 | 69,6 | 24,1 | 24,1 | traces | 5,2 | 0 | 0,24 | 21 | 162 | 0,05 | 0,48 | 40 | 33 | 0,02 | 230 | 410 | 50 | 95,6 | 0,39 | 1,27 | 2,1 | 1,59 | - | 3,17 | 0,02 | 0,032 | 3,47 | 0,41 | 0,17 | 14,3 | 3,74 |
| Rouget-barbet, filet avec peau, surgelé, cru (Thaïlande, Sénégal…) | 407 | 96,7 | 407 | 96,7 | - | 16,3 | 16,3 | traces | 3,5 | 0 | 0,2 | 14 | - | - | 0,3 | - | - | - | - | - | - | 78 | - | - | - | - | - | - | - | - | - | - | - | - | - |
| Roussette ou petite roussette ou saumonette, crue | 415 | 97,8 | 415 | 97,8 | 75,2 | 23,3 | 23,3 | traces | 0,5 | 0 | 0,31 | 11,7 | 0 | < 0,1 | 0,37 | 26,2 | 20,9 | < 0,1 | 147 | 258 | 28,8 | 125 | 0,87 | < 0,5 | 0,65 | 0,1 | - | 0 | 0,025 | 0,083 | 3,14 | 0,2 | 0,25 | 3,07 | 1,95 |
| Roussette ou petite roussette ou saumonette, cuite | 1070 | 256 | 1070 | 256 | - | 25,4 | 25,4 | traces | 17,2 | 0 | 0,2 | - | - | < 0,1 | - | 24,8 | - | < 0,1 | - | - | 42 | - | 0,67 | < 0,5 | - | - | - | - | - | - | - | 0,24 | 0,36 | 3 | - |
| Sabre, cru | 524 | 125 | 524 | 125 | 74,8 | 18 | 18 | traces | 5,9 | 0 | 0,28 | 12 | - | - | 0,6 | - | - | - | 150 | 270 | - | 110 | - | - | - | - | - | 0 | 0,14 | 0,13 | 3 | - | - | - | - |
| Saint-Pierre, cru | 368 | 86,7 | 368 | 86,7 | 77,8 | 19,9 | 19,9 | traces | 0,8 | 0 | 0,17 | 31,5 | 70 | 0,27 | 3,6 | 20 | 22,5 | 0,1 | 180 | 307 | - | 60 | 0,95 | 0 | 0 | - | - | 0 | 0 | 0 | 0 | 0 | 0,95 | - | 0 |
| Salami | 1870 | 451 | 1870 | 451 | 33,2 | 17,7 | 17,7 | 1,05 | 41,8 | traces | 4,05 | 7,2 | 2310 | 0,07 | 1 | 20 | 15 | 0,28 | 150 | 300 | < 20 | 1620 | 1,9 | 0,68 | 1,26 | < 0,8 | 9 | 3,61 | 0,3 | 0,17 | 3,52 | 1,02 | 0,24 | 8,54 | 1,58 |
| Salami type danois | 1900 | 459 | 1900 | 459 | 36,8 | 17 | 17 | 1,35 | 42,9 | 0 | 4,13 | - | - | - | - | - | - | - | - | - | - | 1650 | - | - | - | - | - | - | - | - | - | - | - | - | - |
| Sandre, cru | 351 | 82,9 | 351 | 82,9 | 78,4 | 18,3 | 18,3 | traces | 1,08 | 0 | 0,12 | 54 | - | 0,014 | 0,3 | 21,5 | 29 | 0,03 | 220 | 340 | 20 | 57 | 0,4 | 0,7 | 1,35 | 0,13 | 0,78 | 1 | 0,16 | 0,25 | 2,31 | - | - | - | - |
| Sang, boeuf, cru | 349 | 82,1 | 349 | 82,1 | 80,5 | 19,5 | 19,5 | 0,8 | 0,12 | 0 | 0,6 | 4 | - | - | 46,5 | - | - | - | 14,5 | 107 | - | 240 | 0,2 | 0,1 | 0,4 | - | - | - | 0,055 | 0,065 | 0,85 | - | 0 | 7 | - |
| Sanglier, cru | 616 | 147 | 616 | 147 | 70 | 20,7 | 20,7 | 0 | 7,13 | 0 | 0,2 | 9,65 | - | 0,1 | 1,23 | 1 | 21 | 0,014 | 155 | 325 | 6,9 | 80 | 3,6 | 0,12 | 0,1 | - | - | 0 | 0,56 | 0,13 | 5,1 | 0,77 | 0,3 | 2 | 0,7 |
| Sanglier, rôti/cuit au four | 643 | 153 | 643 | 153 | 63,9 | 28,3 | 28,3 | 0 | 4,38 | 0 | 0,15 | 16 | - | 0,056 | 1,12 | - | 27 | - | 134 | 396 | - | 60 | 3,01 | 0 | 0,38 | 1,4 | - | 0 | 0,31 | 0,14 | 4,21 | - | 0,42 | 6 | 0,7 |
| Sardine, à l'huile d'olive, appertisée, égouttée | 845 | 202 | 845 | 202 | 60,1 | 24,3 | 24,3 | 1,06 | 11,2 | 0 | 0,73 | 798 | 510 | < 0,1 | 3,3 | 15 | - | 0,16 | 530 | 357 | 11 | 291 | 2,2 | 10,8 | 1,9 | - | - | 0 | < 0,04 | 0,15 | 7,1 | 0,69 | 0,36 | 12 | 13,7 |
| Sardine, à l'huile, appertisée, égouttée | 867 | 207 | 867 | 207 | 60,5 | 24,4 | 24,4 | 0,49 | 12 | 0 | 0,75 | 333 | 477 | 0,062 | 1,97 | 80,1 | 38,5 | < 0,1 | 306 | 368 | 35 | 300 | 1,89 | 7,56 | 1,44 | 0 | - | 0 | < 0,04 | 0,16 | 7,1 | 0,68 | 0,28 | 16 | 13,6 |
| Sardine, crue | 683 | 163 | 683 | 163 | 64 | 19,5 | 19,5 | traces | 9,48 | 0 | 0,22 | 57,5 | 130 | 0,41 | 1,67 | 26,8 | 36 | 0,33 | 286 | 584 | 52,8 | 101 | 1,5 | 14 | 0,28 | - | - | 2,5 | 0,038 | 0,23 | 7,56 | 0,81 | 0,47 | 3,18 | 8,6 |
| Sardine, filets sans arêtes à l'huile d'olive, appertisés, égouttés | 889 | 213 | 889 | 213 | 61,4 | 21,5 | 21,5 | 0,62 | 13,9 | 0 | 0,7 | 108 | 637 | 0,1 | 2,3 | 40 | 37 | < 0,1 | 262 | 357 | 46 | 282 | 2 | 7,4 | 2,1 | - | - | - | < 0,04 | 0,23 | 7,3 | 0,75 | 0,4 | - | 13,6 |
| Sardine, grillée | 812 | 194 | 812 | 194 | 64 | 25,1 | 25,1 | traces | 10,4 | 0 | 0,35 | 130 | 170 | 0,14 | 1,7 | 32 | 34 | 0,08 | 320 | 400 | 38 | 140 | 1,4 | 12,3 | 0,31 | - | - | traces | traces | 0,25 | 6,9 | 0,88 | 0,41 | 4 | 12 |
| Sardine, sauce tomate, appertisée, égouttée | 804 | 193 | 804 | 193 | 64,9 | 20,2 | 20,2 | 1,38 | 11,6 | 1,05 | 0,7 | 378 | 590 | 0,24 | 2,6 | 45,1 | 32,5 | 0,24 | 352 | 376 | 37,7 | 280 | 1,8 | 12,5 | 2,3 | 0 | - | 0 | < 0,04 | 0,17 | 7,38 | 0,7 | 0,15 | 16 | 9,88 |
| Saucisse (aliment moyen) | 1170 | 281 | 1170 | 281 | 55,8 | 17,3 | 17,3 | 0,88 | 23 | 0,29 | 2,05 | 20,8 | | 0,066 | 1,15 | 11,4 | 19,1 | 0,039 | 149 | 438 | 7,94 | 834 | 2,1 | 0,31 | 0,6 | | | 5,3 | 0,31 | 0,17 | 4,06 | 0,52 | 0,18 | 6,18 | 0,7 |
| Saucisse alsacienne fumée ou Gendarme | - | - | - | - | 49,5 | 19 | 19 | - | 24,9 | traces | 2,75 | 18,3 | 1500 | 0,1 | 2,2 | 2 | 21,3 | < 0,1 | 202 | 283 | - | 1100 | 1,5 | < 0,5 | 0,64 | - | - | 16,8 | 0,37 | 0,15 | 5,35 | 0,54 | 0,25 | < 16 | 0,95 |
| Saucisse cocktail | 1200 | 290 | 1200 | 290 | 59,5 | 12,7 | 12,7 | 1,46 | 25,9 | 0,27 | 2,1 | 31,6 | 1210 | < 0,1 | 0,65 | 14 | 20 | < 0,1 | 112 | 305 | 5 | 849 | 1,1 | < 0,5 | 0,1 | 0 | - | 8,8 | 0,2 | 0,14 | 2,43 | 0,3 | 0,13 | 24,2 | 0,8 |
| Saucisse de bière | - | - | - | - | 53,6 | 13 | 13 | - | 29,7 | traces | - | - | - | - | - | - | - | - | - | - | - | - | - | - | - | - | - | - | - | - | - | - | - | - | - |
| Saucisse de foie | 1550 | 374 | 1550 | 374 | 50,7 | 11,7 | 11,7 | 0,9 | 35,9 | 0,3 | 2,03 | 41 | 1340 | - | 5,3 | - | - | - | 154 | 143 | 58 | 810 | 2,7 | 1,09 | 2,15 | - | - | 2 | 0,24 | 0,92 | 3,6 | - | 0,17 | 23 | - |
| Saucisse de Francfort | 1120 | 271 | 1120 | 271 | 56,7 | 13,7 | 13,7 | 1,24 | 23,4 | < 0,5 | 2 | 16,5 | - | 0,18 | 0,91 | 14 | 11,9 | < 0,2 | 152 | 164 | 5,48 | 790 | 1,4 | 0,26 | 0,1 | - | - | - | 0,29 | 0,2 | 2,47 | - | 0,11 | 1,6 | 0,42 |
| Saucisse de langue à la pistache | - | - | - | - | 55,6 | 15,3 | 15,3 | - | 25,8 | traces | - | - | - | - | - | - | - | - | - | - | - | - | - | - | - | - | - | - | - | - | - | - | - | - | - |
| Saucisse de Montbéliard | 1320 | 319 | 1320 | 319 | 50,5 | 17,3 | 17,3 | 1 | 27,3 | 0 | 2,33 | 29,9 | 1340 | 0,1 | 0,5 | 29,7 | 26,8 | < 0,1 | 146 | 311 | < 5 | 931 | 2 | < 0,5 | 0,34 | - | - | < 0,5 | 0,53 | 0,18 | 5,28 | 0,74 | 0,34 | 28,5 | 0,63 |
| Saucisse de Morteau | 1640 | 397 | 1640 | 397 | 43,7 | 16 | 16 | 1 | 36,6 | 0 | 2,6 | - | - | - | - | 7 | - | - | - | - | - | 1040 | - | - | - | - | - | - | - | - | - | - | - | - | - |
| Saucisse de Morteau, bouillie/cuite à l'eau | 1340 | 323 | 1340 | 323 | 49,6 | 15,2 | 15,2 | 0 | 29,1 | traces | 2,02 | 8,4 | - | < 0,1 | 0,8 | 9 | 15 | < 0,1 | 126 | 230 | 7 | 809 | 1,9 | < 0,5 | 0,13 | - | - | < 1 | 0,41 | 0,15 | 3,75 | 0,61 | 0,27 | < 20 | 0,49 |
| Saucisse de Strasbourg ou Knack | 1200 | 291 | 1200 | 291 | 57,9 | 12,4 | 12,4 | 1,39 | 25,9 | 1,37 | 2,21 | 23,7 | 1140 | 0,065 | 1,12 | 7 | 13,4 | 0,015 | 152 | 204 | < 10 | 942 | 1,13 | 0,26 | 0,38 | - | - | 9,4 | 0,29 | 0,2 | 2,47 | - | 0,11 | 1,6 | 0,42 |
| Saucisse de Toulouse, crue | 1350 | 327 | 1350 | 327 | 56,2 | 13,2 | 13,2 | 1 | 30 | traces | 1,89 | 28,9 | - | 0,069 | 0,9 | 7 | 21,7 | 0,023 | 137 | - | 11,2 | 747 | 2,22 | - | - | - | - | - | - | - | - | - | - | - | - |
| Saucisse de Toulouse, cuite | 1140 | 274 | 1140 | 274 | 54,9 | 18,8 | 18,8 | 0 | 22,1 | 0 | 1,71 | 11,5 | - | < 0,1 | 0,9 | 8 | 18,2 | < 0,1 | 141 | 650 | 9 | 686 | 2,2 | < 0,5 | 0,35 | - | - | < 1 | 0,41 | 0,22 | 4,4 | 0,56 | 0,24 | < 20 | 0,41 |
| Saucisse de volaille, façon charcutière | 960 | 231 | 960 | 231 | 59,8 | 16,8 | 16,8 | 1,19 | 17,7 | traces | 1,75 | 8 | - | - | - | - | - | - | - | - | - | - | - | - | - | - | - | - | - | - | - | - | - | - | - |
| Saucisse de volaille, type Knack | 978 | 236 | 978 | 236 | 64 | 14 | 14 | 1,82 | 18,9 | 0 | 1,87 | - | - | - | - | - | - | - | - | - | - | 746 | - | - | - | - | - | - | - | - | - | - | - | - | - |
| Saucisse fumée, à cuire | - | - | - | - | 49,5 | 16,8 | 16,8 | - | 30,7 | traces | 1,89 | - | 1430 | - | - | - | - | - | - | - | - | 755 | - | - | - | - | - | - | - | - | - | - | - | - | - |
| Saucisse sèche | 1870 | 451 | 1870 | 451 | 26,2 | 27,3 | 27,3 | 1,7 | 37,1 | 0,59 | 4,74 | - | - | - | - | 7 | - | - | - | - | - | 1900 | - | - | - | - | - | - | - | - | - | - | - | - | - |
| Saucisse végétale au blé ou seitan, préemballé | 1050 | 250 | 1020 | 243 | 47,7 | 27,1 | 29,1 | 7,53 | 10,5 | 4,6 | 1,82 | 37 | 1100 | 0,27 | 1,7 | < 20 | 32 | 0,75 | 110 | 200 | < 20 | 728 | 1,8 | < 0,25 | 5,92 | 2,29 | - | < 0,5 | 0,11 | 0,16 | 0,49 | 0,46 | 0,078 | 40,6 | - |
| Saucisse végétale au tofu (convient aux véganes ou végétaliens), préemballée | 963 | 231 | 941 | 226 | 60,4 | 13,4 | 14,6 | 5,69 | 16,3 | < 3 | 1,68 | 140 | 1040 | 0,23 | 2 | < 20 | 58 | 0,86 | 210 | 120 | < 20 | 670 | 1,4 | < 0,25 | 7,78 | 12,8 | - | < 0,5 | 0,042 | 0,026 | 0,18 | 0,19 | 0,027 | 17,3 | - |
| Saucisse viennoise, crue | - | - | - | - | 56,8 | 10,3 | 10,3 | - | 21,3 | traces | - | - | - | - | - | - | - | - | - | - | - | - | - | - | - | - | - | - | - | - | - | - | - | - | - |
| Saucisson à l'ail | 1160 | 280 | 1160 | 280 | 57,2 | 15 | 15 | 1,63 | 23,6 | 0,63 | 2 | 9 | 1220 | - | 2,3 | 7 | 9 | - | 174 | 230 | - | 783 | - | - | - | - | - | 5 | 0,16 | 0,22 | 2,6 | - | - | - | - |
| Saucisson brioché, cuit | 1290 | 309 | 1290 | 309 | 42,4 | 14,7 | 14,7 | 22,6 | 17,4 | 2 | 1,38 | 19 | 1230 | 0,09 | 1,1 | < 20 | 21 | 0,22 | 160 | 240 | < 50 | 550 | 1,7 | < 0,5 | 0,76 | < 0,8 | - | - | 0,3 | 0,14 | 2,66 | 1,02 | 0,18 | 12 | 0,54 |
| Saucisson de cheval type cervelas | - | - | - | - | 57,6 | 14,2 | 14,2 | - | 20 | 0 | 2,46 | 15 | 1160 | 0,09 | 1,9 | < 20 | 13 | 0,05 | 290 | 200 | < 20 | 985 | 1,9 | < 0,25 | 0,9 | - | - | - | 0,059 | 0,064 | 2,04 | 0,41 | 0,15 | 6,18 | 2,18 |
| Saucisson de Paris | 1430 | 346 | 1430 | 346 | 50,6 | 12,5 | 12,5 | 0,6 | 32,7 | traces | 1,73 | - | 1400 | - | - | - | - | - | 140 | - | - | 691 | - | - | - | - | - | - | - | - | - | - | - | - | - |
| Saucisson de Paris, fumé | 1190 | 288 | 1190 | 288 | 57,3 | 13,7 | 13,7 | 0,6 | 25,6 | traces | - | - | - | - | - | - | - | - | 90 | - | - | - | - | - | - | - | - | - | - | - | - | - | - | - | - |
| Saucisson sec | 1730 | 418 | 1730 | 418 | 30 | 34,2 | 34,2 | 2,41 | 34,5 | 0,73 | 4,75 | 14,4 | 3000 | 0,099 | 1,27 | 7 | 32 | 0,12 | 237 | 583 | < 10 | 1900 | 2,7 | 0,5 | 0,16 | - | - | 5 | 0,62 | 0,37 | 8,55 | - | 0,43 | 1,64 | 0,67 |
| Saucisson sec aux noix et/ou noisettes | 1980 | 477 | 1980 | 477 | - | 27,7 | 27,7 | 1,78 | 39,7 | 0,75 | 4,31 | - | - | - | - | - | - | - | - | - | - | 1660 | - | - | - | - | - | - | - | - | - | - | - | - | - |
| Saumon fumé | 743 | 178 | 743 | 178 | 63,1 | 22 | 22 | 0,91 | 9,49 | < 0,25 | 3,51 | 5,97 | 2000 | 0,035 | 0,16 | 40 | 38,3 | 0,0098 | 214 | 551 | < 10 | 1410 | 0,24 | 5,45 | 2,85 | 0 | - | 0 | 0,23 | 0,1 | 10 | 0,71 | 1 | 26 | 3,35 |
| Saumon, appertisé, égoutté | 751 | 180 | 751 | 180 | 68,5 | 20,5 | 20,5 | 1,5 | 10,2 | 0 | 1,22 | 270 | 730 | 0,038 | 0,45 | 30 | 42 | 0,048 | 440 | 300 | 42 | 488 | 0,7 | 13 | 1,5 | 0 | - | 0 | 0,04 | 0,18 | 7 | 0,55 | 0,45 | 12 | 3,1 |
| Saumon, bouilli/cuit à l'eau, élevage | 805 | 193 | 805 | 193 | 63,4 | 25 | 25 | traces | 10,3 | 0 | 0,084 | 4,6 | - | < 0,1 | 0,3 | - | 29,7 | < 0,1 | 150 | 347 | < 5 | 33,6 | 0,4 | 3,45 | 2,63 | - | - | - | 0,25 | 0,056 | 6,25 | 1,5 | 0,58 | 28 | - |
| Saumon, cru, élevage | 807 | 194 | 807 | 194 | 66,5 | 20,5 | 20,5 | traces | 12,4 | < 0,25 | 0,16 | 5,84 | 73,2 | < 0,1 | 0,48 | 8,21 | 27,4 | < 0,1 | 181 | 358 | 16,5 | 45,4 | 0,36 | 3,69 | 1,89 | 0,3 | 0,5 | 1,8 | 0,21 | 0,076 | 8,25 | 0,95 | 0,58 | 20,8 | 3,95 |
| Saumon, cru, sauvage | 695 | 166 | 695 | 166 | 68,4 | 21 | 21 | traces | 9,17 | 0 | 0,12 | 13,5 | 83 | 0,16 | 0,66 | 12,2 | 29,2 | 0,016 | 225 | 436 | 27 | 46 | 0,58 | < 8,6 | 1,11 | 0,3 | - | 0 | 0,21 | 0,24 | 8,83 | 1,48 | 0,49 | 13,1 | 4,84 |
| Saumon, cuit à la vapeur | 816 | 195 | 816 | 195 | 63,5 | 23 | 23 | traces | 11,5 | 0 | 0,13 | 9,23 | 63 | 0,026 | 0,4 | 17 | 21,4 | 0,007 | 270 | 390 | 19 | 51,5 | 0,39 | 8,7 | 2,05 | - | - | 0 | 0,22 | 0,11 | 7 | 1,09 | 0,49 | 11,7 | 3,05 |
| Saumon, cuit au micro-ondes, élevage | 834 | 200 | 834 | 200 | 63 | 24,2 | 24,2 | traces | 11,4 | 0 | 0,098 | 5 | - | < 0,1 | 0,4 | - | 31,6 | < 0,1 | 250 | 402 | < 5 | 39 | 0,4 | 3,3 | 2,4 | - | - | - | 0,21 | 0,054 | 7,6 | 1,45 | 0,71 | 19 | - |
| Saumon, cuit, sans précision (aliment moyen) | 854 | 205 | 854 | 205 | 62,6 | 23 | 23 | 0 | 12,5 | 0 | 0,13 | 11,5 | | 0,045 | 0,35 | | 28,6 | 0,023 | 243 | 383 | | 53,3 | 0,42 | | | | | 2,79 | 0,27 | 0,099 | 7,6 | 1,42 | 0,46 | 19,9 | 2,78 |
| Saumon, élevage, rôti/cuit au four | 874 | 210 | 874 | 210 | 62,2 | 22,1 | 22,1 | traces | 13,5 | 0 | 0,15 | 15 | - | 0,049 | 0,34 | - | 30 | 0,016 | 252 | 384 | - | 61 | 0,43 | - | - | - | - | 3,7 | 0,3 | 0,11 | 8,05 | 1,49 | 0,39 | 21,1 | 2,65 |
| Saumon, grillé/poêlé | 933 | 223 | 933 | 223 | 60,1 | 25,5 | 25,5 | traces | 13,5 | 0 | 0,11 | 8,4 | 86,2 | 0,05 | 0,43 | 13,5 | 31,7 | 0,05 | 265 | 414 | 11 | 44,2 | 0,48 | 5,82 | 3,52 | - | - | - | 0,23 | 0,086 | 8,15 | 1,7 | 0,34 | 18 | 3,4 |
| Saupe, crue | 399 | 94,5 | 399 | 94,5 | 78 | 18 | 18 | traces | 2,5 | 0 | - | 28 | - | 0,6 | 4,3 | - | 27 | 0,1 | - | 540 | - | - | 3,3 | - | - | - | - | - | - | - | - | - | - | - | - |
| Sébaste du nord, ou grand sébaste, ou dorade sébaste, ou daurade sébaste, crue | 378 | 89,5 | 378 | 89,5 | 78,4 | 18,8 | 18,8 | traces | 1,6 | 0 | 0,18 | 12,5 | 0 | < 0,1 | 0,26 | 21,7 | 27,5 | < 0,1 | 168 | 333 | 28,3 | 72,8 | 0,33 | 3,96 | 0,84 | 0 | - | traces | 0,063 | 0,07 | 2,16 | 0,17 | 0,19 | 9 | 1,78 |

| | | | | | | | | | | | | | | | | | | | | | | | | | | | | | | | | | | | |
|---|---|---|---|---|---|---|---|---|---|---|---|---|---|---|---|---|---|---|---|---|---|---|---|---|---|---|---|---|---|---|---|---|---|---|---|
| Seiche, crue | 323 | 76,3 | 323 | 76,3 | 80,6 | 16,2 | 16,2 | 0,51 | 1,05 | 0 | 0,93 | 37,1 | 0 | 0,76 | 2,74 | 21 | 31 | 0,082 | 236 | 351 | 65 | 372 | 1,81 | 0 | 2,4 | - | - | 5,3 | 0,057 | 0,37 | 2,1 | 0,5 | 0,33 | 14 | 2,5 |
| Seitan, préemballé | 566 | 134 | 566 | 134 | 67,8 | 20,6 | 20,6 | 6,74 | 2,5 | 0,9 | 1,13 | 20 | 700 | 0,23 | 1,5 | < 20 | 18 | 0,61 | 62 | 75 | 30 | 451 | 1,2 | < 0,25 | 0,51 | < 0,8 | - | < 0,5 | 0,036 | 0,03 | 6,92 | 0,16 | 0,022 | 14,7 | 0,038 |
| Sole tropicale ou Sole langue, crue | 289 | 68,2 | 289 | 68,2 | 82,3 | 15,7 | 15,7 | traces | 0,58 | 0 | 0,35 | 15,6 | - | < 0,1 | 0,22 | 21,7 | 20,1 | < 0,1 | 80,4 | 125 | 30,6 | 140 | 0,29 | < 0,5 | 0,22 | - | - | - | 0,038 | 0,034 | 1,14 | 0,1 | 0,1 | - | 0,92 |
| Sole, bouillie/cuite à l'eau | 292 | 69 | 292 | 69 | 84 | 15 | 15 | traces | 1 | 0 | 0,21 | 90 | - | - | 0,9 | - | 22 | - | 195 | 137 | - | 82 | - | - | - | - | - | - | 0,07 | 0,2 | 1,7 | - | 1 | - | - |
| Sole, crue | 328 | 77,3 | 328 | 77,3 | 80,1 | 18 | 18 | traces | 0,6 | 0 | 0,19 | 22,5 | 142 | < 0,1 | 0,35 | 88,2 | 26,9 | < 0,1 | 170 | 349 | 23 | 76,3 | 0,42 | 0,75 | 0,47 | 0 | - | traces | 0,046 | 0,077 | 3,51 | 0,24 | 0,26 | 7,5 | 1,91 |
| Sole, cuite à la vapeur | 400 | 94,6 | 400 | 94,6 | 77 | 20,3 | 20,3 | traces | 1,48 | 0 | 0,41 | 96 | - | 0,026 | 0,72 | 14,5 | 23,1 | 0,007 | 128 | 175 | 103 | 166 | 0,39 | - | - | - | - | - | 0,064 | 0,2 | 2,39 | - | 0,22 | - | 1,12 |
| Sole, frite | 306 | 72,2 | 306 | 72,2 | 83 | 15,8 | 15,8 | traces | 1 | 0 | 0,36 | - | - | - | 0,86 | - | 28 | - | 250 | 240 | 42,3 | 144 | - | - | - | - | - | - | 0,07 | 0,18 | 1,53 | - | - | - | - |
| Sole, poêlée | 410 | 96,7 | 410 | 96,7 | 75,6 | 23,6 | 23,6 | traces | < 0,5 | 0 | 0,37 | 39 | 228 | 0,02 | 0,14 | 20 | 27 | < 0,01 | 180 | 370 | 30 | 147 | 0,45 | 0,47 | 0,5 | < 0,8 | - | 0,87 | 0,058 | 0,094 | 2,18 | 0,37 | 0,34 | 15 | 1,4 |
| Sole, rôtie/cuite au four | 330 | 78,2 | 330 | 78,2 | 82,4 | 15,7 | 15,7 | traces | 1,69 | 0 | 0,29 | 47,7 | - | 0,027 | 0,57 | 30 | 31,3 | 0,035 | 285 | 224 | 17,4 | 117 | 0,58 | 3,5 | 0,77 | 0,1 | - | 0 | 0,048 | 0,11 | 1,49 | 0,23 | 0,56 | 6 | 1,31 |
| Spécialité végétale type jambon cuit, préemballée | 1010 | 242 | 1010 | 242 | 50,2 | 29,7 | 29,7 | 4,13 | 11,1 | 3,2 | 1,62 | 32 | 969 | 0,26 | 1,4 | < 20 | 24 | 0,75 | 87 | 100 | < 20 | 646 | 1,8 | < 0,25 | 7,47 | < 0,8 | - | < 0,5 | 0,088 | 0,098 | 0,9 | 0,4 | 0,027 | 27,7 | - |
| Spécialité végétale type pâté, préemballée | 816 | 197 | 816 | 197 | 66,7 | 3,56 | 3,56 | 9,45 | 15,4 | 2,96 | 1,27 | 30 | 793 | 0,11 | 1,3 | 40 | 25 | 0,33 | 72 | 170 | < 20 | 509 | 0,96 | < 0,25 | 4,87 | 8,46 | - | < 0,5 | 0,05 | 0,096 | 1,31 | 0,35 | 0,053 | 14,7 | - |
| Sprat, cru | 777 | 186 | 777 | 186 | 66,7 | 19,8 | 19,8 | traces | 11,9 | 0 | - | - | - | 0,13 | 0,59 | - | 20 | 0,13 | 220 | 230 | 10 | - | 1,02 | 0 | 2,5 | 0 | - | 0 | 0,04 | 0,26 | 5 | - | - | - | - |
| Surimi, bâtonnets, tranche ou râpé saveur crabe | 525 | 125 | 525 | 125 | 72,5 | 8,31 | 8,31 | 11,8 | 4,92 | < 0,33 | 1,58 | 19 | 1010 | 0,04 | 0,4 | 10 | 11,5 | 0,0089 | 44 | 51,8 | 9,23 | 632 | 0,19 | < 0,5 | 1,1 | 0,1 | - | 0 | 0,02 | 0,021 | 0,22 | 0,07 | 0,03 | 2 | 0,71 |
| Surimi, fourré au fromage | 619 | 149 | 619 | 149 | 72,3 | 6,6 | 6,6 | 9,24 | 9,3 | < 1 | 1,72 | 19 | 1020 | 0,02 | 0,08 | < 20 | 9,5 | 0,01 | 36 | 49 | < 20 | 688 | 0,2 | < 0,25 | 0,98 | 4,15 | - | < 0,5 | < 0,015 | 0,031 | < 0,1 | 0,091 | 0,018 | 10,5 | 0,47 |
| Tacaud, cru | 346 | 81,6 | 346 | 81,6 | 78,9 | 19,7 | 19,7 | traces | 0,33 | 0 | 0,2 | 26,7 | - | < 0,1 | 0,21 | 214 | 26,5 | < 0,1 | 148 | 306 | 35,4 | 81,5 | 0,34 | 0,32 | 0,34 | 0 | - | 0 | 0,062 | 0,082 | 1,49 | 0,17 | 0,13 | - | 1,47 |
| Tarama, préemballé | 2200 | 534 | 2200 | 534 | 30 | 6,75 | 6,75 | 3,19 | 54,8 | 0,74 | 1,69 | 8,5 | 935 | < 0,1 | 0,4 | 10 | 6,6 | < 0,1 | 79 | 80,9 | < 5 | 676 | 0,7 | 0,96 | 9 | - | - | < 0,5 | 0,1 | 0,095 | 0,57 | 0,64 | 0,084 | < 16 | 2,7 |
| Terrine de canard | 1330 | 320 | 1330 | 320 | 50,7 | 13,2 | 13,2 | 4,44 | 27,1 | 1,75 | 1,39 | 25 | 900 | 0,62 | 5,2 | < 20 | 16 | 0,16 | 170 | 210 | 40 | 554 | 2,6 | 0,75 | 0,89 | 1,43 | - | 12,9 | 0,17 | 0,41 | 7,97 | 2,49 | 0,14 | 343 | 13,6 |
| Terrine de fruits de mer, avec ou sans poisson, préemballée | 782 | 188 | 782 | 188 | 68,4 | 8,4 | 8,4 | 6,5 | 14 | 1,3 | 1,48 | 183 | - | 0,05 | 2,3 | 310 | 22,7 | 0,05 | 66 | 157 | 13 | 590 | 1,2 | 1,2 | 4 | - | - | 0 | 0,045 | 0,11 | 1,6 | 0,32 | 0,12 | 23 | 1,05 |
| Terrine de lapin | 1170 | 281 | 1170 | 281 | 53,8 | 16,5 | 16,5 | 5,67 | 21,2 | 1 | 1,84 | - | 1100 | - | - | - | - | - | - | - | - | 738 | - | - | - | - | - | - | - | - | - | - | - | - | - |
| Terrine de poisson, préemballée | 714 | 171 | 714 | 171 | 66,5 | 11 | 11 | 5,5 | 11,4 | 1,4 | 1,25 | 30,5 | 877 | < 0,1 | 0,4 | 25 | 19 | < 0,1 | 66 | 210 | 13 | 500 | 0,4 | 1,2 | 4 | - | - | < 0,5 | 0,045 | 0,11 | 1,6 | 0,32 | 0,12 | 23 | 1,05 |
| Thon à la tomate, miettes, appertisées, égouttées | 545 | 130 | 545 | 130 | - | 14,6 | 14,6 | 4,33 | 5,93 | 0,58 | 1,22 | - | - | - | - | - | - | - | - | - | - | 489 | - | - | - | - | - | - | - | - | - | - | - | - | - |
| Thon à l'huile de tournesol, entier, appertisé, égoutté | 861 | 206 | 861 | 206 | 62,1 | 23,3 | 23,3 | 0,76 | 12,2 | 0 | 0,93 | 12 | 632 | 0,06 | 0,77 | < 20 | 32 | < 0,01 | 180 | 220 | 90 | 373 | 0,65 | 0,81 | 7,65 | < 0,8 | - | - | < 0,015 | 0,028 | 10,9 | 0,16 | 0,14 | 9,99 | 3,05 |
| Thon à l'huile de tournesol, miettes, appertisées, égouttées | 946 | 227 | 946 | 227 | 60,9 | 23,4 | 23,4 | 0 | 14,8 | 0 | 0,85 | 11 | 569 | 0,06 | 1 | < 20 | 27 | < 0,01 | 160 | 210 | 100 | 339 | 0,71 | 1,21 | 8,68 | < 0,8 | - | - | < 0,015 | 0,027 | 12,2 | 0,19 | 0,11 | 6,71 | 4,46 |
| Thon albacore ou thon jaune, au naturel, appertisé, égoutté | 542 | 128 | 542 | 128 | 68,5 | 26,6 | 26,6 | 0,58 | 1,98 | < 1,5 | 0,71 | 1,5 | 573 | < 0,1 | 1,3 | 1 | 26,1 | < 0,1 | 166 | 245 | 68 | 283 | 0,4 | 6,1 | 1,08 | 2,3 | - | - | < 0,04 | 0,045 | 10,6 | < 0,16 | 0,4 | - | 2,05 |
| Thon albacore ou thon jaune, cru | 459 | 108 | 459 | 108 | 71,6 | 25 | 25 | traces | 0,94 | 0 | 0,11 | 7,77 | - | 0,089 | 0,98 | 23,7 | 40,8 | 0,021 | 257 | 410 | 79,6 | 40 | 0,23 | 2,3 | 0,72 | 0,1 | - | 2,6 | 0,14 | 0,098 | 13,5 | 0,47 | 0,97 | 8,5 | 3,44 |
| Thon germon ou thon blanc, à l'huile d'olive, appertisé, égoutté | 807 | 192 | 807 | 192 | 59,3 | 31,3 | 31,3 | 1,01 | 6,95 | 0 | 0,67 | 1 | 240 | < 0,1 | 0,5 | 6 | 35,9 | < 0,1 | 196 | 331 | 11 | 270 | 0,4 | 0,86 | 1,75 | - | - | 0 | < 0,04 | 0,051 | 15,6 | < 0,16 | 0,63 | 5 | 2,45 |
| Thon germon ou thon blanc, cru | 514 | 121 | 514 | 121 | 69,5 | 27,2 | 27,2 | traces | 1,37 | 0 | 0,29 | 3,5 | 195 | < 0,1 | 0,95 | 14 | 37,6 | < 0,1 | 242 | 355 | 188 | 77,7 | 0,38 | 2,23 | 0,44 | - | - | - | 0,062 | 0,047 | 19,4 | 0,14 | 0,95 | 15,3 | 2,78 |
| Thon germon ou thon blanc, cuit à la vapeur sous pression | 697 | 165 | 697 | 165 | 64 | 30 | 30 | 0 | 5,06 | 0 | 0,1 | 1,33 | - | 0,22 | 1,01 | 40 | 34,3 | 0,02 | 307 | 390 | - | 41,1 | 0,54 | - | - | - | - | - | - | - | - | - | - | - | - |
| Thon listao ou Bonite à ventre rayé, cru | 411 | 97,1 | 411 | 97,1 | - | 22 | 22 | traces | 1,01 | 0 | 0,093 | 29 | - | 0,086 | 1,25 | - | 34 | 0,015 | 222 | 407 | - | 37 | 0,82 | - | - | - | - | 1 | 0,033 | 0,1 | 15,4 | 0,42 | 0,85 | 9 | 1,9 |
| Thon rouge, cru | 649 | 155 | 649 | 155 | 68,1 | 23,3 | 23,3 | traces | 6,81 | 0 | 0,098 | 8 | - | 0,086 | 1,02 | - | 50 | 0,015 | 254 | 252 | 73,4 | 39 | 0,6 | 5,7 | 1 | 0 | - | 0 | 0,24 | 0,25 | 8,65 | 1,05 | 0,46 | 2 | 9,43 |
| Thon, à la catalane ou à l'escabèche (sauce tomate), appertisé | 594 | 142 | 594 | 142 | - | 10,8 | 10,8 | 5,05 | 8,6 | < 1,5 | 1 | - | - | - | - | - | - | - | - | - | - | 402 | - | - | - | - | - | - | - | - | - | - | - | - | - |
| Thon, au naturel, appertisé, égoutté | 471 | 111 | 471 | 111 | 72,7 | 26,8 | 26,8 | 0 | 0,4 | 0 | 0,74 | 6,26 | 428 | 0,031 | 0,76 | 16 | 24 | 0,0037 | 155 | 207 | 305 | 251 | 0,45 | 5,08 | 1,2 | 0 | - | 0 | 0,03 | 0,053 | 13,6 | 0,094 | 0,44 | 16,3 | 2,53 |
| Thon, cru | 607 | 144 | 607 | 144 | 66,6 | 24 | 24 | traces | 5,38 | 0 | 0,12 | 17,7 | 0 | 0,13 | 1,77 | 26,9 | 34,9 | 0,023 | 229 | 429 | 97,1 | 49 | 0,55 | 7,8 | 1,2 | - | - | 1,37 | 0,13 | 0,12 | 11 | 0,66 | 0,54 | 13,6 | 3,79 |
| Thon, rôti/cuit au four | 576 | 136 | 576 | 136 | 66,9 | 29,9 | 29,9 | traces | 1,83 | 0 | 0,14 | 18,7 | 0,1 | 0,047 | 1,8 | 150 | 46 | 0,0076 | 273 | 445 | 81,2 | 56 | 0,44 | 1,83 | 0,32 | - | - | - | 0,052 | 0,062 | 16,4 | 0,13 | 0,47 | 40,6 | 2,57 |
| Tilapia, cru | 387 | 91,6 | 387 | 91,6 | 78,5 | 18,1 | 18,1 | traces | 2,13 | 0 | 0,071 | 8,16 | - | < 0,1 | 0,27 | 6,38 | 25,4 | < 0,1 | 131 | 282 | 17,8 | 28,3 | 0,32 | 19,6 | 0,4 | 1,3 | - | 0 | < 0,04 | 0,051 | 3,28 | 0,68 | 0,23 | 34 | 1,07 |
| Tofu fumé, préemballé | 685 | 164 | 661 | 158 | 70,3 | 14,9 | 16,4 | 2,91 | 9,5 | < 0,5 | 0,74 | 62 | 567 | 0,25 | 2 | - | 140 | 1,1 | 190 | 190 | < 20 | 295 | 1,4 | < 0,25 | 0,61 | 13 | - | < 0,5 | 0,026 | < 0,01 | 0,11 | 0,21 | 0,044 | 16,4 | 0,019 |
| Tofu nature, préemballé | 617 | 148 | 595 | 143 | 73,8 | 13,4 | 14,7 | 2,87 | 8,5 | < 0,5 | 0,025 | 100 | 74,7 | 0,24 | 2,4 | < 20 | 100 | 0,93 | 210 | 140 | < 20 | 10 | 1,3 | < 0,25 | 0,71 | 12,7 | - | < 0,5 | 0,024 | < 0,01 | 0,13 | 0,14 | 0,028 | 25,1 | 0 |
| Tortilla espagnole aux oignons (omelette aux pommes de terre et oignons), préemballée | 686 | 165 | 686 | 165 | 67,8 | 5,56 | 5,56 | 11,7 | 9,9 | 2,7 | 1,06 | 26 | 686 | 0,12 | 0,64 | < 20 | 34 | 0,12 | 90 | 450 | < 20 | 425 | 0,55 | < 0,25 | 4,46 | 1,92 | - | 1,98 | 0,051 | 0,098 | 0,29 | 0,97 | 0,1 | 31,1 | 0,34 |
| Tripes, boeuf, crues | 342 | 81,5 | 342 | 81,5 | 84,2 | 12,1 | 12,1 | 0 | 3,69 | 0 | 0,24 | 69 | - | 0,07 | 0,59 | - | 13 | 0,085 | 64 | 67 | - | 97 | 1,42 | 0 | 0,09 | 0 | - | 3 | 0 | 0,064 | 0,88 | 0,23 | 0,014 | 5 | 1,39 |
| Truite arc en ciel, crue, élevage | 553 | 132 | 553 | 132 | 74 | 19 | 19 | traces | 6,24 | 0 | 0,11 | 41,7 | 57 | 0,06 | 0,46 | 8,1 | 28,7 | 0,026 | 246 | 410 | 17,6 | 45,5 | 0,61 | 12,1 | 1,55 | 0,33 | 3,39 | traces | 0,13 | 0,11 | 4,36 | 1,58 | 0,38 | 12,8 | 4,33 |
| Truite arc en ciel, élevage, cuite à la vapeur | 493 | 117 | 493 | 117 | 75 | 20 | 20 | traces | 4,14 | 0 | - | - | - | 0,03 | - | 8 | - | 0,027 | - | - | 3 | - | 0,56 | - | - | - | - | - | - | - | - | - | - | - | - |
| Truite arc en ciel, élevage, rôtie/cuite au four | 600 | 143 | 600 | 143 | 71 | 21,5 | 21,5 | traces | 6,33 | 0 | 0,13 | 60,5 | 65 | 0,03 | 0,35 | 9,5 | 28,3 | 0,027 | 263 | 391 | 2,85 | 53,3 | 0,59 | 15 | 1,9 | 0,1 | - | 2,9 | 0,17 | 0,11 | 5,42 | 1,79 | 0,37 | 11 | 4,56 |
| Truite de mer, crue | 467 | 111 | 467 | 111 | 74,5 | 19,6 | 19,6 | traces | 3,63 | 0 | 0,14 | 12,7 | 0 | 0,04 | 1,23 | 20 | 29,3 | 0,027 | 270 | 458 | 22,9 | 55 | 0,95 | 2,1 | 0,65 | 0 | - | 0,55 | 0,1 | 0,14 | 3,5 | 1,95 | 0,54 | 16 | 5 |
| Truite d'élevage, crue | 558 | 133 | 558 | 133 | 73,9 | 19,3 | 19,3 | traces | 6,22 | 0 | 0,1 | 17,1 | - | < 0,1 | 0,5 | 9,23 | 27,5 | < 0,1 | 188 | 410 | 9,75 | 41,8 | 0,44 | 5,92 | 1,94 | - | - | - | 0,14 | 0,1 | 5,54 | 1,51 | 0,34 | 9,23 | 2,54 |
| Truite d'élevage, fumée | 753 | 180 | 753 | 180 | 61,7 | 23,4 | 23,4 | 0,69 | 9,27 | 0 | 3,18 | 2,7 | 2460 | < 0,1 | 0,4 | 40 | 31,4 | < 0,1 | 227 | 412 | 23 | 1270 | 0,4 | 5,2 | 2,3 | - | - | 0 | 0,21 | 0,081 | 7,1 | 2,05 | 0,67 | 26 | 3,2 |
| Truite saumonée, crue | 600 | 143 | 600 | 143 | 74,2 | 19,2 | 19,2 | traces | 7,4 | 0 | 0,2 | 46,7 | - | 0,043 | 0,52 | 8,89 | 30,5 | traces | 210 | 346 | 14,8 | 92,8 | 0,61 | 18,7 | - | 0 | - | 0 | 0,16 | 0,09 | - | - | - | 22 | 4,7 |
| Truite, cuite à la vapeur | 419 | 99,4 | 419 | 99,4 | 75 | 19 | 19 | traces | 2,6 | 0 | - | - | - | 0,03 | - | 6 | - | 0,027 | - | - | 3 | - | 0,56 | - | - | - | - | - | - | - | - | - | - | - | - |
| Truite, rôtie/cuite au four | 766 | 183 | 766 | 183 | 63,4 | 26,6 | 26,6 | traces | 8,47 | 0 | 0,17 | 55 | - | 0,14 | 1,92 | 6 | 28 | 0,56 | 314 | 463 | 3 | 67 | 0,71 | - | - | - | - | 0,5 | 0,43 | 0,42 | 5,77 | 2,24 | 0,23 | 15 | 7,49 |
| Turbot d'élevage, cru | 457 | 109 | 457 | 109 | 76,5 | 18,3 | 18,3 | traces | 3,94 | 0 | 0,19 | 10,2 | - | < 0,1 | 0,16 | 24,2 | 24,9 | < 0,1 | 134 | 306 | 17,7 | 77,9 | 0,65 | 0,71 | 3,63 | - | - | - | 0,057 | 0,042 | 3,49 | 0,42 | 0,19 | - | 1,25 |
| Turbot sauvage, cru | 386 | 91,5 | 386 | 91,5 | 78,1 | 17,2 | 17,2 | traces | 2,54 | 0 | 0,24 | 38,7 | 140 | 0,039 | 0,45 | 35 | 49 | 0,017 | 176 | 253 | 47,3 | 95,3 | 0,21 | traces | 0 | - | - | 0,85 | 0,055 | 0,1 | 2,13 | 0,57 | 0,21 | 8 | 2,07 |
| Turbot, cru | 361 | 85,5 | 361 | 85,5 | 79,2 | 17,9 | 17,9 | traces | 1,55 | 0 | 0,23 | 49 | - | - | 0,25 | 35 | 23 | - | 203 | 415 | - | 84 | - | - | - | 0 | - | 1,7 | 0,043 | 0,085 | 3,6 | 0,25 | 0,15 | 16 | 1,05 |
| Turbot, rôti/cuit au four | 497 | 118 | 497 | 118 | 72,7 | 21,3 | 21,3 | traces | 3,64 | 0 | 0,25 | 42,5 | 180 | 0,049 | 0,53 | 30 | 63,5 | 0,022 | 213 | 323 | - | 192 | 0,29 | 0 | 0 | - | - | 1,7 | 0,068 | 0,12 | 2,54 | 0,66 | 0,24 | 9 | 2,54 |
| Veau, carré, cru | 606 | 145 | 606 | 145 | 74,1 | 19,5 | 19,5 | 0,4 | 7,25 | 0 | 0,22 | 15,3 | - | 0,12 | 1,45 | - | 25 | 0,028 | 208 | 327 | - | 89,3 | 3,2 | 0 | 0,27 | 4,6 | - | 0 | 0,08 | 0,27 | 7,63 | 1,26 | 0,44 | 15 | 1,37 |
| Veau, carré, sauté/poêlé | 720 | 171 | 720 | 171 | - | 28 | 28 | traces | 6,6 | 0 | - | - | - | - | 1 | - | - | - | - | - | - | - | 7,6 | - | 0,46 | - | - | - | 0,1 | - | 3 | 0,44 | 0,37 | - | 1,7 |
| Veau, collier, braisé ou bouilli | 978 | 233 | 978 | 233 | 53,9 | 34,9 | 34,9 | traces | 10,4 | 0 | - | - | - | - | 0,65 | - | - | - | - | - | 12,7 | - | 7,54 | - | 0,34 | - | - | - | 0,7 | - | 4,87 | 0,37 | 0,13 | - | 2,01 |
| Veau, collier, cru | 556 | 133 | 556 | 133 | 73,8 | 19,8 | 19,8 | traces | 5,92 | 0 | 0,15 | - | - | - | 1,01 | - | - | - | - | - | 7,21 | 59,9 | 4,29 | - | - | - | - | - | - | - | 4,87 | - | 0,21 | - | 2,02 |
| Veau, côte, crue | 787 | 189 | 787 | 189 | 68,9 | 18,3 | 18,3 | 0 | 12,9 | 0 | 0,11 | 11,5 | - | 0,1 | 1,26 | 2,8 | 22 | 0,027 | 184 | 290 | 5,62 | 43,4 | 4,29 | - | 0,21 | - | - | 0 | 0,07 | 0,23 | 2,69 | 1,13 | 0,2 | 12 | 2,82 |
| Veau, côte, grillée/poêlée | 657 | 156 | 657 | 156 | 69,4 | 25,1 | 25,1 | 0 | 6,24 | 0 | 0,25 | 29 | - | < 0,1 | 1,3 | 31 | 22,5 | < 0,1 | 196 | 305 | 8,98 | 100 | 6,54 | < 0,5 | 0,49 | - | - | < 1 | 0,1 | 0,11 | 3,41 | 0,44 | 0,26 | < 20 | 2,82 |
| Veau, épaule, braisée ou bouillie | 920 | 219 | 920 | 219 | 55,9 | 36,4 | 36,4 | traces | 8,14 | 0 | - | - | - | - | 0,74 | - | - | - | - | - | 12,1 | - | 6,99 | - | - | - | - | - | - | - | 6 | - | 0,25 | - | 2,69 |
| Veau, épaule, crue | 523 | 124 | 523 | 124 | 74,9 | 20,7 | 20,7 | 0 | 4,63 | 0 | 0,15 | 16 | - | 0,12 | 1,16 | 2,8 | 24 | 0,029 | 208 | 311 | 6,9 | 61,1 | 3,98 | 0,5 | 0,26 | - | - | 0 | 0,09 | 0,29 | 6 | 1,45 | 0,38 | 11 | 2,69 |
| Veau, épaule, grillée/poêlée | 698 | 166 | 698 | 166 | 66,5 | 27,6 | 27,6 | traces | 6,17 | 0 | - | - | - | - | 1,16 | - | - | - | - | - | 9,2 | - | 5,3 | - | - | - | - | - | - | - | 6 | - | 0,38 | - | 2,69 |
| Veau, escalope panée, cuite | 1130 | 271 | 1130 | 271 | 47,8 | 22,9 | 22,9 | 12,6 | 14,1 | 0,94 | 0,98 | 28,7 | - | 0,07 | 1,11 | 7,97 | 24,3 | 0,12 | 183 | 328 | 6,95 | 394 | 2,26 | 0,13 | 3,32 | - | - | 0,0063 | 0,12 | 0,23 | 5,89 | 0,53 | 0,28 | 15,3 | 0,72 |
| Veau, escalope, crue | 457 | 108 | 457 | 108 | 73,7 | 20,7 | 20,7 | 0,58 | 2,6 | 0 | 0,097 | 10 | 73 | 0,09 | 2 | 2,52 | 21 | 0,013 | 171 | 380 | 7,2 | 38,7 | 2,56 | - | 0,15 | - | - | - | 0,13 | 0,28 | 5,7 | 0,85 | 0,44 | 5 | 2 |
| Veau, escalope, cuite | 620 | 147 | 620 | 147 | 65 | 31 | 31 | traces | 2,5 | 0 | 0,15 | 10,8 | - | 0,067 | 1 | 6 | 29,5 | 0,012 | - | - | 6,22 | 58,9 | 3,21 | - | 0,11 | - | - | - | 0,2 | - | 10 | 0,38 | 0,72 | - | 2 |
| Veau, filet, cru | 402 | 95 | 402 | 95 | - | 20,6 | 20,6 | 0 | 1,4 | 0 | 0,24 | 12 | - | - | 2 | 2,4 | - | - | - | - | - | 95 | - | - | - | - | - | - | - | - | - | - | - | - | - |
| Veau, filet, rôti/cuit au four | 838 | 201 | 838 | 201 | 60,7 | 23,4 | 23,4 | 0,5 | 11,7 | 0 | 0,23 | 19 | - | 0,11 | 0,87 | 6 | 25 | 0,029 | 212 | 325 | 13 | 93 | 3,03 | 0 | 0,44 | 5,5 | - | 0 | 0,05 | 0,28 | 8,86 | 1,2 | 0,34 | 15 | 1,24 |
| Veau, jarret, braisé ou bouilli | 873 | 207 | 873 | 207 | 56,7 | 37,4 | 37,4 | traces | 6,43 | 0 | - | - | - | - | 0,77 | - | - | - | - | - | 11,1 | - | 8,81 | - | 0,33 | - | - | - | 0,4 | - | 5,92 | 0,39 | 0,19 | - | 2,3 |
| Veau, jarret, cru | 496 | 118 | 496 | 118 | 75,4 | 21,3 | 21,3 | 0 | 3,66 | 0 | 0,21 | 9 | - | - | 1,2 | - | - | - | - | - | 6,29 | 82,8 | 5,01 | - | - | - | - | - | - | - | 5,92 | - | 0,3 | - | 2,3 |
| Veau, noix, crue | 467 | 111 | 467 | 111 | 75,4 | 21,8 | 21,8 | 0 | 2,58 | 0 | 0,19 | 9 | - | 0,11 | 0,92 | - | 26 | 0,029 | 220 | 367 | 6,95 | 74,5 | 2,64 | 0 | 0,3 | 3,9 | - | 0 | 0,08 | 0,27 | 7,41 | 1,07 | 0,53 | 14 | 2,08 |
| Veau, noix, grillée/poêlée | 622 | 147 | 622 | 147 | 67,2 | 29,1 | 29,1 | traces | 3,44 | 0 | - | - | - | - | 0,92 | - | - | - | - | - | 9,26 | - | 3,52 | - | - | - | - | - | - | - | 7,4 | - | 0,53 | - | 2,08 |
| Veau, noix, rôtie | 622 | 147 | 622 | 147 | 67,2 | 29,1 | 29,1 | traces | 3,44 | 0 | - | - | - | - | 0,73 | - | - | - | - | - | 9,26 | - | 3,52 | - | - | - | - | - | - | - | 6,91 | - | 0,42 | - | 2,08 |
| Veau, pied, cru | 769 | 184 | 769 | 184 | - | 19,1 | 19,1 | traces | 12 | 0 | 0,12 | 11 | - | - | 2,9 | - | 15 | - | 206 | 320 | - | 48 | - | - | - | - | - | 0 | 0,18 | 0,27 | 6,3 | - | 0,2 | 6,9 | 1,4 |
| Veau, poitrine, crue | 733 | 176 | 733 | 176 | 67,7 | 18,7 | 18,7 | 0 | 11,2 | 0 | 0,21 | 8 | - | 0,094 | 1,22 | 2,8 | 18 | 0,01 | 172 | 286 | 8 | 82,5 | 2,33 | - | - | - | - | - | - | - | - | - | - | - | - |
| Veau, rôti, cru | 611 | 145 | 611 | 145 | 66,5 | 27,3 | 27,3 | 0 | 3,96 | 0 | 0,2 | 7 | - | 0,11 | 1,35 | < 5 | 27 | 0,029 | 223 | 372 | 9,11 | 79 | 2,34 | 1,3 | 0,29 | - | - | 0 | 0,08 | 0,28 | 9,56 | 1,09 | 0,47 | 14 | 1,05 |
| Veau, rôti, cuit | 603 | 143 | 603 | 143 | 67 | 28,1 | 28,1 | 0 | 3,39 | 0 | 0,17 | 6 | - | 0,13 | 0,9 | - | 28 | 0,031 | 236 | 393 | - | 68 | 3,08 | - | 0,55 | - | - | 0 | 0,06 | 0,33 | 10,1 | 1 | 0,31 | 16 | 1,18 |
| Veau, steak haché 15% MG, cru | 864 | 208 | 864 | 208 | - | 18,2 | 18,2 | traces | 15 | 0 | 0,17 | - | - | - | - | - | - | - | - | - | - | - | - | - | - | - | - | - | - | - | - | - | - | - | - |
| Veau, steak haché 20% MG, cru | 1010 | 243 | 1010 | 243 | - | 16,7 | 16,7 | 0,4 | 19,5 | 0 | 0,19 | 11 | - | - | 0,84 | - | - | - | - | - | - | 74,5 | - | - | - | - | - | - | - | - | - | - | - | - | - |

| | | | | | | | | | | | | | | | | | | | | | | | | | | | | | | | | | | | |
|---|---|---|---|---|---|---|---|---|---|---|---|---|---|---|---|---|---|---|---|---|---|---|---|---|---|---|---|---|---|---|---|---|---|---|---|
| Veau, tête, bouillie/cuite à l'eau | 779 | 187 | 779 | 187 | - | 21 | 21 | traces | 11,4 | 0 | - | - | - | - | 0,7 | - | - | - | - | - | - | - | 1,6 | - | 0,57 | - | - | - | 0,04 | - | 1,8 | 0,75 | 0,06 | - | 2,3 |
| Veau, viande, cuite (aliment moyen) | 712 | 169 | 712 | 169 | 63,7 | 29 | 29 | 0,74 | 5,55 | 0,051 | 0,25 | 14,9 | | 0,085 | 0,95 | | 27 | 0,035 | | | 8,78 | 101 | 4,46 | | 0,57 | | | | 0,14 | | 7,38 | 0,61 | 0,41 | | 1,92 |
| Viande blanche, cuite (aliment moyen) | 728 | 173 | 728 | 173 | 64,2 | 28,1 | 28,1 | 0,26 | 6,55 | 0,0009 | 0,17 | 9,43 | | 0,059 | 0,84 | 7,23 | 31,3 | 0,014 | 233 | 376 | 15,6 | 69,7 | 2,01 | 0,29 | 0,2 | 1,27 | | 0,81 | 0,27 | 0,14 | 8,66 | 1,06 | 0,35 | 7,74 | 0,61 |
| Viande cuite (aliment moyen) | 765 | 182 | 765 | 182 | 63,5 | 27,2 | 27,2 | 0,24 | 8,04 | 0,0083 | 0,2 | 10,4 | | 0,084 | 1,69 | 6,72 | 30,1 | 0,023 | 222 | 361 | 12,3 | 81,6 | 3,45 | 0,22 | 0,22 | 1,41 | | 0,69 | 0,21 | 0,17 | 7,45 | 1,01 | 0,37 | 8,38 | 1,31 |
| Viande des Grisons | 881 | 209 | 881 | 209 | 49,5 | 38,9 | 38,9 | 1 | 5,47 | 0 | 3,83 | 30,8 | 2530 | 0,15 | 1,8 | 53,8 | 46,6 | < 0,1 | 278 | 660 | 9 | 1530 | 5,9 | < 0,5 | < 0,01 | - | - | < 0,5 | 0,062 | 0,26 | 11,8 | 1,33 | 0,7 | 33,9 | 2,39 |
| Viande rouge, cuite (aliment moyen) | 816 | 195 | 816 | 195 | 62,6 | 26 | 26 | 0,027 | 10,1 | 0 | 0,22 | 11,8 | | 0,12 | 2,84 | 6,17 | 28,3 | 0,038 | 201 | 338 | 7,77 | 96,4 | 5,44 | 0,094 | 0,25 | | | | 0,084 | 0,24 | 5,76 | 0,93 | 0,4 | 9,44 | 2,31 |
| Vivaneau, cru | 398 | 94,1 | 398 | 94,1 | 78 | 20,5 | 20,5 | traces | 1,34 | 0 | 0,16 | 32 | - | 0,028 | 0,69 | - | 32 | 0,013 | 198 | 417 | - | 64 | 0,36 | 10,2 | 0,96 | 0,1 | - | 1,6 | 0,046 | 0,003 | 0,28 | 0,75 | 0,4 | 5 | 3 |
| Vivaneau, cuit | 511 | 121 | 511 | 121 | 71,7 | 26,3 | 26,3 | traces | 1,72 | 0 | 0,14 | 40 | - | 0,046 | 0,82 | - | 37 | 0,017 | 201 | 522 | - | 57 | 0,44 | - | - | - | - | 1,6 | 0,053 | 0,004 | 0,35 | 0,87 | 0,46 | 6 | 3,5 |
| Volaille, croquette panée ou nuggets | 1000 | 239 | 1000 | 239 | 51,5 | 13,1 | 13,1 | 16,2 | 13,2 | 1,7 | 1,2 | 16,4 | - | 0,056 | 0,6 | 13,6 | 24 | 0,15 | 225 | 215 | 5,88 | 676 | 0,66 | < 0,5 | 2,59 | - | - | 0 | 0,11 | 0,086 | 5,65 | 0,72 | 0,27 | 19,3 | 0,21 |
| Volaille, cuite (aliment moyen) | 698 | 166 | 698 | 166 | 65,5 | 27,6 | 27,6 | 0,28 | 5,96 | 0,0013 | 0,24 | 8,45 | 127 | 0,086 | 1,05 | 7,77 | 33 | 0,015 | 239 | 377 | 14,8 | 97,9 | 1,57 | 0,24 | 0,2 | 1,84 | | 1,15 | 0,095 | 0,17 | 9,82 | 1,37 | 0,38 | 8,1 | 0,58 |

# Appendix 2

| produits laitiers et assimilés | Energie, Règlement UE N° 1169/2011 (kJ/100 g) | Energie, Règlement UE N° 1169/2011 (kcal/100 g) | Energie, N x facteur Jones, avec fibres (kJ/100 g) | Energie, N x facteur Jones, avec fibres (kcal/100 g) | Eau (g/100 g) | Protéines, N x facteur de Jones (g/100 g) | Protéines, N x 6.25 (g/100 g) | Glucides (g/100 g) | Lipides (g/100 g) | Fibres alimentaires (g/100 g) | Sel chlorure de sodium (g/100 g) | Calcium (mg/100 g) | Chlorure (mg/100 g) | Cuivre (mg/100 g) | Fer (mg/100 g) | Iode (µg/100 g) | Magnésium (mg/100 g) | Manganèse (mg/100 g) | Phosphore (mg/100 g) | Potassium (mg/100 g) | Sélénium (µg/100 g) | Sodium (mg/100 g) | Zinc (mg/100 g) | Vitamine D (µg/100 g) | Vitamine E (mg/100 g) | Vitamine K1 (µg/100 g) | Vitamine K2 (µg/100 g) | Vitamine C (mg/100 g) | Vitamine B1 ou Thiamine (mg/100 g) | Vitamine B2 ou Riboflavine (mg/100 g) | Vitamine B3 ou PP ou Niacine (mg/100 g) | Vitamine B5 ou Acide pantothénique (mg/100 g) | Vitamine B6 (mg/100 g) | Vitamine B9 ou Folates totaux (µg/100 g) | Vitamine B12 (µg/100 g) |
|---|---|---|---|---|---|---|---|---|---|---|---|---|---|---|---|---|---|---|---|---|---|---|---|---|---|---|---|---|---|---|---|---|---|---|---|
| Abondance | 1630 | 393 | 1640 | 395 | 37,2 | 27,1 | 26,6 | traces | 31,6 | 0 | 1,73 | 760 | 1090 | 0,49 | 0,09 | < 20 | 28 | 0,02 | 570 | 120 | < 20 | 690 | 3,5 | < 0,25 | 0,75 | 2,25 | - | < 0,5 | < 0,015 | 0,27 | < 0,1 | 0,34 | 0,092 | 39,6 | 2,19 |
| Asiago | 1480 | 356 | 1490 | 359 | - | 32,1 | 31,4 | traces | 25,6 | 0 | 1,9 | 770 | - | - | 0,7 | - | - | - | 530 | 97 | - | 760 | 2,2 | 0,19 | 0,48 | - | - | 0 | 0,03 | 0,35 | 0,1 | - | 0,09 | 40 | - |
| Beaufort | 1700 | 410 | 1710 | 412 | 34,3 | 26,5 | 26 | traces | 34 | 0 | 1,27 | 745 | - | 0,09 | 0,24 | 37,3 | - | 0,03 | 788 | 118 | 7,22 | 506 | 4,75 | - | - | - | - | - | - | - | - | - | - | - | - |
| Bleu de Gex ou Fromage bleu du Haut-jura ou Bleu de septmoncel (AOC) | 1510 | 364 | 1520 | 366 | 43,6 | 22,5 | 22,1 | traces | 30,7 | 0 | 1,44 | 600 | - | - | - | - | - | - | - | - | - | 575 | - | - | - | - | - | - | - | - | - | - | - | - | - |
| Boisson lactée, lait fermenté ou yaourt à boire, aromatisé, avec édulcorants, allégé en sucres, 0% MG, au L Casei | - | - | - | - | - | 2,9 | 2,84 | 3,64 | 0,1 | 0,35 | 0,11 | - | - | - | - | - | - | - | - | - | - | 42 | - | 0,75 | - | - | - | - | - | - | - | - | 0,21 | - | - |
| Boisson lactée, lait fermenté ou yaourt à boire, aromatisé, sucré | 312 | 73,7 | 313 | 73,9 | 82,4 | 2,98 | 2,92 | 11,6 | 1,4 | 0,028 | 0,08 | 115 | 88,3 | 0,01 | 0,12 | < 20 | 11 | 0,01 | 82 | 140 | < 20 | 31,3 | 0,42 | 0,75 | < 0,08 | < 0,8 | - | < 0,5 | 0,038 | 0,14 | < 0,1 | 0,33 | 0,029 | 5,12 | 0,2 |
| Boisson lactée, lait fermenté ou yaourt à boire, aromatisé, sucré, au L Casei | - | - | - | - | - | 2,81 | 2,75 | 12,7 | 1,52 | 0,033 | 0,1 | 120 | - | - | - | - | - | - | - | - | - | 40,5 | - | 0,81 | - | - | - | - | - | - | - | - | 0,21 | - | - |
| Boisson lactée, lait fermenté ou yaourt à boire, aromatisé, sucré, enrichi en vitamine D | - | - | - | - | - | 3,08 | 3,02 | 12,6 | 1,2 | traces | 0,091 | 134 | - | - | - | - | - | - | - | - | - | 36,1 | - | 0,83 | - | - | - | - | - | 0,21 | - | - | - | - | 0,38 |
| Boisson lactée, lait fermenté ou yaourt à boire, aux fruits, sucré | 350 | 82,8 | 351 | 83,1 | - | 3,09 | 3,03 | 13,1 | 1,69 | 0,079 | 0,12 | 133 | - | 0,004 | 0,13 | 11 | 10 | 0,002 | 82 | 116 | 1 | 46,9 | 0,3 | 0,75 | 0,05 | - | - | < 0,5 | 0,031 | 0,16 | 0,11 | 0,29 | 0,05 | 28 | 0,07 |
| Boisson lactée, lait fermenté ou yaourt à boire, aux fruits, sucré, enrichie en vitamine D | - | - | - | - | - | 3,03 | 2,97 | 13,9 | 1,5 | 0,11 | 0,1 | 172 | - | - | - | - | - | - | - | - | - | 40 | - | 1,14 | - | - | - | - | - | 0,21 | - | - | - | - | 0,38 |
| Boisson lactée, lait fermenté ou yaourt à boire, nature, sucré, au L Casei | - | - | - | - | 83,4 | 2,76 | 2,7 | 12,9 | 1,6 | < 0,05 | 0,1 | 99,1 | 70 | < 0,1 | < 0,1 | 9,6 | 7,8 | < 0,1 | 66 | 112 | < 5 | 40 | 0,3 | < 0,5 | 0,015 | - | - | < 0,5 | 0,021 | 0,14 | 0,07 | 0,23 | 0,036 | 8,9 | 0,095 |
| Boulette d'Avesne | 1470 | 355 | 1480 | 356 | 44,1 | 20,9 | 20,5 | traces | 30 | 0 | 1,99 | 600 | 1320 | 0,07 | 0,47 | < 20 | 26 | 0,1 | 460 | 150 | < 20 | 797 | 2,9 | < 0,25 | 1 | 8,25 | - | - | < 0,015 | 0,38 | < 0,1 | 1,08 | 0,36 | 103 | 1,31 |
| Brie de Meaux | 1130 | 271 | 1130 | 273 | 53,4 | 21,4 | 21 | traces | 20,7 | 0 | 2,02 | 310 | 1320 | < 0,01 | 0,15 | 50 | 15 | 0,02 | 310 | 120 | < 20 | 806 | 2,3 | < 0,25 | 0,44 | < 0,8 | - | < 0,5 | < 0,015 | 0,58 | - | 0,36 | 0,094 | 115 | 1,28 |
| Brie de Melun | 1200 | 288 | 1200 | 290 | 52,1 | 22 | 21,6 | traces | 22,2 | 0 | 1,89 | 180 | 1360 | 0,03 | 0,08 | 70 | 17 | 0,02 | 280 | 210 | 20 | 755 | 0,84 | < 0,25 | 0,57 | 1,91 | - | < 0,5 | 0,029 | 0,57 | - | 0,66 | 0,26 | 140 | 1,89 |
| Brie, sans précision | 1230 | 297 | 1240 | 299 | 52,5 | 17,3 | 16,9 | traces | 25,5 | 0 | 1,51 | 424 | 1020 | 0,045 | 0,36 | 13,8 | 20 | < 0,1 | 282 | 145 | 6,41 | 599 | 2,38 | 0,39 | 0,47 | 2,3 | - | 0 | 0,06 | 0,43 | 0,38 | 0,69 | 0,24 | 77,7 | 0,76 |
| Camembert au lait cru | 1110 | 267 | 1120 | 269 | 55,9 | 20,7 | 20,3 | traces | 20,3 | 0 | 1,45 | 380 | 981 | 0,02 | 0,09 | 20 | 18 | 0,03 | 370 | 160 | < 20 | 580 | 2,8 | < 0,25 | 0,46 | 3,1 | - | - | < 0,015 | 0,6 | 10,9 | 0,44 | 0,26 | 88 | 1,61 |
| Camembert, sans précision | 1160 | 280 | 1170 | 281 | 54,8 | 19,5 | 19,1 | traces | 22,5 | 0 | 1,45 | 449 | 932 | 0,03 | 0,18 | 13,8 | 19,3 | 0,016 | 370 | 148 | 5,76 | 579 | 2,5 | 0,24 | 0,57 | 0 | - | 0 | 0,045 | 0,45 | 1,8 | 1,56 | 0,25 | 62 | 1,13 |
| Cancoillotte (spécialité fromagère fondue) | 630 | 151 | 635 | 152 | 72,5 | 13,8 | 13,6 | 0,5 | 10,5 | 0 | 1,73 | 101 | 550 | 0,1 | 0,13 | 20 | 8 | 0,02 | 425 | 59 | 4,5 | 692 | 1,4 | 0,2 | 0,15 | - | - | 0 | 0,08 | 0,31 | 0,49 | 0,4 | 0,04 | 12 | 0,39 |
| Cantal entre-deux | 1590 | 383 | 1600 | 385 | 37,1 | 26,6 | 26,1 | traces | 30,5 | 0 | 1,85 | 791 | - | - | - | - | - | - | - | - | - | 741 | - | - | - | - | - | - | - | - | - | - | - | - | - |
| Cantal, Salers ou Laguiole | 1570 | 378 | 1580 | 380 | 39,6 | 25,3 | 24,8 | traces | 31 | 0 | 2,15 | 772 | - | 0,09 | 0,54 | 21,8 | 25,7 | 0,04 | 487 | 97,8 | 5,44 | 768 | 4,1 | - | - | - | - | - | - | - | - | - | - | 19,4 | - |
| Carré de l'Est | 1360 | 327 | 1360 | 329 | - | 21,4 | 21 | traces | 27 | 0 | 1,58 | 450 | - | 0,06 | - | 21 | 20 | 0,02 | 350 | 140 | 5 | 630 | 6 | - | - | - | - | - | - | - | - | - | - | - | - |
| Chabichou (fromage de chèvre) | 1220 | 294 | 1230 | 296 | 50,4 | 20,5 | 20,1 | traces | 23,8 | 0 | 1,85 | 104 | 1110 | 0,1 | 0,35 | 30 | 14,1 | - | 194 | 238 | 11,5 | 742 | 0,52 | < 0,5 | 0,31 | - | - | - | < 0,04 | 0,82 | 1,61 | 0,51 | 0,23 | 46 | - |
| Chaource | 1140 | 276 | 1150 | 277 | 56 | 17,3 | 17 | traces | 22,9 | 0 | 1,9 | 110 | 1170 | 0,03 | 0,06 | < 20 | 12 | 0,03 | 220 | 130 | < 20 | 760 | 0,56 | < 0,25 | 0,59 | 0,89 | - | < 0,5 | 0,032 | 0,53 | - | 0,36 | 0,2 | 63,2 | 1,6 |
| Cheddar | 1650 | 399 | 1660 | 401 | 37,1 | 24 | 23,6 | traces | 33,8 | 0 | 1,61 | 675 | 1030 | 0,056 | 0,16 | 37,5 | 27 | 0,033 | 473 | 76 | 11,1 | 644 | 3,43 | 0,6 | 0,78 | 2,9 | - | 0 | 0,027 | 0,43 | 0,039 | 0,48 | 0,049 | 26 | 0,88 |
| Cheesecake ou Gâteau au fromage frais, préemballé | 1380 | 330 | 1380 | 330 | 39,2 | 4,57 | 4,57 | 31,7 | 20,3 | 0,9 | 0,38 | 55 | 259 | 0,04 | 0,34 | < 20 | 12 | 0,14 | 84 | 110 | < 20 | 154 | 0,41 | < 0,25 | 0,9 | 3,39 | - | < 0,5 | 0,059 | 0,14 | < 0,1 | 0,42 | 0,031 | 12 | 0,2 |
| Chevrot (fromage de chèvre) | 1150 | 277 | 1160 | 279 | 55,4 | 18,5 | 18,1 | traces | 22,8 | 0 | 1,12 | 98,4 | 984 | < 0,1 | 0,26 | 30 | 12,9 | - | 160 | - | 10 | 449 | 0,28 | < 0,5 | 0,34 | - | - | - | < 0,04 | 0,83 | 1,24 | 1,24 | 0,11 | 77 | - |
| Clafoutis aux fruits, préemballé | 806 | 192 | 807 | 192 | 58,5 | 5,22 | 5,11 | 27,2 | 6,51 | 1,88 | 0,18 | 84 | 91 | 0,069 | 0,57 | 11,9 | 15 | 0,087 | 76,2 | 182 | < 10 | 76,7 | 0,39 | < 0,5 | 0,43 | - | - | < 0,5 | 0,043 | 0,15 | 0,32 | 0,49 | 0,1 | 23 | 0,24 |
| Comté | 1730 | 418 | 1740 | 420 | 33,2 | 27,2 | 26,7 | traces | 34,6 | 0 | 0,8 | 993 | - | 0,61 | 0,49 | 24,4 | 43,7 | 0,03 | 681 | 116 | 7,02 | 322 | 3,84 | - | 0,8 | - | - | - | - | 0,38 | - | - | - | 5 | 2,59 |
| Coulommiers | 1160 | 279 | 1160 | 280 | 56 | 18,8 | 18,4 | traces | 22,8 | 0 | 1,37 | 400 | - | 0,07 | 0,23 | 28 | 21,7 | traces | 185 | 165 | 5 | 534 | 3,28 | - | 0,6 | - | - | - | - | - | - | - | - | - | - |
| Crème aux oeufs (petit pot de crème chocolat, vanille, etc.), rayon frais | 736 | 176 | 737 | 176 | 66,5 | 4,32 | 4,23 | 19,4 | 8,88 | 0,67 | 0,14 | 100 | 109 | 0,08 | 0,79 | < 20 | 19 | 0,09 | 110 | 210 | < 20 | 54,2 | 0,53 | < 0,25 | 0,46 | 0,81 | - | < 0,5 | 0,04 | 0,17 | < 0,1 | 0,68 | 0,029 | 12,4 | 0,26 |
| Crème brûlée, rayon frais | 1130 | 272 | 1130 | 272 | 54,7 | 4,46 | 4,37 | 16,6 | 20,8 | 0,61 | 0,11 | 64,3 | 90 | < 0,1 | 0,8 | 17,5 | 7,6 | < 0,1 | 97 | 94 | < 5 | 46,1 | 0,6 | < 0,5 | 1,17 | - | - | < 0,5 | 0,04 | 0,19 | < 0,16 | 0,59 | 0,042 | 26,2 | 0,35 |
| Crème caramel, rayon frais | 558 | 132 | 560 | 133 | 69 | 4,45 | 4,35 | 20,6 | 3,6 | traces | 0,14 | 84,9 | 100 | 0,0084 | 0,32 | 14,4 | 12,8 | 0,004 | 77 | 150 | < 10 | 57,1 | 0,31 | 0,07 | 0,16 | - | - | 0 | 0,03 | 0,2 | 0,1 | 0,5 | 0,03 | 8 | 0,3 |
| Crème chantilly, sous pression, UHT | 1270 | 308 | 1270 | 308 | 57,1 | 2,11 | 2,06 | 10,6 | 28,2 | <3 | 0,065 | 101 | - | 0,01 | 0,05 | 11 | 11 | 0,001 | 89 | 147 | - | 26 | 0,37 | 0,4 | 0,64 | 1,9 | - | 0 | 0,037 | 0,065 | 0,07 | 0,31 | 0,041 | 3 | 0,29 |
| Crème de lait ou spécialité à base de crème légère, teneur en matière grasse inconnue (aliment moyen) | 1110 | 269 | 1110 | 269 | 68 | 2,88 | 2,82 | 2,87 | 26 | 1,39 | 0,08 | 82,2 | 89 | 0,0075 | 0,098 | 10,6 | 8,13 | 0,0031 | 66,2 | 105 | 1,18 | 31,9 | 0,27 | 0,17 | 0,52 | 0,47 | | 0,3 | 0,028 | 0,18 | 0,15 | 0,25 | 0,035 | 22,3 | 0,18 |
| Crème de lait, 15 à 20% MG, légère, épaisse, rayon frais | 684 | 166 | 685 | 166 | 76,2 | 2,81 | 2,75 | 3 | 15,3 | <3 | 0,085 | 98 | 80 | 0,007 | 0,22 | 10,6 | 9 | 0,007 | 73 | 117 | 1,29 | 34 | 0,34 | < 0,2 | 0,28 | 0 | - | traces | 0,037 | 0,23 | 0,2 | 0,3 | 0,044 | 34 | 0,24 |
| Crème de lait, 15 à 20% MG, légère, fluide, rayon frais | - | - | - | - | 72,4 | 2,96 | 2,9 | 4,3 | 13,5 | 0 | 0,15 | 105 | - | - | - | - | 9 | - | 71 | 127 | - | 60 | - | - | - | - | - | - | - | - | - | - | - | - | - |
| Crème de lait, 15 à 20% MG, légère, semi-épaisse, UHT | 798 | 194 | 799 | 194 | 73,6 | 2,62 | 2,56 | 2,89 | 18,7 | <3 | 0,083 | 99 | 80 | 0,007 | 0,04 | 12 | 10 | 0,007 | 84 | 128 | 1,3 | 33 | 0,35 | 0,18 | 0,34 | 0 | - | 0,5 | 0,036 | 0,19 | 0,08 | 0,3 | 0,047 | 12 | 0,37 |

| | | | | | | | | | | | | | | | | | | | | | | | | | | | | | | | | | | | |
|---|---|---|---|---|---|---|---|---|---|---|---|---|---|---|---|---|---|---|---|---|---|---|---|---|---|---|---|---|---|---|---|---|---|---|---|
| Crème de lait, 30% MG, épaisse, rayon frais | 1250 | 304 | 1250 | 304 | 65,5 | 3 | 2,94 | 2,8 | 30,7 | < 3 | 0,07 | 76,9 | 100 | 0,0075 | 0,08 | 10,6 | 7,92 | 0,0018 | 64 | 101 | < 22 | 28 | 0,24 | < 0,2 | 0,5 | 0 | - | < 0,5 | 0,024 | 0,17 | 0,16 | 0,24 | 0,034 | 21,5 | 0,12 |
| Crème de lait, 30% MG, semi-épaisse, UHT | 1220 | 297 | 1220 | 297 | 62,9 | 2,49 | 2,44 | 1,94 | 30,7 | < 3 | 0,073 | 67 | 59 | 0,0083 | 0,041 | 10,6 | 6,93 | 0,0025 | 59,3 | 89,5 | 12,9 | 29 | 0,24 | 0,51 | 0,97 | 2,95 | - | 0,7 | 0,027 | 0,14 | 0,055 | 0,23 | 0,024 | 7,4 | 0,32 |
| Crème dessert à la vanille, appertisée | 558 | 132 | 559 | 133 | 73,4 | 3,7 | 3,63 | 21,1 | 3,62 | 0,5 | 0,17 | 115 | - | 0,05 | 0,8 | 14 | 10 | 0,05 | 80 | 134 | 1,1 | 66,5 | 0,6 | 0,1 | 0,1 | - | - | 0,2 | 0,04 | 0,22 | 0,3 | 0,52 | 0,06 | 22 | 0,1 |
| Crème dessert à la vanille, rayon frais | 443 | 105 | 444 | 106 | 76,6 | 3,06 | 3 | 14,4 | 3,6 | < 3 | 0,12 | 133 | - | - | - | - | - | - | - | 131 | 1 | 48 | - | - | - | - | - | - | - | - | - | - | - | - | - |
| Crème dessert au café, rayon frais | 488 | 116 | 489 | 116 | 76,1 | 2,93 | 2,87 | 17,8 | 3,69 | 0,042 | 0,16 | - | - | - | - | - | - | - | - | - | - | 67,1 | - | - | - | - | - | - | - | - | - | - | - | - | - |
| Crème dessert au caramel, rayon frais | 494 | 117 | 495 | 118 | 75,9 | 2,92 | 2,86 | 18,6 | 3,47 | 0,12 | 0,18 | - | - | - | - | - | - | - | - | - | - | 78 | - | - | - | - | - | - | - | - | - | - | - | - | - |
| Crème dessert au chocolat, appertisée | 580 | 138 | 581 | 138 | 69 | 3,25 | 3,18 | 21,8 | 3,42 | 3,6 | 0,1 | 89,7 | - | 0,17 | 1,35 | 6,1 | 27,9 | 0,19 | 80 | 247 | < 5 | 41,6 | 0,58 | 0,03 | 0,13 | - | - | 2 | 0,03 | 0,16 | 0,3 | 0,18 | 0,03 | 3 | traces |
| Crème dessert au chocolat, rayon frais | 546 | 130 | 548 | 130 | 72,5 | 3,39 | 3,32 | 19,9 | 3,93 | 0,72 | 0,15 | 123 | 111 | 0,12 | 1,6 | < 20 | 25 | 0,14 | 100 | 260 | < 20 | 59,2 | 0,48 | - | 0,1 | < 0,8 | - | < 0,5 | 0,038 | 0,14 | < 0,1 | 0,5 | 0,025 | 6,99 | 0,075 |
| Crème dessert, allégée en MG, rayon frais | 334 | 79,2 | 335 | 79,4 | 80 | 3,64 | 3,56 | 12,4 | 1,3 | < 3 | 0,14 | 127 | - | - | - | - | - | - | - | - | - | 54 | - | - | - | - | - | - | - | - | - | - | - | - | - |
| Crème dessert, appertisée (aliment moyen) | 569 | 135 | 570 | 135 | 71,3 | 3,49 | 3,42 | 21,4 | 3,52 | 1,98 | 0,14 | 103 | | 0,11 | 1,06 | 10,2 | 18,6 | 0,12 | 80 | 188 | 1,77 | 54,6 | 0,59 | 0,067 | 0,11 | | | 1,06 | 0,035 | 0,19 | 0,3 | 0,35 | 0,046 | 12,9 | 0,052 |
| Crème dessert, rayon frais (aliment moyen) | 558 | 133 | 559 | 133 | 72,5 | 3,42 | 3,35 | 17,9 | 5,1 | 0,85 | 0,14 | 119 | 108 | 0,1 | 1,37 | 11 | 22 | 0,12 | 99,7 | 201 | 6,51 | 56,4 | 0,49 | | 0,25 | | | 0,24 | 0,038 | 0,15 | 0,055 | 0,53 | 0,027 | 9,54 | 0,13 |
| Crème dessert, rayon frais ou appertisée (aliment moyen) | 558 | 133 | 559 | 133 | 72,5 | 3,42 | 3,35 | 18 | 5,07 | 0,88 | 0,14 | 119 | | 0,1 | 1,36 | 10,9 | 21,9 | 0,12 | 99 | 201 | 6,4 | 56,3 | 0,49 | | 0,24 | | | 0,27 | 0,038 | 0,15 | 0,064 | 0,52 | 0,028 | 9,66 | 0,13 |
| Crème d'Isigny AOP, >= 35% MG | - | - | - | - | 54,8 | 2,33 | 2,29 | 3 | 40 | 0 | 0,076 | - | - | - | - | - | - | - | - | - | - | 30,5 | - | - | - | - | - | - | - | - | - | - | - | - | - |
| Crème pâtissière | 504 | 120 | 504 | 120 | 69,7 | 3,79 | 3,79 | 18 | 3,6 | 0 | 0,2 | 105 | 130 | 0,03 | 0,64 | 25,5 | 11,6 | 0,1 | 124 | 150 | 4 | 79,1 | 0,7 | 0,24 | 0,34 | - | - | 0,93 | 0,065 | 0,19 | 0,21 | 0,68 | 0,055 | 15 | 1 |
| Crottin de Chavignol (fromage de chèvre) | 1230 | 296 | 1240 | 298 | 50,8 | 19,7 | 19,3 | traces | 24,4 | 0 | 1,3 | 147 | 905 | < 0,1 | 0,28 | 30 | 17,2 | - | 207 | 300 | 11 | 518 | 0,83 | < 0,5 | 0,54 | - | - | - | < 0,04 | 0,88 | 1,3 | 2,15 | 0,25 | 89,5 | - |
| Crottin de chèvre, au lait cru | 1590 | 384 | 1600 | 386 | 41,4 | 23,2 | 22,7 | traces | 32,6 | 0 | 1,01 | 141 | 640 | < 0,1 | 0,4 | 30 | 18,7 | - | 239 | - | 12 | 405 | 0,57 | < 0,5 | 1,05 | - | - | - | 0,025 | 1,09 | 1,67 | 1,7 | 0,48 | 91,9 | - |
| Crottin de chèvre, sans précision | 1410 | 340 | 1420 | 342 | - | 22,4 | 21,9 | traces | 28 | 0 | 1,17 | 107 | 959 | < 0,1 | 0,22 | 30 | 12,7 | - | 186 | 238 | 13 | 467 | 0,59 | < 0,5 | 0,44 | - | - | - | < 0,04 | 0,65 | 1,12 | 0,79 | 0,16 | 77,5 | 0,1 |
| Dessert au soja, aromatisé, sucré, enrichi en calcium, préemballé | 401 | 95 | 396 | 93,8 | 77,5 | 3,25 | 3,56 | 15,3 | 1,9 | 1,1 | 0,13 | 120 | 54 | 0,16 | 1,1 | < 20 | 21 | 0,24 | 99 | 170 | < 20 | 51 | 0,32 | < 0,25 | < 0,08 | 3,08 | - | < 0,5 | < 0,015 | < 0,01 | < 0,1 | 0,13 | 0,017 | 9,32 | 0 |
| Dessert au soja, aromatisé, sucré, non enrichi, préemballé | 393 | 93,1 | 389 | 92 | 78,2 | 2,74 | 3 | 16 | 1,7 | 0,8 | 0,11 | 16 | 55 | 0,15 | 0,79 | < 20 | 22 | 0,24 | 46 | 160 | < 20 | 42 | 0,42 | < 0,25 | < 0,08 | 3,2 | - | < 0,5 | < 0,015 | < 0,01 | < 0,1 | 0,14 | 0,011 | 12,2 | - |
| Dessert au soja, aux amandes, fermenté, préemballé | 234 | 56 | 228 | 54,6 | 88,7 | 3,65 | 4 | 2,85 | 2,8 | 0,6 | 0,19 | 58 | 32 | 0,15 | 0,42 | - | 22 | 0,25 | 59 | 140 | < 20 | 74 | 0,36 | 0,63 | 1,64 | 4,83 | - | < 0,5 | < 0,015 | < 0,01 | < 0,1 | 0,07 | 0,044 | 18,1 | 0,26 |
| Dessert au soja, aux fruits, sucré, enrichi en calcium, fermenté, préemballé | 366 | 86,7 | 361 | 85,6 | 80,7 | 2,74 | 3 | 12,9 | 2,3 | < 0,5 | 0,07 | 87 | 20,4 | 0,07 | 0,43 | < 20 | 13 | 0,27 | 63 | 89 | < 20 | 28 | 0,21 | < 0,25 | 0,19 | 4,19 | - | < 0,5 | < 0,015 | < 0,01 | < 0,1 | 0,046 | < 0,01 | 12,9 | 0 |
| Dessert au soja, aux fruits, sucré, non enrichi, fermenté, préemballé | 329 | 78 | 324 | 76,8 | 81,8 | 3,31 | 3,63 | 11 | 1,7 | 1,1 | 0,085 | 36 | < 20 | 0,08 | 0,41 | < 20 | 14 | 0,28 | 53 | 91 | < 20 | 34 | 0,33 | < 0,25 | 0,27 | 4,11 | - | < 0,5 | < 0,015 | < 0,01 | < 0,1 | 0,058 | 0,017 | 17,7 | - |
| Dessert au soja, nature, non sucré, enrichi en calcium, fermenté, préemballé | 176 | 42 | 170 | 40,6 | 91,4 | 3,59 | 3,94 | 1,1 | 2,1 | 0,6 | 0,053 | 120 | 27,2 | 0,12 | 0,47 | < 20 | 16 | 0,22 | 87 | 130 | < 20 | 21 | 0,3 | < 0,25 | 0,23 | 3,96 | - | < 0,5 | < 0,015 | < 0,01 | < 0,1 | 0,066 | 0,025 | 15,1 | - |
| Dessert au soja, nature, non sucré, non enrichi, fermenté, préemballé | 188 | 44,7 | 181 | 43,3 | 90,7 | 3,76 | 4,13 | 2,05 | 1,9 | 0,8 | 0,085 | 12 | 24,4 | 0,09 | 0,4 | < 20 | 16 | 0,22 | 47 | 110 | < 20 | 34 | 0,29 | < 0,25 | 0,23 | 5,02 | - | < 0,5 | < 0,015 | < 0,01 | < 0,1 | 0,044 | < 0,01 | 18,6 | - |
| Dessert végétal sans soja (amande, avoine, chanvre, coco, riz), aromatisé, sucré, non enrichi, préemballé | 617 | 148 | 611 | 146 | 71,7 | 1,66 | 2 | 15,8 | 8 | 2,1 | 0,065 | 17 | 46,8 | 0,2 | 1,3 | < 20 | 28 | 0,26 | 53 | 200 | < 20 | 26 | 0,39 | < 0,25 | 0,79 | < 0,8 | - | < 0,5 | 0,019 | 0,099 | 0,18 | 0,11 | 0,01 | 6,72 | - |
| Dessert végétal sans soja (coco, riz), aux fruits, sucré, enrichi en calcium, fermenté, préemballé | 324 | 76,6 | 324 | 76,4 | 81,8 | < 0,42 | < 0,5 | 16 | 1 | < 0,5 | 0,095 | 50 | 23 | 0,02 | 0,15 | < 20 | 8,2 | 0,15 | 18 | 46 | < 20 | 38 | 0,12 | < 0,25 | < 0,08 | < 0,8 | - | < 0,5 | < 0,015 | < 0,01 | 0,13 | 0,097 | < 0,01 | < 5 | - |
| Edam | 1370 | 329 | 1380 | 331 | 41,6 | 25,5 | 25 | traces | 25,5 | 0 | 2,22 | 802 | - | 0,056 | 0,12 | 28,4 | 41,2 | 0,018 | 536 | 106 | < 10 | 890 | 3,09 | 0,5 | 0,24 | 2,3 | 47,5 | 0 | 0,037 | 0,39 | 0,082 | 0,28 | 0,076 | 16 | 1,54 |
| Emmental ou emmenthal | 1550 | 373 | 1560 | 375 | 39,1 | 27,9 | 27,3 | traces | 28,8 | 0 | 0,61 | 898 | 340 | 0,043 | 0,15 | 13,8 | 48,9 | 0,024 | 610 | 97,1 | < 10 | 245 | 3,48 | 0,28 | 0,74 | 6,59 | 5,23 | 0 | 0,04 | 0,5 | 0,1 | 0,3 | 0,08 | 20 | 1,5 |
| Emmental ou emmenthal râpé | 1530 | 367 | 1540 | 370 | 38 | 28,1 | 27,5 | traces | 28,6 | 0 | 0,75 | 979 | 454 | < 0,1 | 1,5 | 21,4 | 38,3 | 0,05 | 635 | 97,4 | 16,4 | 297 | 4,4 | < 0,5 | 0,43 | - | - | < 0,5 | 0,019 | 0,35 | 0,095 | 0,14 | 0,049 | 33,4 | 2,02 |
| Époisses | 1200 | 290 | 1210 | 292 | 55,6 | 17,8 | 17,4 | traces | 24,5 | traces | 1,76 | 120 | 1170 | 0,02 | 0,21 | 21 | 10 | 0,01 | 220 | 140 | < 20 | 702 | 0,5 | < 0,25 | 0,64 | 1,73 | - | - | 0,021 | 1,24 | 0,57 | 1,1 | 0,18 | - | 1,7 |
| Faisselle au coulis de fruits | - | - | - | - | - | 3,57 | 3,5 | 8,5 | 5 | - | 0,077 | - | - | - | - | - | - | - | - | - | - | 30,7 | - | - | - | - | - | - | - | - | - | - | - | - | - |
| Faisselle, 0% MG | - | - | - | - | 89,9 | 4,47 | 4,38 | 4,44 | 0 | traces | 0,11 | 124 | - | - | - | - | - | - | 98,2 | 151 | - | 42,5 | - | - | - | - | - | - | - | 0,15 | - | - | - | - | 0,39 |
| Faisselle, 6% MG environ | 351 | 84,2 | 353 | 84,5 | 84,7 | 4,47 | 4,38 | 3,57 | 5,5 | traces | 0,088 | 130 | 99,5 | < 0,01 | < 0,05 | < 20 | 11 | < 0,01 | 110 | 150 | < 20 | 35 | 0,44 | < 0,25 | 0,13 | < 0,8 | - | < 0,5 | < 0,015 | 0,12 | 0,17 | 0,43 | 0,024 | 9,22 | 0,33 |
| Feta AOP | 1180 | 285 | 1190 | 286 | 55,2 | 15,1 | 14,8 | 0,65 | 24,3 | 0 | 2,27 | 220 | 1510 | 0,04 | 0,08 | 80 | 12 | 0,02 | 190 | 66 | < 20 | 908 | 1,2 | < 0,25 | 0,4 | 4,22 | - | - | 0,016 | 0,27 | 0,18 | 0,18 | < 0,01 | 18,9 | 0,53 |
| Flan aux oeufs, rayon frais | 467 | 111 | 469 | 111 | 74,1 | 4,08 | 4 | 18,5 | 2,2 | < 0,5 | 0,14 | 86,3 | - | 0,023 | 0,35 | - | 12,9 | 0,048 | - | 150 | < 10 | 56 | 0,51 | - | - | - | - | - | - | 0,14 | - | - | - | - | - |
| Fondant au chocolat noir et crème anglaise, préemballé | - | - | - | - | 61,4 | 5,05 | 5,05 | 20 | 14,3 | - | - | - | - | - | - | - | - | - | - | - | - | - | - | - | - | - | - | - | - | - | - | - | - | - | - |
| Fontina | 1580 | 381 | 1590 | 383 | 37,9 | 25,6 | 25,1 | traces | 31,1 | 0 | 2 | 550 | - | 0,025 | 0,23 | - | 14 | 0,014 | 346 | 64 | - | 800 | 3,5 | 0,6 | 0,27 | 2,6 | - | 0 | 0,021 | 0,2 | 0,15 | 0,43 | 0,083 | 6 | 1,68 |
| Fourme d'Ambert | 1350 | 326 | 1360 | 327 | 50 | 19,9 | 19,5 | traces | 27,6 | 0 | 2,2 | 442 | - | 0,11 | 0,36 | 27 | 16 | 0,02 | 1040 | 111 | 3,7 | 862 | 3,37 | - | - | - | - | - | - | - | - | - | - | - | - |
| Fourme de Montbrison | 1510 | 365 | 1520 | 367 | 42,8 | 22,5 | 22,1 | traces | 30,8 | 0 | 1,63 | - | - | - | - | - | - | - | - | - | - | 652 | - | - | - | - | - | - | - | - | - | - | - | - | - |
| Fromage (aliment moyen) | 1400 | 338 | 1410 | 340 | 46,2 | 21,8 | 21,3 | 0,56 | 27,4 | 0 | 1,28 | 626 | 815 | 0,13 | 0,27 | 23,1 | 29,2 | 0,028 | 454 | 122 | 7,75 | 510 | 2,87 | 0,29 | 0,55 | | | 0,057 | 0,035 | 0,41 | 1 | 0,51 | 0,11 | 34,4 | 1,36 |
| Fromage 100% brebis (Feta AOP ou type Feta) | - | - | - | - | 54,4 | 14,8 | 14,5 | 2,5 | 22,8 | 0 | 2,59 | 318 | 1580 | < 0,1 | 0,2 | - | 18,3 | traces | 207 | 95 | - | 1040 | 1,03 | 0,5 | 0,37 | - | - | traces | 0,055 | 0,29 | 0,19 | 0,27 | 0,07 | 29,5 | 0,77 |
| Fromage à pâte ferme environ 14% MG type Masdaam à teneur réduite en MG | 993 | 238 | 1000 | 240 | 52,2 | 29 | 28,4 | traces | 13,8 | 0 | 2,06 | 717 | - | 0,086 | 0,21 | 13,8 | 31,6 | 0,059 | 568 | 76,3 | 9,63 | 824 | 4,23 | 0,16 | 0,35 | 0 | - | 0 | 0,058 | 0,32 | 0,1 | 0,34 | 0,085 | 59,5 | 1,56 |
| Fromage à pâte ferme environ 27% MG type Maasdam | 1490 | 359 | 1500 | 361 | 41,1 | 25,5 | 25 | traces | 28,4 | 0 | 1,29 | 848 | 607 | < 0,1 | 0,12 | 20 | 33,5 | < 0,1 | 519 | 81 | 8 | 517 | 3,06 | 0,2 | 0,5 | - | - | 0 | 0,04 | 0,28 | 0,1 | 0,3 | 0,07 | 20 | 1,9 |
| Fromage à pâte ferme, enrobé de cire | 1260 | 303 | 1260 | 304 | 49,6 | 21,7 | 21,3 | traces | 23,7 | 0 | 1,7 | 608 | - | 0,03 | 0,15 | 27 | 29,3 | 0,014 | 546 | 74,5 | < 10 | 679 | 2,57 | - | - | - | - | 0 | 0,03 | 0,33 | 0,06 | - | 0,08 | 21 | 0 |
| Fromage à pâte molle à croûte lavée, au lait pasteurisé (type Vieux Pané) | 1290 | 311 | 1300 | 313 | 51,8 | 20,3 | 19,9 | traces | 25,7 | 0 | 1,57 | 450 | - | - | - | - | 18,1 | - | 357 | 99,3 | - | 626 | - | 0,95 | - | - | - | - | - | 0,43 | - | - | - | - | 1,3 |
| Fromage à pâte molle et croûte fleurie (type camembert) | 1490 | 359 | 1490 | 361 | 46,7 | 17,5 | 17,1 | traces | 32,3 | 0 | 1,18 | 523 | - | - | - | - | 23 | - | 354 | 69 | - | 470 | - | 0,4 | - | - | - | - | - | 0,1 | - | - | - | - | 1,1 |
| Fromage à pâte molle et croûte fleurie double crème environ 30% MG | 1380 | 333 | 1380 | 334 | 51,2 | 17,5 | 17,2 | traces | 29,4 | 0 | 1,28 | 200 | 789 | < 0,1 | 0,3 | 13,8 | 6 | < 0,1 | 160 | 112 | 3,5 | 511 | 0,54 | 0,31 | 0,8 | 0 | - | 0 | 0,04 | 0,2 | 0,1 | 0,27 | 0,055 | 13 | 0,22 |
| Fromage à pâte molle et croûte lavée (aliment moyen) | 1320 | 320 | 1330 | 322 | 50,6 | 20,6 | 20,2 | 0 | 26,5 | 0 | 1,66 | 535 | | | | | 22,4 | | 396 | 113 | | 649 | | 0,68 | | | | | | 0,6 | | | | | 1,54 |
| Fromage à pâte molle et croûte lavée, allégé environ 13% MG | 825 | 198 | 832 | 199 | 59,9 | 21,6 | 21,2 | traces | 12,6 | 0 | 1,63 | 420 | - | - | - | 31,5 | 19 | - | 268 | 94 | - | 650 | - | - | - | - | - | - | - | 0,39 | - | - | - | - | 2,6 |
| Fromage à pâte molle et croûte mixte (lavée et fleurie) colorée | 1380 | 332 | 1380 | 334 | 47,8 | 18,2 | 17,8 | traces | 29 | 0 | - | 450 | 920 | - | 0,2 | - | - | - | 338 | - | - | - | - | - | - | - | - | - | - | 4,2 | - | - | - | 40 | 0,88 |
| Fromage à pâte molle triple crème environ 40% MG | 1550 | 376 | 1550 | 377 | 49,4 | 9,89 | 9,69 | traces | 37,5 | 0 | 1,09 | 230 | - | 0,07 | 0,3 | 13,8 | 6 | 0,035 | 140 | 119 | 3,2 | 436 | 0,54 | 0,36 | 0,94 | 0 | - | 0 | 0,035 | 0,2 | 0,1 | 0,27 | 0,055 | 13 | 0,22 |
| Fromage à pâte pressée cuite (aliment moyen) | 1620 | 390 | 1630 | 392 | 36,6 | 28 | 27,4 | 0 | 30,8 | 0 | 0,8 | 935 | 512 | 0,25 | 0,32 | 22,9 | 45,2 | 0,028 | 630 | 103 | 7,53 | 310 | 3,79 | 0,28 | 0,66 | 5,52 | | 0,051 | 0,036 | 0,43 | 0,21 | 0,3 | 0,088 | 17,1 | 1,86 |
| Fromage à pate pressée cuite type emmental ou emmenthal, allégé en matière grasse | 1180 | 282 | 1190 | 284 | 46,2 | 30,6 | 30 | traces | 18 | 0 | 1 | 950 | - | 0,09 | 0,2 | 32,3 | 45 | 0,09 | 629 | 110 | 10 | 400 | 4,1 | 0,48 | 0,22 | 7,2 | - | 0 | 0,05 | 0,25 | 0,13 | 0,21 | 0,09 | 9,2 | 1,2 |
| Fromage blanc et crème fouettée sur lit de fruits, sucré | - | - | - | - | 72,1 | 4,59 | 4,5 | 16 | 7 | traces | 0,09 | 88,2 | - | 0,015 | - | - | 10,1 | 0,057 | - | - | < 3,35 | 36 | 0,36 | 0,1 | - | - | - | 0 | 0,03 | - | - | - | 0,04 | 32 | - |
| Fromage blanc nature ou aux fruits (aliment moyen) | 368 | 87,7 | 371 | 88,5 | 81,5 | 7,1 | 6,95 | 5,68 | 4,31 | 0,0048 | 0,11 | 122 | | 0,019 | 0,29 | 29,3 | 10,3 | 0,024 | 107 | 137 | 1,91 | 47,1 | 0,47 | 0,71 | 0,078 | | | 0,1 | 0,038 | 0,22 | 0,16 | 0,38 | 0,04 | 19,8 | 0,36 |

| | | | | | | | | | | | | | | | | | | | | | | | | | | | | | | | | | | | |
|---|---|---|---|---|---|---|---|---|---|---|---|---|---|---|---|---|---|---|---|---|---|---|---|---|---|---|---|---|---|---|---|---|---|---|---|
| Fromage blanc nature, 0% MG | 210 | 49,4 | 213 | 50,1 | 86,5 | 7,95 | 7,78 | 3,89 | 0,044 | traces | 0,11 | 134 | 89 | 0,011 | 0,16 | 10,4 | 12,2 | 0,0051 | 116 | 144 | <3,35 | 44,3 | 0,49 | <0,2 | traces | - | - | traces | 0,044 | 0,24 | 0,2 | 0,33 | 0,058 | 24 | 0,4 |
| Fromage blanc nature, 0% MG, enrichi en vitamine D | - | - | - | - | - | 7,55 | 7,4 | 5 | 0,1 | 0 | 0,15 | 164 | - | - | - | - | - | - | - | - | - | 60 | - | 1,25 | - | - | - | - | - | - | - | - | - | - | - |
| Fromage blanc nature, 3% MG environ | 323 | 76,9 | 326 | 77,5 | 83,7 | 8,03 | 7,86 | 3,46 | 3,26 | traces | 0,11 | 130 | - | 0,01 | 0,13 | 14,5 | 10,2 | <0,1 | 115 | 132 | <4,5 | 42,4 | 0,52 | 1,5 | 0,14 | 0 | - | <0,5 | 0,037 | 0,24 | 0,19 | 0,38 | 0,047 | 26 | 0,4 |
| Fromage blanc nature, 3% MG environ, au bifidus | - | - | - | - | - | 7,19 | 7,04 | 3,94 | 3,28 | 0 | 0,11 | - | - | - | - | - | - | - | - | - | - | 42 | - | - | - | - | - | - | - | - | - | - | - | - | - |
| Fromage blanc nature, 3% MG environ, enrichi en vitamine D | - | - | - | - | 84 | 7,04 | 6,9 | 5,4 | 3,2 | 0 | 0,2 | 200 | - | - | - | - | - | - | 120 | - | - | 80 | - | 1,25 | - | - | - | - | - | - | - | - | - | - | - |
| Fromage blanc nature, gourmand, 8% MG environ | 485 | 116 | 487 | 117 | 79,5 | 6,19 | 6,07 | 5,25 | 7,65 | traces | 0,1 | 104 | 61 | 0,051 | <1 | 56 | 9,6 | 0,0039 | 94 | 140 | <3,35 | 40,8 | 0,41 | <0,2 | <0,1 | - | - | traces | 0,038 | 0,18 | 0,1 | 0,39 | <0,05 | 10,8 | 0,29 |
| Fromage blanc ou spécialité laitière nature et crème fouetté, 10% MG environ | - | - | - | - | 79,4 | 6,89 | 6,75 | 3,74 | 10,2 | traces | 0,13 | 95 | - | - | - | 10 | - | - | - | - | - | 51,5 | - | - | - | - | - | - | - | - | - | - | - | - | - |
| Fromage blanc ou spécialité laitière, aromatisé, sucré, 0% MG | - | - | - | - | 78,2 | 7,12 | 6,98 | 12,5 | <0,025 | traces | 0,1 | - | - | - | - | - | - | - | - | - | - | 40 | - | - | - | - | - | - | - | - | - | - | - | - | - |
| Fromage blanc ou spécialité laitière, aromatisé, sucré, 3% MG environ | - | - | - | - | 77,3 | 6,39 | 6,26 | 13,1 | 2,65 | traces | 0,099 | 120 | - | - | - | - | - | - | - | - | - | 84,3 | - | - | - | - | - | - | - | - | - | - | - | - | - |
| Fromage blanc ou spécialité laitière, aromatisé, sucré, 3% MG environ, au bifidus | - | - | - | - | - | 5,58 | 5,47 | 14 | 3,2 | 0,28 | 0,11 | - | - | - | - | - | - | - | - | - | - | 45,3 | - | - | - | - | - | - | - | - | - | - | - | - | - |
| Fromage blanc ou spécialité laitière, aux copeaux de chocolat, sucré, 7% MG environ | - | - | - | - | - | 5,34 | 5,23 | 16,2 | 5,97 | 0,13 | 0,1 | - | - | - | - | - | - | - | - | - | - | 40 | - | - | - | - | - | - | - | - | - | - | - | - | - |
| Fromage blanc ou spécialité laitière, aux fruits, avec édulcorants, allégé en sucres, 0% MG | - | - | - | - | - | 6,24 | 6,11 | 6,76 | 0,12 | 0,58 | 0,14 | 121 | - | 0,01 | 0,12 | 10 | 11 | 0,005 | 115 | 145 | 2 | 57,6 | 0,5 | 1,25 | 0,095 | - | - | traces | 0,032 | 0,21 | 0,29 | 0,55 | 0,045 | 30 | 0,32 |
| Fromage blanc ou spécialité laitière, aux fruits, avec édulcorants, allégé en sucres, 3% MG environ | - | - | - | - | - | 3,16 | 3,1 | 12,2 | 2,5 | 0,2 | 0,13 | 320 | - | - | - | - | - | - | - | - | - | 50 | - | 4 | - | - | - | - | - | - | - | - | - | - | - |
| Fromage blanc ou spécialité laitière, aux fruits, sucré, 0% MG | - | - | - | - | - | 7,93 | 7,77 | 13,5 | traces | 0 | 0,075 | - | - | - | - | - | - | - | - | - | - | 30 | - | - | - | - | - | - | - | - | - | - | - | - | - |
| Fromage blanc ou spécialité laitière, aux fruits, sucré, 3% MG environ | - | - | - | - | 76,6 | 5,23 | 5,13 | 14,7 | 3,31 | 0,12 | 0,087 | 180 | - | - | - | - | - | - | - | - | - | 35,5 | - | 1 | - | - | - | - | - | - | - | - | - | - | - |
| Fromage blanc ou spécialité laitière, aux fruits, sucré, 3% MG environ, au bifidus | - | - | - | - | - | 5,36 | 5,25 | 15,5 | 2,98 | 0,36 | 0,11 | - | - | - | - | - | - | - | - | - | - | 42,5 | - | - | - | - | - | - | - | - | - | - | - | - | - |
| Fromage blanc ou spécialité laitière, aux fruits, sucré, gourmand, 7% MG environ | - | - | - | - | 73 | 5,18 | 5,08 | 13,4 | 5,53 | 0,68 | 0,13 | 93,3 | 93 | - | <1 | 54 | 9,5 | - | 98 | 121 | - | 52,2 | - | <0,2 | <0,1 | - | - | <0,5 | <0,04 | 0,17 | 0,1 | 0,4 | <0,05 | 15,4 | 0,44 |
| Fromage bleu au lait de vache | 1380 | 333 | 1390 | 334 | 46,2 | 19,9 | 19,5 | traces | 28,3 | 0 | 2,89 | 494 | 1750 | <0,1 | 0,23 | 13,8 | 23 | <0,1 | 369 | 130 | 5,5 | 850 | 3,88 | 0,42 | 0,59 | 2,4 | - | 0 | 0,035 | 0,43 | 1,01 | 1,73 | 0,2 | 36 | 1,23 |
| Fromage bleu d'Auvergne | 1420 | 343 | 1430 | 345 | 44,1 | 22,4 | 22 | traces | 28,4 | 0 | 2,85 | 551 | 1730 | 0,07 | 0,3 | 27,1 | 18,1 | <0,1 | 301 | 126 | 3,73 | 1050 | 2,69 | - | - | - | - | - | - | 0,55 | - | 0,66 | - | 54,8 | 0,85 |
| Fromage bleu de Bresse | 1420 | 342 | 1420 | 344 | 51,9 | 17,4 | 17 | traces | 30,5 | 0 | 1,58 | 434 | 960 | <0,1 | <1 | 48 | 17 | <0,1 | 320 | 128 | 16,1 | 450 | 2,5 | <0,2 | <0,1 | - | - | 0 | <0,05 | 0,46 | 1,5 | 0,49 | 0,07 | 47 | 0,74 |
| Fromage bleu de Bresse allegé environ 15% MG | 1020 | 246 | 1030 | 248 | - | 27,1 | 26,5 | traces | 15,5 | 0 | 2,05 | 420 | 1240 | - | - | - | - | - | - | - | - | 800 | - | - | - | - | - | - | - | 0,45 | - | - | - | - | 0,5 |
| Fromage bleu des Causses | 1450 | 350 | 1460 | 352 | 45,2 | 20,4 | 20 | traces | 30 | 0 | 3,3 | 582 | 2000 | - | - | 27 | 35 | - | 400 | 132 | - | 1300 | 3 | - | - | - | - | - | - | - | - | - | - | 25 | - |
| Fromage de brebis à pâte molle et croûte fleurie | 1140 | 276 | 1150 | 277 | 55,6 | 16,8 | 16,5 | traces | 23,3 | 0 | 1,69 | 487 | - | - | - | - | - | - | 364 | - | - | 676 | - | - | - | - | - | - | - | 0,5 | - | - | - | - | 2,05 |
| Fromage de brebis à pâte pressée | 1590 | 384 | 1600 | 386 | 38 | 23,9 | 23,4 | traces | 32,3 | 0 | 1,85 | 722 | - | - | - | - | - | - | 498 | - | - | 738 | - | - | - | - | - | - | - | 0,4 | - | - | - | - | 1,2 |
| Fromage de brebis Corse à pâte molle | 1900 | 457 | 1900 | 460 | 28,7 | 28,5 | 27,9 | traces | 38,4 | 0 | 2,39 | 576 | - | 0,086 | 0,21 | - | 34,2 | - | - | 105 | - | 957 | 2 | - | 0,21 | - | - | - | 0,036 | 0,75 | - | 0,23 | 0,085 | 86,9 | 0,95 |
| Fromage de brebis des Pyrénées | 1640 | 397 | 1650 | 399 | 36,9 | 24,1 | 23,6 | traces | 33,6 | 0 | 1,62 | 750 | 1,5 | <0,1 | 0,3 | 124 | 35,9 | <0,1 | 505 | 71,9 | 19,9 | 648 | 2,4 | 1,08 | 0,67 | - | - | - | 0,032 | 0,49 | 0,19 | <0,16 | 0,064 | 26,8 | 0,73 |
| Fromage de chèvre à pâte molle et croûte fleurie type camembert | 1210 | 292 | 1220 | 294 | 51,7 | 19,9 | 19,4 | traces | 23,8 | 0 | 1 | 534 | 764 | <0,1 | 0,32 | 26,7 | 32 | traces | 326 | 153 | 10,7 | 400 | 2,4 | <0,5 | 0,4 | - | - | traces | <0,04 | 0,4 | 0,8 | 0,35 | 0,2 | 60,8 | 2 |
| Fromage de chèvre à pâte molle non pressée non cuite croûte naturelle, au lait pasteurisé | 1160 | 279 | 1170 | 281 | 54,7 | 19,3 | 19 | traces | 22,6 | 0 | 1,56 | 90 | - | - | - | - | - | - | - | - | - | 624 | - | - | - | - | - | - | - | - | - | - | - | - | - |
| Fromage de chèvre à tartiner, nature | - | - | - | - | 72,9 | 10,8 | 10,6 | 2,41 | 12,2 | 0 | 0,92 | 78,1 | 778 | 0,07 | 0,14 | 32,9 | 9,73 | - | 101 | 93 | <8 | 369 | 0,37 | <0,5 | 0,21 | - | - | <0,5 | <0,04 | 0,2 | 0,34 | 0,13 | 0,03 | 26,5 | 0,08 |
| Fromage de chèvre bûche | 1180 | 285 | 1190 | 287 | 54,8 | 18,8 | 18,4 | traces | 23,3 | 0 | 1,58 | 107 | 1040 | <0,1 | 0,22 | 30 | 12,7 | - | 183 | 167 | 10,1 | 630 | 0,61 | <0,5 | 0,29 | - | - | - | 0,022 | 0,47 | 1,05 | 0,52 | 0,16 | 76,4 | - |
| Fromage de chèvre bûche, allégé en matière grasse | 747 | 179 | 754 | 180 | 60,5 | 22,6 | 22,1 | traces | 10 | 0 | 1,24 | 183 | - | - | 0,1 | - | 15 | - | 275 | 210 | - | 801 | - | - | - | - | - | - | - | - | - | - | - | - | - |
| Fromage de chèvre demi-sec | 1460 | 353 | 1470 | 355 | 45,5 | 21,6 | 21,1 | traces | 29,8 | 0 | 1,81 | 145 | 1100 | 0,068 | 1,62 | 69,5 | 18,2 | 0,036 | 375 | 165 | 8,85 | 471 | 0,64 | 0,5 | 0,26 | 2,5 | - | 0 | 0,072 | 0,68 | 1,15 | 0,19 | 0,06 | 2 | 0,22 |
| Fromage de chèvre frais, au lait cru (type palet ou crottin frais) | - | - | - | - | 65,6 | 13,1 | 12,9 | 2,03 | 16,9 | 0 | 0,77 | 91,6 | 670 | <0,1 | 0,28 | 29,1 | 12,1 | - | 154 | - | 4,36 | 309 | 0,37 | <0,5 | 0,38 | - | - | - | <0,04 | 0,43 | 0,55 | 0,29 | 0,042 | 43,1 | - |
| Fromage de chèvre frais, au lait pasteurisé (type bûchette fraîche) | - | - | - | - | 60 | 16,1 | 15,8 | 2,08 | 20 | 0 | 0,97 | 96,7 | 767 | <0,1 | 0,22 | 28,3 | 11,2 | - | 161 | - | 7,58 | 387 | 0,79 | <0,5 | 0,32 | - | - | - | <0,04 | 0,34 | 0,6 | 0,18 | 0,045 | 50,8 | - |
| Fromage de chèvre frais, au lait pasteurisé ou cru (type crottin frais ou bûchette fraîche) | - | - | - | - | 62,8 | 14,6 | 14,3 | 2,05 | 18,4 | 0 | 1,55 | 94,2 | 940 | <0,1 | 0,25 | 28,7 | 11,6 | 0,1 | 158 | 159 | 6,04 | 349 | 0,59 | <0,5 | 0,35 | 1,8 | - | 0 | <0,04 | 0,39 | 0,58 | 0,24 | 0,044 | 47 | 0,19 |
| Fromage de chèvre lactique affiné (type bûchette, crottin, Sainte-Maure) | 1310 | 315 | 1310 | 317 | 50,3 | 20,4 | 20 | traces | 26,2 | 0 | 1,69 | 117 | 1030 | 0,055 | 0,28 | 30,6 | 14,9 | - | 199 | 161 | 10,8 | 505 | 0,58 | <0,5 | 0,45 | - | - | - | 0,021 | 0,74 | 1,3 | 1,03 | 0,22 | 81,5 | - |
| Fromage de chèvre lactique affiné, au lait cru (type Crottin de Chavignol, Picodon, Rocamadour, Sainte-Maure de Touraine) | 1340 | 324 | 1350 | 326 | 49,7 | 20,7 | 20,3 | traces | 27 | 0 | 1,28 | 123 | 895 | 0,057 | 0,31 | 31 | 16,2 | - | 208 | 230 | 10,9 | 513 | 0,56 | <0,5 | 0,53 | - | - | - | 0,021 | 0,89 | 1,43 | 1,3 | 0,25 | 84,2 | - |
| Fromage de chèvre lactique affiné, au lait pasteurisé (type bûchette ou crottin) | 1240 | 300 | 1250 | 302 | 52,8 | 19,8 | 19,4 | traces | 24,7 | 0 | 1,23 | 107 | 984 | <0,1 | 0,22 | 30 | 12,7 | - | 183 | 260 | 10,6 | 490 | 0,61 | <0,5 | 0,32 | - | - | - | 0,022 | 0,5 | 1,06 | 0,57 | 0,16 | 76,6 | - |
| Fromage de chèvre sec | 1830 | 440 | 1840 | 442 | 29 | 30,5 | 29,9 | traces | 35,6 | 0 | 1,06 | 895 | - | 0,07 | 1,88 | 70 | 54 | - | 729 | 48 | - | 423 | 1,59 | 0,7 | 0,31 | 3 | - | 0 | 0,14 | 1,19 | 2,4 | 0,41 | 0,08 | 53 | 0,12 |
| Fromage de lactosérum de brebis | 787 | 190 | 791 | 190 | 68,7 | 10,1 | 9,91 | 4,3 | 14,8 | 0 | 0,63 | 216 | 413 | - | 0,25 | - | 21,7 | - | - | - | - | 253 | 0,24 | - | - | - | - | - | 0,08 | 0,26 | - | 0,32 | - | 25 | 0,4 |
| Fromage fondu double crème, environ 31% MG | 1280 | 311 | 1290 | 312 | 56,3 | 9,06 | 8,88 | 1,78 | 29,6 | 0 | 1,37 | 50 | 400 | <0,1 | 0,2 | 20 | 18 | <0,1 | 250 | 117 | 6,7 | 547 | 2 | - | - | - | - | - | - | 0,32 | - | 0,47 | - | - | 0,9 |
| Fromage fondu en portions ou en cubes environ 20% MG | 1010 | 242 | 1010 | 243 | 59,4 | 10,7 | 10,4 | 6,38 | 19,3 | 0 | 1,59 | 513 | 530 | 0,35 | 0,34 | 13,8 | 26 | 0,041 | 800 | 199 | 8 | 634 | 2,6 | 0,25 | 0,63 | 0 | - | 0 | 0,09 | 0,38 | 0,1 | 0,4 | 0,025 | 7 | 0,85 |
| Fromage fondu en portions ou en cubes environ 8% MG | - | - | - | - | 68,1 | 11,5 | 11,3 | - | 7,48 | - | 0,74 | 327 | 450 | <0,1 | - | 25 | - | <0,1 | 695 | 204 | - | 79 | 1,1 | - | - | - | - | - | 0,06 | 0,34 | - | 0,3 | - | - | 0,5 |
| Fromage fondu en tranchettes | 1020 | 246 | 1030 | 247 | 56,1 | 14,1 | 13,8 | 5,34 | 18,5 | 0 | 2,46 | 490 | 1080 | 0,02 | 0,12 | 24 | 25 | 0,01 | 840 | 370 | <20 | 983 | 1,8 | <0,25 | 0,53 | 1,39 | - | - | 0,1 | 0,37 | <0,1 | 0,44 | 0,031 | - | 0,56 |
| Fromage frais type petit suisse, aromatisé chocolat, sucré | - | - | - | - | 62,4 | 3,95 | 3,87 | 20,9 | 5,18 | 0,2 | 0,27 | 144 | - | - | - | - | 8 | - | 66 | 321 | - | 108 | - | - | - | - | - | - | - | - | - | - | - | - | - |
| Fromage frais type petit suisse, aromatisé ou aux fruits, 2-3% MG, enrichi en calcium et vitamine D | 381 | 90,4 | 383 | 90,8 | 78,9 | 5,74 | 5,63 | 10,2 | 2,5 | <2 | 0,075 | 170 | 78 | <0,01 | 0,07 | <20 | 9,4 | 0,02 | 110 | 110 | <20 | 30 | 0,43 | 1,96 | <0,08 | <0,8 | - | <0,5 | <0,015 | 0,13 | <0,1 | 0,37 | 0,03 | 25 | 0,36 |
| Fromage frais type petit suisse, aux fruits, 2-3% MG | 390 | 92,7 | 392 | 93,2 | 78,8 | 6,7 | 6,56 | 9,34 | 2,8 | 1 | 0,053 | 110 | 83,6 | 0,01 | 0,06 | <20 | 9,4 | 0,03 | 110 | 140 | <20 | 21 | 0,37 | <0,25 | 0,13 | <0,8 | - | <0,5 | 0,015 | 0,18 | <0,1 | 0,45 | 0,024 | 14,7 | 0,24 |
| Fromage frais type petit suisse, aux fruits, 2-3% MG, enrichi en calcium et vitamine D | - | - | - | - | 78,7 | 6,42 | 6,29 | 12,2 | 2,63 | 0,2 | 0,076 | 155 | 89,2 | <0,01 | 0,05 | <20 | 9,2 | 0,02 | 120 | 110 | <20 | 29,2 | 0,41 | 1,16 | 0,15 | <0,8 | - | <0,5 | 0,044 | 0,17 | <0,1 | 0,24 | 0,029 | 151 | 0,35 |
| Fromage frais type petit suisse, nature, 0% MG | - | - | - | - | 84,3 | 9,89 | 9,69 | 3,8 | 0,24 | 0 | 0,11 | 127 | 115 | - | <1 | 54 | 11,4 | - | 129 | 166 | 9,3 | 44,8 | - | <0,2 | <0,1 | - | - | - | - | 0,21 | 0,1 | 0,52 | <0,05 | 14,2 | 1,6 |
| Fromage frais type petit suisse, nature, 10% MG environ | 621 | 149 | 624 | 150 | 76 | 9,7 | 9,5 | 3,17 | 10,4 | <3 | 0,058 | 110 | 103 | 0,01 | 0,2 | <20 | 9,4 | <0,01 | 125 | 130 | <20 | 23 | 0,43 | - | 0,23 | <0,8 | - | <0,5 | 0,029 | 0,17 | 0,14 | 0,34 | 0,044 | 20,2 | 0,57 |
| Fromage frais type petit suisse, nature, 4% MG environ | 373 | 88,8 | 376 | 89,6 | 81,6 | 9,95 | 9,75 | 2,84 | 4 | 0 | 0,083 | 103 | - | 0,01 | 0,14 | 14 | 40 | - | 126 | 114 | 2 | 33 | 0,48 | - | - | - | - | - | - | 0,24 | 0,26 | 0,34 | - | - | 0,35 |
| Fromage rond à pâte molle et croûte fleurie 5 à 11% MG type camembert allégé en matière grasse | 725 | 173 | 733 | 175 | 63,7 | 22,6 | 22,2 | traces | 9,3 | 0 | 1,44 | 720 | - | 0,07 | 0,18 | 13,8 | 20 | 0,048 | 370 | 187 | 8,5 | 576 | 2,38 | 0,14 | 0,28 | 0 | - | 0 | 0,045 | 0,45 | 1,8 | 1,36 | 0,25 | 62 | 1,13 |
| Fromage rond à pâte molle et croûte fleurie environ 11% MG type coulommiers allégé en matière grasse | 781 | 187 | 789 | 189 | 62,5 | 22,5 | 22 | traces | 11 | 0 | 0,9 | - | - | - | 0,05 | - | - | - | - | - | - | - | - | - | - | - | - | - | - | - | - | - | - | - | - |

| | | | | | | | | | | | | | | | | | | | | | | | | | | | | | | | | | | | |
|---|---|---|---|---|---|---|---|---|---|---|---|---|---|---|---|---|---|---|---|---|---|---|---|---|---|---|---|---|---|---|---|---|---|---|---|
| Fromage rond à pâte molle et croûte fleurie environ 5% MG type camembert allégé en matière grasse | 632 | 150 | 641 | 152 | - | 25,7 | 25,2 | traces | 5,5 | 0 | - | 570 | - | - | - | - | - | - | 440 | - | - | - | - | - | - | - | - | - | - | - | - | - | - | - | - |
| Fromage type feta, au lait de vache | 1120 | 270 | 1130 | 271 | 54 | 16,4 | 16 | 1,42 | 21,7 | 0 | 1,94 | 557 | - | 0,051 | 0,42 | 13,8 | 19,5 | 0,044 | 349 | 125 | 62 | 917 | 2,63 | 0,32 | 0,37 | 1,8 | - | 0 | 0,1 | 0,65 | 1,4 | 1,16 | 0,34 | 47 | 1,41 |
| Fromage type feta, au lait de vache, à l'huile et aux aromates | 1180 | 283 | 1180 | 285 | 53,4 | 17,6 | 17,3 | 0,16 | 23,4 | 0 | 2,52 | 440 | 1840 | 0,04 | 0,13 | < 20 | 15 | 0,03 | 280 | 65 | < 20 | 1010 | 1,9 | 0,84 | 1,31 | 8,3 | - | - | < 0,015 | 0,22 | < 0,1 | 0,18 | 0,022 | 32,3 | 1,3 |
| Gâteau au chocolat, coeur fondant, préemballé (rayon frais) | 1520 | 364 | 1520 | 364 | - | 6,86 | 6,86 | 41,9 | 18,1 | 3,24 | 0,073 | - | - | - | - | - | - | - | - | - | - | 29 | - | - | - | - | - | - | - | - | - | - | - | - | - |
| Gâteau de riz au caramel, rayon frais | 604 | 143 | 604 | 143 | 67,5 | 3,19 | 3,19 | 25,4 | 3,1 | < 0,5 | 0,073 | 63 | 90 | < 0,1 | 0,3 | 2 | 7,8 | < 0,1 | 73 | 84 | < 3,35 | 29 | 0,5 | < 0,5 | 0,2 | - | - | < 0,5 | < 0,04 | 0,14 | < 0,16 | 0,34 | 0,075 | 17 | < 0,08 |
| Gâteau de riz, appertisé | 484 | 115 | 486 | 115 | 71 | 4,29 | 4,2 | 20 | 1,9 | 0,3 | 0,15 | 130 | - | - | - | - | 130 | - | - | 210 | - | 60 | - | - | - | - | - | - | - | - | - | - | - | - | - |
| Gâteau de semoule aux raisins et caramel, rayon frais | 615 | 146 | 615 | 146 | 67,1 | 3,81 | 3,81 | 24,8 | 3,4 | < 0,5 | 0,08 | 64 | 82,4 | 0,06 | 0,35 | < 20 | 13 | 0,07 | 71 | 270 | < 20 | 32 | 0,38 | < 0,25 | 0,26 | < 0,8 | - | < 0,5 | 0,044 | 0,17 | < 0,1 | 0,45 | 0,042 | 14,3 | 0,31 |
| Gâteau de semoule, appertisé | 634 | 150 | 634 | 150 | 68 | 2,85 | 2,85 | 27,4 | 3,03 | 1 | 0,069 | 67,9 | 73 | < 0,1 | 0,3 | 2 | 9,2 | < 0,1 | 63 | 93 | < 3,35 | 27,5 | 0,3 | < 0,5 | 0,34 | - | - | < 0,5 | < 0,04 | 0,14 | 0,23 | 0,21 | 0,038 | 18,9 | < 0,08 |
| Gorgonzola | 1290 | 312 | 1300 | 314 | 51,1 | 19 | 18,6 | traces | 26,4 | 0 | 1,77 | 390 | 1180 | 0,02 | 0,08 | 70 | 18 | 0,01 | 310 | 110 | < 20 | 710 | 2,1 | 0,24 | 0,16 | 0,9 | - | < 0,5 | 0,07 | 0,32 | 2,93 | 0,81 | 0,13 | 41,2 | 0,73 |
| Gouda | 1550 | 374 | 1560 | 376 | 38,4 | 23,2 | 22,8 | traces | 31,5 | 0 | 2 | 728 | 1440 | 0,036 | 0,24 | 29,7 | 29 | < 0,1 | 546 | 121 | 9,02 | 795 | 3,9 | 0,5 | 0,24 | 2,3 | - | 0 | 0,05 | 0,33 | 0,063 | 0,34 | 0,08 | 21 | 1,54 |
| Grana Padano | 1650 | 396 | 1660 | 399 | - | 34,1 | 33,4 | traces | 29,2 | 0 | 1,56 | 1170 | - | - | - | - | - | - | - | - | - | 610 | - | - | - | - | - | - | - | - | - | - | - | - | - |
| Gruyère | 1760 | 423 | 1770 | 426 | 32,5 | 28,4 | 27,9 | traces | 34,6 | 0 | 0,8 | 1090 | 1040 | 0,41 | 0,29 | 13,8 | 39,2 | 0,031 | 608 | 101 | 5,33 | 320 | 4,89 | 0,44 | 0,52 | 2,7 | - | 0 | 0,05 | 0,39 | 0,1 | 0,43 | 0,081 | 15 | 1,55 |
| Gruyère IGP France | 1640 | 396 | 1650 | 398 | 36,9 | 27,3 | 26,7 | traces | 32 | 0 | 0,95 | 871 | 576 | - | - | - | - | - | - | - | - | - | - | - | - | - | - | - | - | - | - | - | - | - | - |
| Ile flottante, rayon frais | 543 | 129 | 544 | 129 | 70,1 | 4,98 | 4,87 | 19,8 | 3,31 | 0,047 | 0,11 | 97 | - | < 0,1 | 0,4 | 11,7 | 9,8 | < 0,1 | 27,8 | 126 | 6 | 45 | 0,9 | < 0,5 | 0,31 | - | - | < 0,5 | 0,04 | 0,21 | < 0,1 | 0,58 | 0,059 | 10,5 | 0,26 |
| Kéfir de lait | - | - | - | - | 87,6 | 3,19 | 3,13 | 4,5 | 3,5 | 0 | 0,12 | 127 | - | 0,01 | 0,05 | 24,3 | 12,7 | 0,009 | 95 | 155 | 1,3 | 49 | 0,44 | 0,1 | 0,092 | 0 | - | 0,3 | 0,045 | 0,2 | 0,087 | 0,31 | 0,038 | 21 | 0,18 |
| Lait à 1,2% de matière grasse, UHT, enrichi en plusieurs vitamines | 186 | 44,1 | 187 | 44,4 | 89,7 | 3,38 | 3,31 | 4,79 | 1,25 | 0 | 0,088 | 120 | 96,9 | < 0,01 | < 0,05 | < 20 | 9,9 | < 0,01 | 86 | 160 | < 20 | 35 | 0,38 | 0,81 | 2,02 | 9,24 | - | < 0,5 | 0,17 | 1,82 | 2,12 | 0,9 | 0,23 | 16,1 | 0,5 |
| Lait concentré non sucré, entier | 510 | 122 | 513 | 123 | 74,5 | 6,44 | 6,31 | 9,91 | 5,9 | <3 | 0,24 | 273 | 220 | - | 0,06 | 23 | 25,7 | 0,011 | 225 | 339 | 2,1 | 95 | 0,8 | 0,1 | 0,2 | - | - | 1,9 | 0,055 | 0,34 | 0,2 | 2,16 | 0,28 | 30,7 | 0,16 |
| Lait concentré sucré, entier | 1360 | 323 | 1370 | 324 | 27,2 | 7,7 | 7,54 | 55,6 | 7,87 | 0 | 0,21 | 290 | 264 | 0,013 | 0,19 | 24,5 | 25,2 | 0,0056 | 253 | 371 | < 4,5 | 85,2 | 0,96 | 0,2 | 0,16 | 0,6 | - | 2,6 | 0,09 | 0,42 | 0,21 | 0,75 | 0,051 | 11 | 0,44 |
| Lait de brebis, entier | 429 | 103 | 431 | 103 | 82,2 | 5,68 | 5,56 | 4,5 | 6,97 | 0 | 0,11 | 199 | 101 | 0,011 | 0,46 | 23,3 | 17,1 | 0,018 | 158 | 103 | 3 | 44 | 0,54 | 0,2 | 0,15 | - | - | 4,2 | 0,057 | 0,34 | 0,42 | 0,41 | 0,06 | 9,19 | 0,71 |
| Lait de chèvre, demi-écrémé, UHT | 193 | 45,8 | 194 | 46,1 | 89,6 | 3,77 | 3,69 | 4,13 | 1,57 | 0 | 0,15 | 110 | 154 | < 0,01 | 0,02 | < 20 | 13 | < 0,01 | 110 | 190 | < 20 | 58,6 | 0,33 | < 0,03 | < 0,08 | < 0,8 | - | < 0,5 | 0,026 | 0,088 | 0,23 | 0,29 | 0,013 | < 5 | 0,05 |
| Lait de chèvre, entier, cru | 240 | 57,4 | 241 | 57,7 | 88,7 | 3,22 | 3,15 | 4,01 | 3,2 | 0 | 0,091 | 107 | 181 | < 0,1 | 0,17 | 25 | 12,9 | - | 77,6 | - | <4 | 36,6 | 0,33 | < 0,2 | 0,056 | - | - | < 0,2 | 0,041 | 0,11 | 0,2 | 0,22 | 0,025 | 6,83 | - |
| Lait de chèvre, entier, UHT | 235 | 56,1 | 236 | 56,4 | 87 | 3,39 | 3,33 | 4,35 | 2,83 | 0 | 0,13 | 134 | 157 | 0,05 | 0,05 | 15,3 | 14 | 0,0053 | 111 | 204 | 2 | 50 | 0,3 | 0,06 | 0,04 | - | - | 1,3 | 0,048 | 0,14 | 0,28 | 0,31 | 0,046 | 1 | 0,07 |
| Lait de jument, entier | 210 | 50 | 211 | 50,2 | 89,2 | 2,41 | 2,36 | 6,18 | 1,73 | 0 | 0,05 | 87,4 | - | 0,03 | 0,065 | - | 7,95 | - | 54 | 63,3 | - | 20 | 0,15 | - | 0,026 | - | - | 15 | 0,03 | 0,034 | 0,14 | 0,3 | 0,03 | - | 0,3 |
| Lait de poule, sans alcool | 368 | 87,8 | 369 | 88,1 | 82,5 | 4,55 | 4,46 | 8,05 | 4,19 | 0 | 0,14 | 130 | - | 0,013 | 0,2 | 35 | 19 | 0,005 | 109 | 165 | - | 54 | 0,46 | 1,2 | 0,21 | 0,3 | - | 1,5 | 0,034 | 0,19 | 0,11 | 0,42 | 0,05 | 9 | 0,45 |
| Lait demi-écrémé (ou à teneur en matière grasse légèrement inférieure) à teneur réduite en lactose | 187 | 44,4 | 188 | 44,7 | 89,7 | 3,45 | 3,38 | 4,69 | 1,29 | 0 | 0,088 | 120 | - | - | - | - | - | - | - | - | - | 35 | - | - | - | - | - | - | - | - | - | - | - | - | - |
| Lait demi-écrémé, pasteurisé | 195 | 46,4 | 197 | 46,7 | 89,4 | 3,19 | 3,13 | 4,9 | 1,54 | 0 | 0,078 | 119 | 100 | 0,0031 | 0,031 | 17,9 | 11,1 | 0,002 | 96 | 153 | < 3,35 | 31 | 0,41 | 0,085 | 0,041 | 0 | - | 1,3 | 0,047 | 0,18 | 0,097 | 0,34 | 0,048 | 11,5 | 0,49 |
| Lait demi-écrémé, UHT | 198 | 47 | 199 | 47,3 | 89,4 | 3,38 | 3,31 | 4,83 | 1,55 | 0 | 0,09 | 117 | 100 | 0,0087 | 0,045 | 12,1 | 12,1 | 0,0031 | 89,1 | 167 | < 10 | 36 | 0,39 | < 0,5 | 0,13 | 0,2 | - | 1,5 | 0,06 | 0,18 | 0,093 | 0,41 | 0,037 | 9,49 | 0,38 |
| Lait demi-écrémé, UHT, enrichi en vitamine D seulement | 196 | 46,6 | 197 | 46,9 | 89,4 | 3,38 | 3,31 | 4,8 | 1,52 | 0 | 0,09 | 120 | 97,6 | < 0,01 | < 0,05 | < 20 | 9,8 | < 0,01 | 84 | 160 | < 20 | 36 | 0,37 | 0,78 | < 0,08 | < 0,8 | - | < 0,5 | 0,023 | 1,16 | < 0,1 | 0,28 | 0,023 | 5,39 | 0,21 |
| Lait écrémé, pasteurisé | 146 | 34,5 | 148 | 34,8 | 90,8 | 3,28 | 3,21 | 4,99 | 0,19 | 0 | 0,11 | 123 | 100 | 0,012 | 0,03 | 24,3 | 11,6 | 0,0055 | 99 | 157 | 1,66 | 43,2 | 0,42 | 0,075 | 0,009 | 0 | - | 1,3 | 0,043 | 0,18 | 0,092 | 0,36 | 0,044 | 4,9 | 0,49 |
| Lait écrémé, UHT | 142 | 33,4 | 143 | 33,7 | 90,9 | 3,51 | 3,44 | 4,64 | 0,06 | 0 | 0,098 | 105 | 100 | 0,007 | 0,028 | 13,5 | 12,3 | 0,0028 | 92,1 | 166 | < 10 | 39 | 0,37 | traces | traces | - | - | 0,89 | 0,049 | 0,19 | 0,09 | 0,42 | 0,041 | 5,11 | 0,39 |
| Lait emprésuré aromatisé, rayon frais | - | - | - | - | 72,1 | 5,63 | 5,52 | 17,6 | 2,5 | 0 | 0,15 | 208 | - | - | - | - | 9,5 | - | 71 | 235 | - | 98,8 | - | - | - | - | - | - | - | - | - | - | - | - | - |
| Lait en poudre, demi-écrémé | 1930 | 459 | 1940 | 461 | 2,47 | 31,5 | 30,8 | 43,4 | 17,5 | 0 | 0,78 | 1030 | 1010 | - | 0,35 | 77,5 | 97 | 0,048 | 829 | 1330 | 8,8 | 313 | 3,7 | 0,2 | 0,4 | - | - | - | 0,21 | 1,71 | 0,8 | 3,07 | 0,32 | 69 | 2,8 |
| Lait en poudre, écrémé | 1530 | 361 | 1540 | 364 | 1,95 | 35,3 | 34,6 | 54 | 0,7 | 0 | 1,21 | 1250 | 1190 | 0,07 | 0,31 | 7 | 112 | 0,05 | 966 | 1670 | 13 | 482 | 4,47 | 0,018 | 0,027 | 0,1 | - | 6,1 | 0,43 | 1,5 | 0,91 | 3,5 | 0,42 | 35,5 | 4,02 |
| Lait en poudre, entier | 2090 | 499 | 2090 | 501 | 2,4 | 27,4 | 26,9 | 37,5 | 26,8 | 0 | 0,9 | 960 | 886 | 0,058 | 0,62 | 7 | 93 | 0,061 | 810 | 1190 | 14 | 360 | 3,91 | 1,2 | 0,68 | - | - | 10 | 0,33 | 1,42 | 0,65 | 2,27 | 0,3 | 37 | 3,25 |
| Lait entier, pasteurisé | 236 | 56,5 | 237 | 56,8 | 89,1 | 3,3 | 3,23 | 3,47 | 3,3 | 0 | 0,2 | 117 | 121 | < 0,1 | 0,04 | 24,3 | 10,9 | 0,002 | 93 | 140 | < 2,2 | 79 | 0,37 | 0,1 | 0,089 | 0 | - | 1,2 | 0,042 | 0,17 | 0,092 | 0,34 | 0,047 | 11 | 0,32 |
| Lait entier, UHT | 272 | 65,1 | 273 | 65,4 | 87,5 | 3,32 | 3,25 | 4,85 | 3,63 | 0 | 0,11 | 120 | 98 | < 0,01 | 0,01 | < 20 | 9,8 | < 0,01 | 97 | 160 | < 50 | 44,2 | 0,37 | < 0,25 | 0,089 | < 0,8 | 0,9 | < 0,5 | 0,041 | 0,17 | < 0,1 | 0,43 | 0,02 | < 2,5 | 0,24 |
| Lait fermenté à boire, nature, au lait entier | 232 | 55,7 | 233 | 56 | 88,3 | 3,38 | 3,31 | 1,99 | 3,2 | <3 | 0,088 | 127 | - | 0,01 | 0,05 | 24,3 | 12,7 | 0,009 | 95 | 155 | 1,3 | 35 | 0,44 | 0,1 | 0,092 | 0 | 8,15 | 0,3 | 0,045 | 0,2 | 0,087 | 0,31 | 0,038 | 6 | 0,18 |
| Lait fermenté à boire, nature, maigre | 141 | 33,6 | 143 | 33,9 | 90,6 | 3,51 | 3,44 | 2,18 | 0,6 | <3 | 0,088 | 160 | 139 | 0,0095 | <1 | 17 | 10,2 | <1 | 91 | 120 | <5 | 35 | <1 | 0 | < 0,1 | 0,1 | - | 1,25 | 0,05 | 0,09 | < 0,05 | 0,38 | < 0,05 | 7,8 | 0,24 |
| Lait fermenté ou spécialité laitière type yaourt, aromatisé, sucré, au bifidus | 397 | 94,4 | 398 | 94,8 | 78,4 | 3,83 | 3,75 | 11,5 | 3,1 | <3 | 0,1 | 122 | 109 | 0,007 | <1 | 8,1 | 11,7 | 0,001 | 101 | 139 | 1,1 | 40 | 0,41 | 0,3 | < 0,1 | - | - | 1 | 0,04 | 0,19 | 0,12 | 0,47 | < 0,05 | 5,3 | 0,22 |
| Lait fermenté ou spécialité laitière type yaourt, aux fruits, 0% MG, avec édulcorants, aux esters de stérol | - | - | - | - | - | 3,88 | 3,8 | 6,6 | 0,5 | 0,45 | 0,16 | 130 | - | - | - | - | - | - | - | - | - | 65 | - | - | - | - | - | - | - | - | - | - | - | - | - |
| Lait fermenté ou spécialité laitière type yaourt, aux fruits, avec édulcorants, 0% MG, au bifidus | - | - | - | - | - | 4,67 | 4,58 | 6,71 | 0,094 | 1,08 | 0,18 | 155 | - | - | - | - | - | - | - | - | - | 71,1 | - | - | - | - | - | - | - | - | - | - | - | - | 0,31 |
| Lait fermenté ou spécialité laitière type yaourt, aux fruits, sucré, au bifidus | 409 | 97,2 | 410 | 97,5 | 78 | 3,57 | 3,5 | 13 | 3,01 | 0,65 | 0,14 | 122 | - | < 0,1 | - | 12,6 | 12 | < 0,1 | 92 | 153 | <5 | 53,8 | 0,42 | - | - | - | - | - | - | - | - | - | - | - | 0,24 |
| Lait fermenté ou spécialité laitière type yaourt, nature, 0% MG, au bifidus | - | - | - | - | - | 4,39 | 4,3 | 5,75 | 0,3 | 0 | 0,14 | 149 | - | - | - | - | - | - | - | - | - | 55 | - | - | - | - | - | - | - | - | - | - | - | - | - |
| Lait fermenté ou spécialité laitière type yaourt, nature, au bifidus | 270 | 64,5 | 271 | 64,8 | 87,2 | 3,84 | 3,77 | 3,47 | 3,6 | < 0,18 | 0,12 | 130 | - | < 0,1 | 0,07 | 11,5 | 12,1 | < 0,1 | 97,3 | 145 | < 2,2 | 47,6 | 0,45 | < 0,5 | 0,3 | - | - | < 0,5 | < 0,04 | 0,22 | 0,088 | 0,3 | 0,049 | 12,3 | 0,23 |
| Lait fermenté ou spécialité laitière type yaourt, sur lit de fruits, sucré, au bifidus | - | - | - | - | - | 3,81 | 3,73 | 13,3 | 2,72 | 0,78 | 0,15 | 129 | - | - | - | - | - | - | - | - | - | 58,3 | - | - | - | - | - | - | - | - | - | - | - | - | - |
| Lait gélifié aromatisé, allégé en matière grasse et en sucre, rayon frais | - | - | - | - | - | 4,17 | 4,08 | 11,8 | 1,08 | 2,22 | 0,19 | 141 | - | - | - | - | - | - | - | - | - | 77,7 | - | - | - | - | - | - | - | - | - | - | - | - | - |
| Lait gélifié aromatisé, nappé caramel, rayon frais | 388 | 91,9 | 389 | 92,1 | 77,2 | 2,55 | 2,5 | 16,7 | 1,3 | <3 | 0,1 | 78,5 | - | 0,011 | - | 7,9 | 8,13 | 0,0016 | 74 | 148 | < 4,5 | 41 | 0,31 | - | - | - | - | - | - | - | - | - | - | - | - |
| Lait gélifié aromatisé, rayon frais | - | - | - | - | 75,2 | 3,39 | 3,32 | 19,5 | 3,53 | 0,52 | 0,14 | 120 | - | 0,11 | 0,61 | 7,8 | 14,5 | 0,1 | 81 | 181 | <5 | 55,4 | 0,55 | < 0,2 | 0,043 | - | - | < 0,5 | 0,03 | 0,17 | 0,19 | 0,33 | 0,037 | 21,5 | 0,045 |
| Lait, teneur en matière grasse inconnue, UHT (aliment moyen) | 199 | 47,3 | 200 | 47,6 | 89,4 | 3,38 | 3,32 | 4,82 | 1,99 | 0 | 0,092 | 117 | 99,7 | 0,0082 | 0,041 | 11,9 | 11,8 | 0,0033 | 89,4 | 166 | 6,47 | 36,6 | 0,39 | 0,27 | 0,11 | 0,23 | | 1,31 | 0,056 | 0,24 | 0,087 | 0,41 | 0,035 | 8,52 | 0,36 |
| Langres | 1170 | 281 | 1170 | 283 | 56,2 | 16,7 | 16,4 | traces | 23,8 | traces | 0,99 | 160 | 698 | 0,03 | 0,05 | 30 | 12 | 0,01 | 220 | 140 | < 20 | 397 | 1,2 | < 0,25 | 0,65 | 1,37 | - | < 0,5 | < 0,015 | 0,84 | 0,1 | 2,18 | 0,6 | 67,6 | 1,78 |
| Liégeois aux fruits, préemballé | 612 | 147 | 612 | 147 | 72,3 | 1,5 | 1,5 | 16 | 8,1 | 1,2 | 0,045 | 36 | 32,5 | 0,04 | 0,35 | < 20 | 7,8 | 0,12 | 32 | 120 | < 20 | 17 | 0,17 | < 0,25 | 0,41 | 2,22 | - | 6,51 | 0,096 | 0,049 | < 0,1 | 0,24 | 0,023 | 20,6 | 0,15 |
| Liégeois ou viennois (chocolat, café, caramel ou vanille), rayon frais | - | - | - | - | 68,5 | 3 | 2,94 | 18,8 | 5,78 | 0,27 | 0,13 | 100 | 100 | 0,14 | 1,4 | < 20 | 27 | 0,15 | 100 | 280 | < 20 | 50,9 | 0,48 | 0,16 | 0,21 | 1,17 | - | < 0,5 | 0,039 | 0,13 | < 0,1 | 0,17 | 0,049 | 12,2 | 0,14 |
| Livarot | 1250 | 301 | 1260 | 303 | 48,9 | 24,6 | 24,1 | traces | 22,6 | 0 | 1,5 | 730 | 958 | 0,03 | 0,07 | 30 | 26 | 0,02 | 470 | 94 | < 20 | 601 | 3,1 | < 0,25 | 0,57 | 1,02 | - | < 0,5 | < 0,015 | 0,55 | 0,1 | 1,16 | 0,12 | 140 | 1,68 |
| Maroilles fermier | 1450 | 349 | 1450 | 351 | 45,6 | 21,2 | 20,8 | traces | 29,5 | 0 | 2,07 | - | - | - | - | - | - | - | - | - | - | 821 | - | - | - | - | - | - | - | - | - | - | - | - | - |

| | | | | | | | | | | | | | | | | | | | | | | | | | | | | | | | | | | | |
|---|---|---|---|---|---|---|---|---|---|---|---|---|---|---|---|---|---|---|---|---|---|---|---|---|---|---|---|---|---|---|---|---|---|---|---|
| Maroilles laitier | 1400 | 338 | 1410 | 339 | 45,9 | 22,3 | 21,8 | traces | 27,8 | 0 | 1,87 | - | - | - | - | - | - | - | - | - | - | 766 | - | - | - | - | - | - | - | - | - | - | - | - | - |
| Maroilles, sans précision | 1350 | 325 | 1360 | 327 | 48,3 | 21,9 | 21,5 | traces | 26,4 | 0 | 2,09 | 580 | 1390 | 0,03 | 0,09 | <20 | 23 | 0,02 | 410 | 100 | <20 | 834 | 2,8 | <0,25 | 0,57 | 0,96 | - | <0,5 | <0,015 | 0,34 | 0,53 | 0,99 | 0,026 | 144 | 2,57 |
| Mascarpone | - | - | - | - | 51,5 | 4,38 | 4,29 | 4 | 39 | 0 | 0,081 | 130 | 79,8 | <0,01 | 0,06 | 21 | 13 | <0,01 | 110 | 120 | <50 | 32,2 | 0,53 | <0,5 | 1 | 1,17 | - | 0 | 0,028 | 0,15 | 0,13 | 0,31 | 0,028 | <5 | 0,48 |
| Milk-shake, provenant de fast food | 530 | 126 | 531 | 126 | 71,9 | 3,38 | 3,32 | 17,4 | 4,66 | 0,65 | 0,21 | 114 | 91 | 0,063 | 0,29 | - | 13 | 0,025 | 99 | 174 | 1,7 | 82 | 0,47 | 0 | 0,25 | 0,4 | - | 0,8 | 0,036 | 0,43 | 0,19 | 0,53 | 0,052 | 3 | 0,27 |
| Mimolette demi-vieille | 1520 | 365 | 1530 | 368 | 34,3 | 31,5 | 30,8 | traces | 26,9 | 0 | - | - | - | - | - | - | - | - | - | - | - | - | - | - | - | - | - | - | - | - | - | - | - | - | - |
| Mimolette extra-vieille | 1610 | 387 | 1620 | 389 | 31 | 33,9 | 33,2 | traces | 28,2 | 0 | - | - | - | - | - | - | - | - | - | - | - | - | - | - | - | - | - | - | - | - | - | - | - | - | - |
| Mimolette jeune | 1380 | 331 | 1390 | 333 | 40,1 | 28,9 | 28,3 | traces | 24,2 | 0 | - | - | - | - | - | - | - | - | - | - | - | - | - | - | - | - | - | - | - | - | - | - | - | - | - |
| Mimolette vieille | 1650 | 397 | 1660 | 400 | 30 | 34 | 33,3 | traces | 28,8 | 0 | 3,68 | 910 | 1810 | 0,05 | 0,15 | <20 | 36 | 0,03 | 660 | 88 | <20 | 1470 | 4,5 | 0,33 | 0,6 | 1,3 | - | <0,5 | <0,015 | 0,3 | 1,15 | 0,38 | 0,066 | 33 | 2,06 |
| Mimolette, sans précision | 1280 | 308 | 1290 | 310 | 45,5 | 24,9 | 24,4 | traces | 23,4 | 0 | 2,84 | 816 | 1730 | 0,07 | <1 | <10 | 37,1 | 0,03 | 514 | 103 | 4,9 | 1140 | 3,1 | <0,2 | 0,3 | - | - | 0 | 0,03 | 0,33 | 0,1 | 0,32 | <0,05 | 18,4 | 0,57 |
| Mont d'or ou Vacherin du Haut-Doubs (produit en France) ou Vacherin-Mont d'Or (produit en Suisse) | 1250 | 302 | 1260 | 304 | 52,9 | 18,9 | 18,5 | traces | 25,4 | 0 | 1,38 | 491 | - | 0,07 | - | 30 | 30 | 0,02 | 430 | 120 | 5,8 | 523 | 8 | - | - | - | - | - | - | - | - | - | - | - | - |
| Morbier | 1470 | 355 | 1480 | 356 | 43,6 | 22,8 | 22,4 | traces | 29,2 | 0 | 1,41 | 600 | 903 | 0,14 | 0,06 | <20 | 23 | 0,02 | 430 | 83 | <20 | 562 | 2,8 | <0,25 | 0,55 | 2,25 | - | <0,5 | <0,015 | 0,21 | 0,88 | 0,48 | 0,036 | 29 | 1,79 |
| Mousse à la crème de marrons, préemballée | 872 | 207 | 872 | 207 | 55,4 | 2,17 | 2,13 | 34,4 | 6,5 | 1,1 | 0,04 | 36 | 25,9 | 0,06 | 0,19 | <20 | 9,3 | 0,48 | 33 | 130 | <20 | 16 | 0,22 | <0,25 | 0,21 | <0,8 | - | <0,5 | 0,019 | 0,031 | 0,21 | 0,15 | 0,02 | 24,4 | 0,029 |
| Mousse au chocolat (base laitière), rayon frais | 631 | 150 | 633 | 150 | 65,8 | 5,42 | 5,31 | 21,3 | 4,4 | <3 | 0,16 | 149 | 86 | 0,29 | 4,64 | 13,5 | 59,5 | 0,3 | 122 | 371 | <10 | 62 | 0,59 | <0,2 | 0,17 | - | - | <0,5 | 0,03 | 0,17 | 0,28 | 0,39 | 0,035 | 22 | 0,07 |
| Mousse au chocolat traditionnelle, préemballée | 1260 | 303 | 1270 | 304 | - | 6,52 | 6,39 | 25,6 | 18,6 | 3,8 | 0,25 | - | 86 | - | - | - | - | - | - | - | - | 100 | - | - | - | - | - | - | - | - | - | - | - | - | - |
| Mousse au chocolat végétale, préemballée | 1090 | 262 | 1090 | 262 | 48,1 | 6,69 | 6,69 | 24,3 | 14 | 5,1 | 0,36 | 34 | 81,5 | 0,64 | 6 | <20 | 81 | 0,71 | 150 | 370 | <20 | 143 | 1,2 | 0,82 | 0,36 | 2,72 | - | <0,5 | 0,049 | <0,01 | 5,23 | 0,27 | 0,016 | 15,6 | - |
| Mousse aux fruits, rayon frais | 225 | 53,1 | 227 | 53,6 | 85,4 | 6,25 | 6,13 | 5,51 | <0,3 | 1,2 | 0,09 | 140 | 81,8 | 0,01 | 0,08 | <20 | 9,1 | 0,05 | 98 | 130 | <20 | 36 | 0,33 | 1,16 | 0,095 | <0,8 | - | <0,5 | <0,015 | 0,14 | <0,1 | 0,37 | 0,02 | 7,39 | 0,4 |
| Mousse de fromage blanc sur lit de fruits, 0% MG, avec édulcorants, enrichie en calcium et vitamine D | - | - | - | - | - | 6,48 | 6,35 | 6 | 0,45 | 1,35 | 0,15 | 120 | - | - | - | - | - | - | - | - | - | 60 | - | 0,75 | - | - | - | - | - | - | - | - | - | - | - |
| Mousse liégeoise (chocolat, café, caramel ou vanille), rayon frais | 779 | 186 | 780 | 187 | - | 3,62 | 3,55 | 20 | 10,1 | 0,75 | 0,14 | - | - | - | - | - | - | - | - | - | - | 57,6 | - | - | - | - | - | - | - | - | - | - | - | - | - |
| Mozzarella au lait de bufflonne ou buflesse ("di bufala") | 1080 | 260 | 1080 | 261 | 61,3 | 14,5 | 14,3 | 0,24 | 22,3 | 0 | 0,49 | 260 | 354 | 0,03 | 0,11 | 30 | 9,8 | 0,01 | 210 | 17 | <20 | 197 | 2,1 | <0,25 | 0,28 | 2,35 | - | 0 | <0,015 | 0,096 | <0,1 | 0,046 | 0,011 | 23,6 | 0,45 |
| Mozzarella au lait de vache | 943 | 227 | 949 | 229 | 62,7 | 16,5 | 16,1 | 0,75 | 17,7 | 0 | 0,6 | 545 | 990 | 0,011 | 0,44 | 37,2 | 20 | 0,03 | 354 | 76 | 16,3 | 240 | 2,92 | 0,4 | 0,19 | 2,3 | - | 0 | 0,03 | 0,28 | 0,1 | 0,14 | 0,037 | 7 | 2,28 |
| Munster | 1420 | 344 | 1430 | 345 | 49,4 | 21,2 | 20,7 | traces | 29 | 0 | 1,78 | 717 | - | 0,031 | 0,41 | 30,4 | 27 | 0,008 | 468 | 134 | 4,32 | 670 | 2,81 | 0,6 | 0,26 | 2,5 | - | 0 | 0,013 | 0,32 | 0,1 | 0,19 | 0,056 | 12 | 1,47 |
| Neufchâtel | 1190 | 287 | 1200 | 289 | 54,8 | 18,5 | 18,1 | traces | 23,9 | 0 | 1,87 | 74,5 | 1190 | 0,02 | 0,07 | 32 | 9,17 | 0,01 | 198 | 117 | <50 | 748 | 0,55 | <0,3 | 0,4 | 1,95 | - | - | 0,084 | 0,5 | 1,36 | 0,18 | 0,32 | 94,5 | 0,65 |
| Ossau-Iraty | 1770 | 428 | 1780 | 430 | 33,8 | 24,2 | 23,8 | traces | 37 | 0 | 1,68 | 764 | - | 0,057 | 0,32 | - | 35,6 | - | - | 45,7 | - | 673 | 2,43 | - | 0,15 | - | - | - | 0,036 | 0,5 | - | 0,16 | 0,028 | 35,3 | 1,06 |
| Pain perdu | 916 | 218 | 916 | 218 | 53,7 | 6 | 6 | 29 | 8,45 | 1,1 | 1,22 | 104 | - | 0,071 | 1,94 | 10,2 | 17 | 0,22 | 128 | 134 | 2,39 | 487 | 0,72 | 0,28 | 0,39 | - | - | 0,3 | 0,24 | 0,35 | 2,18 | 0,74 | 0,29 | 23,5 | 1 |
| Panna cotta, avec préparations de fruits ou caramel, rayon frais | 826 | 198 | 827 | 199 | 66,1 | 2,74 | 2,69 | 14,9 | 13 | <3 | 0,085 | 75,8 | - | - | - | - | - | - | - | - | - | 34 | - | - | - | - | - | - | - | - | - | - | - | - | - |
| Parmesan | 1690 | 406 | 1700 | 409 | 30,4 | 31,1 | 30,5 | traces | 31 | 0 | 1,57 | 980 | 973 | 0,71 | 0,1 | 90 | 40 | 0,02 | 630 | 110 | 20 | 628 | 3,9 | <0,25 | 0,17 | 1,53 | - | <0,5 | <0,015 | 0,26 | 1,23 | 0,34 | 0,18 | 16 | 2,64 |
| Pecorino | - | - | - | - | 32,2 | - | - | traces | 28,8 | 0 | 4,73 | 1160 | - | - | 1,4 | - | - | - | 675 | 94 | - | 1890 | 3,5 | 0,5 | 0,73 | - | - | 0 | 0,03 | 0,47 | 0,2 | - | 0,09 | 7 | - |
| Pélardon (fromage de chèvre) | 1420 | 343 | 1430 | 345 | 43,6 | 23,8 | 23,3 | traces | 27,8 | 0 | 1,43 | 147 | 909 | <0,1 | 0,37 | 30 | 20,5 | - | 242 | - | 9,5 | 572 | 0,57 | <0,5 | 0,58 | - | - | - | <0,04 | 1,01 | 1,28 | 1,2 | 0,28 | 95,5 | - |
| Picodon (fromage de chèvre) | 1310 | 315 | 1320 | 317 | 49,4 | 20,7 | 20,3 | traces | 26 | 0 | 1,25 | 126 | 1230 | <0,1 | 0,28 | 20 | 16,9 | - | 203 | - | 13 | 500 | 0,67 | <0,5 | 0,44 | - | - | - | <0,04 | 0,78 | 1,24 | 1,23 | 0,14 | 80 | - |
| Pont l'Évêque | 1350 | 326 | 1360 | 328 | 47,9 | 22,7 | 22,2 | traces | 26,2 | 0 | 1,3 | 660 | 886 | 0,03 | 0,07 | 30 | 24 | 0,02 | 430 | 110 | <20 | 520 | 2,8 | <0,25 | 0,46 | 1,82 | - | <0,5 | <0,015 | 0,49 | <0,1 | 0,88 | 0,13 | 88,9 | 1,74 |
| Pouligny Saint-Pierre (fromage de chèvre) | 1220 | 294 | 1230 | 296 | 52,7 | 18,9 | 18,5 | traces | 24,5 | 0 | 1,08 | 100 | 702 | <0,1 | 0,33 | 30 | 14,8 | - | 189 | - | 14 | 452 | 0,53 | <0,5 | 0,29 | - | - | - | <0,04 | 0,86 | 1,8 | 1,24 | 0,17 | 108 | - |
| Profiteroles (crème pâtissière et sauce chocolat), préemballées | 1150 | 273 | 1150 | 273 | 44,6 | 5,51 | 5,51 | 34,3 | 12,2 | 2,31 | 0,3 | - | - | - | - | - | - | - | - | - | - | 120 | - | - | - | - | - | - | - | - | - | - | - | - | - |
| Provolone | 1410 | 340 | 1420 | 342 | 41 | 25,6 | 25,1 | traces | 26,6 | 0 | 2,19 | 756 | - | 0,026 | 0,52 | - | 28 | 0,01 | 496 | 138 | - | 876 | 3,23 | 0,5 | 0,23 | 2,2 | - | 0 | 0,019 | 0,32 | 0,16 | 0,48 | 0,073 | 10 | 1,46 |
| Raclette (fromage) | 1420 | 342 | 1430 | 345 | 44,7 | 23,4 | 22,9 | traces | 27,5 | 0 | 1,69 | 630 | 668 | 0,06 | - | 20 | 24,5 | <0,1 | 460 | 75,9 | 8 | 676 | 3 | - | 0,51 | - | - | - | - | - | - | - | - | 51 | - |
| Reblochon | 1350 | 326 | 1360 | 328 | 49,1 | 20,4 | 19,9 | traces | 27,4 | 0 | 1,27 | 471 | - | 0,11 | 0,32 | 19,4 | 20,2 | 0 | 322 | 104 | 5,1 | 509 | 4,44 | - | - | - | - | - | - | - | - | - | - | 25 | - |
| Ricotta | - | - | - | - | 71,7 | 8,79 | 8,61 | 4 | 11,9 | 0 | 0,3 | 314 | - | 0,021 | 0,38 | 18 | 11 | 0,006 | 158 | 105 | - | 109 | 1,16 | 0,2 | 0,11 | 1,1 | - | 0 | 0,013 | 0,2 | 0,1 | 0,21 | 0,043 | 12 | 0,34 |
| Riz au lait, rayon frais | 532 | 126 | 533 | 126 | 71,5 | 3,34 | 3,27 | 21,4 | 3,01 | 0,28 | 0,13 | 61,6 | 120 | <0,1 | 0,2 | 8,83 | 10,9 | <0,1 | 75 | 96 | <5 | 51,3 | 0,4 | <0,5 | 0,095 | - | - | <0,5 | <0,04 | 0,14 | 0,21 | 0,32 | 0,043 | 22 | <0,08 |
| Rocamadour (fromage de chèvre) | 1140 | 274 | 1140 | 275 | 57,2 | 17,9 | 17,5 | traces | 22,6 | 0 | 1,09 | 102 | 651 | 0,1 | 0,2 | 30 | 13,8 | - | 197 | 160 | <8 | 373 | 0,37 | <0,5 | 0,39 | - | - | - | <0,04 | 0,79 | 1,8 | 1,4 | 0,31 | 80 | - |
| Roquefort | 1590 | 384 | 1600 | 386 | 41,9 | 19,5 | 19,1 | traces | 33,9 | 0 | 3,22 | 660 | 2600 | 0,024 | 0,1 | 51,2 | 31,9 | 0,019 | 392 | 115 | <10 | 1290 | 1,83 | - | 0,12 | - | - | 0 | 0,044 | 0,67 | 0,73 | 0,62 | 0,081 | 29,1 | 0,57 |
| Sainte-Maure (fromage de chèvre) | 1480 | 357 | 1490 | 359 | 42,9 | 22,1 | 21,7 | traces | 30 | 0 | 1,81 | 122 | 1100 | <0,1 | 0,3 | 40 | 14,5 | - | 233 | 168 | 10,5 | 543 | 0,62 | <0,5 | 0,41 | - | - | - | <0,04 | 0,9 | 1,25 | 1,04 | 0,19 | 80,2 | 0,1 |
| Sainte-Maure de Touraine (fromage de chèvre) | 1220 | 294 | 1230 | 296 | 53,4 | 19,6 | 19,2 | traces | 24,2 | 0 | 1,41 | - | - | - | - | - | - | - | - | - | - | 526 | - | - | - | - | - | - | - | - | - | - | - | - | - |
| Saint-Félicien | 1160 | 281 | 1170 | 283 | 58,6 | 13,4 | 13,1 | traces | 25,2 | 0 | 1,28 | 110 | 758 | 0,02 | 0,04 | 50 | 9,6 | <0,01 | 160 | 130 | <20 | 511 | 0,44 | <0,25 | 0,65 | 2,41 | - | <0,5 | 0,053 | 0,61 | 1,57 | 2,11 | 0,31 | 57,7 | 1,23 |
| Saint-Marcellin | 1160 | 280 | 1160 | 281 | 58,3 | 15,1 | 14,8 | traces | 24,3 | 0 | 1,08 | 140 | 783 | 0,04 | 0,05 | <20 | 12 | 0,01 | 210 | 180 | <20 | 433 | 0,54 | <0,25 | 0,61 | 2,15 | - | <0,5 | <0,015 | 0,91 | 1,53 | 2,49 | 0,53 | 93,4 | 1,83 |
| Saint-Nectaire, fermier | 1470 | 355 | 1480 | 357 | - | 22,3 | 21,8 | traces | 29,8 | 0 | 1,08 | - | - | - | - | - | - | - | - | - | - | - | - | - | - | - | - | - | - | - | - | - | - | - | - |
| Saint-Nectaire, laitier | 1380 | 332 | 1390 | 334 | - | 22,8 | 22,3 | traces | 27 | 0 | 1,7 | - | - | - | - | - | - | - | - | - | - | - | - | - | - | - | - | - | - | - | - | - | - | - | - |
| Saint-Nectaire, sans précision | 1430 | 344 | 1430 | 346 | 44,6 | 22,5 | 22 | traces | 28,4 | 0 | 1,41 | 540 | - | 0,067 | 0,22 | 31,4 | 24 | 0,035 | 227 | 83,8 | 5,11 | 620 | 2,83 | - | - | - | - | - | - | 0,42 | - | - | - | - | 1,03 |
| Saint-Paulin (fromage à pâte pressée non cuite demi-ferme) | 1360 | 327 | 1370 | 329 | 47,5 | 22,3 | 21,8 | traces | 26,4 | 0 | 1,61 | 600 | 1040 | 0,03 | 0,11 | <20 | 25 | 0,02 | 400 | 71 | <20 | 642 | 2,7 | <0,25 | 0,67 | 3,1 | - | - | <0,015 | 0,17 | <0,1 | 0,4 | 0,02 | 32,5 | 0,46 |
| Salers | 1610 | 388 | 1620 | 390 | 38,6 | 25,9 | 25,4 | traces | 31,4 | 0 | 2,19 | 680 | 1470 | 0,03 | 0,11 | <20 | 25 | 0,02 | 480 | 96 | <20 | 875 | 2,9 | 0,31 | 0,72 | 2,03 | - | <0,5 | 0,063 | 0,37 | <0,1 | 0,31 | 0,079 | 37,5 | 1,86 |
| Selles-sur-Cher (fromage de chèvre) | 1210 | 291 | 1210 | 292 | 53,1 | 19,2 | 18,9 | traces | 23,9 | 0 | 1,4 | 167 | - | - | - | 56 | - | - | - | - | - | 562 | - | - | 0,6 | - | - | - | - | - | - | - | - | - | 0,1 |
| Semoule au lait, rayon frais | 511 | 121 | 512 | 122 | 73,4 | 3,46 | 3,39 | 19,4 | 3,23 | 0,55 | 0,095 | 83,7 | 89 | 0,036 | <1 | 8,4 | 11 | 0,038 | 177 | 135 | <22 | 37,8 | 0,39 | 0,7 | <0,1 | - | - | 0,5 | <0,05 | 0,14 | 0,13 | 0,39 | <0,05 | <5 | 0,2 |
| Snack pour enfants à base de fromage fondu et de gressins | - | - | - | - | 42,6 | 10,8 | 10,6 | 24,1 | 18,9 | 1,2 | 1,36 | 145 | - | 0,4 | 0,63 | 14,8 | 21 | - | 383 | 131 | 5,45 | 543 | 3,79 | 0,27 | 0,99 | - | - | 0 | 0,15 | 0,3 | 0,49 | 0,36 | 0,07 | 11,5 | 0,63 |
| Spécialité à base de crème légère 8% MG, fluide ou épaisse | 318 | 76,2 | 319 | 76,4 | 84,8 | 3,25 | 3,19 | 5,28 | 4,3 | <3 | 0,11 | 105 | - | 0,007 | 0,04 | 10,6 | 10 | 0,007 | 80 | 140 | 1,3 | 43 | 0,34 | 0,14 | 0,23 | 0 | - | 0,9 | 0,04 | 0,17 | 0,08 | 0,3 | 0,04 | 13 | 0,25 |
| Spécialité fromagère fondante au fromage blanc et aux noix | - | - | - | - | 50,8 | 10,7 | 10,4 | 4,2 | 31,6 | <2 | 1,66 | 260 | 467 | 0,17 | 0,45 | <20 | 31 | 0,02 | 590 | 160 | <50 | 665 | 1,3 | <0,5 | 0,7 | 4,32 | - | - | 0,056 | 0,11 | 0,29 | 0,18 | 0,054 | 17,9 | 0,3 |
| Spécialité fromagère non affinée à tartiner environ 30-40 % MG aromatisée (ex: ail et fines herbes) | 1510 | 366 | 1510 | 367 | 50,8 | 7,66 | 7,5 | 4,24 | 35,3 | 0 | 1,2 | 93,8 | - | <0,1 | 0,21 | 15,5 | 7,7 | <0,1 | 130 | 111 | 4,1 | 479 | 1,05 | 0,4 | 0,66 | - | - | <1 | - | 0,17 | 0,56 | - | 0,05 | 28 | 0,24 |
| Spécialité fromagère non affinée environ 20% MG, type fromage en barquette à tartiner ou coque fromagère | - | - | - | - | 67,5 | 7,65 | 7,49 | 1,4 | 21,2 | 0 | 1,24 | 104 | 728 | - | - | 20 | 14 | - | 111 | 119 | 7 | 496 | 2 | 0,2 | 0,5 | - | - | - | 0,04 | 0,22 | 0,15 | 0,47 | 0,08 | 3,4 | 0,39 |

| | | | | | | | | | | | | | | | | | | | | | | | | | | | | | | | | | | | |
|---|---|---|---|---|---|---|---|---|---|---|---|---|---|---|---|---|---|---|---|---|---|---|---|---|---|---|---|---|---|---|---|---|---|---|---|
| Spécialité fromagère non affinée environ 25% MG, type fromage en barquette à tartiner ou coque fromagère | 1060 | 255 | 1060 | 256 | 63,4 | 5,36 | 5,25 | 4,24 | 23,9 | 0 | 1,1 | 772 | - | 0,027 | 0,61 | 20 | 29 | 0,014 | 205 | 216 | 6,7 | 440 | 3,61 | 0,48 | 0,34 | 2,2 | - | 0 | 0,014 | 0,28 | 0,038 | 0,26 | 0,036 | 6 | 1,23 |
| Spécialité laitière type encas, riche en protéines, sur lit de fruits, sucrée | 415 | 98,3 | 418 | 98,9 | 76,3 | 7,46 | 7,31 | 11,5 | 2,29 | 0 | 0,083 | 110 | 85,5 | 0,02 | 0,06 | < 20 | 9,5 | 0,04 | 120 | 120 | < 20 | 33 | 0,39 | 2,43 | 0,16 | < 0,8 | - | < 0,5 | < 0,015 | 0,13 | < 0,1 | 0,37 | 0,03 | 36,7 | 0,58 |
| Spécialité végétale type fromage à tartiner, au soja, préemballée | 726 | 175 | 707 | 170 | 71,9 | 11,7 | 12,8 | 1,05 | 13,1 | < 0,5 | 1,15 | 32 | 734 | 0,28 | 1,3 | < 20 | 32 | 0,75 | 120 | 170 | < 20 | 459 | 1 | < 0,25 | 1,6 | 23,2 | - | < 0,5 | 0,052 | 0,047 | 0,16 | 0,2 | 0,056 | 74,3 | - |
| Spécialité végétale type fromage en tranche ou râpé, sans soja, préemballée | 1130 | 272 | 1130 | 272 | 54,1 | < 0,5 | < 0,5 | 21,2 | 20,1 | 2,3 | 1,88 | 46 | 1070 | 0,02 | 0,15 | < 20 | 5,2 | 0,03 | 16 | 34 | < 20 | 751 | 0,1 | < 0,25 | 0,26 | < 0,8 | - | < 0,5 | < 0,015 | < 0,01 | < 0,1 | 0,059 | < 0,01 | 5,63 | - |
| Spécialité végétale type fromage, à la noix de cajou, préemballée | 1830 | 441 | 1830 | 441 | 29,5 | 14,5 | 14,5 | 15,2 | 35,3 | 2,6 | 0,88 | 51 | 568 | 1,3 | 3,9 | < 20 | 120 | 1,1 | 260 | 290 | < 20 | 351 | 3,4 | < 0,25 | 3,79 | 18,3 | - | < 0,5 | 0,3 | 0,07 | 0,52 | 0,42 | 0,27 | 34,3 | - |
| Tiramisu, préemballé | 1010 | 241 | 1010 | 241 | 51,6 | 4,25 | 4,25 | 30,4 | 9,8 | 1,2 | 0,19 | 94 | 98,1 | 0,06 | 0,68 | < 20 | 18 | 0,1 | 110 | 230 | < 20 | 77 | 0,46 | < 0,25 | 0,86 | - | - | < 0,5 | 0,022 | 0,095 | 0,24 | 0,7 | 0,32 | 5,52 | 0,078 |
| Tome des Bauges | 1580 | 381 | 1590 | 383 | 39,5 | 26,6 | 26,1 | traces | 30,8 | 0 | 1,15 | - | - | - | - | - | - | - | - | - | - | 459 | - | - | - | - | - | - | - | - | - | - | - | - | - |
| Tomme ou tome de montagne ou de Savoie | 1510 | 364 | 1520 | 366 | 41,6 | 25,6 | 25 | traces | 29,1 | 0 | 1,23 | 600 | 910 | 0,31 | 0,07 | < 20 | 20 | 0,02 | 510 | 140 | < 20 | 493 | 2,9 | < 0,25 | 0,85 | 1,48 | - | < 0,5 | < 0,015 | 0,26 | 3,23 | 0,32 | 0,049 | 47,6 | 2,03 |
| Tomme ou tome de vache | 1450 | 350 | 1460 | 352 | 42,3 | 21,6 | 21,1 | traces | 29,4 | 0 | 1,28 | 953 | - | 0,11 | - | 20 | 33,8 | 0,022 | 478 | 130 | 3,73 | 510 | 4,36 | - | - | - | - | - | - | - | - | - | - | - | - |
| Tomme ou tome, allégée en matière grasse, environ 13% MG | 965 | 231 | 975 | 233 | 51 | 31 | 30,4 | traces | 12,1 | 0 | 1,84 | 852 | - | - | - | - | - | - | - | - | - | 737 | - | - | - | - | - | - | - | - | - | - | - | - | - |
| Valençay (fromage de chèvre) | 1270 | 306 | 1270 | 307 | 51,8 | 19,3 | 18,9 | traces | 25,5 | 0 | 1,32 | 120 | 1100 | < 0,1 | 0,19 | 40 | 15,3 | - | 181 | 150 | 9 | 509 | 0,55 | < 0,5 | 0,32 | - | - | - | < 0,04 | 0,64 | 1,12 | 0,95 | 0,11 | 84,8 | - |
| Yaourt à la grecque, aromatisé, sucré | - | - | - | - | - | 4,29 | 4,2 | 12,8 | 8,2 | 0 | 0,18 | 158 | - | - | - | - | - | - | - | - | - | 70 | - | - | - | - | - | - | - | - | - | - | - | - | - |
| Yaourt à la grecque, au lait de brebis | 290 | 69,1 | 291 | 69,4 | 85,7 | 3,84 | 3,76 | 4,99 | 3,4 | 0 | 0,0079 | 170 | 84,8 | < 0,01 | 0,04 | 40 | 15 | < 0,01 | 130 | 110 | < 20 | 3,15 | 0,43 | < 0,25 | < 0,08 | < 0,8 | - | < 0,5 | 0,049 | 0,22 | 0,31 | 0,29 | 0,013 | 31,9 | 0,53 |
| Yaourt à la grecque, nature | - | - | - | - | 80,4 | 3,32 | 3,25 | 4,21 | 9,22 | < 0,14 | 0,097 | 139 | - | - | - | - | - | - | - | - | - | 34,5 | - | - | - | - | - | - | - | - | - | - | - | - | - |
| Yaourt à la grecque, sur lit de fruits | - | - | - | - | 75,2 | 2,31 | 2,27 | 16,3 | 6,3 | 0,38 | 0,098 | 104 | 69,8 | 0,02 | 0,07 | < 20 | 8,4 | 0,12 | 71 | 110 | < 20 | 38,7 | 0,27 | - | 0,25 | 1,17 | - | < 0,5 | 0,028 | 0,1 | < 0,1 | 0,24 | 0,031 | 5,14 | 0,22 |
| Yaourt au lait de brebis, aromatisé, sucré | 394 | 93,7 | 396 | 94,1 | 80 | 5,04 | 4,94 | 9,33 | 3,7 | 0 | 0,12 | 180 | 91,6 | < 0,01 | 0,06 | 70 | 16 | < 0,01 | 120 | 110 | < 20 | 47 | 0,43 | < 0,25 | < 0,08 | 1,42 | - | < 0,5 | 0,05 | 0,23 | < 0,1 | 0,35 | 0,02 | 17,6 | 0,27 |
| Yaourt au lait de brebis, nature, 3% MG environ | - | - | - | - | - | 5,72 | 5,6 | 4,6 | 3 | 0,13 | 0,12 | - | - | - | - | - | - | - | - | - | - | 47,5 | - | - | - | - | - | - | - | - | - | - | - | - | - |
| Yaourt au lait de brebis, nature, 6% MG environ | - | - | - | - | - | 5,14 | 5,04 | 4,56 | 6,08 | < 1 | 0,1 | 183 | - | - | - | - | - | - | - | - | - | 40 | - | - | - | - | - | - | - | - | - | - | - | - | - |
| Yaourt au lait de chèvre, nature, 5% MG environ | 305 | 73,4 | 307 | 73,8 | 86,2 | 4,15 | 4,06 | 1,12 | 5,2 | < 3 | 0,09 | 160 | 130 | 0,01 | 0,2 | - | 14 | traces | 110 | 170 | - | 36 | 0,4 | - | 0,03 | - | - | 1 | 0,04 | 0,17 | 0,27 | 0,23 | 0,06 | 7 | traces |
| Yaourt ou spécialité laitière aux fruits (aliment moyen) | 392 | 93,1 | 393 | 93,3 | 79,1 | 3,2 | 3,14 | 12,8 | 2,89 | 0,43 | 0,11 | 109 | 102 | 0,014 | 0,092 | 9,69 | 10,3 | 0,062 | 78,3 | 149 | 5,25 | 43,6 | 0,34 | 0,16 | 0,12 | 0,66 | | 0,31 | 0,032 | 0,23 | 0,13 | 0,42 | 0,056 | 20,1 | 0,16 |
| Yaourt ou spécialité laitière nature (aliment moyen) | 238 | 56,8 | 240 | 57,1 | 86,5 | 3,88 | 3,8 | 4,73 | 2,3 | 0,86 | 0,11 | 127 | 157 | 0,02 | 0,2 | 20,8 | 12,1 | 0,018 | 96,4 | 167 | 2,17 | 44,8 | 0,4 | 0,61 | 0,095 | 0,21 | 0,67 | 0,26 | 0,033 | 0,23 | 0,16 | 0,31 | 0,048 | 20,7 | 0,28 |
| Yaourt ou spécialité laitière nature ou aux fruits (aliment moyen) | 292 | 69,4 | 293 | 69,7 | 84 | 3,65 | 3,57 | 7,47 | 2,5 | 0,71 | 0,11 | 121 | 138 | 0,018 | 0,16 | 16,9 | 11,5 | 0,033 | 90,2 | 161 | 3,24 | 44,4 | 0,38 | 0,48 | 0,1 | 0,37 | | 0,28 | 0,033 | 0,23 | 0,15 | 0,35 | 0,051 | 20,5 | 0,24 |
| Yaourt, lait fermenté ou spécialité laitière sur lit de fruits, sucré | - | - | - | - | 79,8 | 2,95 | 2,89 | 15,2 | 3,91 | 0,42 | 0,099 | 88,5 | - | - | - | - | - | - | - | - | - | 39,2 | - | - | - | - | - | - | - | - | - | - | - | - | - |
| Yaourt, lait fermenté ou spécialité laitière, aromatisé ou aux fruits (aliment moyen) | 376 | 89,3 | 377 | 89,5 | 79,5 | 3,32 | 3,25 | 12,4 | 2,63 | 0,33 | 0,12 | 116 | 83,4 | 0,027 | 0,13 | 9,14 | 10,6 | 0,047 | 83,2 | 145 | 4,01 | 45 | 0,38 | 0,54 | 0,078 | 0,32 | | 0,33 | 0,089 | 0,21 | 0,12 | 0,61 | 0,048 | 17,5 | 0,19 |
| Yaourt, lait fermenté ou spécialité laitière, aromatisé ou aux fruits, 0% MG (aliment moyen) | 180 | 42,4 | 181 | 42,8 | 87,7 | 4,29 | 4,2 | 4,92 | 0,13 | 1,12 | 0,12 | 129 | | 0,024 | 0,19 | 12,9 | 14 | 0,078 | 97,2 | 164 | 2,84 | 49,7 | 0,44 | | 0,08 | | | 0,25 | 0,097 | 0,26 | 0,25 | 0,64 | 0,14 | 28,5 | 0,17 |
| Yaourt, lait fermenté ou spécialité laitière, aromatisé ou aux fruits, avec édulcorants (aliment moyen) | 180 | 42,4 | 181 | 42,8 | 87,7 | 4,29 | 4,2 | 4,92 | 0,13 | 1,12 | 0,12 | 129 | | 0,024 | 0,19 | 12,9 | 14 | 0,078 | 97,2 | 164 | 2,84 | 49,7 | 0,44 | | 0,08 | | | 0,25 | 0,097 | 0,26 | 0,25 | 0,64 | 0,14 | 28,5 | 0,17 |
| Yaourt, lait fermenté ou spécialité laitière, aromatisé ou aux fruits, non allégé en MG (aliment moyen) | 392 | 93,2 | 394 | 93,4 | 79 | 3,24 | 3,17 | 12,9 | 2,81 | 0,27 | 0,12 | 115 | 83 | 0,027 | 0,12 | 8,81 | 10,4 | 0,044 | 82,4 | 142 | 4,11 | 44,1 | 0,37 | 0,52 | 0,078 | 0,32 | | 0,34 | 0,088 | 0,21 | 0,11 | 0,61 | 0,042 | 16,7 | 0,19 |
| Yaourt, lait fermenté ou spécialité laitière, aromatisé ou aux fruits, sucré (aliment moyen) | 393 | 93,3 | 394 | 93,5 | 78,8 | 3,24 | 3,18 | 13 | 2,83 | 0,27 | 0,12 | 115 | 83 | 0,027 | 0,12 | 8,83 | 10,4 | 0,044 | 82,3 | 144 | 4,11 | 44,6 | 0,37 | 0,52 | 0,078 | 0,32 | | 0,34 | 0,088 | 0,21 | 0,11 | 0,61 | 0,042 | 16,7 | 0,19 |
| Yaourt, lait fermenté ou spécialité laitière, aromatisé ou aux fruits, sucré, non allégé en MG (aliment moyen) | 392 | 93,2 | 394 | 93,4 | 78,8 | 3,24 | 3,18 | 13 | 2,83 | 0,27 | 0,12 | 115 | 83 | 0,027 | 0,12 | 8,81 | 10,4 | 0,044 | 82,2 | 144 | 4,11 | 44,6 | 0,37 | 0,52 | 0,078 | 0,32 | | 0,34 | 0,088 | 0,21 | 0,11 | 0,61 | 0,042 | 16,7 | 0,19 |
| Yaourt, lait fermenté ou spécialité laitière, aromatisé, avec édulcorants, 0% MG | 232 | 54,5 | 233 | 54,8 | 85,2 | 4,53 | 4,44 | 8,29 | 0,06 | traces | 0,15 | 141 | - | < 0,1 | - | 15,1 | - | < 0,1 | - | - | < 5 | 57,5 | 0,52 | 1,25 | - | - | - | - | - | - | - | - | - | - | - |
| Yaourt, lait fermenté ou spécialité laitière, aromatisé, sucré | 375 | 89,1 | 377 | 89,3 | 79,8 | 3,42 | 3,35 | 12,2 | 2,6 | 0,17 | 0,15 | 120 | 61 | < 0,1 | 0,1 | 8,17 | 10,9 | < 0,1 | 86 | 140 | < 2,78 | 50,9 | 0,4 | 0,76 | 0,041 | 0 | - | < 0,5 | 0,17 | 0,22 | 0,13 | 0,9 | 0,05 | 20 | 0,18 |
| Yaourt, lait fermenté ou spécialité laitière, aromatisé, sucré, à la crème | - | - | - | - | 77,6 | 3,08 | 3,02 | 13,4 | 5,45 | 0,083 | 0,093 | 105 | 89 | 0,01 | 0,04 | < 20 | 9,2 | < 0,01 | 82 | 120 | < 20 | 35,5 | 0,38 | 0,75 | 0,15 | < 0,8 | - | < 0,5 | 0,1 | 0,15 | < 0,1 | 0,62 | 0,034 | 14,7 | 0,19 |
| Yaourt, lait fermenté ou spécialité laitière, aromatisé, sucré, enrichi en vitamine D | - | - | - | - | 79,5 | 3,43 | 3,36 | 13,6 | 1,79 | 0,075 | 0,11 | 156 | - | - | - | - | 12 | - | 90 | - | - | 50 | - | 1,24 | - | - | - | - | - | 0,21 | - | - | - | - | 0,38 |
| Yaourt, lait fermenté ou spécialité laitière, aux céréales | - | - | - | - | 74,1 | 3,64 | 3,57 | 11,3 | 3,43 | 1 | 0,11 | 125 | - | < 0,1 | 0,085 | < 5 | 12,6 | 0,01 | 95 | 145 | < 5 | 42,3 | 0,4 | 0,093 | 0,084 | 0 | - | 0,1 | 0,036 | 0,16 | 0,12 | 0,31 | 0,042 | 23 | 0,15 |
| Yaourt, lait fermenté ou spécialité laitière, aux céréales, 0% MG | - | - | - | - | 86,1 | 4,63 | 4,53 | 8,25 | 1,08 | 0,3 | 0,17 | 130 | - | 0,2 | 0,42 | 19 | 13,9 | 0,07 | 103 | 177 | 1,9 | 67 | 0,5 | 0 | 0,2 | - | - | 3 | 0,1 | 0,29 | 0,93 | 0,73 | 0,14 | 36,6 | 0,3 |
| Yaourt, lait fermenté ou spécialité laitière, aux copeaux de chocolat, à la crème, sucré | 510 | 122 | 511 | 122 | 74,2 | 3,51 | 3,44 | 14,1 | 5,1 | < 3 | 0,085 | 120 | - | 0,02 | 0,3 | 17 | 14 | 0,02 | 98 | 178 | 1,3 | 34 | 0,4 | 0,1 | 0,16 | - | - | 0,2 | 0,04 | 0,2 | 0,15 | 0,34 | 0,05 | 20 | 0,19 |
| Yaourt, lait fermenté ou spécialité laitière, aux fruits, avec édulcorants, 0% MG | 165 | 38,9 | 166 | 39,3 | 88,5 | 4,21 | 4,13 | 3,72 | < 0,3 | < 3 | 0,12 | 122 | - | 0,016 | 0,2 | 11,6 | 14,1 | 0,089 | 95 | 159 | < 4,5 | 46 | 0,41 | - | 0,08 | - | - | < 0,5 | 0,1 | 0,27 | 0,26 | 0,65 | 0,15 | 30,5 | 0,16 |
| Yaourt, lait fermenté ou spécialité laitière, aux fruits, avec édulcorants, 0% MG, enrichi en vitamine D | - | - | - | - | 87 | 4,54 | 4,45 | 7,45 | 0,097 | 0,66 | 0,16 | 163 | 136 | 0,02 | 0,06 | 20 | 14 | 0,04 | 120 | 220 | < 20 | 64,8 | 0,52 | 1,44 | - | < 0,8 | - | < 0,5 | 0,049 | 0,21 | 0,16 | 0,55 | 0,044 | 7,88 | 0,32 |
| Yaourt, lait fermenté ou spécialité laitière, aux fruits, sucré | 404 | 95,7 | 405 | 96 | 77,8 | 3,61 | 3,54 | 14,2 | 2,4 | 0,22 | 0,12 | 127 | 121 | 0,01 | 0,13 | 7,3 | 10,4 | < 0,1 | 78,5 | 157 | < 3,35 | 46,8 | 0,35 | < 0,2 | 0,051 | 0,1 | - | 0,43 | 0,032 | 0,33 | 0,17 | 0,55 | 0,057 | 22 | 0,09 |
| Yaourt, lait fermenté ou spécialité laitière, aux fruits, sucré, à la crème | 457 | 109 | 458 | 109 | 77,1 | 2,42 | 2,38 | 14,5 | 4,28 | 0,21 | 0,093 | 83 | 84,2 | 0,01 | < 0,05 | < 20 | 8,5 | 0,07 | 68 | 140 | < 20 | 37 | 0,3 | < 0,25 | 0,2 | 1,13 | - | < 0,5 | < 0,015 | 0,13 | < 0,1 | 0,24 | 0,026 | 15 | 0,2 |
| Yaourt, lait fermenté ou spécialité laitière, aux fruits, sucré, enrichi en vitamine D | - | - | - | - | 78,9 | 3,51 | 3,44 | 14 | 2,35 | 0,1 | 0,11 | 129 | 110 | 0,01 | 0,04 | < 20 | 12 | 0,02 | 94 | 160 | < 20 | 44,8 | 0,44 | 0,93 | 0,21 | < 0,8 | - | < 0,5 | 0,046 | 0,18 | 0,13 | 0,43 | 0,038 | < 5 | 0,25 |
| Yaourt, lait fermenté ou spécialité laitière, nature | 191 | 45,5 | 192 | 45,8 | 88,6 | 3,96 | 3,88 | 2,65 | 1,5 | < 3 | 0,11 | 128 | 170 | 0,0078 | 0,25 | 26,3 | 12,5 | 0,0053 | 98 | 176 | < 3,35 | 43 | 0,39 | 0,8 | 0,076 | 0,2 | 0,68 | < 0,5 | 0,04 | 0,26 | 0,19 | 0,31 | 0,057 | 25 | 0,33 |
| Yaourt, lait fermenté ou spécialité laitière, nature, 0% MG | 168 | 39,4 | 169 | 39,8 | 88,9 | 4,82 | 4,73 | 4,1 | 0,064 | traces | 0,14 | 129 | 135 | 0,0083 | 0,1 | 27,8 | 13,6 | 0,0026 | 105 | 182 | < 4,5 | 57,3 | 0,46 | 0,8 | 0 | 0,2 | 0,1 | < 0,5 | 0,042 | 0,24 | 0,26 | 0,44 | 0,065 | 28 | 0,29 |
| Yaourt, lait fermenté ou spécialité laitière, nature, 0% MG, enrichi en vitamine D | - | - | - | - | - | 4,59 | 4,5 | 5 | 0 | < 0,5 | 0,14 | 120 | - | - | - | - | - | - | - | - | - | 54 | - | 0,8 | - | - | - | - | - | - | - | - | - | - | - |
| Yaourt, lait fermenté ou spécialité laitière, nature, 0% MG, sucré, enrichi en vitamine D | - | - | - | - | - | 4,41 | 4,33 | 5,4 | 1,17 | < 0,15 | 0,17 | 184 | - | - | - | - | - | - | - | - | - | 66 | - | 1,83 | - | - | - | - | - | - | - | - | - | - | 0,25 |
| Yaourt, lait fermenté ou spécialité laitière, nature, à la crème | 476 | 115 | 478 | 115 | 82,6 | 3,06 | 3 | 3,08 | 9,8 | traces | 0,088 | 110 | 88,4 | < 0,01 | < 0,05 | < 20 | 9,4 | < 0,01 | 86 | 150 | < 20 | 35 | 0,35 | < 0,25 | 0,22 | < 0,8 | - | < 0,5 | < 0,015 | 0,075 | < 0,1 | 0,48 | 0,024 | < 5 | 0,21 |
| Yaourt, lait fermenté ou spécialité laitière, nature, sucré | 356 | 84,3 | 357 | 84,6 | 80,2 | 3,48 | 3,41 | 12,5 | 1,95 | traces | 0,12 | 120 | - | < 0,1 | 0,23 | 8 | 11,4 | < 0,1 | 92,5 | 157 | < 5 | 48,8 | 0,41 | < 0,5 | 0,07 | - | - | < 0,5 | < 0,04 | 0,21 | < 0,16 | 0,27 | < 0,04 | 14,6 | 0,2 |

# Appendix 3

| pâtes , pains, céréales et assimilés | Energie, Règlement UE N° 1169/2011 (kJ/100 g) | Energie, Règlement UE N° 1169/2011 (kcal/100 g) | Energie, N x facteur Jones, avec fibres (kJ/100 g) | Energie, N x facteur Jones, avec fibres (kcal/100 g) | Eau (g/100 g) | Protéines, N x facteur de Jones (g/100 g) | Protéines, N x 6.25 (g/100 g) | Glucides (g/100 g) | Lipides (g/100 g) | Fibres alimentaires (g/100 g) | Sel chlorure de sodium (g/100 g) | Calcium (mg/100 g) | Chlorure (mg/100 g) | Cuivre (mg/100 g) | Fer (mg/100 g) | Iode (µg/100 g) | Magnésium (mg/100 g) | Manganèse (mg/100 g) | Phosphore (mg/100 g) | Potassium (mg/100 g) | Sélénium (µg/100 g) | Sodium (mg/100 g) | Zinc (mg/100 g) | Vitamine D (µg/100 g) | Vitamine E (mg/100 g) | Vitamine K1 (µg/100 g) | Vitamine K2 (µg/100 g) | Vitamine C (mg/100 g) | Vitamine B1 ou Thiamine (mg/100 g) | Vitamine B2 ou Riboflavine (mg/100 g) | Vitamine B3 ou PP ou Niacine (mg/100 g) | Vitamine B5 ou Acide pantothénique (mg/100 g) | Vitamine B6 (mg/100 g) | Vitamine B9 ou Folates totaux (µg/100 g) | Vitamine B12 (µg/100 g) |
|---|---|---|---|---|---|---|---|---|---|---|---|---|---|---|---|---|---|---|---|---|---|---|---|---|---|---|---|---|---|---|---|---|---|---|---|
| Amarante, crue | 1560 | 369 | 1540 | 365 | 11,3 | 13,5 | 14,5 | 58,6 | 7,02 | 6,7 | 0,01 | 159 | - | 0,53 | 7,61 | - | 248 | 3,33 | 557 | 508 | - | 4 | 2,87 | 0 | 1,19 | 0 | - | 4,2 | 0,12 | 0,2 | 0,92 | 1,46 | 0,59 | 82 | 0 |
| Avoine, crue | 1590 | 378 | 1570 | 374 | 8,22 | 16,9 | 18,1 | 55,7 | 6,9 | 10,6 | 0,005 | 54 | 70 | 0,63 | 4,72 | - | 177 | 4,92 | 523 | 429 | 9 | 2 | 3,97 | 0 | 1,09 | - | - | 0 | 0,76 | 0,14 | 0,96 | 1,35 | 0,12 | 56 | 0 |
| Bagel | 1150 | 272 | 1140 | 269 | 34,3 | 9,26 | 10,2 | 47,3 | 3,98 | 3,42 | 1,08 | 37,3 | - | 0,17 | 1,83 | - | 36,3 | 0,84 | 111 | 122 | - | 410 | 0,95 | 0 | 0,32 | 1,5 | - | 0 | 0,24 | 0,1 | 2,3 | 0,37 | 0,083 | 25 | 0 |
| Biscotte briochée | 1760 | 417 | 1740 | 413 | - | 10,9 | 11,9 | 69 | 9,28 | 4,85 | 0,99 | - | - | - | - | - | - | - | - | - | - | 394 | - | - | - | - | - | - | - | - | - | - | - | - | - |
| Biscotte classique | 1730 | 409 | 1710 | 405 | 2,8 | 9,8 | 10,8 | 76,6 | 5,9 | 3,1 | 1,27 | 29 | 886 | 0,16 | 1,1 | < 20 | 24 | 0,62 | 100 | 160 | < 20 | 509 | 0,79 | < 0,25 | 0,83 | 2,22 | - | 0 | 0,12 | < 0,01 | 0,7 | 0,52 | 0,081 | 22,3 | - |
| Biscotte complète ou riche en fibres | 1660 | 393 | 1640 | 388 | 4,2 | 11,7 | 12,8 | 67 | 6,17 | 8,84 | 1,25 | 40,6 | 813 | < 1 | 3,1 | < 10 | 115 | 2 | 260 | 350 | < 5 | 492 | 1,9 | - | 0,68 | - | - | - | 0,15 | 0,13 | 1,9 | 0,74 | 0,14 | 26 | 0 |
| Biscotte multicéréale | 1680 | 398 | 1660 | 393 | 5,1 | 11,8 | 12,9 | 66,7 | 7,22 | 7,08 | 1,51 | 39,5 | 947 | < 1 | 2,3 | < 10 | 57 | 1,1 | 233 | 233 | < 5 | 597 | 1,5 | 0 | 0,66 | - | - | 0 | 0,16 | 0,12 | 1,6 | 0,58 | 0,16 | 29 | 0 |
| Biscotte sans adjonction de sel | 1710 | 405 | 1690 | 401 | 3,98 | 10,4 | 11,4 | 74 | 5,88 | 5,16 | 0,046 | 23,5 | - | 0,18 | 1,3 | 6,3 | 22,5 | 0,7 | 121 | 161 | < 3,35 | 18,1 | 7 | - | 4,7 | - | - | 0 | 0,66 | 0,22 | 8,5 | - | 0,94 | 94 | 0 |
| Blé de Khorasan, cru | 1440 | 342 | 1430 | 337 | 11,1 | 14,5 | 15,6 | 59,5 | 2,13 | 11,1 | 0,013 | 22 | - | 0,51 | 3,77 | - | 130 | 2,74 | 364 | 403 | - | 5 | 3,68 | - | 0,61 | 1,8 | - | 0 | 0,57 | 0,18 | 6,38 | 0,95 | 0,26 | - | - |
| Blé dur entier, cru | 1450 | 343 | 1440 | 340 | 10,7 | 12,1 | 13 | 62,4 | 2,24 | 11 | 0,0088 | 30,2 | - | 0,44 | 3,51 | 2,7 | 130 | 3,31 | 390 | 408 | 4,1 | 3,5 | 3,48 | 0 | 1,4 | 30 | - | 0 | 0,38 | 0,12 | 6,17 | 1,02 | 0,39 | 32 | 0 |
| Blé dur précuit, entier, cru | 1520 | 359 | 1500 | 355 | 8,81 | 13,7 | 14,7 | 67,7 | 1,83 | 6,67 | 0,01 | 55 | - | 0,4 | 0 | 0 | 50 | 3,4 | 238 | 950 | - | 4 | 0 | 0 | - | - | - | 0 | 1,11 | 0 | 8,4 | 0,6 | 0,3 | 331 | 0 |
| Blé dur précuit, grains entiers, cuit, non salé | 630 | 149 | 623 | 147 | 61,9 | 5,54 | 5,94 | 27,4 | 0,9 | 3,8 | < 0,013 | 21 | < 20 | 0,17 | 0,78 | < 20 | 28 | 0,8 | 96 | 99 | < 20 | < 5 | 0,71 | 0 | < 0,08 | - | - | < 0,5 | 0,061 | < 0,01 | 0,26 | 0,29 | 0,052 | 10,2 | 0 |
| Blé germé, cru | 897 | 211 | 887 | 209 | 47,8 | 7,49 | 8,03 | 41,4 | 1,27 | 1,1 | 0,04 | 28 | - | 0,26 | 2,14 | - | 82 | 1,86 | 200 | 169 | - | 16 | 1,65 | 0 | - | - | - | 2,6 | 0,23 | 0,16 | 3,09 | 0,95 | 0,27 | 38 | 0 |
| Blé tendre entier ou froment, cru | - | - | - | - | - | - | - | - | - | - | - | - | - | - | - | - | - | - | - | - | - | - | - | - | - | - | - | - | - | - | - | - | - | - | - |
| Blini | - | - | - | - | 45,9 | 5,98 | 5,98 | 33,7 | 11,2 | 2,35 | 1,32 | 38,9 | 762 | < 0,1 | 0,6 | 30 | 14,2 | 0,3 | 116 | 137 | < 5 | 521 | 0,45 | < 0,5 | 2,4 | - | - | 0 | 0,11 | 0,12 | 0,73 | 0,35 | 0,07 | 33 | 0,28 |
| Boulgour de blé, cru | 1490 | 351 | 1470 | 348 | 9,05 | 11,5 | 12,4 | 66,8 | 1,71 | 9,57 | 0,043 | 34,3 | - | 0,33 | 2,46 | 2,7 | 140 | 3,32 | 285 | 410 | 4,1 | 17 | 2,37 | 0 | 0,73 | 16 | - | 0 | 0,23 | 0,12 | 5,11 | 1,07 | 0,35 | 24 | 0 |
| Boulgour de blé, cuit, non salé | 471 | 111 | 467 | 110 | 71,5 | 3,73 | 4 | 21,7 | 0,5 | 2,2 | 0,018 | 14 | 21,1 | 0,16 | 0,55 | < 20 | 22 | 0,46 | 66 | 95 | < 20 | 7 | 0,73 | 0 | < 0,08 | 0,5 | - | < 0,5 | 0,047 | < 0,01 | 0,34 | 0,19 | 0,039 | 10,8 | 0 |
| Bretzel, pain frais | 1390 | 330 | 1380 | 327 | 21,5 | 9,18 | 10,1 | 54,6 | 7,1 | 3,8 | 3,35 | 53 | 1830 | 0,13 | 0,84 | - | 25 | 0,55 | 170 | 170 | < 20 | 1340 | 0,88 | - | 1,37 | 5,54 | - | < 0,5 | 0,15 | 0,038 | 0,68 | 0,6 | 0,075 | 26,4 | - |
| Chapelure | 1580 | 374 | 1560 | 369 | 7,29 | 11,6 | 12,7 | 70 | 3,77 | 4,5 | 1,82 | 72,8 | 1250 | < 0,1 | 1,8 | < 10 | 55,5 | 1,2 | 196 | 270 | < 5 | 728 | 1,5 | 0 | 0,2 | 0 | - | - | 0,2 | 0,09 | 1 | 0,71 | 0,12 | 22 | 0 |
| Crackers de table au froment | 1900 | 450 | 1880 | 447 | - | 7,75 | 8,5 | 75 | 12 | 4,1 | 1,5 | - | - | - | - | - | - | - | - | - | - | 600 | - | - | - | - | - | - | - | - | - | - | - | - | - |
| Croûton à l'ail aux fines herbes ou aux oignons, préemballé | 2060 | 492 | 2060 | 492 | 3,4 | 8,81 | 8,81 | 58,9 | 24,1 | 2 | 2,44 | 31 | 1460 | 0,11 | 1,1 | < 20 | 27 | 0,58 | 110 | 200 | < 20 | 976 | 0,67 | < 0,25 | 10,1 | 8,15 | - | - | 0,071 | 0,065 | < 0,1 | 0,63 | 0,063 | 57,1 | - |
| Croûton à tartiner | 1580 | 374 | 1560 | 369 | - | 12,3 | 13,5 | 70 | 2,85 | 7 | 1,3 | - | - | - | - | - | - | - | - | - | - | 500 | - | - | - | - | - | - | - | - | - | - | - | - | - |
| Croûtons nature, préemballés | 2070 | 495 | 2060 | 492 | 3,6 | 8,38 | 9,19 | 59,1 | 24,1 | 2,6 | 1,7 | 27 | 1080 | 0,11 | 1,5 | < 20 | 24 | 0,57 | 110 | 190 | < 20 | 680 | 0,67 | < 0,25 | 7,86 | 2,04 | - | - | 0,099 | 0,056 | 0,22 | 0,67 | 0,054 | 37,1 | - |
| Épeautre, cru | 1450 | 344 | 1430 | 340 | 11 | 14,6 | 15,6 | 59,5 | 2,43 | 10,7 | 0,02 | 27 | - | 0,51 | 4,44 | - | 136 | 2,98 | 401 | 388 | - | 8 | 3,28 | - | 0,79 | 3,6 | - | 0 | 0,36 | 0,11 | 6,84 | 1,07 | 0,23 | 45 | 0 |
| Flocons d'avoine, bouillis/cuits à l'eau | 318 | 75,5 | 315 | 74,7 | - | 2,54 | 2,72 | 11,9 | 1,52 | 1,7 | 0,01 | 9 | - | 0,074 | 0,9 | 0,3 | 27 | 0,58 | 77 | 70 | - | 4 | 1 | 0 | 0,08 | 0,3 | - | 0 | 0,076 | 0,016 | 0,23 | 0,31 | 0,005 | 6 | 0 |
| Frik (blé dur immature concassé), cru | 1360 | 323 | 1350 | 321 | 11,3 | 9,62 | 10,3 | 55,8 | 2,25 | 19,3 | 0,011 | 50,3 | 89 | < 1 | 5,2 | < 10 | 114 | 3,6 | 321 | 512 | < 5 | 4,2 | 3,6 | - | < 0,1 | - | - | - | 0,24 | 0,09 | 3,49 | 0,82 | 0,14 | 68,2 | - |
| Frik (blé dur immature concassé), cuit, non salé | 321 | 76 | 318 | 75,4 | 79,6 | 2,04 | 2,19 | 14 | 0,48 | 3,5 | 0,005 | 19,6 | 50 | < 1 | 1 | < 10 | 30,4 | < 1 | 55 | 95,5 | < 5 | 2 | < 1 | - | < 0,1 | - | - | - | < 0,05 | < 0,05 | 1,6 | 0,24 | 0,06 | 18 | - |
| Galette de maïs soufflé | - | - | - | - | 2,6 | 7,08 | 7,44 | - | 1,7 | - | 0,2 | - | - | - | - | - | - | - | - | - | - | 79,8 | - | - | - | - | - | - | - | - | - | - | - | - | - |
| Galette de riz soufflé complet | 1630 | 385 | 1630 | 384 | 4,8 | 7,32 | 7,69 | 80,5 | 3 | 2,9 | 0,18 | 12 | 159 | 0,23 | 1,1 | < 20 | 160 | 2,8 | 410 | 300 | < 20 | 72,6 | 1,8 | - | 1,01 | < 0,8 | - | - | 0,088 | 0,02 | 1,26 | 0,74 | 0,18 | 41,5 | - |
| Galette multicéréales soufflée | 1670 | 394 | 1660 | 392 | 2,45 | 8,43 | 8,86 | 82,4 | 2,5 | 3 | 0,23 | - | - | - | - | - | - | - | - | - | - | 90,1 | - | - | - | - | - | - | - | - | - | - | - | - | - |
| Gnocchi à la pomme de terre, cru | 744 | 176 | 744 | 176 | - | 4,66 | 4,66 | 35 | 1,42 | 2,2 | 1,18 | - | - | - | - | - | - | - | - | - | - | 468 | - | - | - | - | - | - | - | - | - | - | - | - | - |
| Gnocchi à la pomme de terre, cuit | 748 | 177 | 748 | 177 | 55,5 | 5,01 | 5,01 | 33,6 | 2,05 | 2 | 1,01 | 5,1 | 695 | 0,14 | 0,7 | 60 | 18,8 | 1,4 | 39 | 184 | 5 | 402 | 0,5 | < 0,5 | 0,92 | - | - | - | 0,084 | < 0,04 | 1,6 | 0,24 | 0,12 | 27,1 | 0,15 |
| Gnocchi à la semoule, cru | 846 | 200 | 846 | 200 | - | 6,2 | 6,2 | 37 | 2,6 | 1,9 | 2,21 | - | - | - | - | - | - | - | - | - | - | 870 | - | - | - | - | - | - | - | - | - | - | - | - | - |
| Gnocchi à la semoule, cuit | 783 | 185 | 783 | 185 | - | 5,6 | 5,6 | 34,4 | 2,4 | 1,8 | 1,2 | - | - | - | - | - | - | - | - | - | - | - | - | - | - | - | - | - | - | - | - | - | - | - | - |
| Gnocchi, cuit (aliment moyen) | 766 | 181 | 766 | 181 | 54,8 | 6,34 | 6,34 | 34 | 7,88 | 1,9 | 1,06 | 95,6 | | | 0,75 | | 17 | | | 158 | | 395 | 0,8 | | | | | | | | | | | | |
| Graine de couscous (semoule de blé dur précuite), crue | 1550 | 365 | 1530 | 360 | 8,56 | 12 | 13,2 | 72,7 | 1,45 | 4,33 | 0,013 | 24 | - | 0,25 | 1,08 | - | 44 | 0,78 | 170 | 166 | - | 5,5 | 0,83 | 0 | - | - | - | 0 | 0,16 | 0,078 | 3,49 | 1,24 | 0,11 | 20 | 0 |
| Graine de couscous (semoule de blé dur précuite), cuite, non salée | 667 | 157 | 659 | 156 | 60,6 | 4,56 | 5 | 31 | 1 | 2,3 | < 0,013 | 15 | 48,7 | 0,16 | 0,53 | < 20 | 19 | 0,31 | 73 | 110 | < 20 | < 5 | 0,57 | 0 | < 0,08 | < 0,8 | - | < 0,5 | 0,064 | < 0,01 | 0,24 | 0,32 | 0,43 | 12,4 | 0 |
| Gressin | 1820 | 433 | 1800 | 429 | 2,48 | 12,7 | 13,9 | 65,3 | 11,8 | 5,16 | 1,8 | 48 | 1210 | 0,25 | 1,2 | 20 | 66,3 | 0,36 | 188 | 213 | < 5 | 715 | 1,4 | - | 3 | - | - | 7 | 0,42 | 0,082 | 1,83 | 0,47 | 0,6 | 70 | 0,26 |
| Maïs entier, cru | 1460 | 346 | 1460 | 346 | 13,6 | 8,1 | 8,1 | 67,2 | 3,7 | 5,7 | 0,01 | 40 | 50 | 0,2 | 3,2 | - | 100 | 0,8 | 260 | 320 | 12,8 | 4 | 1,9 | - | 1,7 | - | - | - | 0,4 | 0,14 | 2,1 | 0,6 | 0,5 | - | 0 |
| Mélange de céréales et légumineuses, cru | 1490 | 354 | 1470 | 349 | - | 13,7 | 15 | 59 | 4 | 11 | 0,99 | - | - | - | - | - | 70 | - | - | - | - | 397 | - | - | - | - | - | - | - | - | - | - | - | - | - |
| Mil entier, cru | 1520 | 359 | 1500 | 356 | 8,69 | 11 | 11,8 | 64,3 | 4,21 | 8,5 | 0,0068 | 8,75 | - | 0,62 | 3,91 | 5 | 107 | 1,17 | 263 | 208 | 16 | 2,7 | 2,54 | 0 | 0,05 | 0,9 | - | 0 | 0,42 | 0,29 | 4,72 | 0,85 | 0,57 | 57,5 | 0 |
| Mil, cuit, non salé | 492 | 116 | 487 | 115 | 71,4 | 3,51 | 3,76 | 22,4 | 1 | 1,3 | 0,005 | 3 | - | 0,16 | 0,63 | - | 44 | 0,27 | 100 | 62 | - | 2 | 0,91 | 0 | 0,02 | 0,3 | - | 0 | 0,11 | 0,082 | 1,33 | 0,17 | 0,11 | 19 | 0 |
| Muffin anglais, complet, petit pain spécial, préemballé | 917 | 217 | 896 | 212 | 45,6 | 12,5 | 13,7 | 32,9 | 1,87 | 7,07 | 0,98 | - | - | - | - | - | - | - | - | - | - | 393 | - | - | - | - | - | - | - | - | - | - | - | - | - |
| Muffin anglais, petit pain spécial, préemballé | 966 | 228 | 949 | 224 | 42,3 | 10,6 | 11,6 | 39,6 | 1,88 | 3,18 | 1,05 | - | - | - | - | - | - | - | - | - | - | 418 | - | - | - | - | - | - | - | - | - | - | - | - | - |
| Nouilles asiatiques cuites, aromatisées | 616 | 147 | 611 | 146 | 67,5 | 4,78 | 5,12 | 18,8 | 4,3 | 6,3 | 1,2 | - | - | - | - | - | - | - | - | - | - | 470 | - | - | - | - | - | - | - | - | - | - | - | - | - |
| Orge entière, crue | 1410 | 334 | 1390 | 330 | 9,44 | 12,5 | 13,4 | 56,2 | 2,3 | 17,3 | 0,03 | 33 | 110 | 0,5 | 3,6 | - | 133 | 1,94 | 264 | 452 | 6,15 | 12 | 2,77 | 0 | 0,57 | 2,2 | - | 0 | 0,65 | 0,29 | 4,6 | 0,28 | 0,32 | 19 | 0 |
| Orge perlée, bouilli/cuite à l'eau, non salée | 503 | 119 | 500 | 118 | 68,8 | 2,26 | 2,42 | 34,4 | 0,44 | 3,8 | 0,0075 | 11 | - | 0,11 | 1,33 | 0,5 | 22 | 0,26 | 54 | 93 | - | 3 | 0,82 | 0 | 0,01 | 0,8 | - | 0 | 0,083 | 0,062 | 2,06 | 0,14 | 0,12 | 16 | 0 |
| Orge perlée, crue | 1460 | 346 | 1450 | 343 | 10,1 | 9,91 | 10,6 | 68,6 | 1,16 | 9,1 | 0,023 | 29 | - | 0,42 | 2,5 | - | 79 | 1,32 | 221 | 280 | - | 9 | 2,13 | 0 | 0,02 | 2,2 | - | 0 | 0,19 | 0,11 | 4,6 | 0,28 | 0,26 | 23 | 0 |
| Pain (aliment moyen) | 1170 | 276 | 1150 | 273 | 30 | 8,21 | 9,01 | 54,4 | 1,61 | 3,84 | 1,28 | 31,1 | 891 | 0,14 | 1,25 | 4,5 | 27,7 | 0,67 | 108 | 165 | 10,3 | 512 | 0,75 | 0,16 | 0,28 | 0,6 | | 0,3 | 0,098 | 0,01 | 0,63 | 0,41 | 0,023 | 23 | |
| Pain au son | 1050 | 249 | 1040 | 246 | 33,3 | 8,05 | 8,8 | 44,4 | 2,6 | 6,21 | 1,65 | 34,3 | - | < 0,5 | 1,8 | - | 52,2 | 1,5 | 198 | 253 | < 5 | 410 | 1,1 | - | 0,2 | - | - | - | 0,15 | 0,04 | 0,67 | 0,58 | 0,16 | 3,5 | - |
| Pain blanc maison (avec farine pour machine à pain) | 1090 | 256 | 1070 | 253 | 33,9 | 7,81 | 8,56 | 52,1 | 0,9 | 2,6 | 1,45 | 21 | 880 | 0,09 | 0,85 | 18,5 | 20 | 0,47 | 94 | 130 | < 50 | 578 | 0,62 | - | 0,18 | < 0,8 | - | 0,6 | 0,079 | 0,04 | 0,69 | 0,31 | 0,082 | 30,3 | - |
| Pain complet ou intégral (à la farine T150) | 1030 | 244 | 1020 | 241 | 36,6 | 8,38 | 9,19 | 44,3 | 1,8 | 6,9 | 1,12 | 39 | 780 | 0,21 | 2,1 | < 20 | 51 | 1,6 | 180 | 240 | < 20 | 448 | 1,4 | 0,86 | 0,29 | < 0,8 | - | < 0,5 | 0,11 | < 0,01 | < 0,1 | 0,44 | 0,087 | 35,6 | 0 |
| Pain courant, 400g ou boule | 1130 | 266 | 1110 | 263 | 36,9 | 7,9 | 8,67 | 52 | 2 | 2,6 | 1,33 | 37 | - | 0,12 | 1 | 20,7 | 21,3 | 0,6 | 124 | 128 | 2,65 | 499 | 0,63 | 0 | 0,5 | 0 | - | 0 | 0,12 | 0,067 | 1 | 0,5 | 0,053 | 50,3 | 0 |
| Pain de campagne maison (avec farine pour machine à pain) | 1020 | 240 | 1000 | 237 | 37,3 | 7,98 | 8,75 | 46,7 | 0,9 | 5,1 | 1,34 | 20 | 899 | 0,13 | 1,3 | 22 | 34 | 0,84 | 130 | 210 | < 50 | 537 | 1,1 | - | 0,26 | < 0,8 | - | 1,33 | 0,16 | 0,06 | 0,64 | 0,31 | 0,062 | 25,6 | - |
| Pain de mie brioché, préemballé | 1310 | 309 | 1290 | 306 | 28,3 | 7,87 | 8,63 | 52,4 | 6,5 | 3,1 | 1,12 | 65 | 730 | 0,09 | 0,7 | < 20 | 17 | 0,39 | 95 | 110 | < 20 | 447 | 0,67 | < 0,25 | 1,19 | 5,67 | - | < 0,5 | 0,11 | 0,056 | 0,24 | 0,76 | 0,028 | 14,5 | 0,26 |
| Pain de mie, au son | 1170 | 278 | 1160 | 275 | 34,3 | 6,84 | 7,5 | 46 | 5,5 | 7 | 1,35 | - | - | < 0,5 | 3,5 | - | 76,6 | 1,6 | - | - | < 5 | 423 | 1,4 | - | - | - | - | - | - | - | 1,35 | - | - | < 2 | - |

| | | | | | | | | | | | | | | | | | | | | | | | | | | | | | | | | | | | |
|---|---|---|---|---|---|---|---|---|---|---|---|---|---|---|---|---|---|---|---|---|---|---|---|---|---|---|---|---|---|---|---|---|---|---|---|
| Pain de mie, complet | 1110 | 262 | 1090 | 259 | 35,4 | 8,51 | 9,33 | 43,7 | 4,06 | 6,23 | 1,18 | 141 | 787 | <1 | 1,6 | <10 | 51,5 | 1,5 | 160 | 226 | <5 | 471 | 1,1 | - | 0,5 | - | - | - | 0,17 | 0,1 | <0,05 | 0,56 | 0,11 | 25 | - |
| Pain de mie, courant | 1180 | 278 | 1160 | 275 | 33,3 | 7,13 | 7,81 | 52,3 | 3,6 | 1,8 | 1,15 | 88 | 760 | 0,11 | 0,7 | <20 | 17 | 0,43 | 75 | 120 | <20 | 459 | 0,47 | 0 | 0,12 | 2,22 | - | <0,5 | 0,12 | 0,015 | 0,31 | 0,43 | 0,03 | 13,3 | - |
| Pain de mie, multicéréale | 1160 | 274 | 1140 | 271 | 34,4 | 8,81 | 9,66 | 43,6 | 5,41 | 5,74 | 1,29 | 77,6 | 728 | <0,5 | 1,9 | 6 | 40,9 | 1,52 | 140 | 157 | 8 | 515 | 1,5 | <0,5 | 3 | - | - | <1 | 0,2 | 0,12 | 4,2 | 0,55 | 0,13 | 52 | - |
| Pain de mie, sans croûte, préemballé | 1150 | 271 | 1130 | 268 | 34,8 | 7,52 | 8,25 | 49,5 | 3,8 | 3 | 1,06 | 87 | 712 | 0,12 | 0,74 | <20 | 20 | 0,42 | 86 | 110 | <20 | 425 | 0,49 | - | <0,08 | 3,07 | - | <0,5 | 0,091 | 0,013 | 0,37 | 0,42 | 0,034 | 18,6 | - |
| Pain de seigle, et froment | 1100 | 260 | 1090 | 257 | 32,8 | 8,27 | 9,06 | 51,5 | 1 | 4,5 | 1,35 | 33 | 923 | 0,14 | 1,2 | <20 | 27 | 0,74 | 110 | 170 | <20 | 538 | 0,86 | <0,25 | 0,26 | <0,8 | - | <0,5 | 0,11 | <0,01 | 0,28 | 0,44 | <0,01 | 49,3 | 0 |
| Pain grillé brioché, tranché, préemballé | 1770 | 419 | 1750 | 415 | 4,45 | 10,6 | 11,6 | 70,8 | 9,19 | 3,58 | 1,03 | - | - | - | - | - | - | - | 120 | - | - | 409 | - | - | - | - | - | - | - | - | - | - | - | - | - |
| Pain grillé suédois au blé complet | 1680 | 398 | 1660 | 394 | 5,8 | 10,5 | 11,5 | 66,4 | 8 | 7 | 1,53 | 26,8 | 1020 | <1 | 1,6 | <10 | 60,9 | 1,7 | 213 | 284 | <5 | 610 | 1,4 | - | 0,68 | - | - | - | 0,19 | 0,11 | 1,4 | 0,69 | 0,25 | 26 | - |
| Pain grillé suédois au froment | 1690 | 402 | 1680 | 398 | 5,5 | 8,89 | 9,75 | 68,5 | 8,25 | 7,3 | 1,06 | 42,2 | 765 | - | 1 | <10 | 23,4 | <1 | 120 | 163 | <5 | 422 | <1 | - | 0,86 | - | - | - | 0,2 | 0,07 | 1,4 | 0,52 | 0,14 | 21 | - |
| Pain grillé suédois aux fruits | 1690 | 401 | 1680 | 397 | 6,7 | 7,98 | 8,75 | 70,8 | 8 | 5,25 | 0,86 | - | - | - | - | - | - | - | - | - | - | 340 | - | - | - | - | - | - | - | - | - | - | - | - | - |
| Pain grillé suédois aux graines de lin | 1670 | 396 | 1650 | 392 | - | 10,3 | 11,3 | 63 | 9 | 9 | 1,57 | - | - | - | - | - | - | - | - | - | - | 613 | - | - | - | - | - | - | - | - | - | - | - | - | - |
| Pain grillé, domestique | 1340 | 317 | 1330 | 313 | 19,6 | 9,46 | 10,4 | 64 | 1,3 | 3,7 | 1,51 | 20 | 972 | 0,11 | 0,97 | <20 | 25 | 0,51 | 85 | 140 | <20 | 602 | 0,65 | <0,25 | 0,24 | <0,8 | - | <0,5 | 0,11 | 0,018 | 0,33 | 0,58 | 0,044 | 19,6 | - |
| Pain grillé, tranches, au froment | 1710 | 404 | 1690 | 401 | 3,13 | 9,76 | 10,7 | 74,4 | 6,16 | 4,34 | 1,56 | 29,3 | - | 0,18 | 1,9 | 4 | 33,8 | 0,7 | 135 | 175 | 2 | 616 | - | - | - | - | - | - | - | - | - | 0,1 | - | - | 0 |
| Pain grillé, tranches, multicéréale | 1770 | 419 | 1750 | 415 | 3 | 11,1 | 12,1 | 68,8 | 9,3 | 5,95 | 1,09 | 30 | - | 0,18 | 9 | 4 | 66 | 0,7 | 194 | 300 | 1,7 | 437 | 1,5 | 0 | 1,5 | - | - | 0 | 0,2 | 0,15 | 1,5 | 0,1 | 0,1 | 57 | 0 |
| Pain panini | 1150 | 272 | 1140 | 269 | 30,9 | 7,66 | 8,4 | 55,9 | 1,1 | 2,6 | 1,55 | 22 | - | - | 0,9 | - | 29 | - | 118 | 148 | - | 619 | - | traces | 0,4 | - | - | - | 0,1 | 0,1 | 0,8 | 0,4 | 0,17 | 39 | 0,1 |
| Pain pita | 1100 | 259 | 1080 | 256 | 34,1 | 7,53 | 8,26 | 53 | 0,92 | 2,68 | 1,38 | 52,7 | 982 | - | 1 | <10 | 18,1 | <1 | 84 | 134 | <5 | 543 | <1 | - | <0,1 | - | - | - | 0,11 | 0,08 | 2,6 | 0,49 | 0,05 | 21 | - |
| Pain pour hamburger ou hot dog (bun), complet, préemballé | 1200 | 285 | 1190 | 282 | - | 8,89 | 9,75 | 46,3 | 5,75 | 4,75 | 1,11 | - | - | - | - | - | - | - | - | - | - | 413 | - | - | - | - | - | - | - | - | - | - | - | - | - |
| Pain pour hamburger ou hot dog (bun), préemballé | 1240 | 293 | 1220 | 290 | 30,9 | 9,07 | 9,95 | 49,7 | 5,35 | 3,3 | 1,14 | 89 | 757 | 0,13 | 0,85 | <20 | 27 | 0,56 | 110 | 140 | <50 | 453 | 0,84 | - | 0,22 | 2,81 | - | <0,5 | 0,19 | 0,03 | 1 | 0,37 | 0,097 | 13,5 | 0,036 |
| Pain, baguette ou boule, au levain | 1110 | 261 | 1090 | 258 | 34,4 | 7,41 | 8,13 | 53,1 | 1,3 | 2,1 | - | 18,7 | - | 0,11 | 1,3 | - | 18,7 | 0,57 | 87,3 | 124 | - | - | 0,53 | - | 0,16 | 1 | - | - | 0,07 | 0,02 | 1,15 | 0,4 | 0,06 | 13,8 | - |
| Pain, baguette ou boule, aux céréales et graines, artisanal | 1140 | 269 | 1120 | 265 | 33,2 | 9,58 | 10,5 | 46,9 | 3,2 | 5,2 | 1,13 | 40 | 791 | 0,25 | 1,6 | <20 | 53 | 0,78 | 150 | 230 | <20 | 450 | 1,1 | <0,25 | 0,85 | <0,8 | - | <0,5 | 0,098 | <0,01 | 0,71 | 0,43 | 0,059 | 30,5 | - |
| Pain, baguette ou boule, bio (à la farine T55 jusqu'à T110) | 1090 | 257 | 1080 | 254 | 34,5 | 8,68 | 9,52 | 49,9 | 1,19 | 4,33 | 1,37 | 24,2 | - | 0,19 | 1,8 | - | 36,6 | 1,27 | 145 | 190 | - | 538 | 1,1 | - | 0,27 | - | - | - | 0,13 | 0,04 | 2,09 | 0,51 | 0,09 | 19,3 | - |
| Pain, baguette ou boule, bis (à la farine T80 ou T110) | 1120 | 265 | 1110 | 262 | 30,2 | 8,6 | 9,43 | 54 | 0,33 | 4,2 | - | 27,5 | - | <0,98 | 1,31 | - | 33 | 1 | 140 | 178 | - | - | 1,2 | - | 0,23 | - | - | - | 0,13 | 0,06 | 1,66 | 0,4 | 0,27 | 22,4 | - |
| Pain, baguette ou boule, de campagne | 1070 | 253 | 1060 | 250 | 34,2 | 7,52 | 8,25 | 50 | 1,3 | 4,3 | 1,35 | 23 | 905 | 0,15 | 1,5 | <20 | 33 | 0,82 | 120 | 190 | <20 | 539 | 0,88 | <0,25 | <0,08 | <0,8 | - | <0,5 | 0,097 | <0,01 | <0,1 | 0,39 | 0,051 | 30,2 | - |
| Pain, baguette, courante | 1220 | 287 | 1200 | 284 | 27,5 | 8,27 | 9,06 | 58,3 | 1,4 | 2,7 | 1,3 | 22 | 938 | 0,13 | 1,2 | <20 | 23 | 0,59 | 110 | 180 | <20 | 518 | 0,73 | <0,25 | 0,32 | <0,8 | - | <0,5 | 0,17 | <0,01 | 2,88 | 0,46 | <0,01 | 26,8 | - |
| Pain, baguette, de tradition française | 1180 | 279 | 1170 | 276 | 28,6 | 8,15 | 8,94 | 56,6 | 1 | 3,8 | 1,33 | 20 | 924 | 0,12 | 1,2 | <0,2 | 24 | 0,58 | 96 | 150 | <20 | 530 | 0,65 | <0,25 | 0,2 | <0,8 | - | <0,5 | 0,068 | <0,01 | <0,1 | 0,38 | <0,01 | 19,8 | - |
| Pain, baguette, sans sel | 1180 | 279 | 1170 | 276 | 26,9 | 8,3 | 9,1 | 56,5 | 1,3 | 2,68 | 0,013 | 42 | - | 0,12 | - | 6 | 21,3 | 0,52 | - | 129 | 4,7 | - | 0,75 | - | - | - | - | - | 0,069 | - | - | - | 0,067 | 23 | - |
| Pain, sans gluten | 1100 | 261 | 1100 | 260 | 33,8 | 2,62 | 2,75 | 51,8 | 3,3 | 6,5 | 1,56 | 60 | 778 | 0,06 | 0,5 | <20 | 16 | 0,19 | 95 | 75 | <20 | 624 | 0,36 | - | 0,98 | <0,8 | - | 38,6 | 0,045 | 0,02 | 0,15 | 0,4 | 0,025 | 10,7 | 0,059 |
| Pâtes fraîches, aux oeufs, crues | 1200 | 283 | 1180 | 279 | - | 10,1 | 11,1 | 53,8 | 1,87 | 3,17 | 0,35 | 35 | - | - | 1,29 | - | - | - | - | - | - | 156 | - | - | - | - | - | - | - | - | - | - | - | - | - |
| Pâtes fraîches, aux oeufs, cuites, non salées | 712 | 168 | 702 | 166 | 59,1 | 5,76 | 6,31 | 32 | 1,3 | 1,5 | 0,042 | 24,1 | 37 | <1 | 1,1 | - | 20,9 | <1 | 88 | 61 | - | 16,8 | <1 | <0,2 | <0,1 | - | - | - | 0,07 | 0,06 | 0,21 | 0,37 | <0,05 | 9,2 | 0,24 |
| Pâtes ou nouilles asiatiques au blé et aux oeufs, crues, nature | 1570 | 371 | 1550 | 366 | - | 11,9 | 13 | 71 | 3 | 4 | 0,99 | - | - | - | - | - | - | - | - | - | - | 397 | - | - | - | - | - | - | - | - | - | - | - | - | - |
| Pâtes ou nouilles asiatiques au blé, cuites, nature, non salées | 690 | 165 | 686 | 164 | 66,9 | 3,5 | 3,75 | 21,4 | 6,9 | 1 | 0,23 | 11 | 113 | 0,04 | 0,32 | <20 | 9,1 | 0,24 | 43 | 26 | <50 | 93,6 | 0,19 | - | - | <0,8 | - | - | 0,023 | <0,01 | 0,1 | 0,91 | <0,01 | <5 | 0,073 |
| Pâtes sèches standard, crues | 1420 | 336 | 1400 | 331 | 17,1 | 11,5 | 12,6 | 65,8 | 1,79 | 3 | 0,031 | 20,5 | - | 0,27 | 1,5 | 0,6 | 57,5 | 0,75 | 165 | 219 | 4,8 | 9,2 | 1,31 | 0 | 0,2 | 0,1 | - | 0 | 0,12 | 0,048 | 1,35 | 0,37 | 0,096 | 23,7 | 0 |
| Pâtes sèches standard, cuites, non salées | 535 | 126 | 529 | 125 | 68,3 | 3,99 | 4,38 | 25 | 0,55 | 1,9 | <0,01 | 17 | 13,3 | 0,12 | 0,43 | <20 | 17 | 0,31 | 59 | 34 | <20 | <5 | 0,47 | 0 | <0,08 | <0,8 | - | <0,5 | 0,025 | <0,01 | 0,52 | 0,28 | 0,022 | 23,5 | 0,014 |
| Pâtes sèches, au blé complet, crues | 1490 | 353 | 1480 | 350 | 11 | 11,8 | 12,6 | 67,6 | 2,2 | 6,1 | 0,016 | 32,1 | 64 | <1 | 3,2 | 10 | 82,2 | 1,8 | 160 | 378 | 5,8 | 6,2 | 1,9 | 0 | <0,1 | 0 | - | 0 | 0,25 | 0,07 | 0,87 | 0,66 | 0,13 | 22,4 | 0 |
| Pâtes sèches, au blé complet, cuites, non salées | 541 | 128 | 535 | 127 | 67,5 | 4,55 | 4,88 | 25,4 | 0,9 | 3,3 | <0,01 | 22 | 11,5 | 0,19 | 0,98 | <20 | 36 | 0,76 | 110 | 65 | <20 | <5 | 0,99 | 0 | <0,08 | <0,8 | - | <0,5 | 0,07 | 0,021 | 0,16 | 0,19 | 0,032 | 10,5 | 0,021 |
| Pâtes sèches, aux oeufs, crues | 1610 | 381 | 1590 | 377 | 9,01 | 14 | 15,2 | 68 | 4,72 | 3,15 | 0,085 | 35 | - | 0,3 | 1,9 | 7,04 | 58 | 0,86 | 241 | 244 | 32,5 | 34 | 1,92 | 0,3 | 0,37 | 0,5 | - | 0 | 0,17 | 0,09 | 2,1 | 0,91 | 0,22 | 29 | 0,29 |
| Pâtes sèches, aux oeufs, cuites, non salées | 566 | 134 | 559 | 132 | 68,1 | 4,57 | 4,94 | 23 | 2 | 2 | <0,01 | 23 | 11,9 | 0,11 | 0,57 | <20 | 16 | 0,3 | 68 | 25 | <20 | <5 | 0,53 | <0,25 | 0,19 | <0,8 | - | <0,5 | 0,02 | 0,014 | <0,1 | 0,2 | 0,033 | 24,1 | 0,14 |
| Pâtes sèches, sans gluten, crues | 1510 | 356 | 1500 | 354 | - | 6,17 | 6,61 | 79 | 1,24 | 1,22 | 0,0078 | - | - | - | - | - | - | - | - | - | - | 3,61 | - | - | - | - | - | - | - | - | - | - | - | - | - |
| Pâtes sèches, sans gluten, cuites, non salées | 690 | 163 | 687 | 162 | 60,1 | 3,09 | 3,25 | 34,4 | 1,1 | 1 | 0,015 | 9,6 | 13,3 | 0,07 | 0,48 | <20 | 23 | 0,34 | 62 | 32 | <20 | 5,9 | 0,46 | - | <0,08 | <0,8 | - | <0,5 | 0,04 | <0,01 | 0,15 | 0,17 | 0,019 | 12,7 | 0,09 |
| Pâtes, sans gluten, à base de lentilles corail, cuites à l'eau, non salées | 709 | 168 | 695 | 164 | 56,8 | 11,7 | 12,6 | 25,1 | 0,8 | 4,9 | <0,013 | 29 | <20 | 0,48 | 2,7 | - | 59 | 0,57 | 130 | 240 | <20 | <5 | 1,8 | - | <0,08 | 2,12 | - | - | 0,12 | <0,01 | 0,21 | 0,29 | 0,066 | 23,1 | - |
| Pâtes, sans gluten, à base de riz et maïs, cuites à l'eau, non salées | 720 | 170 | 716 | 169 | 58 | 3,15 | 3,38 | 36,6 | 0,8 | 1,3 | <0,013 | 9,6 | <20 | 0,06 | 0,13 | - | 8,6 | 0,1 | 27 | 15 | <20 | <5 | 0,19 | - | 0,12 | <0,8 | - | - | <0,015 | <0,01 | <0,1 | 0,058 | 0,019 | 6,32 | - |
| Polenta ou semoule de maïs, cuite, non salée | 328 | 77,5 | 328 | 77,5 | 79,9 | 1,38 | 1,38 | 16,9 | <0,3 | 1,6 | <0,01 | 4,7 | 10,2 | 0,07 | 0,21 | <20 | 4 | 0,03 | 15 | 24 | <20 | <5 | 0,11 | - | <0,08 | <0,8 | - | <0,5 | <0,015 | <0,01 | <0,1 | 0,049 | 0,023 | <5 | 0,03 |
| Polenta ou semoule de maïs, précuite, sèche | 1480 | 350 | 1480 | 350 | 12,6 | 7,88 | 7,88 | 74 | 1,8 | 3,2 | 0,0043 | 1,55 | 94 | - | <1 | <10 | 35 | <1 | 104 | 144 | <5 | 1,7 | <1 | - | <1 | - | - | - | 0,07 | 0,06 | <0,05 | 0,39 | 0,08 | 35 | - |
| Quinoa, bouilli/cuit à l'eau, non salé | 631 | 149 | 625 | 148 | 61,6 | 4,66 | 5 | 27,9 | 1,1 | 3,8 | 0,015 | 23 | 26,6 | 0,21 | 1,6 | - | 71 | 0,7 | 180 | 220 | <20 | 6 | 1,2 | 0 | 0,88 | <0,8 | - | 0 | 0,14 | 0,014 | 0,2 | 0,61 | 0,089 | 49,9 | 0 |
| Quinoa, cru | 1510 | 358 | 1490 | 354 | 13,3 | 13,2 | 14,1 | 58,1 | 6,07 | 7 | 0,013 | 47 | - | 0,59 | 4,57 | - | 197 | 2,03 | 457 | 563 | - | 5 | 3,1 | 0 | 2,44 | 0 | - | - | 0,36 | 0,32 | 1,52 | 0,77 | 0,49 | 184 | 0 |
| Riz basmati, cuit, non salé | 495 | 117 | 492 | 116 | 71,1 | 2,74 | 2,88 | 24,4 | 0,6 | 1 | <0,013 | 11 | <20 | 0,08 | 0,17 | <20 | 7,1 | 0,23 | 32 | 18 | <20 | <5 | 0,4 | <0,25 | - | <0,8 | - | <0,5 | <0,015 | <0,01 | <0,1 | 0,13 | <0,01 | 8,77 | - |
| Riz blanc étuvé, cru | 1510 | 356 | 1500 | 354 | 11,3 | 7,11 | 7,47 | 78,6 | 0,98 | 1,32 | 0,012 | 101 | - | 0,24 | 0,97 | 2,2 | 31 | 2,47 | 162 | 162 | 6 | 5,08 | 1,36 | 0 | 0,04 | 0,1 | - | 0 | 0,42 | 0,043 | 5,02 | 0,79 | 0,43 | 13 | 0 |
| Riz blanc étuvé, cuit, non salé | 621 | 146 | 618 | 146 | 63,5 | 2,95 | 3,1 | 31,7 | 0,56 | 1,1 | 0,0059 | 11,1 | - | <0,1 | <0,1 | <10 | 10 | <0,1 | 35 | 41,7 | <5 | 2,35 | <0,1 | 0 | 0,01 | 0 | - | 0 | 0,07 | 0,019 | 0,4 | 0,15 | <0,05 | 12,1 | 0 |
| Riz blanc, cru | 1490 | 352 | 1490 | 350 | 12,5 | 7,04 | 7,4 | 78 | 0,91 | 1,05 | 0,0061 | 33 | - | 0,19 | 1,57 | 2,2 | 31,3 | 0,99 | 118 | 121 | 6 | 2,28 | 1,41 | 0 | 0,08 | 0,1 | - | 0 | 0,15 | 0,042 | 1,92 | 0,88 | 0,13 | 19,3 | 0 |
| Riz blanc, cuit, non salé | 615 | 145 | 612 | 144 | 63,9 | 2,92 | 3,06 | 31,8 | 0,41 | 0,8 | <0,01 | 14 | 5,91 | 0,07 | 0,04 | <20 | 7,1 | 0,17 | 35 | 16 | <20 | <5 | 0,26 | 0 | <0,08 | <0,8 | - | <0,5 | 0,03 | <0,01 | 0,16 | 0,21 | 0,029 | 9,9 | 0,032 |
| Riz complet, cru | 1480 | 350 | 1480 | 349 | 12,6 | 7,02 | 7,38 | 71,4 | 2,8 | 5 | 0,01 | 11,1 | 62 | 0,27 | <1 | <10 | 118 | 2,2 | 163 | 219 | <5 | 4,1 | 2 | 0 | 0,2 | 1,9 | - | 0 | 0,26 | 0,05 | <0,05 | 1,1 | 0,09 | 48 | 0 |
| Riz complet, cuit, non salé | 668 | 158 | 665 | 157 | 60,4 | 3,21 | 3,38 | 32,6 | 1 | 2,3 | <0,01 | 13 | 3,69 | 0,1 | 0,32 | <20 | 49 | 1,1 | 120 | 43 | <20 | <5 | 0,62 | 0 | 0,08 | <0,8 | - | <0,5 | 0,065 | <0,01 | 0,25 | 0,38 | 0,049 | 29,4 | 0,094 |
| Riz rouge, cru | 1490 | 352 | 1480 | 350 | 12,4 | 7,97 | 8,38 | 70,6 | 3 | 4,5 | 0,028 | 13,5 | 61 | <1 | <1 | <10 | 130 | 2 | 372 | 288 | <5 | 11 | 2,1 | - | 0,7 | - | - | - | 0,26 | 0,06 | 1,7 | 1,1 | 0,12 | 15 | - |
| Riz rouge, cuit, non salé | 598 | 141 | 595 | 141 | 63,1 | 3,45 | 3,63 | 28,2 | 0,69 | 4 | 0,0088 | 19,9 | 38 | - | <1 | <10 | 54,5 | <1 | 156 | 75,4 | <5 | 3,5 | <1 | - | <0,1 | - | - | - | 0,11 | <0,05 | <0,05 | 0,29 | 0,44 | 9 | - |
| Riz sauvage, cru | 1460 | 344 | 1450 | 342 | 11,5 | 11,1 | 11,7 | 69,2 | 0,94 | 5,9 | 0,0068 | 6,95 | 78 | <1 | <1 | <10 | 90,1 | 1,1 | 317 | 298 | <5 | 2,7 | 3,7 | 0 | <0,1 | 1,9 | - | 0 | 0,41 | 0,15 | 3,5 | 1,1 | 0,34 | 49 | 0 |
| Riz sauvage, cuit, non salé | 430 | 102 | 427 | 101 | 73,9 | 3,8 | 3,99 | 19,7 | 0,34 | 1,8 | 0,0075 | 3 | - | 0,12 | 0,6 | - | 32 | 0,28 | 82 | 101 | - | 3 | 1,34 | 0 | 0,24 | 0,5 | - | 0 | 0,052 | 0,087 | 1,29 | 0,15 | 0,14 | 26 | 0 |
| Riz thaï ou basmati, cru | 1500 | 353 | 1490 | 351 | 12,4 | 7,67 | 8,06 | 77,8 | 0,84 | 0,92 | 0,014 | - | - | - | - | - | - | - | - | - | - | 6,39 | - | - | - | - | - | - | - | - | - | - | - | - | - |
| Riz thaï ou basmati, cuit, non salé | - | - | - | - | 67,8 | 2,86 | 3 | 28,3 | - | 0,8 | <0,01 | 10 | 9,38 | 0,07 | 0,07 | <20 | 7,2 | 0,26 | 29 | 11 | <20 | <5 | 0,44 | - | <0,08 | <0,8 | - | <0,5 | 0,01 | <0,01 | <0,1 | 0,16 | 0,05 | 10,7 | 0,048 |
| Riz thaï, cuit, non salé | 605 | 143 | 602 | 142 | 64,7 | 2,92 | 3,06 | 30,5 | 0,7 | 1,1 | <0,013 | 5,7 | <20 | 0,1 | <0,05 | <20 | 4 | 0,25 | 24 | 12 | <20 | <5 | 0,55 | <0,25 | - | <0,8 | - | <0,5 | <0,015 | <0,01 | <0,1 | 0,068 | <0,01 | 26,3 | - |
| Riz, mélange de variétés (blanc, complet, rouge, sauvage, etc.), cru | 1500 | 355 | 1500 | 353 | 12,2 | 7,14 | 7,5 | 78,1 | 1 | 1,5 | 0,025 | - | - | - | - | - | - | - | - | - | - | 10 | - | - | - | - | - | - | - | - | - | - | - | - | - |
| Sarrasin entier, cru | 1530 | 356 | 1500 | 356 | 10,1 | 12,9 | 12,9 | 64,6 | 3,55 | 7 | 0,0025 | 16,3 | - | 1,1 | 2,2 | 2,5 | 231 | 1,3 | 362 | 460 | - | 1 | 2,4 | 0 | - | - | - | 0 | 0,1 | 0,43 | 7,02 | 1,23 | 0,21 | 30 | 0 |
| Seigle entier, cru | 1410 | 334 | 1400 | 331 | 10,7 | 9,83 | 10,5 | 61 | 1,97 | 15 | 0,0073 | 26,4 | 60 | 0,34 | 3,32 | 2,7 | 104 | 3,14 | 346 | 469 | 1,5 | 2,9 | 2,78 | 0 | 1 | 5,9 | - | 0 | 0,34 | 0,21 | 2,99 | 1,4 | 0,29 | 47 | 0 |
| Semoule de blé dur, crue | 1490 | 352 | 1480 | 348 | 12,1 | 10,9 | 12 | 71,6 | 1,25 | 3,37 | 0,003 | 16,9 | - | 0,3 | 0,92 | 0,3 | 26,5 | 2,31 | 102 | 152 | 1 | 1,2 | 2,18 | 0 | 0,1 | 30 | - | 0 | 0,25 | 0,063 | 2,16 | 0,54 | 0,094 | 57 | 0 |

| | | | | | | | | | | | | | | | | | | | | | | | | | | | | | | | | | | | |
|---|---|---|---|---|---|---|---|---|---|---|---|---|---|---|---|---|---|---|---|---|---|---|---|---|---|---|---|---|---|---|---|---|---|---|---|
| Semoule de blé dur, cuite, non salée | 516 | 122 | 510 | 120 | 69,6 | 3,42 | 3,75 | 24 | 0,8 | 1,9 | < 0,015 | 13 | 23,2 | 0,12 | 1 | < 20 | 16 | 3 | 62 | 96 | < 20 | < 5 | 0,48 | < 0,25 | - | < 0,8 | - | < 0,5 | 0,053 | < 0,01 | 0,24 | 0,24 | < 0,01 | 19,5 | - |
| Sorgho entier, cru | 1470 | 349 | 1470 | 349 | 12,4 | 10,6 | 10,6 | 65,4 | 3,46 | 6,7 | 0,005 | 13 | 60 | 0,28 | 3,36 | - | 165 | 1,61 | 289 | 363 | 43 | 2 | 1,67 | 0 | 0,5 | - | - | 0 | 0,33 | 0,096 | 3,69 | 0,37 | 0,44 | 20 | 0 |
| Tartine craquante, extrudée et grillée | 1610 | 379 | 1590 | 375 | 6,4 | 9,97 | 10,9 | 73,7 | 3,57 | 4,37 | 1,43 | 42,3 | 871 | - | 1 | < 10 | 29,8 | 1 | 131 | 254 | < 5 | 576 | < 1 | - | 0,35 | - | - | 0 | 0,77 | 0,8 | 9,82 | 0,71 | 0,8 | 87 | 0 |
| Tortilla souple (à garnir), à base de blé | 1380 | 327 | 1360 | 323 | 26 | 8,34 | 9,15 | 52,1 | 8,2 | 3,85 | 1,25 | 45 | 219 | 0,1 | 0,76 | < 20 | 19 | 0,4 | 360 | 150 | < 20 | 650 | 0,57 | - | < 0,08 | 4,78 | - | - | 0,11 | < 0,01 | 0,17 | 0,35 | 0,027 | 39,4 | - |
| Tortilla souple (à garnir), à base de maïs | 1320 | 313 | 1320 | 313 | 26 | 9,06 | 9,06 | 54,7 | 5,8 | 2,8 | 1,08 | 23 | 303 | 0,09 | 0,81 | < 20 | 20 | 0,31 | 280 | 120 | < 20 | 450 | 0,63 | - | 1,04 | 4,49 | - | < 0,5 | 0,086 | < 0,01 | < 0,1 | 0,24 | 0,042 | 34,5 | - |
| Vermicelle de riz, cuite, non salée | 379 | 89,3 | 378 | 89,1 | 77,8 | 1,43 | 1,5 | 20,1 | < 0,5 | < 0,5 | 0,024 | 5,2 | 6,22 | 0,03 | 0,08 | < 20 | 1,3 | 0,05 | 13 | 2,2 | < 20 | 9,5 | 0,15 | < 0,25 | 0,03 | < 0,8 | - | < 0,5 | < 0,015 | < 0,01 | < 0,1 | 0,054 | < 0,01 | 6,57 | 0 |
| Vermicelle de riz, sèche | 1550 | 365 | 1540 | 363 | 9,5 | 7,44 | 7,81 | 80,5 | 1 | 1,2 | 0,03 | 14,2 | 48 | < 1 | < 1 | < 10 | 13 | < 1 | 60 | 20,3 | < 5 | 12 | 1,3 | - | < 0,1 | - | - | - | < 0,05 | < 0,05 | < 0,05 | 0,22 | < 0,05 | 10 | - |
| Vermicelle de soja, cuite, non salée | 260 | 61,2 | 259 | 61,1 | 84,4 | < 0,57 | < 0,63 | 14,4 | 0,1 | 0,8 | 0,0063 | 14,2 | 39 | - | < 1 | < 10 | 2,9 | < 1 | 10 | 1,2 | < 5 | 2,5 | < 1 | - | < 0,1 | - | - | - | < 0,05 | < 0,05 | < 0,05 | 0,11 | < 0,05 | 10 | - |
| Vermicelle de soja, sèche | 1460 | 344 | 1460 | 344 | 13,1 | < 0,57 | < 0,63 | 84,7 | 0,14 | 1,6 | 0,039 | 25,2 | 47 | < 1 | < 1 | < 10 | 9,6 | < 1 | 40 | 13,2 | < 5 | 15,6 | < 1 | - | < 0,1 | - | - | - | < 0,05 | < 0,05 | < 0,05 | 1,1 | < 0,05 | 7 | - |

# Appendix 4

| fruits, légumes, légumineuses et oléagineux | Energie, Règlement UE N° 1169/2011 (kJ/100 g) | Energie, Règlement UE N° 1169/2011 (kcal/100 g) | Energie, N x facteur Jones, avec fibres (kJ/100 g) | Energie, N x facteur Jones, avec fibres (kcal/100 g) | Eau (g/100 g) | Protéines, N x facteur de Jones (g/100 g) | Protéines, N x 6.25 (g/100 g) | Glucides (g/100 g) | Lipides (g/100 g) | Fibres alimentaires (g/100 g) | Sel chlorure de sodium (g/100 g) | Calcium (mg/100 g) | Chlorure (mg/100 g) | Cuivre (mg/100 g) | Fer (mg/100 g) | Iode (µg/100 g) | Magnésium (mg/100 g) | Manganèse (mg/100 g) | Phosphore (mg/100 g) | Potassium (mg/100 g) | Sélénium (µg/100 g) | Sodium (mg/100 g) | Zinc (mg/100 g) | Vitamine D (µg/100 g) | Vitamine E (mg/100 g) | Vitamine K1 (µg/100 g) | Vitamine K2 (µg/100 g) | Vitamine C (mg/100 g) | Vitamine B1 ou Thiamine (mg/100 g) | Vitamine B2 ou Riboflavine (mg/100 g) | Vitamine B3 ou PP ou Niacine (mg/100 g) | Vitamine B5 ou Acide pantothénique (mg/100 g) | Vitamine B6 (mg/100 g) | Vitamine B9 ou Folates totaux (µg/100 g) | Vitamine B12 (µg/100 g) |
|---|---|---|---|---|---|---|---|---|---|---|---|---|---|---|---|---|---|---|---|---|---|---|---|---|---|---|---|---|---|---|---|---|---|---|---|
| Abricot au sirop (sans précision sur léger ou classique), appertisé, égoutté (aliment moyen) | | | | | 81 | 0,5 | 0,5 | 16,2 | 0,25 | 1,4 | 0,03 | 16 | 1,8 | 0,04 | 0,13 | 10 | 5,3 | 0,02 | 10 | 130 | 10 | 12 | 0,06 | | | 0,4 | | 0,25 | 0,7 | 0,005 | 0,11 | 0,071 | 0,02 | 9,3 | |
| Abricot au sirop léger, appertisé, égoutté | - | - | - | - | 83 | 0,94 | 0,94 | 13,7 | < 0,5 | 1,4 | 0,024 | 18 | 4,63 | 0,04 | 0,12 | < 20 | 4,9 | 0,03 | 11 | 140 | < 20 | 9,7 | 0,06 | - | - | < 0,8 | - | 2,19 | 0,7 | < 0,01 | < 0,1 | 0,12 | 0,017 | 6,85 | - |
| Abricot au sirop léger, appertisé, non égoutté | - | - | - | - | 83,5 | 0,81 | 0,81 | 14 | 0,043 | 0,83 | 0,025 | 14,3 | 4,45 | 0,036 | 0,1 | < 20 | 4,74 | 0,026 | 9,25 | 136 | < 20 | 9,86 | 0,06 | 0 | 0,6 | < 0,8 | - | 2,28 | 0,68 | < 0,01 | < 0,1 | 0,11 | 0,017 | 6,98 | 0 |
| Abricot au sirop, appertisé, égoutté | - | - | - | - | 81 | 0,5 | 0,5 | 16,2 | < 0,5 | 1,4 | 0,03 | 16 | < 3,6 | 0,04 | 0,13 | < 20 | 5,3 | 0,02 | 10 | 130 | < 20 | 12 | 0,06 | - | - | < 0,8 | - | < 0,5 | 0,7 | < 0,01 | 0,11 | 0,071 | 0,02 | 9,3 | - |
| Abricot au sirop, appertisé, non égoutté | - | - | - | - | 81,4 | 0,29 | 0,29 | 16,6 | < 0,5 | 0,81 | 0,03 | 11,3 | < 3,6 | 0,036 | 0,11 | < 20 | 4,8 | 0,02 | 8,66 | 114 | < 20 | 12 | 0,035 | - | - | < 0,8 | - | < 0,5 | - | < 0,01 | 0,12 | 0,073 | 0,036 | 9,89 | - |
| Abricot pays, pulpe, au sirop, appertisé, non égoutté, prélevé à la Martinique | - | - | - | - | 84,5 | 0,28 | 0,28 | 13,3 | 0,1 | 1,8 | - | - | - | - | - | - | - | - | - | - | - | - | - | - | - | - | - | 0 | - | - | - | - | - | - | - |
| Abricot pays, pulpe, cru, prélevé à la Martinique | - | - | - | - | 85,4 | 0,48 | 0,48 | 10,8 | 0,05 | 3,1 | - | - | - | 0,095 | traces | - | - | - | - | - | - | - | - | - | - | - | - | - | - | - | - | - | - | 12,7 | - |
| Abricot, dénoyauté, cru | 194 | 45,9 | 194 | 45,9 | 87,1 | 0,81 | 0,81 | 9,01 | < 0,5 | 1,7 | < 0,013 | 15 | < 20 | 0,06 | 0,19 | < 20 | 8,4 | 0,07 | 22 | 260 | < 20 | < 5 | 0,09 | 0 | 0,7 | < 0,8 | - | 2,55 | 0,03 | 0,013 | < 0,1 | 0,19 | 0,054 | 7,6 | 0 |
| Abricot, dénoyauté, sec | 1010 | 239 | 1010 | 239 | 24,7 | 2,88 | 2,88 | 59,1 | 0,5 | 8,3 | < 0,013 | 71 | < 20 | 0,45 | 1,4 | < 20 | 41 | 0,27 | 79 | 1400 | < 20 | < 5 | 0,35 | < 0,25 | 5,52 | < 0,8 | - | < 0,5 | < 0,015 | < 0,01 | 1,12 | 0,69 | 0,48 | 8,85 | 0 |
| Abricot, dénoyauté, sec, moelleux (réhydraté à 35-45%) | 846 | 200 | 846 | 200 | 39,3 | 2,31 | 2,31 | 51,1 | 0,4 | 3,5 | < 0,013 | 53 | < 20 | 0,29 | 0,86 | < 20 | 32 | 0,19 | 59 | 980 | < 20 | < 5 | 0,3 | < 0,25 | 3,45 | < 0,8 | - | < 0,5 | < 0,015 | < 0,01 | 0,46 | 0,43 | 0,51 | 7,59 | - |
| Amande (avec peau) | - | - | - | - | 4,8 | 18,8 | 22,6 | 9,51 | 51,3 | 12,5 | < 0,013 | 260 | < 20 | 0,96 | 3,4 | < 20 | 270 | 1,9 | 510 | 800 | < 20 | < 5 | 3,5 | < 0,25 | 22,3 | < 0,8 | - | < 0,5 | 0,15 | 0,29 | 1,97 | 0,62 | 0,07 | 120 | 0 |
| Amande, grillée, salée | - | - | - | - | 2,41 | 21,1 | 24,1 | 8,72 | 53,9 | 10,9 | 1,25 | 268 | - | 1,1 | 3,73 | - | 279 | 2,23 | 471 | 713 | - | 498 | 3,31 | 0 | 23,9 | 0 | - | 0 | 0,077 | 1,2 | 3,64 | 0,32 | 0,14 | 55 | 0 |
| Amande, mondée, émondée ou blanchie | - | - | - | - | 4,51 | 21,4 | 25,8 | 8,76 | 52,5 | 9,9 | 0,048 | 236 | - | 1,03 | 3,28 | - | 268 | 1,84 | 481 | 659 | - | 19 | 2,97 | 0 | 23,8 | 0 | - | 0 | 0,19 | 0,71 | 3,5 | 0,31 | 0,12 | 49 | 0 |
| Ananas au jus d'ananas, appertisé, non égoutté | - | - | - | - | 84,9 | < 0,5 | < 0,5 | 13,6 | < 0,5 | 0,92 | 0,00049 | 10,3 | 18 | 0,033 | 0,19 | < 20 | 11,7 | 1,13 | 7,14 | 127 | < 20 | 0,2 | 0,11 | - | - | < 0,8 | - | 4,98 | 0,063 | < 0,01 | 1,44 | 0,061 | 0,064 | 12,9 | - |
| Ananas au jus d'ananas, égoutté, appertisé | - | - | - | - | 84,5 | < 0,5 | < 0,5 | 13,6 | < 0,5 | 1,4 | - | 11 | 18 | 0,04 | 0,19 | < 20 | 12 | 1,2 | 7,8 | 130 | < 20 | < 1 | 0,11 | - | - | < 0,8 | - | 4,99 | 0,063 | < 0,01 | 2,2 | 0,067 | 0,076 | 13,5 | - |
| Ananas au sirop léger, appertisé, égoutté | - | - | - | - | 82 | < 0,5 | < 0,5 | 15,9 | < 0,5 | 1,6 | 0,0048 | 6,4 | 13,1 | 0,04 | 0,13 | < 20 | 11 | 1,2 | 4,8 | 95 | < 20 | 1,9 | 0,05 | - | - | < 0,8 | - | 5,22 | 0,034 | < 0,01 | < 0,1 | 0,04 | 0,046 | 11,9 | - |
| Ananas au sirop léger, appertisé, non égoutté | - | - | - | - | 82,4 | < 0,5 | < 0,5 | 16 | 0,078 | 1,01 | 0,0047 | 5,77 | 13,9 | 0,033 | 0,12 | < 20 | 11,4 | 1,16 | 4,06 | 101 | < 20 | 1,86 | 0,031 | 0 | 0,01 | < 0,8 | - | 5,11 | 0,037 | < 0,01 | < 0,1 | 0,03 | 0,029 | 11,8 | 0 |
| Ananas Victoria ou ananas Queen Victoria, pulpe crue, prélevé à La Réunion Ananas comosus (L.) merr var. Queen) | 304 | 71,8 | 304 | 71,8 | 81,3 | 0,94 | 0,94 | 15,1 | < 0,3 | 2,4 | < 0,013 | 6,3 | < 20 | 0,07 | 0,17 | < 20 | 15 | 0,2 | 9,6 | 130 | < 20 | < 5 | 0,11 | - | < 0,08 | < 0,8 | - | 18,3 | 0,084 | 0,022 | 0,23 | 0,19 | 0,055 | 25,4 | - |
| Ananas, pulpe, cru | 231 | 54,4 | 231 | 54,4 | 85,5 | < 0,5 | < 0,5 | 11,7 | < 0,5 | 1,2 | < 0,013 | 8 | 21,6 | 0,06 | 0,17 | < 20 | 15 | 0,84 | 8,1 | 140 | < 20 | < 5 | 0,08 | 0 | < 0,08 | < 0,8 | - | 46,1 | 0,056 | 0,033 | 0,31 | 0,17 | 0,052 | 19,6 | 0 |
| Ananas, pulpe, cru , prélevé à la Martinique | - | - | - | - | 83,4 | 0,56 | 0,56 | - | 0 | - | - | - | - | - | - | - | - | - | - | - | - | - | - | - | - | - | - | 38,3 | - | - | - | - | - | - | - |
| Anone ou chérimole, pulpe, crue | - | - | - | - | 79,4 | 1,72 | 1,72 | 15,4 | 0,64 | 2,65 | 0,018 | 9 | - | 0,069 | 0,29 | 0,4 | 17 | 0,093 | 26 | 287 | - | 7 | 0,16 | 0 | 0,27 | - | - | 12,1 | 0,096 | 0,13 | 0,61 | 0,35 | 0,26 | 23 | 0 |
| Arachide, bouillie/cuite à l'eau, salée | - | - | - | - | 41,8 | 15,5 | 15,5 | 9,54 | 22 | 8,8 | 1,88 | 55 | - | 0,5 | 1,01 | - | 102 | 1,02 | 198 | 180 | - | 751 | 1,83 | 0 | 4,1 | 0 | - | 0 | 0,26 | 0,063 | 5,26 | 0,83 | 0,15 | 75 | 0 |
| Artichaut, appertisé, égoutté | - | - | - | - | 89,9 | 1,5 | 1,5 | 6,5 | 0,2 | 3 | 0,8 | - | - | - | - | - | - | - | - | - | - | 320 | - | - | - | - | - | - | - | - | - | - | - | - | - |
| Artichaut, coeur, appertisé, égoutté | 109 | 26 | 109 | 26 | 90,5 | 1,8 | 1,8 | 2,31 | < 0,5 | 3,2 | 0,78 | 25 | 499 | 0,07 | 0,4 | < 20 | 18 | 0,13 | 34 | 150 | < 20 | 310 | 0,28 | < 0,25 | 1,12 | 6,29 | - | 12,2 | 0,021 | 0,03 | 0,39 | 0,13 | 0,026 | 54,3 | - |
| Artichaut, cru | - | - | - | - | 84,9 | 3,2 | 3,2 | 4,92 | 0,18 | 5,4 | 0,15 | 52 | - | 0,23 | 1,19 | 0,5 | 47,8 | 0,26 | 81,2 | 387 | 0,23 | 60,5 | 0,49 | 0 | 0,27 | 14,8 | - | 11,7 | 0,081 | 0,063 | 1,05 | 0,32 | 0,11 | 68 | 0 |
| Artichaut, cuit | - | - | - | - | 85,8 | 2,53 | 2,53 | 0,99 | 0,28 | 8,3 | 0,15 | 42,9 | - | 0,082 | 0,67 | < 5 | 44,6 | 0,14 | 73 | 427 | < 10 | 60 | 0,29 | 0 | 0,19 | 14,8 | - | 9,1 | 0,05 | 0,089 | 1,11 | 0,24 | 0,081 | 89 | 0 |
| Artichaut, cuit à la vapeur sous pression | 195 | 47,3 | 195 | 47,3 | 81,4 | 2,63 | 2,63 | 3,24 | < 0,3 | 10,9 | 0,068 | 42 | 92,4 | 0,05 | 0,7 | < 20 | 41 | 0,17 | 56 | 480 | < 20 | 27 | 0,27 | - | < 0,08 | 1,47 | - | < 0,5 | 0,03 | 0,023 | 0,82 | 0,35 | 0,086 | 52,6 | - |
| Artichaut, fond, appertisé, égoutté | 106 | 25,2 | 106 | 25,2 | 91,3 | 1,2 | 1,2 | 3,02 | < 0,5 | 2,6 | 0,73 | 36 | 464 | 0,06 | 0,36 | < 20 | 20 | 0,05 | 26 | 140 | < 20 | 293 | 0,2 | < 0,25 | 0,42 | 1,57 | - | 18 | 0,041 | < 0,01 | 0,2 | 0,09 | 0,22 | 53,6 | - |
| Artichaut, fond, surgelé, cru | - | - | - | - | 84,7 | 2,08 | 2,08 | 8,09 | 0,43 | 7,54 | 0,18 | - | - | - | - | - | - | - | - | - | - | 45 | - | - | - | - | - | - | - | - | - | - | - | - | - |
| Asperge blanche, bouillie/cuite à l'eau | 78,1 | 18,6 | 78,1 | 18,6 | 94,9 | 1,44 | 1,44 | 1,63 | 0,3 | 1,3 | 0,013 | 15 | 31,6 | 0,09 | 0,26 | < 20 | 7,1 | 0,07 | 29 | 120 | < 20 | 5 | 0,3 | - | 0,39 | < 0,8 | - | 5,16 | 0,04 | 0,018 | 0,15 | 0,1 | 0,029 | 66,8 | - |
| Asperge verte, bouillie/cuite à l'eau | 112 | 26,6 | 112 | 26,6 | 92,7 | 2,69 | 2,69 | 1,75 | 0,3 | 2,2 | < 0,013 | 21 | 54,4 | 0,11 | 0,64 | < 20 | 10 | 0,13 | 49 | 220 | < 20 | < 5 | 0,49 | - | 1,5 | 12,5 | - | 4,73 | 0,067 | 0,056 | 0,7 | 0,24 | 0,046 | 69,9 | - |
| Asperge, appertisée, égouttée | - | - | - | - | 94,2 | 1,58 | 1,58 | 1,43 | 0,27 | 1,5 | 0,73 | 15,8 | - | 0,076 | 1,09 | 0,7 | 8,5 | 0,16 | 35 | 142 | 0,2 | 291 | 0,42 | 0 | 0,74 | 41,3 | - | 18,4 | 0,046 | 0,07 | 0,63 | 0,1 | 0,11 | 72 | 0 |
| Asperge, blanche ou violette, pelée, crue | - | - | - | - | 92,4 | 2,5 | 2,5 | 2,5 | 0,3 | 1,8 | 0,019 | 16,6 | - | - | - | - | 7,3 | - | 42,5 | 169 | - | 7,7 | - | 0 | - | - | - | 18 | - | - | - | - | 0,072 | 150 | 0 |
| Asperge, bouillie/cuite à l'eau | - | - | - | - | 93,4 | 2,68 | 2,68 | 0,81 | 0,32 | 1,8 | 0,021 | 19,9 | - | 0,074 | 0,74 | 0,25 | 6,31 | 0,049 | 51,5 | 198 | < 3,35 | 8,5 | 0,2 | 0 | 1,35 | 65,3 | - | 16,1 | 0,11 | 0,12 | 1,06 | 0,19 | 0,05 | 142 | 0 |
| Asperge, pelée, crue | - | - | - | - | 93,3 | 2,04 | 2,04 | 2,4 | 0,21 | 1,95 | 0,011 | 22 | - | 0,12 | 1,32 | 0,15 | 11,5 | 0,16 | 52 | 202 | 1 | 4,2 | 0,49 | 0 | 0,72 | 40,3 | - | 8,8 | 0,13 | 0,12 | 0,99 | 0,17 | 0,076 | 101 | 0 |
| Asperge, verte, crue | - | - | - | - | 92,5 | 2,46 | 2,46 | 2,03 | 0,27 | 2,15 | 0,014 | 22,4 | - | - | 1 | - | 12,2 | - | 85,5 | 267 | - | 3,85 | - | 0 | - | - | - | 15,5 | - | - | - | - | 0,15 | 150 | 0 |
| Aubergine, crue | - | - | - | - | 92,9 | 1,12 | 1,12 | 2,39 | 0,14 | 2,7 | 0,0063 | 8,7 | - | 0,081 | 0,32 | 0,15 | 12 | 0,24 | 27,2 | 235 | 0,2 | 2,5 | 0,16 | 0 | 0,17 | 3,5 | - | 2,1 | 0,045 | 0,034 | 0,72 | 0,25 | 0,082 | 24,5 | 0 |
| Aubergine, cuite | - | - | - | - | 88,6 | 1,33 | 1,33 | 4,17 | 0,28 | 4,3 | 0,015 | 20,1 | - | 0,058 | 0,25 | 1 | 15 | 0,12 | 15 | 123 | < 3,35 | 5,95 | 0,098 | 0 | 0,41 | 2,9 | - | 1,3 | 0,076 | 0,02 | 0,6 | 0,075 | 0,086 | 14 | 0 |

| | | | | | | | | | | | | | | | | | | | | | | | | | | | | | | | | | | | |
|---|---|---|---|---|---|---|---|---|---|---|---|---|---|---|---|---|---|---|---|---|---|---|---|---|---|---|---|---|---|---|---|---|---|---|---|
| Aubergine, pulpe et peau, rôtie/cuite au four | 143 | 33,8 | 143 | 33,8 | 91 | 1,31 | 1,31 | 3,83 | < 0,3 | 1,9 | < 0,013 | 13 | 55,9 | 0,08 | 0,59 | < 20 | 20 | 0,15 | 31 | 310 | < 20 | < 5 | 0,14 | - | < 0,08 | < 0,8 | - | < 0,5 | < 0,015 | < 0,01 | 0,42 | 0,52 | 0,053 | < 5 | - |
| Avocat, pulpe, cru | 843 | 205 | 843 | 205 | 70,3 | 1,56 | 1,56 | 0,83 | 20,6 | 3,6 | 0,015 | 9,4 | < 20 | 0,18 | 0,34 | < 20 | 21 | 0,2 | 38 | 430 | < 20 | 6 | 0,43 | 0 | 2,23 | 14,5 | - | < 0,5 | 0,052 | 0,037 | 1,56 | 1,07 | 0,17 | 70,4 | 0 |
| Bambou, pousse, crue | 148 | 35,1 | 148 | 35,1 | 89,8 | 2,52 | 2,52 | 5,08 | 0,2 | 1,45 | 0,01 | 15,5 | - | 0,19 | 0,4 | - | 3 | 0,26 | 54,5 | 517 | - | 4 | 1,1 | 0 | 1 | 0 | - | 7,5 | 0,095 | 0,095 | 0,55 | 0,16 | 0,24 | 7 | 0 |
| Bambou, pousses, appertisées, égouttées | - | - | - | - | 95 | 1,61 | 1,61 | 1,9 | 0,15 | 1,25 | 0,19 | 17 | - | 0,098 | 0,24 | - | 3,5 | 0,14 | 26,8 | 196 | - | 74,1 | 0,56 | 0 | 0,63 | 0 | - | 1,1 | 0,017 | 0,039 | 0,16 | 0,079 | 0,12 | 2,5 | 0 |
| Banane jaune, pulpe, cuite à la vapeur, prélevée à la Martinique | - | - | - | - | 64,1 | 1,06 | 1,06 | - | 0,057 | - | - | - | - | - | - | - | - | - | - | - | - | - | - | - | - | - | - | 6,07 | - | - | - | - | - | - | - |
| Banane plantain, crue | - | - | - | - | 65,3 | 1,28 | 1,28 | 29,6 | 0,39 | 2,3 | 0,01 | 3 | - | 0,081 | 0,6 | 2,5 | 37 | traces | 34 | 499 | - | 4 | 0,14 | 0 | 0,14 | 0,7 | - | 18,4 | 0,052 | 0,054 | 0,69 | 0,26 | 0,3 | 22 | 0 |
| Banane plantain, cuite | 529 | 125 | 529 | 125 | 67,3 | 0,79 | 0,79 | 28,9 | 0,18 | 2,3 | 0,013 | 2 | - | 0,066 | 0,58 | - | 32 | - | 28 | 465 | - | 5 | 0,13 | 0 | 0,13 | 0,7 | - | 20,9 | 0,046 | 0,052 | 0,76 | 0,23 | 0,24 | 26 | 0 |
| Banane, pulpe, crue | 383 | 90,5 | 383 | 90,5 | 75,8 | 1,06 | 1,06 | 19,7 | < 0,5 | 2,7 | < 0,013 | 5,1 | 79,8 | 0,06 | 0,2 | < 20 | 28 | 0,36 | 29 | 320 | < 20 | < 5 | 0,14 | 0 | < 0,08 | < 0,8 | - | 7,16 | 0,054 | < 0,01 | 0,39 | 0,31 | 0,18 | 19 | 0 |
| Banane, pulpe, sèche | - | - | - | - | 3 | 3,89 | 3,89 | 78,4 | 1,81 | 9,9 | 0,0075 | 22 | - | 0,39 | 1,15 | 3 | 108 | 0,57 | 74 | 1490 | - | 3 | 0,61 | 0 | 0,39 | 2 | - | 7 | 0,18 | 0,24 | 2,8 | - | 0,44 | 14 | 0 |
| Batavia, crue | 74,6 | 17,9 | 74,6 | 17,9 | 95,6 | 1,25 | 1,25 | 1,78 | < 0,5 | 2,1 | 0,015 | 26 | 60,6 | 0,04 | 0,39 | < 20 | 8,7 | 0,28 | 17 | 200 | < 20 | 6 | 0,16 | - | < 0,08 | 20,6 | - | 4,38 | 0,042 | < 0,01 | 0,2 | 0,079 | 0,051 | 65,6 | - |
| Bette ou blette, côte et feuille, bouillie/cuite à l'eau | 70 | 16,9 | 70 | 16,9 | 95,2 | 0,88 | 0,88 | 1,43 | 0,3 | 2,5 | 0,19 | 60 | 79,6 | 0,04 | 0,52 | < 20 | 10 | 0,34 | 18 | 160 | < 20 | 77 | 0,11 | - | < 0,08 | 8,29 | - | < 0,5 | < 0,015 | < 0,01 | < 0,1 | 0,1 | 0,098 | 18,3 | - |
| Bette ou blette, crue | 68,3 | 16,4 | 68,3 | 16,4 | 95,4 | 1 | 1 | 1,63 | < 0,5 | 1,8 | 0,57 | 25 | 221 | 0,03 | 0,85 | < 20 | 17 | 0,16 | 10 | 300 | < 20 | 53,2 | 0,07 | < 0,25 | 0,25 | < 0,8 | - | 3,08 | < 0,015 | 0,015 | < 0,1 | 0,2 | 0,036 | 24,7 | 0 |
| Bette ou blette, cuite | 61,1 | 14,7 | 61,1 | 14,7 | 95,3 | 0,7 | 0,7 | 1,23 | < 0,5 | 2,5 | 0,25 | 56 | 114 | 0,04 | 0,35 | < 20 | 18 | 0,16 | 11 | 240 | < 20 | 99,6 | 0,08 | < 0,25 | 0,41 | 327 | - | < 0,5 | < 0,015 | < 0,01 | < 0,1 | 0,17 | 0,024 | 43,2 | 0 |
| Betterave rouge, crue | - | - | - | - | 86,7 | 1,74 | 1,74 | 9,1 | 0,24 | 2,55 | 0,15 | 22,1 | - | 0,098 | 0,7 | 0,5 | 19 | 0,46 | 38,1 | 328 | 0,2 | 60,5 | 0,54 | 0 | 0,06 | 0,2 | - | 6,45 | 0,028 | 0,043 | 0,27 | 0,15 | 0,059 | 100 | 0 |
| Betterave rouge, cuite | 180 | 42,8 | 180 | 42,8 | 88,3 | 1,44 | 1,44 | 7,13 | 0,4 | 2,5 | 0,23 | 24 | 137 | 0,07 | 0,29 | < 20 | 17 | 0,31 | 18 | 320 | < 20 | 90 | 0,29 | 0 | < 0,08 | < 0,8 | - | < 0,5 | < 0,015 | < 0,01 | < 0,1 | 0,14 | 0,031 | 12,4 | 0 |
| Beurre de cacahuète ou Pâte d'arachide | - | - | - | - | 1,23 | 22,2 | 25,4 | 16,1 | 52,5 | 5 | 0,97 | 43 | - | 0,56 | 1,92 | 0,5 | 174 | 1,59 | 333 | 629 | 6,9 | 388 | 2,76 | 0 | 6,9 | 0,3 | - | 0 | 0,16 | 0,15 | 14,1 | 1,09 | 0,47 | 70 | 0 |
| Brèdes chou de Chine ou bok choy ou pak choï, tiges et feuilles, cuites à la vapeur, prélevées à La Réunion (Brassica rapa subsp. Chinensis) | 74,4 | 17,7 | 74,4 | 17,7 | 94,1 | 1,69 | 1,69 | 1,7 | < 0,3 | 1,3 | 0,038 | 88 | 48,3 | 0,04 | 0,93 | < 20 | 17 | 0,25 | 28 | 330 | < 20 | 15 | 0,28 | - | 1,5 | 53,8 | - | 9,54 | 0,026 | 0,061 | 0,36 | 0,18 | 0,073 | 71,9 | - |
| Brocoli, bouilli/cuit à l'eau, croquant | 98,1 | 23,5 | 98,1 | 23,5 | 92,9 | 2,5 | 2,5 | 1,23 | 0,4 | 2,6 | 0,043 | 42 | < 20 | 0,11 | 0,41 | < 20 | 12 | 0,18 | 44 | 110 | < 20 | 17 | 0,23 | - | 2,42 | 23,3 | - | 22,7 | < 0,015 | < 0,01 | < 0,1 | 0,27 | 0,06 | 67,1 | - |
| Brocoli, bouilli/cuit à l'eau, fondant | 96,1 | 23,1 | 96,1 | 23,1 | 93,8 | 2,19 | 2,19 | 1,05 | 0,5 | 3 | 0,033 | 43 | < 20 | 0,15 | 0,33 | < 20 | 11 | 0,17 | 39 | 90 | < 20 | 13 | 0,18 | - | 1,69 | 49,6 | - | 18,1 | < 0,015 | < 0,01 | < 0,1 | 0,25 | 0,048 | 24,7 | - |
| Brocoli, cru | - | - | - | - | 88,9 | 3,95 | 3,95 | 1,7 | 0,48 | 2,9 | 0,048 | 45,9 | 89 | 0,059 | 0,76 | 0,8 | 21,9 | 0,4 | 76,5 | 357 | 0,6 | 19 | 0,4 | 0 | 1,04 | 181 | - | 106 | 0,074 | 0,12 | 0,82 | 0,78 | 0,17 | 153 | 0 |
| Brocoli, cuit | - | - | - | - | 92,5 | 2,1 | 2,1 | 1,1 | 0,78 | 1,5 | 0,098 | 43,3 | 61 | < 0,1 | < 0,1 | < 10 | 11,2 | < 0,1 | 51 | 125 | < 2,58 | 39 | < 0,1 | 0 | 0,9 | 141 | - | 23,9 | < 0,05 | 0,05 | < 0,05 | 0,22 | < 0,05 | 75,1 | 0 |
| Brocoli, cuit à la vapeur | 158 | 37,6 | 158 | 37,6 | 89,4 | 4,13 | 4,13 | 2,53 | 0,7 | 2,2 | 0,04 | 36 | < 20 | 0,04 | 0,58 | < 20 | 23 | 0,28 | 72 | 340 | < 20 | 16 | 0,53 | - | 2,28 | 23,2 | - | 20,7 | 0,059 | 0,11 | 0,81 | 0,77 | 0,16 | 81,7 | - |
| Brocoli, purée | - | - | - | - | 91,7 | 2,5 | 2,5 | 2,75 | 0,38 | 2,39 | 0,021 | 87 | - | - | 0,9 | - | - | - | - | - | - | 7,93 | - | - | - | - | - | 90 | - | - | - | - | - | - | - |
| Brocoli, surgelé, cru | - | - | - | - | 91,5 | 2,79 | 2,79 | 1,56 | 0,44 | 2,77 | 0,05 | 48,5 | - | 0,036 | 0,77 | - | 17 | 0,26 | 54,5 | 231 | 2,8 | 21,4 | 0,41 | 0 | 1,29 | 101 | - | 62,3 | 0,063 | 0,11 | 0,46 | 0,24 | 0,18 | 80,5 | 0 |
| Brocoli, surgelé, cuit | - | - | - | - | 90,7 | 3,1 | 3,1 | 2,36 | 0,11 | 3 | 0,06 | 51 | - | 0,043 | 0,61 | - | 20 | 0,33 | 55 | 180 | - | 24 | 0,3 | 0 | 1,32 | 88,1 | - | 40,1 | 0,055 | 0,081 | 0,46 | 0,27 | 0,13 | 30 | 0 |
| Cacahuète ou Arachide | - | - | - | - | 2,2 | 22,8 | 26,1 | 14,8 | 49,1 | 8,6 | 0,022 | 57 | 23,6 | 0,46 | 1,6 | < 20 | 190 | 1,4 | 400 | 700 | 30 | 8,6 | 3 | 0 | 2,46 | < 0,8 | - | < 0,5 | 0,43 | 0,024 | 10,6 | 1,76 | 0,4 | 89,3 | 0 |
| Cacahuète, grillée à sec, salée | - | - | - | - | 2 | 21,6 | 24,8 | 18,2 | 47,7 | 7,4 | 1,8 | - | - | - | - | - | - | - | - | - | - | 700 | - | - | - | - | - | - | - | - | - | - | - | - | - |
| Cacahuète, grillée, salée | - | - | - | - | 1,22 | 22,9 | 26,2 | 15 | 50 | 8,04 | 1,33 | 47,4 | 304 | 0,68 | 1,2 | 5,7 | 169 | 1,19 | 244 | 531 | 4,8 | 531 | 2,92 | 0 | 2,5 | 0 | - | < 0,5 | 0,14 | 0,069 | 12,4 | 1,17 | 0,43 | 66,6 | 0 |
| Cacahuète, grillée, sans sel ajouté | - | - | - | - | 1,81 | 23,4 | 26,8 | 11,3 | 51,9 | 8,67 | 0,018 | 58 | - | 0,43 | 1,58 | - | 178 | 1,79 | 363 | 634 | - | 8 | 2,77 | 0 | 4,93 | 0 | - | 0 | 0,15 | 0,2 | 14,4 | 1,01 | 0,47 | 97 | 0 |
| Caïmite, pulpe, cru, prélevé à la Martinique | - | - | - | - | 80,3 | 0,31 | 0,31 | - | 2,87 | - | - | - | - | - | - | - | - | - | - | - | - | - | - | - | - | - | - | 5,6 | - | - | - | - | - | - | - |
| Canneberge ou cranberry, crue | - | - | - | - | 86,8 | 0,75 | 0,75 | 7,6 | 0,13 | 5,13 | 0,02 | 11 | - | 0,061 | 0,44 | 0,15 | 6 | 0,36 | 13,5 | 98,8 | 0 | 8 | 0,1 | 0 | 1,2 | 5,1 | - | 13,7 | 0,012 | 0,018 | 0,3 | 0,3 | 0,057 | 1 | 0 |
| Canneberge ou cranberry, séchée, sucrée | 1410 | 333 | 1410 | 333 | 14,6 | < 0,5 | < 0,5 | 76,4 | 1 | 5,7 | < 0,013 | 8 | < 20 | 0,04 | 0,16 | - | 3,9 | 0,19 | 7,5 | 58 | < 20 | < 5 | < 0,05 | < 0,25 | 1,92 | 5,09 | - | < 0,5 | < 0,015 | < 0,01 | < 0,1 | 0,19 | 0,43 | < 5 | 0 |
| Carambole, pulpe, crue | - | - | - | - | 91,4 | 1,15 | 1,15 | 3,9 | 0,32 | 2,8 | 0,005 | 3 | - | 0,14 | 0,08 | 0,4 | 10 | 0,037 | 12 | 133 | - | 2 | 0,12 | 0 | 0,15 | 0 | - | 34,4 | 0,014 | 0,016 | 0,37 | 0,39 | 0,017 | 12 | 0 |
| Cardon, cru | - | - | - | - | 94 | 0,7 | 0,7 | 1,7 | 0,1 | 1,6 | 0,43 | 70 | - | 0,23 | 0,7 | 0,095 | 42 | 0,26 | 23 | 400 | 0,2 | 170 | 0,17 | 0 | 0,19 | - | - | 2 | 0,02 | 0,03 | 0,3 | 0,34 | 0,12 | 68 | 0 |
| Cardon, cuit | - | - | - | - | 93,5 | 0,76 | 0,76 | 3,63 | 0,11 | 1,7 | 0,44 | 72 | - | - | 0,73 | - | 43 | 0,13 | 23 | 392 | - | 176 | 0,18 | 0 | - | - | - | 1,7 | 0,018 | 0,031 | 0,29 | 0,097 | 0,042 | 22 | 0 |
| Carotte, appertisée, égouttée | 92 | 21,9 | 92 | 21,9 | 93 | 0,67 | 0,67 | 3,4 | 0,2 | 1,75 | 0,58 | 25 | - | 0,1 | 0,64 | 2 | 8 | 0,45 | 24 | 179 | 0,85 | 233 | 0,26 | 0 | 0,74 | 9,8 | - | 2,7 | 0,018 | 0,03 | 0,55 | 0,14 | 0,11 | 9 | 0 |
| Carotte, bouillie/cuite à l'eau, croquante | 150 | 35,7 | 150 | 35,7 | 89,2 | 0,75 | 0,75 | 6,33 | < 0,3 | 3,2 | 0,11 | 27 | 55,1 | 0,07 | 0,19 | < 20 | 8 | 0,17 | 16 | 170 | < 20 | 45 | 0,22 | - | 0,8 | 2,36 | - | 1,71 | 0,016 | < 0,01 | 0,31 | 0,2 | 0,047 | 19,3 | - |
| Carotte, bouillie/cuite à l'eau, fondante | 132 | 31,6 | 132 | 31,6 | 90,1 | < 0,5 | < 0,5 | 5,73 | < 0,3 | 3,3 | 0,12 | 28 | 49,3 | 0,09 | 0,19 | < 20 | 9,4 | 0,14 | 17 | 140 | < 20 | 46 | 0,22 | - | 0,78 | 2,76 | - | 0,52 | < 0,015 | < 0,01 | 0,35 | 0,2 | 0,051 | 17,7 | - |
| Carotte, crue | 169 | 40,2 | 169 | 40,2 | 88,1 | 0,63 | 0,63 | 7,59 | < 0,5 | 2,7 | 0,11 | 25 | 56,8 | 0,05 | 0,24 | < 20 | 10 | 0,1 | 22 | 230 | < 20 | 43 | 0,18 | 0 | 0,27 | 2,96 | - | 2,05 | 0,028 | < 0,01 | < 0,1 | 0,2 | 0,093 | 59,4 | 0 |
| Carotte, cuite | 78,8 | 18,9 | 78,8 | 18,9 | 93 | 0,55 | 0,55 | 2,6 | 0,1 | 2,8 | 0,051 | 31,5 | - | < 0,1 | < 0,1 | < 10 | 7,5 | < 0,1 | 18 | 96,4 | < 2,97 | 20,2 | < 0,1 | 0 | 0,56 | 13,7 | - | < 1 | < 0,05 | < 0,05 | < 0,05 | 0,1 | < 0,05 | 23,3 | 0 |
| Carotte, cuite à la vapeur | 175 | 41,6 | 175 | 41,6 | 87,7 | 0,63 | 0,63 | 7,33 | 0,3 | 3,7 | 0,11 | 26 | 76,1 | 0,04 | 0,22 | < 20 | 10 | 0,18 | 19 | 240 | < 20 | 45 | 0,21 | - | 0,84 | 2,39 | - | 1,28 | 0,018 | < 0,01 | 0,41 | 0,26 | 0,066 | 21,3 | - |
| Carotte, déshydratée | - | - | - | - | 4 | 8,1 | 8,1 | 56 | 1,49 | 23,6 | 0,69 | 212 | - | 0,37 | 3,93 | - | 118 | 1,12 | 346 | 2540 | - | 275 | 1,57 | 0 | 5,45 | 108 | - | 14,6 | 0,53 | 0,42 | 6,57 | 1,47 | 1,04 | 55 | 0 |
| Carotte, purée | 132 | 31,5 | 132 | 31,5 | 90,1 | 0,88 | 0,88 | 5,06 | < 0,3 | 3,1 | 0,13 | 30 | 62,9 | 0,06 | 0,32 | < 20 | 8,4 | 0,1 | 25 | 230 | < 20 | 52 | 0,23 | - | 1,16 | 1,66 | - | < 0,5 | 0,024 | < 0,01 | 0,13 | 0,31 | 0,056 | 14,3 | - |
| Carotte, purée cuisinée à la crème, préemballée | 185 | 44,4 | 185 | 44,4 | 89,6 | 1,44 | 1,44 | 4,57 | 1,6 | 3 | 0,53 | 30 | 324 | 0,04 | 0,19 | < 20 | 6,4 | 0,08 | 21 | 190 | < 20 | 212 | 0,18 | - | 1 | 5,76 | - | < 0,5 | 0,024 | 0,013 | 0,24 | 0,31 | 0,051 | 11,7 | - |
| Carotte, surgelée, crue | - | - | - | - | 90,9 | 0,63 | 0,63 | 4,99 | 0,27 | 2,76 | 0,15 | 34,9 | - | 0,074 | 0,41 | - | 12 | 0,17 | 33 | 235 | - | 51 | 0,33 | 0 | 0,57 | 17,6 | - | 4,32 | 0,044 | 0,037 | 0,46 | 0,19 | 0,095 | 10 | 0 |
| Carotte, surgelée, cuite | 137 | 32,9 | 137 | 32,9 | 90,3 | 0,58 | 0,58 | 4,52 | 0,68 | 3,3 | 0,15 | 35 | - | 0,082 | 0,53 | - | 11 | 0,17 | 31 | 192 | - | 59 | 0,35 | 0 | 1,01 | 13,6 | - | 2,3 | 0,03 | 0,037 | 0,42 | 0,17 | 0,084 | 11 | 0 |
| Cassis, cru | - | - | - | - | 80,5 | 1,33 | 1,33 | 9,68 | 0,86 | 5,8 | 0,0063 | 57,1 | - | 0,093 | 1,17 | 1,5 | 23 | 0,28 | 53,5 | 330 | 1,1 | 2,5 | 0,28 | 0 | 2,1 | - | - | 181 | 0,038 | 0,038 | 0,3 | 0,4 | 0,073 | 8,2 | 0 |
| Céleri branche, appertisé, égoutté | 54,4 | 13,1 | 54,4 | 13,1 | 95 | 0,63 | 0,63 | 1,38 | 0,15 | 2 | 0,54 | 29,2 | - | 0,04 | - | 0,2 | 8 | 0,11 | - | 281 | 1 | 213 | 0,14 | - | - | - | - | - | - | - | - | - | - | - | - |
| Céleri branche, cru | 73,4 | 17,6 | 73,4 | 17,6 | 93,4 | 0,63 | 0,63 | 2,41 | < 0,5 | 2,2 | 0,2 | 46 | 215 | 0,03 | 0,06 | < 20 | 11 | 0,08 | 24 | 390 | < 20 | 79 | 0,09 | 0 | < 0,08 | 4,86 | - | 5,51 | < 0,015 | < 0,01 | < 0,1 | 0,27 | 0,052 | 56,8 | 0 |
| Céleri branche, cuit | - | - | - | - | 94,1 | 0,83 | 0,83 | 1,2 | 0,16 | 1,6 | 0,23 | 53,3 | - | 0,059 | 0,42 | 0,3 | 9 | 0,11 | 25 | 284 | < 2,58 | 91 | 0,077 | 0 | 0,35 | 37,8 | - | 6,1 | 0,023 | 0,07 | 0,37 | 0,32 | 0,086 | 33 | 0 |
| Céleri branche, cuit à l'étouffée | 70 | 16,7 | 70 | 16,7 | 94,3 | 0,75 | 0,75 | 2,48 | < 0,3 | 1,5 | 0,35 | 31 | 170 | 0,04 | 0,12 | < 20 | 8,3 | 0,07 | 24 | 340 | < 20 | 138 | 0,08 | - | 1,58 | 1,26 | - | < 0,5 | < 0,015 | 0,019 | 0,21 | 0,27 | 0,034 | 7,98 | - |
| Céleri rave, bouilli/cuit à l'eau | 110 | 26,2 | 110 | 26,2 | 91 | 0,94 | 0,94 | 4,32 | 0,3 | 2,6 | 0,093 | 41 | 40,1 | 0,11 | 0,27 | < 20 | 10 | 0,16 | 38 | 300 | < 20 | 37 | 0,17 | - | 1,13 | < 0,8 | - | 0,87 | 0,019 | 0,02 | 0,36 | 0,28 | 0,074 | 84,5 | - |
| Céleri-rave, cru | 122 | 29,3 | 122 | 29,3 | 88,8 | 1,38 | 1,38 | 3,98 | < 0,5 | 4,5 | 0,085 | 40 | 104 | 0,11 | 0,28 | < 20 | 13 | 0,2 | 56 | 500 | < 20 | 34 | 0,33 | 0 | 0,12 | < 0,8 | - | 2,87 | 0,059 | 0,041 | 0,3 | 0,59 | 0,2 | 39 | 0 |
| Céleri-rave, cuit | - | - | - | - | 90,1 | 1,3 | 1,3 | 4,6 | 0,86 | 2,6 | 0,15 | 52,4 | - | 0,12 | 0,32 | < 10 | 15,1 | 0,27 | 58 | 306 | - | 58,2 | 0,24 | 0 | 0,22 | - | - | 2,7 | < 0,05 | < 0,05 | 0,92 | 0,41 | 0,07 | 67,4 | 0 |
| Céleri-rave, purée | - | - | - | - | 91,4 | 1,37 | 1,37 | 2,83 | 0,18 | 4,01 | 0,13 | 50 | - | - | 0,5 | - | - | - | - | - | - | 54,5 | - | - | - | - | - | 8,3 | - | - | - | - | - | - | - |

| | | | | | | | | | | | | | | | | | | | | | | | | | | | | | | | | | | | |
|---|---|---|---|---|---|---|---|---|---|---|---|---|---|---|---|---|---|---|---|---|---|---|---|---|---|---|---|---|---|---|---|---|---|---|---|
| Cerise acérola, pulpe, crue, prélevée à la Martinique | - | - | - | - | 93,1 | 0,56 | 0,56 | - | 0,097 | - | - | - | - | - | - | - | - | - | - | - | - | - | - | - | - | - | - | 2850 | - | - | - | - | - | - | - |
| Cerise, dénoyautée, crue | 235 | 55,7 | 235 | 55,7 | 85,7 | 0,81 | 0,81 | 13 | < 0,3 | 1,6 | < 0,013 | 9,9 | < 20 | 0,08 | 0,17 | < 20 | 8,8 | 0,06 | 19 | 190 | < 20 | < 5 | 0,06 | 0 | < 0,08 | < 0,8 | - | 4,09 | < 0,015 | 0,012 | < 0,1 | 0,14 | 0,04 | 6,75 | 0 |
| Chadèque, pulpe, cru, prélevé à la Martinique | - | - | - | - | 90,6 | 0,48 | 0,48 | - | 0,3 | - | - | - | - | - | - | - | - | - | - | - | - | - | - | - | - | - | - | 44,2 | - | - | - | - | - | - | - |
| Champignon de Paris ou champignon de couche, appertisé, égoutté | 80,8 | 19,4 | 80,8 | 19,4 | 92,9 | 2,23 | 2,23 | 0,7 | 0,37 | 2,5 | 0,63 | 22,2 | 425 | 0,17 | 0,9 | < 5 | 4,85 | 0,027 | 47 | 72,1 | < 1 | 251 | 0,44 | 0 | 0 | 0 | - | 6,5 | 0,081 | 0,12 | 0,5 | 0,34 | 0,025 | 29,6 | 0 |
| Champignon de Paris ou champignon de couche, bouilli/cuit à l'eau | 120 | 28,7 | 120 | 28,7 | 91,1 | 2,17 | 2,17 | 3 | 0,47 | 2,2 | 0,005 | 6 | - | 0,5 | 1,74 | - | 12 | 0,12 | 87 | 356 | 11,9 | 2 | 0,87 | 0,2 | 0,01 | 0 | - | 4 | 0,073 | 0,3 | 4,46 | 2,16 | 0,095 | 18 | 0 |
| Champignon de Paris ou champignon de couche, cru | 118 | 28 | 118 | 28 | 93,9 | 2,62 | 2,62 | 3,15 | 0,36 | 1 | 0,098 | 6,03 | 61 | 0,35 | 0,31 | 1 | 10,5 | 0,076 | 96,6 | 364 | 6,47 | 39 | 0,5 | 0,3 | 0,02 | 0 | - | 3,09 | 0,073 | 0,29 | 5 | 1,57 | 0,1 | 34,5 | 0 |
| Champignon de Paris ou champignon de couche, sauté/poêlé, sans matière grasse | 161 | 38,4 | 161 | 38,4 | 88,4 | 4,44 | 4,44 | 4,53 | 0,6 | 0,9 | 0,028 | 2,7 | 132 | 0,39 | 0,31 | < 20 | 14 | 0,05 | 140 | 540 | < 20 | 11 | 0,81 | 0,2 | < 0,08 | < 0,8 | - | < 0,5 | 0,062 | 0,36 | 4,53 | 2,33 | 0,029 | 14,3 | 0 |
| Champignon de Paris ou champignon de couche, surgelé, cru | 66,2 | 15,9 | 66,2 | 15,9 | 93,2 | 1,8 | 1,8 | 0,83 | 0,25 | 1,97 | 0,02 | 10 | - | - | 0,8 | - | - | - | - | - | - | 5 | - | - | - | - | - | 4 | - | - | - | - | - | - | - |
| Champignon noir, séché | - | - | - | - | - | 7,27 | 7,27 | - | 0,075 | - | 0,094 | - | - | - | - | - | - | - | - | - | - | 37,8 | - | - | - | - | - | - | - | - | - | - | - | - | - |
| Champignon, cèpe, cru | 133 | 31,6 | 133 | 31,6 | 90,3 | 3,13 | 3,13 | 3,03 | 0,43 | 1,93 | 0,024 | 23 | - | 0,23 | 1 | - | 12 | 0,17 | 129 | 361 | - | 5 | 0,4 | 3,1 | 0,04 | - | - | 3 | 0,21 | 0,32 | 4,45 | 2,7 | 0,18 | 44 | - |
| Champignon, chanterelle ou girolle, crue | 137 | 33,1 | 137 | 33,1 | 88,7 | 2,28 | 2,28 | 2,3 | 0,78 | 4,23 | 0,016 | 8,33 | 50 | 0,39 | 3,52 | 3,3 | 10 | 0,28 | 53,5 | 449 | 0,9 | 5 | 0,84 | 5,3 | 0,06 | - | - | 5,5 | 0,053 | 0,31 | 5,44 | 1,08 | 0,042 | 2 | 0 |
| Champignon, lentin comestible ou shiitaké, cuit | 258 | 60,9 | 258 | 60,9 | 83,5 | 1,56 | 1,56 | 12,3 | 0,22 | 2,1 | 0,01 | 3 | - | 0,9 | 0,44 | - | 14 | 0,2 | 29 | 117 | - | 4 | 1,33 | 0,7 | 0 | 0 | - | 0,3 | 0,037 | 0,17 | 1,5 | 3,59 | 0,16 | 21 | 0 |
| Champignon, lentin comestible ou shiitaké, séché | 1330 | 316 | 1330 | 316 | 9,5 | 9,58 | 9,58 | 63,9 | 0,99 | 11,5 | 0,033 | 11 | - | 5,17 | 1,72 | - | 132 | 1,18 | 294 | 1530 | - | 13 | 7,66 | 3,9 | 0 | 0 | - | 3,5 | 0,3 | 1,27 | 14,1 | 21,9 | 0,97 | 163 | 0 |
| Champignon, morille, crue | 102 | 24,5 | 102 | 24,5 | 89,6 | 3,16 | 3,16 | 0,1 | 0,54 | 3,35 | 0,064 | 43 | 8 | 0,63 | 1,2 | - | 19 | 0,59 | 194 | 411 | - | 21 | 2,03 | 5,1 | 0,05 | - | - | 5 | 0,069 | 0,21 | 2,25 | 0,44 | 0,14 | 9 | - |
| Champignon, oronge vraie, crue | 97,2 | 23,2 | 97,2 | 23,2 | 92,9 | 2 | 2 | 2,45 | 0,3 | 1,7 | 0,013 | 17 | - | - | 1,1 | - | - | - | 89 | 320 | - | 5 | 0,4 | 2,1 | 0,12 | - | - | 3 | 0,1 | 0,31 | 4 | - | 0,18 | 44 | - |
| Champignon, pleurote, crue | 107 | 25,5 | 107 | 25,5 | 89,2 | 3,06 | 3,06 | 1,5 | 0,36 | 2,4 | 0,034 | 5 | - | 0,24 | 1,17 | - | 18 | 0,11 | 120 | 420 | - | 13,5 | 0,77 | 0,7 | 0 | 0 | - | 4 | 0,13 | 0,35 | 4,96 | 1,29 | 0,11 | 38 | 0 |
| Champignon, rosé des prés, cru | 106 | 25,3 | 106 | 25,3 | 92,1 | 2,3 | 2,3 | 2,45 | 0,4 | 1,7 | 0,013 | 10 | - | - | 1,2 | - | - | - | 102 | 320 | - | 5 | 0,4 | 0 | 0,12 | - | - | 4 | 0,1 | 0,13 | 4,2 | - | 0,18 | 44 | - |
| Champignon, tout type, appertisé, égoutté | 104 | 24,9 | 104 | 24,9 | 91,1 | 1,87 | 1,87 | 2,56 | 0,29 | 2,4 | 1,06 | 18,9 | - | 0,15 | 0,79 | 2 | 6,65 | 0,035 | 66 | 129 | < 2,2 | 425 | 0,51 | 0,2 | 0,01 | 0 | - | 0 | 0,085 | 0,021 | 1,59 | 0,81 | 0,061 | 12 | 0 |
| Champignon, tout type, cru | 91,2 | 21,7 | 91,2 | 21,7 | 92,6 | 2,37 | 2,37 | 1,88 | 0,23 | 1,72 | 0,012 | 7,78 | - | 0,32 | 0,69 | 1 | 10,9 | 0,061 | 85,6 | 341 | < 2,78 | 4,76 | 0,65 | 0,2 | 0,01 | 0 | - | 3,4 | 0,084 | 0,42 | 4,55 | 1,75 | 0,082 | 29 | 0 |
| Champignon, truffe noire, crue | - | - | - | - | 75,9 | 5,77 | 5,77 | - | 0,51 | 8,5 | 0,17 | 24 | 27,7 | - | 3,5 | - | 23,8 | - | 62 | 447 | - | 66 | 3,8 | 0 | 0,12 | - | - | 1 | 0,05 | 0,09 | 2 | - | 0,18 | 44 | - |
| Châtaigne ou Marron, appertisé | - | - | - | - | 62,7 | 1,96 | 2,31 | 28,8 | 1,4 | 4 | 0,13 | 21,1 | 4,5 | - | < 1 | - | 24,1 | - | 51 | 359 | - | 52,1 | - | - | < 0,1 | - | - | - | < 0,05 | 0,07 | 1,5 | 0,44 | 0,22 | 58 | - |
| Châtaigne, bouillie/cuite à l'eau | - | - | - | - | 68,2 | 2 | 2,36 | 23 | 1,38 | 4,5 | 0,068 | 46 | - | 0,47 | 1,73 | 0,1 | 54 | 0,85 | 99 | 715 | 2,2 | 27 | 0,25 | 0 | 1 | - | - | 26,7 | 0,15 | 0,1 | 0,73 | 0,32 | 0,23 | 38 | 0 |
| Châtaigne, crue | - | - | - | - | 51 | 1,81 | 2,13 | 36,5 | 2,23 | 7,45 | 0,0081 | 27 | - | 0,33 | 0,94 | 0,1 | 34,5 | 2,16 | 69,8 | 550 | 0,9 | 3,25 | 0,51 | 0 | 0,5 | - | - | 41,6 | 0,22 | 0,1 | 1,47 | 0,48 | 0,35 | 60 | 0 |
| Châtaigne, grillée | - | - | - | - | 40,5 | 3,17 | 3,74 | 47,9 | 2,2 | 5,1 | 0,005 | 29 | - | 0,51 | 0,91 | 0,12 | 33 | 1,18 | 107 | 592 | - | 2 | 0,57 | 0 | 0,5 | 7,8 | - | 26 | 0,24 | 0,18 | 1,34 | 0,55 | 0,5 | 70 | 0 |
| Chayote ou christophine ou chouchou, bouillie/cuite à l'eau | - | - | - | - | 93,4 | 0,62 | 0,62 | 2,29 | 0,48 | 2,8 | 0,0025 | 13 | - | 0,11 | 0,22 | - | 12 | 0,17 | 29 | 173 | - | 1 | 0,31 | 0 | - | - | - | 8 | 0,026 | 0,04 | 0,42 | 0,41 | 0,12 | 18 | 0 |
| Chayote ou christophine ou chouchou, crue | - | - | - | - | 94,2 | 0,72 | 0,72 | 2,8 | 0,12 | 1,7 | 0,005 | 17 | - | 0,12 | 0,34 | - | 12 | 0,19 | 18 | 125 | - | 2 | 0,74 | 0 | 0,12 | 4,1 | - | 7,7 | 0,025 | 0,029 | 0,47 | 0,25 | 0,076 | 93 | 0 |
| Chayote ou christophine ou chouchou, pulpe avec pépins, cuite à la vapeur, prélevée à La Réunion (Sechium edule) | 93,8 | 22,2 | 93,8 | 22,2 | 93,4 | 0,63 | 0,63 | 3,5 | < 0,3 | 1,4 | 0,018 | 14 | 36,1 | 0,03 | 0,46 | < 20 | 8,5 | 0,14 | 23 | 170 | < 20 | 7 | 0,1 | - | < 0,08 | 1,77 | - | < 0,5 | < 0,015 | < 0,01 | 0,35 | 0,32 | 0,024 | 16,2 | - |
| Chia, graine, séchée | - | - | - | - | 5,8 | 16,5 | 19,5 | 7,72 | 30,7 | 34,4 | 0,04 | 631 | - | 0,92 | 7,72 | - | 335 | 2,72 | 860 | 407 | - | 16 | 4,58 | - | 0,5 | - | - | 1,6 | 0,62 | 0,17 | 8,83 | - | - | 49 | 0 |
| Chicorée rouge, crue | - | - | - | - | 94 | 1,4 | 1,4 | 1,6 | 0,1 | 3 | 0,018 | 36 | - | - | 0,3 | - | - | - | 30 | 180 | - | 7 | 0,2 | 0 | 0,57 | - | - | 10 | 0,07 | 0,05 | 0,3 | - | 0,07 | 13 | - |
| Chicorée verte, crue | - | - | - | - | 91,2 | 1,9 | 1,9 | 0,5 | 0,5 | 1 | 0,041 | 82,5 | - | 0,06 | 4,15 | - | 17 | 0,3 | 45 | 498 | - | 16,5 | 0,32 | 0 | 0,57 | - | - | 46 | 0,06 | 0,53 | 0,3 | - | 0,02 | 29 | - |
| Chips d'abricot pays, pulpe, prélevé à la Martinique | - | - | - | - | 5,13 | 0,96 | 0,96 | 68,7 | 18,7 | 6,2 | - | - | - | - | - | - | - | - | - | - | - | - | - | - | - | - | - | 0,2 | - | - | - | - | - | - | - |
| Chips de giraumon (variété locale), pulpe, prélevé à la Martinique | - | - | - | - | - | - | - | 12,3 | - | 8,4 | - | - | - | - | - | - | - | - | - | 2400 | - | - | - | - | - | - | - | - | - | - | - | - | - | - | - |
| Chips de giraumon (variété phoenix), pulpe, prélevé à la Martinique | - | - | - | - | - | - | - | 13,9 | - | 8,7 | - | - | - | - | - | - | - | - | - | 2580 | - | - | - | - | - | - | - | - | - | - | - | - | - | - | - |
| Chips de pommes de terre et assimilés nature ou aromatisées, allégées en matière grasse | 1940 | 463 | 1940 | 463 | 1,3 | 6,58 | 6,58 | 62,1 | 19,8 | 5,19 | 1,22 | 56 | 802 | 0,25 | 1,2 | < 20 | 67 | 0,26 | 210 | 1300 | < 20 | 487 | 0,87 | - | 3,32 | 2,56 | - | - | 0,32 | 0,029 | 2,56 | 1,57 | 0,43 | - | 0,049 |
| Chips de pommes de terre nature ou aromatisées, à l'ancienne | 2370 | 570 | 2370 | 570 | 1,4 | 5,62 | 5,62 | 49,8 | 37,6 | 4,51 | 1,78 | 21 | 1070 | 0,19 | 0,98 | < 20 | 55 | 0,25 | 120 | 1100 | < 20 | 702 | 0,6 | < 0,25 | 15,3 | 4,56 | - | 5,07 | 0,31 | 0,02 | 4,74 | 0,83 | 0,35 | 27,3 | - |
| Chips de pommes de terre nature ou aromatisées, standard | 2270 | 545 | 2270 | 545 | 1,4 | 5,67 | 5,67 | 51,1 | 34,3 | 4,57 | 1,65 | 27 | 742 | 0,22 | 1,2 | < 20 | 66 | 0,35 | 140 | 1300 | < 20 | 652 | 0,82 | < 0,25 | 12,3 | 5,02 | - | 8,26 | 0,13 | 0,014 | 19,4 | 0,85 | 0,44 | 31,6 | 0 |
| Chou blanc, bouilli/cuit à l'eau | 98,1 | 23,5 | 98,1 | 23,5 | 92,9 | 1 | 1 | 3,23 | < 0,3 | 2,7 | 0,018 | 42 | 23,2 | 0,07 | 0,19 | < 20 | 6,4 | 0,08 | 19 | 120 | < 20 | 7 | 0,07 | - | 0,64 | 13 | - | 13,3 | < 0,015 | < 0,01 | 0,18 | 0,059 | 0,048 | 9,55 | - |
| Chou blanc, cru | 153 | 36,5 | 153 | 36,5 | 90 | 1,38 | 1,38 | 4,63 | 0,6 | 3,5 | 0,033 | 59 | 50 | 0,02 | 0,3 | < 20 | 11 | 0,1 | 33 | 240 | 20 | 13 | 0,16 | 0 | < 0,08 | 9,6 | - | 8,88 | 0,04 | < 0,01 | 0,2 | 0,18 | 0,17 | 69,6 | 0 |
| Chou chinois (pak-choi ou pé-tsai), cuit | - | - | - | - | 95,4 | 1,53 | 1,53 | 0,75 | 0,17 | 1,35 | 0,054 | 62,5 | - | 0,024 | 0,67 | - | 10,5 | 0,15 | 34 | 298 | - | 21,5 | 0,18 | 0 | 0,09 | 34 | - | 20,9 | 0,038 | 0,054 | 0,46 | 0,08 | 0,17 | 47 | 0 |
| Chou chinois ou pak-choi ou pé-tsai, cru | - | - | - | - | 95,2 | 1,38 | 1,38 | 1,65 | 0,2 | 1,05 | 0,13 | 86,4 | - | 0,021 | 0,68 | 0,3 | 17,5 | 0,17 | 37,8 | 249 | 0,29 | 51 | 0,17 | 0 | 0,11 | 45,5 | - | 40,3 | 0,04 | 0,065 | 0,48 | 0,092 | 0,19 | 73,6 | 0 |
| Chou de Bruxelles, appertisé, égoutté | - | - | - | - | 88,9 | 2,5 | 2,5 | 4,1 | 0,53 | 3,3 | 0,57 | 26,3 | - | 0,03 | 0,7 | 1 | 12,2 | 0,1 | 53 | 252 | 1,5 | 222 | 0,2 | 0 | 0,69 | - | - | 18 | - | - | - | - | - | - | 0 |
| Chou de Bruxelles, bouilli/cuit à l'eau | 190 | 45,4 | 190 | 45,4 | 85,5 | 3,19 | 3,19 | 5,4 | < 0,3 | 4,8 | 0,023 | 35 | 28,8 | 0,13 | 0,51 | < 20 | 19 | 0,2 | 62 | 410 | < 20 | 9 | 0,31 | - | 0,44 | 61,8 | - | 55,4 | 0,067 | 0,058 | 0,48 | 0,33 | 0,11 | 49,3 | - |
| Chou de Bruxelles, cru | - | - | - | - | 85 | 3,98 | 3,98 | 5,67 | 0,4 | 3,95 | 0,041 | 32,3 | - | 0,062 | 1,15 | 0,25 | 23,5 | 0,3 | 76,4 | 436 | 1,1 | 16,5 | 0,43 | 0 | 0,89 | 214 | - | 103 | 0,13 | 0,13 | 0,77 | 0,52 | 0,25 | 95,5 | 0 |
| Chou de Bruxelles, cuit | - | - | - | - | 87,6 | 2,6 | 2,6 | 4,2 | 0,11 | 3,2 | 0,024 | 36,2 | - | < 0,1 | < 0,1 | < 10 | 16,9 | < 0,1 | 65 | 324 | - | 9,7 | < 0,1 | 0 | 0,43 | 140 | - | 56,4 | < 0,05 | 0,06 | 0,72 | 0,27 | 0,06 | 113 | 0 |
| Chou de Bruxelles, surgelé, cru | - | - | - | - | 86,3 | 3,43 | 3,43 | 4,5 | 0,33 | 4,28 | 0,029 | 34 | - | 0,043 | 0,82 | 0,25 | 22 | 0,29 | 65,5 | 427 | 1,1 | 14 | 0,37 | 0 | 0,34 | 250 | - | 77,5 | 0,12 | 0,11 | 0,69 | 0,35 | 0,19 | 95,5 | 0 |
| Chou de Bruxelles, surgelé, cuit | - | - | - | - | 86,7 | 3,64 | 3,64 | 4,22 | 0,39 | 4,1 | 0,038 | 26 | - | 0,034 | 0,48 | - | 18 | 0,21 | 56 | 290 | - | 15 | 0,24 | 0 | 0,51 | 194 | - | 45,7 | 0,1 | 0,11 | 0,54 | 0,34 | 0,29 | 101 | 0 |
| Chou dur ou chou caraïbe, pulpe, cuit à la vapeur, prélevé à la Martinique | - | - | - | - | 63,7 | 3,25 | 3,25 | - | 0,0033 | - | - | - | - | - | - | - | - | - | - | - | - | - | - | - | - | - | - | 19,3 | - | - | - | - | - | - | - |
| Chou frisé, cru | - | - | - | - | 82,8 | 4,33 | 4,33 | 4,2 | 1,07 | 4,9 | 0,091 | 185 | - | 0,8 | 1,74 | 1,4 | 33,5 | 0,58 | 83 | 378 | 2,1 | 36,5 | 0,59 | 0 | 3,47 | 477 | - | 145 | 0,13 | 0,21 | 1,9 | 0,55 | 0,31 | 101 | 0 |
| Chou frisé, cuit | - | - | - | - | 91,2 | 1,9 | 1,9 | 3,63 | 0,4 | 2 | 0,058 | 72 | - | 0,16 | 0,9 | - | 18 | 0,42 | 28 | 228 | - | 23 | 0,24 | 0 | 0,85 | 817 | - | 41 | 0,053 | 0,07 | 0,5 | 0,049 | 0,14 | 13 | 0 |
| Chou romanesco ou brocoli à pomme, cru | - | - | - | - | 89,8 | 2,93 | 2,93 | 2,2 | 0,4 | 2,35 | 0,039 | 34,5 | - | 0,041 | 0,92 | - | 20 | 0,25 | 62 | 300 | - | 15,5 | 0,64 | 0 | 0,04 | 20,2 | - | 70,6 | 0,08 | 0,1 | 0,73 | 0,7 | 0,22 | 57 | 0 |
| Chou romanesco ou brocoli à pomme, cuit | 149 | 35,8 | 149 | 35,8 | 91,5 | 3 | 3 | 2,05 | 0,8 | 4,3 | < 0,013 | 32 | < 20 | 0,05 | 0,53 | < 20 | 13 | 0,19 | 55 | 260 | < 20 | < 5 | 0,32 | < 0,25 | 0,74 | 12,4 | - | 15,1 | 0,025 | 0,046 | 0,23 | 0,59 | 0,13 | 49,1 | - |
| Chou rouge, bouilli/cuit à l'eau | - | - | - | - | 90,8 | 1,51 | 1,51 | 3,32 | 0,09 | 2,6 | 0,07 | 54,6 | - | 0,034 | 0,66 | 1,35 | 11,8 | 0,16 | 33 | 262 | < 2,97 | 28 | 0,14 | 0 | 0,12 | 47,6 | - | 34,4 | 0,071 | 0,06 | 0,38 | 0,15 | 0,23 | 24 | 0 |
| Chou rouge, cru | 126 | 30 | 126 | 30 | 90,7 | 1,13 | 1,13 | 4,33 | < 0,5 | 2,8 | 0,035 | 36 | 32,7 | 0,02 | 0,25 | < 20 | 9,7 | 0,12 | 30 | 260 | < 20 | 14 | 0,15 | 0 | < 0,08 | 6,41 | - | 18,7 | 0,036 | < 0,01 | 0,12 | 0,24 | 0,14 | 31,9 | 0 |
| Chou rouge, cuit à l'étouffée | 146 | 34,9 | 146 | 34,9 | 89,1 | 1,19 | 1,19 | 5,7 | < 0,3 | 3 | 0,033 | 44 | 34,8 | 0,02 | 0,29 | < 20 | 13 | 0,16 | 26 | 250 | < 20 | 13 | 0,15 | - | < 0,08 | 0,84 | - | 26,2 | 0,047 | 0,03 | 0,29 | 0,32 | 0,11 | 45,1 | - |
| Chou vert, bouilli/cuit à l'eau | 90 | 21,6 | 90 | 21,6 | 92,7 | 1,63 | 1,63 | 1,83 | < 0,3 | 3,3 | 0,013 | 48 | < 20 | 0,11 | 0,34 | < 20 | 7,3 | 0,11 | 29 | 130 | < 20 | 5 | 0,13 | - | < 0,08 | 9,77 | - | 12 | 0,02 | < 0,01 | < 0,1 | 0,078 | 0,052 | 12,4 | - |
| Chou vert, cru | - | - | - | - | 91,8 | 2,53 | 2,53 | 1,92 | 0,41 | 3,08 | 0,065 | 96,2 | - | 0,049 | 0,93 | 0,3 | 18 | 0,17 | 32,7 | 231 | 1,86 | 25,8 | 0,22 | 0 | 0,16 | 121 | - | 69 | 0,059 | 0,039 | 0,26 | 0,17 | 0,15 | 67,9 | 0 |
| Chou vert, cuit | - | - | - | - | 93,9 | 1,03 | 1,03 | 3,04 | 0,45 | 2,4 | 0,098 | 69,7 | 61 | 0,098 | 0,28 | 1,45 | 15,1 | 0,18 | 33 | 190 | 3,56 | 39 | 0,22 | 0 | 0,14 | 109 | - | 27,3 | 0,056 | 0,029 | 0,14 | 0,17 | 0,13 | 38 | 0 |
| Chou-fleur, cru | 109 | 26,2 | 109 | 26,2 | 92,9 | 1,81 | 1,81 | 2,13 | 0,7 | 2,2 | 0,015 | 23 | 25,4 | 0,02 | 0,27 | < 20 | 9,8 | 0,14 | 40 | 270 | < 20 | 6 | 0,22 | 0 | < 0,08 | 3,31 | - | 4,14 | 0,031 | < 0,01 | 0,26 | 0,72 | 0,22 | 56,2 | 0 |

| | | | | | | | | | | | | | | | | | | | | | | | | | | | | | | | | | | | |
|---|---|---|---|---|---|---|---|---|---|---|---|---|---|---|---|---|---|---|---|---|---|---|---|---|---|---|---|---|---|---|---|---|---|---|---|
| Chou-fleur, cuit | 87,3 | 20,9 | 87,3 | 20,9 | 93,7 | 1,6 | 1,6 | 1,6 | 0,46 | 2 | 0,017 | 19,2 | < 61 | 0,05 | 0,31 | < 5 | 12,9 | 0,12 | 34 | 238 | < 10 | < 39 | 0,22 | 0 | 0,07 | 13,8 | - | 9,7 | < 0,05 | < 0,05 | 0,21 | 0,23 | < 0,05 | 78,8 | 0 |
| Chou-fleur, cuit à la vapeur | 127 | 30,1 | 127 | 30,1 | 91,1 | 2,56 | 2,56 | 3,41 | 0,3 | 1,8 | 0,013 | 26 | < 20 | 0,03 | 0,38 | < 20 | 12 | 0,16 | 46 | 300 | < 20 | 5 | 0,28 | - | < 0,08 | 8,13 | - | 21,4 | 0,034 | 0,056 | 0,44 | 0,94 | 0,13 | 54,3 | - |
| Chou-fleur, surgelé, cru | - | - | - | - | 92,5 | 2,09 | 2,09 | 2,64 | 0,37 | 1,82 | 0,047 | 21,7 | - | 0,031 | 0,51 | 0,3 | 12 | 0,2 | 41,6 | 193 | 0,8 | 20,3 | 0,17 | 0 | 0,07 | 15,4 | - | 51,9 | 0,051 | 0,07 | 0,43 | 0,14 | 0,12 | 64 | 0 |
| Chou-fleur, surgelé, cuit | 74,8 | 18 | 74,8 | 18 | 94 | 1,61 | 1,61 | 1,05 | 0,22 | 2,7 | 0,045 | 17 | - | 0,024 | 0,41 | 0,1 | 9 | 0,15 | 24 | 139 | traces | 18 | 0,13 | 0 | 0,06 | 11,9 | - | 31,3 | 0,037 | 0,053 | 0,31 | 0,098 | 0,088 | 41 | 0 |
| Chou-rave, bouilli/cuit à l'eau | - | - | - | - | 90,3 | 1,8 | 1,8 | 4,4 | 0,11 | 1,1 | 0,053 | 25 | - | 0,13 | 0,4 | - | 19 | 0,14 | 45 | 340 | - | 21 | 0,31 | 0 | 0,52 | 0,1 | - | 54 | 0,04 | 0,02 | 0,39 | 0,16 | 0,15 | 12 | 0 |
| Chou-rave, cru | - | - | - | - | 91 | 1,79 | 1,79 | 2,35 | 0,1 | 3,6 | 0,05 | 24 | - | 0,13 | 0,4 | 0,71 | 19 | 0,14 | 46 | 350 | 0,47 | 20 | 0,03 | 0 | 0,48 | 0,1 | - | 55,1 | 0,05 | 0,02 | 0,4 | 0,17 | 0,15 | 16 | 0 |
| Christophine à peau blanche, pulpe, cuite à la vapeur, prélevée à la Martinique | - | - | - | - | 90,4 | 1,21 | 1,21 | 5,93 | 0,033 | 1,9 | - | - | - | 0,063 | - | - | - | - | - | 192 | - | - | - | - | - | - | - | 7,4 | - | - | - | - | - | 20,8 | - |
| Christophine à peau verte, pulpe, cuite à la vapeur, prélevée à la Martinique | - | - | - | - | 90,7 | 1,23 | 1,23 | 4,6 | 0,023 | 3 | - | - | - | 0,0019 | - | - | - | - | - | 198 | - | - | - | - | - | - | - | 12,5 | - | - | - | - | - | 17,8 | - |
| Christophine blanche, pulpe, appertisée, non égouttée, prélevée à la Martinique | - | - | - | - | 93,2 | - | - | - | - | 1 | - | - | - | - | - | - | - | - | - | - | - | - | - | - | - | - | - | 0,23 | - | - | - | - | - | 23,4 | - |
| Christophine, pulpe, cuite à la vapeur, surgelée, prélevée à la Martinique | - | - | - | - | 91,6 | - | - | - | - | 1,6 | - | - | - | - | - | - | - | - | - | - | - | - | - | - | - | - | - | 6,1 | - | - | - | - | - | 14,9 | - |
| Citron vert ou Lime, pulpe, cru | 170 | 40,2 | 170 | 40,2 | 86,3 | 1,13 | 1,13 | 3,14 | < 0,3 | 4,3 | 0,015 | 57 | < 20 | 0,08 | 0,2 | < 20 | 14 | 0,09 | 24 | 190 | < 20 | 6 | 0,15 | 0 | 0,45 | < 0,8 | - | 29,3 | < 0,015 | 0,023 | 0,25 | 0,23 | 0,018 | 36,5 | 0 |
| Citron, pulpe, cru | 118 | 27,6 | 118 | 27,6 | 91,3 | < 0,5 | < 0,5 | 1,56 | < 0,5 | < 0,5 | < 0,013 | 11 | < 20 | 0,04 | 0,15 | < 20 | 7,9 | 0,02 | 12 | 140 | < 20 | < 5 | 0,33 | 0 | < 0,08 | < 0,8 | - | 45 | 0,043 | < 0,01 | < 0,1 | 0,14 | 0,023 | 28,4 | 0 |
| Citron, zeste, cru | - | - | - | - | 81,6 | 1,38 | 1,38 | 5,4 | 0,3 | 10,6 | 0,015 | 134 | - | 0,092 | 0,9 | - | 15 | - | 12 | 160 | 7 | 6 | 0,25 | 0 | 0,25 | 0 | - | 129 | 0,06 | 0,08 | 0,4 | 0,32 | 0,17 | 13 | 0 |
| Citrouille, pulpe, crue | - | - | - | - | 91,6 | 1 | 1 | 6 | 0,1 | 0,5 | 0,0025 | 21 | - | 0,13 | 0,8 | - | 12 | 0,13 | 44 | 340 | - | 1 | 0,32 | 0 | 1,06 | 1,1 | - | 9 | 0,05 | 0,11 | 0,6 | 0,3 | 0,061 | 16 | 0 |
| Clémentine ou Mandarine, pulpe, crue | 200 | 47,3 | 200 | 47,3 | 87 | 0,81 | 0,81 | 9,17 | < 0,5 | 1,7 | < 0,013 | 23 | < 20 | 0,04 | 0,09 | < 20 | 9,3 | 0,02 | 18 | 140 | < 20 | < 5 | 0,1 | 0 | 0,21 | < 0,8 | - | 49,2 | 0,064 | < 0,01 | 0,23 | 0,2 | 0,079 | 27,6 | 0 |
| Coeur de palmier, appertisé, égoutté | - | - | - | - | 90,2 | 2,45 | 2,45 | 4,03 | 0,5 | 2,32 | 0,81 | 58 | - | 0,13 | 3,13 | 0,5 | 38 | 1,39 | 65 | 177 | - | 316 | 1,15 | 0 | - | - | - | 7,9 | 0,011 | 0,057 | 0,44 | 0,13 | 0,022 | 39 | 0 |
| Coing, cru | - | - | - | - | 83,8 | 0,51 | 0,51 | 13,4 | 0,1 | 1,9 | 0,01 | 11 | - | 0,13 | 0,7 | 0,4 | 8 | - | 17 | 197 | - | 4 | 0,04 | 0 | 0,55 | - | - | 15 | 0,02 | 0,03 | 0,2 | 0,081 | 0,04 | 3 | 0 |
| Compote de pomme | 432 | 102 | 432 | 102 | 72,9 | 0,23 | 0,23 | 24,4 | 0,21 | 1,53 | traces | 4,44 | - | 0,043 | 0,11 | 0,2 | 5,09 | 0,027 | 6 | 104 | < 10 | traces | 0,025 | 0 | 0,18 | 0,6 | - | 14,5 | 0,017 | 0,022 | 0,072 | 0,044 | 0,027 | 1 | 0 |
| Compote de pomme, allégée en sucres | 276 | 65,1 | 276 | 65,1 | 82,9 | < 0,5 | < 0,5 | 15,1 | 0,08 | 1,3 | < 0,013 | 3,8 | < 20 | 0,04 | 0,12 | - | 3,9 | 0,03 | 9,5 | 120 | < 20 | < 5 | < 0,05 | < 0,25 | 0,32 | < 0,8 | - | 7,75 | < 0,015 | < 0,01 | < 0,1 | 0,072 | 0,032 | 15,1 | - |
| Compote ou assimilé, tout type de fruits, teneur en sucre (allégée en sucres ou non, sans sucres ajoutés...) inconnue (aliment moyen) | 340 | 80,1 | 340 | 80,1 | 78,9 | 0,25 | 0,25 | 18,6 | 0,15 | 1,68 | 0,0048 | 7,87 | 9,68 | 0,046 | 0,092 | 3,72 | 5,59 | 0,05 | 10 | 132 | 9,12 | 1,9 | 0,033 | 0,091 | 0,45 | 0,67 | | 19,6 | 0,01 | 0,0096 | 0,12 | 0,087 | 0,02 | 7,3 | 0 |
| Compote, tout type de fruits | 434 | 102 | 434 | 102 | 73,1 | < 0,5 | < 0,5 | 23,9 | 0,092 | 2 | < 0,013 | 18 | < 20 | 0,05 | 0,13 | 0,1 | 5,8 | 0,04 | 13 | 150 | < 20 | < 5 | 0,07 | < 0,25 | 0,87 | 1,52 | - | 42 | < 0,015 | < 0,01 | 0,15 | 0,12 | 0,025 | 13,6 | 0 |
| Compote, tout type de fruits, allégée en sucres | 279 | 66 | 279 | 66 | 82,9 | < 0,5 | < 0,5 | 15,3 | 0,08 | 1,6 | < 0,013 | 6,2 | < 20 | 0,04 | 0,14 | 0,1 | 5,5 | 0,08 | 11 | 140 | < 20 | < 5 | < 0,05 | < 0,25 | 0,53 | < 0,8 | - | 11,5 | < 0,015 | < 0,01 | < 0,1 | 0,095 | 0,027 | 9,03 | 0 |
| Compote, tout type de fruits, allégée en sucres, rayon frais | - | - | - | - | 83,9 | 0,32 | 0,32 | 14,9 | 0,48 | 1,4 | 0,011 | 4,7 | < 3,6 | 0,04 | 0,09 | < 20 | 4,7 | 0,03 | 9,6 | 120 | < 50 | 4,38 | < 0,05 | - | 0,37 | - | - | 10,1 | < 0,01 | < 0,01 | 0,12 | 0,036 | 0,044 | < 5 | - |
| Concombre, pulpe avec graines, cru, prélevé à la Martinique | - | - | - | - | 95,8 | 0,73 | 0,73 | - | 0,58 | - | - | - | - | - | - | - | - | - | - | - | - | - | - | - | - | - | - | 0,1 | - | - | - | - | - | - | - |
| Concombre, pulpe et peau, cru | 65,8 | 15,6 | 65,8 | 15,6 | 96 | 0,64 | 0,64 | 2,54 | 0,11 | 0,6 | 0,012 | 19,2 | - | 0,026 | 0,13 | 0,3 | 13,6 | 0,085 | 24,7 | 157 | < 10 | 4,8 | 0,095 | 0 | 0,09 | 16,4 | - | 8,25 | 0,022 | 0,025 | 0,15 | 0,28 | 0,037 | 12,3 | 0 |
| Concombre, pulpe, cru | 61,6 | 14,7 | 61,6 | 14,7 | 96,3 | 0,56 | 0,56 | 2,23 | < 0,5 | 0,8 | < 0,013 | 16 | < 20 | 0,02 | 0,14 | < 20 | 8,9 | 0,11 | 25 | 140 | < 20 | < 5 | 0,13 | 0 | < 0,08 | 2,75 | - | 3,52 | < 0,015 | < 0,01 | < 0,1 | 0,15 | 0,042 | 7,2 | 0 |
| Concombre, pulpe, cru, prélevé à la Martinique | - | - | - | - | 96,8 | 0,42 | 0,42 | - | 0,27 | - | - | - | - | - | - | - | - | - | - | - | - | - | - | - | - | - | - | 0,3 | - | - | - | - | - | - | - |
| Coulis de fruits rouges (framboises, fraises, groseilles, cassis) | - | - | - | - | - | 0,8 | 0,8 | 25,1 | 0,54 | 3,46 | 0,014 | 25,6 | - | 0,074 | 0,71 | 0,88 | 14,5 | 0,32 | 26,9 | 204 | 0,3 | 7,79 | 0,2 | 0 | 0,45 | - | - | 60,9 | 0,024 | 0,067 | 0,33 | 0,22 | 0,055 | 34,6 | 0 |
| Courge doubeurre (butternut), pulpe, crue | - | - | - | - | 86,4 | 1 | 1 | 5,4 | 0,1 | 2 | 0,01 | 48 | - | 0,072 | 0,7 | - | 34 | 0,2 | 33 | 352 | - | 4 | 0,15 | 0 | 1,44 | 1,1 | - | 21 | 0,1 | 0,02 | 1,2 | 0,4 | 0,15 | 27 | 0 |
| Courge doubeurre (butternut), pulpe, cuite | - | - | - | - | 87,8 | 1,23 | 1,23 | - | 0,07 | - | 0,005 | 19 | - | 0,036 | 0,58 | - | 9 | 0,17 | 14 | 133 | - | 2 | 0,12 | 0 | - | - | - | 3,5 | 0,05 | 0,039 | 0,46 | 0,15 | 0,069 | 16 | 0 |
| Courge hokkaïdo, pulpe, crue | - | - | - | - | 86,1 | 1,4 | 1,4 | 9,98 | 0,6 | 2,6 | - | 24,7 | - | - | 0,4 | - | 10,8 | - | 32,8 | 436 | - | - | - | - | - | - | - | 13,9 | 0,015 | - | - | - | - | - | - |
| Courge melonnette, pulpe, crue | - | - | - | - | 91,2 | 0,75 | 0,75 | 5,98 | 0,13 | 1,35 | - | 18,3 | - | - | 0,48 | - | 16,8 | - | 23,5 | 359 | - | - | - | - | - | - | - | 9,7 | 0,01 | - | - | - | - | - | - |
| Courge musquée, pulpe, crue | - | - | - | - | 93,3 | 0,55 | 0,55 | 4,6 | 0,1 | 1 | - | 18,3 | - | - | 0,43 | 0,5 | 6,25 | - | 18,3 | 235 | - | - | - | - | - | - | - | 4,6 | 0,01 | - | - | - | - | - | - |
| Courge musquée, pulpe, cuite | 129 | 30,7 | 129 | 30,7 | 90,9 | 1,06 | 1,06 | 5,24 | < 0,5 | 1,8 | < 0,013 | 17 | 56,3 | 0,05 | 0,15 | < 20 | 7,4 | 0,04 | 17 | 340 | < 20 | < 5 | 0,13 | < 0,25 | 0,94 | < 0,8 | - | 2,28 | 0,02 | < 0,01 | 0,5 | 0,23 | 0,064 | 30,5 | - |
| Courge spaghetti, pulpe, crue | - | - | - | - | 91,6 | 0,64 | 0,64 | 5,41 | 0,57 | 1,5 | 0,043 | 23 | - | 0,037 | 0,31 | - | 12 | 0,13 | 12 | 108 | - | 17 | 0,19 | 0 | 0,13 | 0,9 | - | 2,1 | 0,037 | 0,018 | 0,95 | 0,36 | 0,1 | 12 | 0 |
| Courge spaghetti, pulpe, cuite | - | - | - | - | 92,3 | 0,66 | 0,66 | 5,06 | 0,26 | 1,4 | 0,045 | 21 | - | 0,035 | 0,34 | - | 11 | 0,11 | 14 | 117 | - | 18 | 0,2 | 0 | 0,12 | 0,8 | - | 3,5 | 0,038 | 0,022 | 0,81 | 0,36 | 0,099 | 8 | 0 |
| Courge, crue | - | - | - | - | 92,3 | 1,1 | 1,1 | 1,6 | 0,17 | 1,3 | 0,011 | 28,5 | - | 0,061 | 0,6 | 0,15 | 16,5 | 0,17 | 30,5 | 312 | 0,24 | 4,5 | 0,26 | 0 | 0,12 | 1,1 | - | 14,7 | 0,028 | 0,1 | 0,49 | 0,17 | 0,13 | 34 | 0 |
| Courgette, pulpe et peau, crue | 69,4 | 16,5 | 69,4 | 16,5 | 94,7 | 1,23 | 1,23 | 1,8 | 0,26 | 1,05 | 0,023 | 18,7 | - | 0,052 | 0,36 | 1,64 | 17,5 | 0,18 | 44,7 | 262 | 0,58 | 9 | 0,31 | 0 | 0,12 | 4,3 | - | 17,5 | 0,047 | 0,12 | 0,47 | 0,18 | 0,19 | 36 | 0 |
| Courgette, pulpe et peau, cuite | 64,9 | 15,5 | 64,9 | 15,5 | 93,8 | 0,93 | 0,93 | 1,4 | 0,36 | 1,5 | 0,098 | 19,3 | 61 | 0,085 | 0,32 | < 5 | 22,3 | 0,089 | 65 | 238 | < 10 | 39 | 0,22 | 0 | 0,12 | 4,2 | - | 2,9 | < 0,05 | < 0,05 | < 0,05 | < 0,05 | < 0,05 | 32,8 | 0 |
| Courgette, pulpe et peau, rôtie/cuite au four | 96,6 | 23 | 96,6 | 23 | 93,1 | 1,5 | 1,5 | 3,13 | < 0,3 | 1,7 | < 0,013 | 24 | 41,8 | 0,07 | 0,4 | < 20 | 23 | 0,14 | 41 | 300 | < 20 | < 5 | 0,27 | - | 0,13 | 2,92 | - | < 0,5 | 0,02 | 0,035 | 0,64 | 0,2 | 0,038 | 16,4 | - |
| Courgette, pulpe et peau, surgelée, crue | - | - | - | - | 94,6 | 1,19 | 1,19 | 2,18 | 0,14 | 1,48 | 0,019 | 20 | - | 0,05 | 0,64 | - | 13 | 0,24 | 28 | 218 | - | 8,36 | 0,21 | 0 | 0,12 | 4,2 | - | 9,32 | 0,048 | 0,042 | 0,43 | 0,3 | 0,05 | 10 | 0 |
| Courgette, purée | - | - | - | - | - | 2,8 | 2,8 | 6,5 | 1,9 | 1,5 | 0,6 | - | - | - | - | - | - | - | - | - | - | 220 | - | - | - | - | - | - | - | - | - | - | - | - | - |
| Crème de marrons | - | - | - | - | 33,3 | 1,28 | 1,51 | 55 | 0,67 | 1,3 | - | - | - | - | - | - | - | - | - | - | - | - | - | - | - | - | - | 10,3 | - | - | - | - | - | - | - |
| Crème de marrons vanillée, appertisée | - | - | - | - | 34,6 | 0,96 | 1,13 | 59,8 | 0,7 | 3,2 | < 0,01 | 11,3 | - | 0,08 | 1,7 | 0,4 | 12,2 | 0,24 | - | 174 | 1 | < 10 | 0,3 | - | - | - | - | 2,27 | - | - | - | - | - | 30 | - |
| Cresson alénois, cru | - | - | - | - | 91,1 | 2,55 | 2,55 | 2,2 | 0,6 | 1,35 | 0,096 | 75,5 | - | 0,13 | 1,8 | 0,9 | 41 | 0,55 | 76 | 368 | - | 38,5 | 0,35 | 0 | 2,15 | 542 | - | 52 | 0,13 | 0,23 | 1,75 | 0,27 | 0,27 | 95 | 0 |
| Cresson de fontaine, cru | - | - | - | - | 94,6 | 2,09 | 2,09 | 0,9 | 0,15 | 0,85 | 0,13 | 101 | - | 0,11 | 0,9 | 13,5 | 19 | 0,37 | 64,2 | 320 | - | 50,5 | 0,16 | 0 | 1 | 250 | - | 51,5 | 0,095 | 0,11 | 0,4 | 0,21 | 0,13 | 9 | 0 |
| Cresson, feuille, cru, prélevé à la Martinique | - | - | - | - | 95 | 2,25 | 2,25 | traces | 0,1 | 2,1 | - | 0,059 | - | - | 1,31 | - | - | 0,41 | - | 432 | - | - | - | - | - | 22,5 | - | 2,67 | - | - | - | - | - | 57,2 | - |
| Crosne, cuit | 215 | 50,8 | 215 | 50,8 | 85,7 | 2,13 | 2,13 | 9,11 | < 0,5 | 1,9 | 0,013 | 20 | < 20 | 0,21 | 0,37 | < 20 | 14 | 0,09 | 58 | 310 | < 20 | 5 | 0,24 | < 0,25 | < 0,08 | < 0,8 | - | - | 0,025 | < 0,01 | 0,32 | 0,38 | 0,068 | 32,3 | - |
| Crosne, surgelé, cru | - | - | - | - | - | 1,4 | 1,4 | 8,7 | 0 | 0 | 0,0025 | - | - | - | - | - | - | - | - | - | - | 1 | - | - | - | - | - | - | - | - | - | - | - | - | - |
| Cucurbitacées, graine | - | - | - | - | 5,23 | 30,2 | 35,6 | 4,71 | 49,1 | 6 | 0,018 | 46 | - | 1,34 | 8,82 | - | 592 | 4,54 | 1230 | 809 | - | 7 | 7,81 | 0 | 2,18 | 7,3 | - | 1,9 | 0,27 | 0,15 | 4,99 | 0,75 | 0,14 | 58 | 0 |
| Dachine, pulpe, cuit à la vapeur, prélevé à la Martinique | - | - | - | - | 65,3 | 1,15 | 1,15 | - | 0,3 | - | - | - | - | - | - | - | - | - | - | - | - | - | - | - | - | - | - | 4,07 | - | - | - | - | - | - | - |
| Datte, pulpe et peau, sèche | - | - | - | - | 22,9 | 1,81 | 1,81 | 64,7 | 0,25 | 7,3 | 0,098 | 44,9 | 61 | 0,22 | 0,9 | 14 | 47,3 | 0,3 | 62 | 696 | 3,05 | 39 | 0,23 | 0 | 0,46 | 2,7 | - | 3 | 0,04 | 0,075 | 1,41 | 0,79 | 0,19 | 18 | 0 |
| Dessert de fruits, tout type de fruits (en taux de sucres : compotes allégées en sucres < desserts de fruits < compotes allégée) | - | - | - | - | 79,4 | 0,32 | 0,32 | 19,5 | 0,2 | 1,61 | 0,0049 | 4 | < 3,6 | 0,04 | 0,15 | < 20 | 4 | 0,03 | 9,3 | 130 | < 20 | 1,71 | < 0,05 | - | - | < 0,8 | - | 11,8 | < 0,01 | < 0,01 | 0,28 | 0,067 | 0,03 | < 5 | - |
| Dourian, pulpe, cru | - | - | - | - | 65 | 1,47 | 1,47 | 23,3 | 5,33 | 3,8 | 0,005 | 6 | - | 0,21 | 0,43 | - | 30 | 0,33 | 39 | 436 | - | 2 | 0,28 | - | - | - | - | 19,7 | 0,37 | 0,2 | 1,07 | 0,23 | 0,32 | 36 | 0 |
| Échalote, crue | 269 | 63,6 | 269 | 63,6 | 82,3 | 1,81 | 1,81 | 12,2 | < 0,5 | 2,6 | 0,013 | 9,5 | 27 | 0,07 | 0,12 | < 20 | 7,5 | 0,08 | 46 | 220 | < 20 | 5 | 0,2 | 0 | < 0,08 | < 0,8 | - | 8 | 0,036 | < 0,01 | 0,2 | 0,19 | 0,1 | 10,2 | 0 |
| Échalote, cuite | - | - | - | - | 80 | 2,13 | 2,13 | 13,5 | < 0,5 | 3,4 | 0,031 | 28 | 29,2 | 0,11 | 0,34 | < 20 | 14 | 0,15 | 76 | 400 | < 50 | 12,5 | 0,32 | - | 0,12 | < 0,8 | - | 2,25 | 0,53 | - | 0,17 | 0,22 | 0,19 | 12,9 | - |
| Échalote, sautée/poêlée sans matière grasse | 282 | 66,9 | 282 | 66,9 | 81,5 | 2 | 2 | 12,3 | 0,4 | 2,9 | < 0,013 | 22 | 26 | 0,09 | 0,25 | < 20 | 11 | 0,12 | 48 | 270 | < 20 | < 5 | 0,27 | - | < 0,08 | < 0,8 | - | 0,84 | 0,032 | < 0,01 | 0,59 | 0,23 | 0,1 | - | - |

| | | | | | | | | | | | | | | | | | | | | | | | | | | | | | | | | | | | |
|---|---|---|---|---|---|---|---|---|---|---|---|---|---|---|---|---|---|---|---|---|---|---|---|---|---|---|---|---|---|---|---|---|---|---|---|
| Endive, crue | 84,8 | 20,2 | 84,8 | 20,2 | 94,3 | 1,19 | 1,19 | 2,83 | < 0,5 | 1,1 | < 0,013 | 20 | 40,7 | 0,04 | 0,12 | < 20 | 8,3 | 0,08 | 23 | 150 | < 20 | < 5 | 0,13 | 0 | < 0,08 | < 0,8 | - | 2,12 | 0,067 | < 0,01 | 0,21 | 0,15 | 0,026 | 25,6 | 0 |
| Endive, rôtie/cuite au four | 99,5 | 23,5 | 99,5 | 23,5 | 93,7 | 1,13 | 1,13 | 4,38 | < 0,3 | < 0,5 | < 0,013 | 27 | 44,7 | 0,07 | 0,23 | < 20 | 10 | 0,09 | 27 | 180 | < 20 | < 5 | 0,18 | - | < 0,08 | 2,09 | - | < 0,5 | 0,032 | < 0,01 | 0,21 | 0,11 | 0,028 | 5,89 | - |
| Épinard, appertisé, égoutté | - | - | - | - | 91,8 | 2,81 | 2,81 | 0,95 | 0,5 | 2,4 | 0,81 | 127 | - | 0,18 | 2,3 | 2,79 | 76 | 0,6 | 44 | 346 | 1,35 | 322 | 0,46 | 0 | 1,25 | 462 | - | 14,3 | 0,016 | 0,14 | 0,39 | 0,047 | 0,1 | 98 | 0 |
| Épinard, bouilli/cuit à l'eau | 117 | 28,1 | 117 | 28,1 | 91,9 | 3,38 | 3,38 | < 0,85 | 0,7 | 3,5 | 0,03 | 240 | < 20 | 0,14 | 1,3 | < 20 | 30 | 0,81 | 30 | 160 | < 20 | 12 | 0,79 | - | 3,98 | 38,1 | - | < 0,5 | < 0,015 | 0,053 | < 0,1 | 0,15 | 0,038 | 62,7 | - |
| Épinard, cru | - | - | - | - | 91,6 | 2,62 | 2,62 | 2,25 | 0,5 | 2,37 | 0,17 | 114 | 98 | 0,1 | 3,61 | 3,4 | 52,5 | 1,3 | 45,2 | 504 | 0,1 | 66,7 | 0,82 | 0 | 2,47 | 521 | - | 41,1 | 0,087 | 0,21 | 0,71 | 0,18 | 0,21 | 207 | 0 |
| Épinard, cuit | - | - | - | - | 92,8 | 3,2 | 3,2 | 0,5 | 0,14 | 2,7 | 0,069 | 140 | - | 0,1 | 2,14 | < 10 | 54,4 | 0,43 | 40 | 396 | < 10 | 27,6 | 0,65 | 0 | 2,4 | 494 | - | 2,1 | < 0,05 | 0,1 | 0,24 | < 0,05 | 0,06 | 125 | 0 |
| Épinard, jeunes pousses pour salades, cru | 76 | 18,3 | 76 | 18,3 | 93,7 | 2,06 | 2,06 | < 0,85 | 0,4 | 2,4 | 0,063 | 110 | 70,1 | 0,08 | 1,8 | 60 | 44 | 0,62 | 32 | 580 | < 20 | 25 | 0,43 | < 0,25 | 0,56 | 75,5 | - | 6,04 | 0,041 | 0,057 | 0,26 | 0,2 | 0,15 | 78,6 | - |
| Épinard, purée | - | - | - | - | - | 3,7 | 3,7 | 5,2 | 2,2 | 2,2 | 0,72 | - | - | - | - | - | - | - | - | - | - | 289 | - | - | - | - | - | - | - | - | - | - | - | 94 | - |
| Épinard, surgelé, cru | - | - | - | - | 91,6 | 2,87 | 2,87 | 0,89 | 0,49 | 2,7 | 0,098 | 159 | - | 0,12 | 2,28 | 3,4 | 50,5 | 0,53 | 47,7 | 398 | 0,1 | 46,5 | 0,49 | 0 | 2,9 | 356 | - | 21,4 | 0,093 | 0,2 | 0,51 | 0,11 | 0,18 | 183 | 0 |
| Épinard, surgelé, cuit | - | - | - | - | 88,9 | 4,01 | 4,01 | 0,51 | 0,87 | 3,7 | 0,24 | 153 | - | 0,16 | 1,96 | - | 82 | 0,72 | 50 | 302 | - | 97 | 0,49 | 0 | 3,54 | 541 | - | 2,2 | 0,078 | 0,18 | 0,44 | 0,075 | 0,14 | 121 | 0 |
| Farine de châtaigne | - | - | - | - | 5,52 | 5,69 | 6,1 | 70,4 | 3,43 | 12,6 | - | 61,6 | - | 0,5 | < 0,1 | - | 67,9 | 3,2 | - | - | - | - | 1,1 | - | - | - | - | 16 | - | - | 2,1 | 1,53 | 0,58 | 215 | - |
| Farine de pulpe de patate douce, prélevée à la Martinique | - | - | - | - | 8,55 | - | - | 50,1 | - | 11,3 | - | - | - | - | - | - | - | - | - | - | - | - | - | - | - | - | - | 0 | - | - | - | - | - | - | - |
| Feijoa, pulpe, crue | - | - | - | - | 83,3 | 0,71 | 0,71 | 8,2 | 0,42 | 6,4 | 0,0075 | 17 | - | 0,036 | 0,14 | - | 9 | 0,084 | 19 | 172 | - | 3 | 0,06 | - | 0,16 | 3,5 | - | 32,9 | 0,006 | 0,018 | 0,3 | 0,23 | 0,067 | 23 | 0 |
| Fenouil, bouilli/cuit à l'eau | - | - | - | - | 94,4 | 1,13 | 1,13 | 0,8 | < 0,1 | 2 | 0,15 | 45 | 82 | 0,16 | 0,2 | 10 | 10,2 | 0,14 | 17,3 | 270 | < 3,6 | 58,1 | 0,18 | 0 | 0,32 | - | - | 1,6 | 0,024 | < 0,04 | 0,36 | 0,15 | 0,12 | 31,8 | 0 |
| Fenouil, cru | 90,9 | 21,8 | 90,9 | 21,8 | 90,9 | 1 | 1 | 2,63 | < 0,5 | 2,6 | 0,055 | 37 | 78,6 | 0,08 | 0,2 | < 20 | 11 | 0,1 | 28 | 320 | < 20 | 22 | 0,19 | 0 | < 0,08 | 62,8 | - | 2,57 | < 0,015 | < 0,01 | < 0,1 | 0,22 | 0,05 | 19,6 | 0 |
| Fenouil, cuit à l'étouffée | 57,1 | 13,7 | 57,1 | 13,7 | 95,1 | < 0,5 | < 0,5 | 1,94 | < 0,3 | 1,8 | 0,028 | 37 | 42,8 | 0,04 | 0,12 | < 20 | 6,7 | 0,07 | 16 | 210 | < 20 | 11 | 0,08 | - | < 0,08 | < 0,8 | - | < 0,5 | < 0,015 | < 0,01 | 0,12 | 0,11 | 0,019 | 16,8 | - |
| Fève à écosser, fraîche | - | - | - | - | 76,8 | 6,76 | 6,76 | 4,2 | 0,67 | 5,85 | 0,094 | 29,5 | 14 | 0,24 | 1,73 | - | 35,5 | 0,49 | 112 | 291 | 0,6 | 37,5 | 0,79 | 0 | 1,16 | 40,9 | - | 18,4 | 0,15 | 0,2 | 1,87 | 0,16 | 0,071 | 122 | 0 |
| Fève, bouillie/cuite à l'eau | 350 | 82,9 | 350 | 82,9 | 77,9 | 8,06 | 8,06 | 9,35 | 0,8 | 3,1 | 0,025 | 37 | < 20 | 0,31 | 1,5 | < 20 | 37 | 0,31 | 160 | 200 | < 20 | 10 | 1 | 0 | 0,28 | 5,52 | - | < 0,5 | 0,043 | 0,064 | 0,69 | 0,12 | 0,037 | 65,3 | 0 |
| Fève, fraîche, surgelée | - | - | - | - | 75,8 | 7,5 | 7,5 | 7,95 | 0,22 | 3,6 | 0,022 | - | - | - | - | - | - | - | - | - | - | - | - | - | - | - | - | - | - | - | - | - | - | - | - |
| Fève, pelée, surgelée, crue | - | - | - | - | - | 14,9 | 14,9 | 14,9 | 0,3 | 6,5 | 0,038 | - | - | - | - | - | - | - | - | - | - | 15 | - | - | - | - | - | - | - | - | - | - | - | - | - |
| Fève, pelée, surgelée, cuite à l'eau | - | - | - | - | 80,7 | 6,4 | 6,4 | 9 | 0,5 | 2,9 | 0,018 | 31 | 4,13 | 0,17 | 1,3 | < 20 | 22 | 0,18 | 97 | 97 | < 20 | 7,2 | 0,62 | < 0,25 | - | 3,2 | - | < 0,5 | 0,043 | 0,038 | 0,57 | 0,28 | 0,042 | 49,1 | - |
| Fève, sèche | - | - | - | - | 11 | 26,1 | 26,1 | 33,3 | 1,53 | 25 | 0,033 | 103 | - | 0,82 | 6,7 | - | 192 | 1,63 | 421 | 1060 | - | 13 | 3,14 | 0 | 0,05 | 9 | - | 1,4 | 0,56 | 0,33 | 2,83 | 0,98 | 0,37 | 423 | 0 |
| Fève, surgelée, bouillie/cuite à l'eau | - | - | - | - | 80,5 | 5,6 | 5,6 | 7,4 | < 0,5 | 5,8 | 0,02 | 61 | 9,8 | 0,21 | 0,93 | < 20 | 26 | 0,23 | 88 | 150 | < 20 | 8 | 0,55 | < 0,25 | - | 4,16 | - | < 0,5 | 0,061 | 0,055 | 0,98 | 0,13 | 0,038 | 37,1 | - |
| Figue de Barbarie, pulpe et graines, crue | - | - | - | - | 88,4 | 0,37 | 0,37 | 6 | 0,31 | 4,45 | 0,011 | 118 | - | 0,019 | 0,25 | 1,5 | 77 | 0,58 | 17,5 | 175 | - | 4,5 | 0,14 | 0 | 0,01 | 2,9 | - | 12,5 | 0,011 | 0,046 | 0,38 | - | 0,079 | 8 | 0 |
| Figue, crue | 293 | 69,4 | 293 | 69,4 | 80,2 | 1,19 | 1,19 | 13,5 | < 0,5 | 4,1 | 0,013 | 57 | < 20 | 0,08 | 0,27 | < 30 | 22 | 0,06 | 21 | 230 | < 20 | 5 | 0,16 | 0 | < 0,08 | 5,22 | - | < 0,5 | < 0,015 | < 0,01 | 0,37 | 0,15 | 0,059 | 24,5 | 0 |
| Figue, sèche | - | - | - | - | 30,1 | 2,99 | 2,99 | 54,3 | 0,87 | 9,72 | 0,22 | 167 | - | 0,27 | 2,12 | 1,1 | 52,5 | 0,4 | 75,1 | 845 | < 2,2 | 86,3 | 0,51 | 0 | 0,35 | 15,6 | - | 1,2 | 0,085 | 0,096 | 0,68 | 0,43 | 0,11 | 9 | 0 |
| Fondue de poireau | - | - | - | - | - | 1,1 | 1,1 | 0 | 5,2 | 3,6 | 0,5 | - | - | - | - | - | - | - | - | - | - | 200 | - | - | - | - | - | - | - | - | - | - | - | - | - |
| Fraise, crue | 162 | 38,6 | 162 | 38,6 | 90,3 | 0,63 | 0,63 | 6,03 | < 0,5 | 3,8 | < 0,013 | 18 | < 20 | 0,02 | 0,19 | < 20 | 12 | 0,26 | 23 | 140 | < 20 | < 5 | 0,11 | 0 | 0,3 | < 0,8 | - | 54 | < 0,015 | < 0,01 | 0,21 | 0,13 | 0,04 | 98,9 | 0 |
| Framboise, crue | 206 | 49,2 | 206 | 49,2 | 86,8 | 1,19 | 1,19 | 5,83 | 0,8 | 4,3 | < 0,013 | 16 | < 20 | 0,04 | 0,4 | < 20 | 20 | 0,44 | 29 | 170 | < 20 | < 5 | 0,24 | 0 | 0,88 | 5,02 | - | 18,7 | < 0,015 | 0,02 | 0,35 | 0,85 | 0,032 | 38,1 | 0 |
| Framboise, surgelée, crue | - | - | - | - | 85,9 | 1,16 | 1,16 | 6,43 | 0,8 | 6,05 | 0,0053 | 29,9 | - | 0,11 | 0,78 | 0,4 | 17 | 1,2 | 38 | 228 | 0,19 | 2,17 | 0,34 | 0 | 1,4 | - | - | 24,5 | 0,03 | 0,05 | 0,5 | 0,24 | 0,09 | 44 | 0 |
| Frites de pommes de terre, surgelées, cuites en friteuse | 1200 | 285 | 1200 | 285 | 42,5 | 3,28 | 3,28 | 39,4 | 11,9 | 3,9 | 0,073 | 23,8 | 0,1 | 0,12 | 0,9 | < 5 | 23,3 | 0,16 | 106 | 684 | < 2,2 | 29 | 0,36 | 0 | 0,44 | 11,2 | - | 19 | 0,11 | 0,059 | 2,57 | 0,43 | 0,37 | 32,3 | 0,04 |
| Frites de pommes de terre, surgelées, préfrites, pour cuisson en friteuse | 1020 | 244 | 1020 | 244 | - | 3,12 | 3,12 | 29,1 | 12,1 | 3,39 | 0,12 | 20 | - | - | 1,8 | - | - | - | - | - | - | 57,9 | - | - | - | - | - | 6 | - | - | - | - | - | - | - |
| Frites de pommes de terre, surgelées, préfrites, pour cuisson micro-ondes | 1070 | 254 | 1070 | 254 | 49,5 | 3,5 | 3,5 | 34 | 10,8 | 3,83 | 0,27 | - | - | - | - | - | - | - | - | - | - | 108 | - | - | - | - | - | - | - | - | - | - | - | - | - |
| Frites de pommes de terre, surgelées, préfrites, pour cuisson rôtie/ au four | 648 | 154 | 648 | 154 | 67,9 | 2,41 | 2,41 | 23,4 | 5,09 | 2,58 | 0,28 | - | - | - | - | - | - | - | - | - | - | 109 | - | - | - | - | - | - | - | - | - | - | - | - | - |
| Frites de pommes de terre, surgelées, rôties/cuites au four | 896 | 213 | 896 | 213 | 51,1 | 3,75 | 3,75 | 32,6 | 6,6 | 4,2 | 0,39 | 11 | 267 | 0,11 | 0,67 | < 20 | 29 | 0,15 | 81 | 590 | < 20 | 156 | 0,45 | < 0,25 | 7,64 | < 0,8 | - | 2,38 | 0,097 | 0,018 | 1,15 | 0,52 | 0,52 | 40,6 | 0 |
| Fruit à pain, cru | 444 | 105 | 444 | 105 | 70,7 | 1,07 | 1,07 | 22,2 | 0,23 | 4,9 | 0,005 | 17 | - | 0,084 | 0,54 | - | 25 | 0,06 | 30 | 490 | - | 2 | 0,12 | 0 | 0,1 | 0,5 | - | 29 | 0,11 | 0,03 | 0,9 | 0,46 | 0,1 | 14 | 0 |
| Fruit à pain, pulpe, cuit à la vapeur, prélevé à la Martinique | - | - | - | - | 70,5 | 1,29 | 1,29 | - | 0,11 | - | - | - | - | - | - | - | - | - | - | - | - | - | - | - | - | - | - | 8,17 | - | - | - | - | - | - | - |
| Fruit cru (aliment moyen) | 251 | 59,5 | 251 | 59,5 | 84,5 | 0,7 | 0,7 | 11,6 | 0,26 | 1,97 | 0,0088 | 14,5 | 33,2 | 0,059 | 0,18 | 6,83 | 12,1 | 0,12 | 20,1 | 197 | 7,53 | 3,45 | 0,11 | 0,003 | 0,29 | 1,98 | - | 20,9 | 0,036 | 0,014 | 0,28 | 0,17 | 0,065 | 22 | 0 |
| Fruit de la passion ou maracudja, pulpe et pépins, cru | 425 | 101 | 425 | 101 | 73,6 | 2,13 | 2,13 | 10,9 | 3 | 6,8 | < 0,013 | 8,1 | < 20 | 0,15 | 0,56 | < 20 | 26 | 0,19 | 46 | 240 | < 20 | < 5 | 0,76 | 0 | 0,5 | < 0,8 | - | 25,6 | < 0,015 | 0,053 | 1,37 | 0,56 | 0,17 | 101 | 0 |
| Fruits rouges, crus (framboises, fraises, groseilles, cassis) | - | - | - | - | 84,5 | 1,11 | 1,11 | 7,88 | 0,31 | 5,3 | 0,008 | 40 | - | 0,088 | 1 | 1,15 | 16,9 | 0,42 | 32,6 | 225 | 0,35 | 3 | 0,22 | 0 | 0,67 | - | - | 87 | 0,031 | 0,084 | 0,38 | 0,21 | 0,064 | 37,6 | 0 |
| Giraumon (variété locale), pulpe, cuit à la vapeur, prélevé à la Martinique | - | - | - | - | 91,8 | 1,1 | 1,1 | 4,51 | 0,097 | 2 | - | - | - | - | - | - | - | - | - | 321 | - | - | - | - | - | - | - | - | - | - | - | - | - | 20,5 | - |
| Giraumon (variété locale), pulpe, cuit à la vapeur, surgelé, prélevé à la Martinique | - | - | - | - | - | - | - | 3,75 | - | 2,4 | - | - | - | - | - | - | - | - | - | 347 | - | - | - | - | - | - | - | - | - | - | - | - | - | - | - |
| Giraumon (variété locale), pulpe, râpé, cru, prélevé à la Martinique | - | - | - | - | - | - | - | 2,42 | - | 2,9 | - | - | - | - | - | - | - | - | - | 319 | - | - | - | - | - | - | - | - | - | - | - | - | - | - | - |
| Giraumon (variété phoenix), pulpe, cuit à la vapeur, prélevé à la Martinique | - | - | - | - | 90,6 | 1,03 | 1,03 | 5,62 | 0,05 | 2 | - | - | - | - | - | - | - | - | - | 442 | - | - | - | - | - | - | - | - | - | - | - | - | - | 31,7 | - |
| Giraumon (variété phoenix), pulpe, râpé, cru, prélevé à la Martinique | - | - | - | - | - | - | - | 3,51 | - | 1,6 | - | - | - | - | - | - | - | - | - | 353 | - | - | - | - | - | - | - | - | - | - | - | - | - | - | - |
| Giraumon (variété phoenix), pulpe, surgelé, cuit à la vapeur, prélevé à la Martinique | - | - | - | - | - | - | - | 4,37 | - | 2,4 | - | - | - | - | - | - | - | - | - | 454 | - | - | - | - | - | - | - | - | - | - | - | - | - | - | - |
| Gombo, entier, appertisé, non égoutté, prélevé à la Martinique | - | - | - | - | 92,3 | - | - | - | - | 2,8 | - | - | - | - | - | - | - | 0,17 | - | - | - | - | - | - | - | - | - | - | - | - | - | - | - | 18,3 | - |
| Gombo, entier, cuit à la vapeur, prélevé à la Martinique | - | - | - | - | 87,5 | 1,8 | 1,8 | 4,69 | 0,3 | 4,8 | - | - | - | - | - | - | 61 | 0,52 | - | 374 | - | - | - | - | - | - | - | 3 | - | - | - | - | - | 107 | - |
| Gombo, fruit, cuit | - | - | - | - | 92,6 | 1,87 | 1,87 | 2,01 | 0,21 | 2,5 | 0,015 | 77 | - | 0,085 | 0,28 | - | 36 | 0,29 | 32 | 135 | - | 6 | 0,43 | 0 | 0,27 | 40 | - | 16,3 | 0,13 | 0,055 | 0,87 | 0,21 | 0,19 | 46 | 0 |
| Gombo, pulpe, blanchi, surgelé, prélevé à la Martinique | - | - | - | - | 88,4 | 0,24 | 0,24 | 5,28 | 0,1 | 5,17 | - | - | - | - | - | - | - | 0,67 | - | - | - | - | - | - | - | - | - | - | - | - | - | - | - | 69,8 | - |
| Goyave, pulpe, crue | - | - | - | - | 83,5 | 1,59 | 1,59 | 9,02 | 0,73 | 5,15 | 0,005 | 14 | - | 0,23 | 0,26 | 0,71 | 16 | 0,15 | 40 | 308 | 0,5 | 2 | 0,21 | 0 | 0,73 | 2,6 | - | 228 | 0,052 | 0,04 | 1,08 | 0,45 | 0,094 | 49 | 0 |
| Goyave, pulpe, purée, prélevée à la Martinique | - | - | - | - | 88,5 | 0,83 | 0,83 | - | 0,0067 | - | - | - | - | - | - | - | - | - | - | - | - | - | - | - | - | - | - | 492 | - | - | - | - | - | - | - |
| Grenade, pulpe et pépins, crue | 340 | 80,6 | 340 | 80,6 | 79,4 | 1,44 | 1,44 | 14,3 | 1,2 | 2,3 | < 0,013 | 9,5 | 44,1 | 0,11 | 0,17 | < 20 | 12 | 0,1 | 27 | 230 | < 20 | < 5 | 0,22 | 0 | < 0,08 | < 0,8 | - | 9,02 | 0,054 | < 0,01 | 0,23 | 0,37 | 0,032 | 8,77 | 0 |
| Griotte, crue | - | - | - | - | 86,1 | 1,1 | 1,1 | 10,5 | 0,3 | 1,65 | 0,0063 | 16 | - | 0,1 | 0,32 | - | 9 | 0,11 | 15 | 173 | - | 3 | 0,1 | 0 | 0,07 | 2,1 | - | 10 | 0,03 | 0,04 | 0,4 | 0,14 | 0,044 | 8 | 0 |
| Groseille à maquereau, crue | - | - | - | - | 87,9 | 0,75 | 0,75 | 4,88 | 0,59 | 4,03 | 0,0038 | 25,4 | - | 0,063 | 0,33 | 0,3 | 9 | 0,16 | 26,2 | 187 | 0,068 | 1,5 | 0,14 | 0 | 0,69 | - | - | 29 | 0,033 | 0,029 | 0,29 | 0,29 | 0,048 | 8,85 | 0 |
| Groseille, crue | 289 | 68,5 | 289 | 68,5 | 82,1 | 1,56 | 1,56 | 7,06 | 0,7 | 4,6 | < 0,013 | 38 | < 20 | 0,09 | 0,35 | < 20 | 12 | 0,13 | 38 | 230 | < 20 | < 5 | 0,15 | 0 | 1,19 | 2,06 | - | 29,8 | 0,041 | < 0,01 | 0,18 | 0,65 | 0,03 | 11,8 | 0 |

| | | | | | | | | | | | | | | | | | | | | | | | | | | | | | | | | | | | |
|---|---|---|---|---|---|---|---|---|---|---|---|---|---|---|---|---|---|---|---|---|---|---|---|---|---|---|---|---|---|---|---|---|---|---|---|
| Haricot beurre, appertisé, égoutté | - | - | - | - | 92,8 | 1,22 | 1,22 | 3,48 | 0,23 | 1,68 | 0,66 | 25 | - | 0,044 | 0,9 | 0,6 | 15 | 0,2 | 19 | 104 | 0,43 | 265 | 0,3 | 0 | 0,29 | 9,9 | - | 4,1 | 0,02 | 0,054 | 0,2 | 0,085 | 0,034 | 66 | 0 |
| Haricot beurre, bouilli/cuit à l'eau | 145 | 34,6 | 145 | 34,6 | 89,6 | 2,19 | 2,19 | 4,45 | 0,3 | 2,7 | < 0,013 | 56 | 20,1 | 0,13 | 0,99 | < 20 | 26 | 0,35 | 43 | 210 | < 20 | < 5 | 0,34 | - | 0,31 | 3,88 | - | 7,57 | 0,04 | 0,053 | 0,36 | 0,12 | < 0,01 | 39,7 | - |
| Haricot beurre, cru | - | - | - | - | 90,1 | 1,85 | 1,85 | 5,6 | 0,16 | 2,45 | 0,015 | 41 | - | 0,07 | 1,07 | 0,6 | 26 | 0,21 | 38 | 209 | - | 6 | 0,32 | 0 | 0,45 | - | - | 18,7 | 0,08 | 0,12 | 0,68 | 0,072 | 0,087 | 68,5 | 0 |
| Haricot beurre, surgelé, cru | - | - | - | - | 89,9 | 2 | 2 | 4,9 | 0,17 | 2,33 | 0,0092 | 51 | - | 0,049 | 0,78 | 0,4 | 22 | 0,39 | 32 | 186 | - | 2,5 | 0,26 | 0 | - | - | - | 16 | 0,099 | 0,092 | 0,5 | 0,085 | 0,042 | 15 | 0 |
| Haricot blanc, appertisé, égoutté | - | - | - | - | 70,1 | 6,57 | 6,57 | 10,9 | 0,4 | 7 | 0,69 | 73 | - | 0,23 | 2,99 | 3 | 51 | 0,52 | 91 | 454 | - | 270 | 1,12 | 0 | 0,79 | 2,9 | - | 0 | 0,096 | 0,037 | 0,11 | 0,19 | 0,075 | 65 | 0 |
| Haricot blanc, bouilli/cuit à l'eau | 469 | 112 | 469 | 112 | 66,9 | 6,75 | 6,75 | 12 | 1,1 | 13,8 | 0,023 | 120 | < 20 | 0,19 | 1,7 | < 20 | 33 | 0,8 | 110 | 260 | < 20 | 9,3 | 0,67 | < 0,25 | < 0,08 | 4,37 | - | < 0,5 | 0,03 | 0,017 | 0,32 | 0,061 | 0,039 | 31,1 | 0 |
| Haricot blanc, sec | - | - | - | - | 10,6 | 19,1 | 19,1 | 43,9 | 1,78 | 16,8 | 0,028 | 183 | 47 | 0,73 | 7,97 | 0,6 | 187 | 1,9 | 363 | 1660 | 8,8 | 11 | 3,23 | 0 | 0,28 | 5,6 | - | 2,04 | 0,44 | 0,15 | 0,48 | 0,86 | 0,41 | 307 | 0 |
| Haricot coco, bouilli/cuit à l'eau | 547 | 131 | 547 | 131 | 61,3 | 9,63 | 9,63 | 13,7 | 0,5 | 15,8 | 0,023 | 72 | < 20 | 0,26 | 2,7 | < 20 | 53 | 0,59 | 180 | 480 | < 20 | 9 | 1,2 | < 0,25 | 0,1 | 3,61 | - | 1,62 | 0,17 | 0,02 | 0,86 | 0,14 | 0,065 | 71,4 | - |
| Haricot de Lima, cru | 444 | 105 | 444 | 105 | 70,2 | 6,84 | 6,84 | 15,3 | 0,86 | 4,9 | 0,02 | 34 | - | 0,32 | 3,14 | - | 58 | 1,22 | 136 | 467 | - | 8 | 0,78 | 0 | 0,32 | 5,6 | - | 23,4 | 0,22 | 0,1 | 1,47 | 0,25 | 0,2 | 34 | 0 |
| Haricot flageolet, appertisé, égouttés | - | - | - | - | 72 | 6,1 | 6,1 | 12,2 | 0,79 | 6,3 | 0,73 | 63,1 | - | < 0,1 | 1,3 | < 10 | 26,5 | < 0,1 | 86 | 281 | < 2,2 | 293 | < 0,1 | 0 | 0,37 | - | - | < 1 | 0,07 | 0,16 | 0,79 | < 0,05 | 0,05 | 23 | 0 |
| Haricot flageolet, bouilli/cuit à l'eau | 469 | 112 | 469 | 112 | 66,9 | 6,75 | 6,75 | 12 | 1,1 | 13,8 | 0,023 | 120 | < 20 | 0,19 | 1,7 | < 20 | 33 | 0,8 | 110 | 260 | < 20 | 9,3 | 0,67 | < 0,25 | < 0,08 | 4,37 | - | < 0,5 | 0,03 | 0,017 | 0,32 | 0,061 | 0,039 | 31,1 | traces |
| Haricot flageolet, surgelé | - | - | - | - | 55,1 | 9,32 | 9,32 | 25,6 | 0,7 | 4,4 | 0,78 | - | - | - | - | - | - | - | - | - | - | - | - | - | - | - | - | - | - | - | - | - | - | - | - |
| Haricot flageolet, vert, bouilli/cuit à l'eau | - | - | - | - | 63 | 5,63 | 5,63 | 15,8 | 1,1 | 16,5 | 0,071 | 68 | 9,36 | 0,18 | 1,8 | < 20 | 32 | 0,47 | 97 | 360 | < 50 | 28,2 | 0,6 | < 0,5 | < 0,08 | 1,45 | - | < 0,5 | 0,072 | 0,02 | 0,31 | 0,065 | 0,064 | 34,6 | 0 |
| Haricot flageolet, vert, sec | - | - | - | - | 8,5 | 19,1 | 19,1 | 42,1 | 2,6 | 23,4 | 0,02 | - | 38,6 | 0,68 | 7,3 | < 20 | 160 | 1,6 | 440 | - | < 50 | 7,9 | 2,3 | < 0,5 | 0,08 | 13,7 | - | < 0,5 | 0,31 | 0,05 | 1,93 | 0,68 | 0,41 | 121 | - |
| Haricot mungo germé ou pousse de "soja", appertisé, égoutté | - | - | - | - | 95,8 | 1,56 | 1,71 | 2,36 | 0,12 | 1,63 | 0,35 | 13,4 | - | 0,047 | 0,26 | 2,55 | 7,53 | 0,039 | 26 | 31,5 | < 10 | 141 | 0,11 | 0 | 0,065 | 13,4 | - | 0,65 | 0,025 | 0,05 | 0,21 | 0,14 | 0,031 | 26 | 0 |
| Haricot mungo germé ou pousse de "soja", cru | - | - | - | - | 91,7 | 2,57 | 2,81 | 3,4 | < 0,5 | 1,7 | 0,043 | 16 | 6,7 | 0,1 | 0,44 | < 20 | 16 | 0,12 | 43 | 120 | < 50 | 17,1 | 0,32 | < 0,5 | 0,095 | 2,1 | - | 4,13 | 0,065 | 0,06 | 0,71 | 0,17 | 0,06 | 61,3 | 0 |
| Haricot mungo, bouilli/cuit à l'eau | - | - | - | - | 72,5 | 7,54 | 7,54 | 11,9 | 0,55 | 6,4 | 0,018 | 53 | - | 0,14 | 1,75 | - | 63 | 0,41 | 156 | 231 | - | 7 | 0,83 | 0 | 0,15 | 2,7 | - | 1 | 0,15 | 0,075 | 1,5 | 0,43 | 0,058 | 94 | 0 |
| Haricot mungo, sec | - | - | - | - | 9,52 | 24,5 | 24,5 | 47,6 | 1,42 | 16,7 | 0,066 | 113 | - | 0,96 | 7,16 | - | 228 | 1,28 | 387 | 1110 | 8,8 | 26,5 | 3,02 | 0 | 1,9 | 170 | - | 12,2 | 0,45 | 0,24 | 1,85 | 0,55 | 0,39 | 421 | 0 |
| Haricot plat, cru | - | - | - | - | 92,2 | 1,93 | 1,93 | 4,47 | 0,12 | 2,97 | 0,049 | 60 | - | - | 0,8 | - | - | - | - | - | - | 2 | - | - | - | - | - | 16 | - | - | - | - | - | - | - |
| Haricot rouge, appertisé, égoutté | - | - | - | - | 68,9 | 8,31 | 8,31 | 13 | 0,97 | 7 | 0,68 | 58 | - | 0,28 | 1,5 | - | 29 | 0,38 | 118 | 250 | - | 269 | 0,74 | 0 | - | - | - | 0,2 | 0,06 | 0,015 | 0,42 | - | - | 23 | 0 |
| Haricot rouge, bouilli/cuit à l'eau | 487 | 116 | 487 | 116 | 65 | 9,63 | 9,63 | 12,3 | 0,6 | 11,6 | 0,021 | 55 | < 20 | 0,27 | 2,3 | < 20 | 39 | 0,44 | 150 | 300 | < 20 | 8,4 | 0,94 | < 0,25 | < 0,08 | 1,06 | - | < 0,5 | 0,12 | 0,022 | 0,25 | 0,083 | 0,067 | 78,3 | 0 |
| Haricot rouge, sec | - | - | - | - | 11,8 | 22,5 | 22,5 | 46,1 | 1,06 | 15,2 | 0,03 | 83 | - | 0,7 | 6,69 | 1,9 | 138 | 1,11 | 406 | 1360 | - | 12 | 2,79 | 0 | 0,21 | 5,6 | - | 4,5 | 0,61 | 0,22 | 2,11 | 0,78 | 0,4 | 394 | 0 |
| Haricot vert, appertisé, égoutté | 96,1 | 23,1 | 96,1 | 23,1 | 92,6 | 1,33 | 1,33 | 2,07 | 0,31 | 3,36 | 0,68 | 42,7 | 518 | 0,13 | 0,81 | 1,6 | 13 | 0,094 | 25,3 | 94 | 0,9 | 261 | 0,23 | 0 | 0,14 | 38,9 | - | 3,1 | 0,053 | 0,053 | 0,15 | 0,15 | 0,03 | 20,5 | 0 |
| Haricot vert, bouilli/cuit à l'eau | 117 | 28 | 117 | 28 | 90,8 | 1,75 | 1,75 | 3,39 | < 0,3 | 3,2 | < 0,013 | 51 | 30 | 0,07 | 0,54 | < 20 | 26 | 0,19 | 38 | 260 | < 20 | < 5 | 0,22 | - | 0,1 | 5,16 | - | 1,16 | 0,031 | 0,038 | 0,28 | 0,12 | 0,031 | 49,5 | - |
| Haricot vert, cru | - | - | - | - | 90,4 | 1,85 | 1,85 | 4,14 | 0,21 | 2,85 | 0,0096 | 48,5 | 16 | 0,063 | 1,02 | 0,8 | 21 | 0,23 | 38,5 | 224 | 0,3 | 3,85 | 0,32 | 0 | 0,36 | 14,4 | - | 13,6 | 0,086 | 0,11 | 0,72 | 0,14 | 0,12 | 48,5 | 0 |
| Haricot vert, cuit | 123 | 29,4 | 123 | 29,4 | 89,3 | 2 | 2 | 3 | 0,17 | 4 | 0,4 | 55,3 | 11 | 0,081 | 0,6 | < 5 | 22,3 | 0,19 | 36 | 174 | < 10 | 5,8 | 0,2 | 0 | 0,45 | 16 | - | 5 | < 0,05 | 0,08 | 0,34 | < 0,05 | < 0,05 | 33 | 0 |
| Haricot vert, surgelé, cru | 142 | 34 | 142 | 34 | 89,4 | 1,97 | 1,97 | 4,35 | 0,17 | 3,68 | 0,015 | 51,5 | - | 0,063 | 0,85 | 0,8 | 22 | 0,29 | 40,8 | 186 | 0,3 | 3,86 | 0,31 | 0 | 0,24 | 38,4 | - | 18 | 0,099 | 0,092 | 0,5 | 0,077 | 0,068 | 39,5 | 0 |
| Haricot vert, surgelé, cuit | 152 | 36,3 | 152 | 36,3 | 91,4 | 1,95 | 1,95 | 5,1 | 0,19 | 3,3 | 0,0038 | 51 | - | 0,059 | 0,68 | 0,4 | 19 | 0,29 | 29 | 159 | 0,4 | 1,5 | 0,24 | 0 | 0,04 | 12,7 | - | 11,6 | 0,035 | 0,09 | 0,38 | 0,049 | 0,06 | 23 | 0 |
| Haricots verts, purée | - | - | - | - | - | 1,7 | 1,7 | 4,2 | 2,9 | 3,3 | 0,36 | - | - | - | - | - | - | - | - | - | - | - | - | - | - | - | - | - | - | - | - | - | - | - | - |
| Igname cousse-couche, pulpe, cuit à la vapeur, prélevé à la Martinique | - | - | - | - | 75,4 | 1,65 | 1,65 | 17,9 | 0,3 | 3,9 | - | - | - | 0 | - | - | - | 0,09 | - | 198 | - | - | - | - | 0,14 | - | - | 0,87 | - | - | < 0,1 | - | - | - | - |
| Igname jaune, pulpe, cuit à la vapeur, prélevé à la Martinique | - | - | - | - | 66,6 | 2,04 | 2,04 | 27,6 | 0,37 | 2,67 | - | - | - | 0,05 | - | - | - | 0,15 | - | 349 | - | - | - | - | 0,27 | - | - | 0,81 | - | - | 0,6 | - | - | - | - |
| Igname saint martin, pulpe, cuit à la vapeur, prélevé à la Martinique | - | - | - | - | 74,2 | 2,25 | 2,25 | 18,4 | 0,3 | 3,9 | - | - | - | 0,01 | - | - | - | 0,05 | - | 25,1 | - | - | - | - | 0,18 | - | - | 22 | - | - | 0,7 | - | - | - | - |
| Igname, épluchée, bouillie/cuite à l'eau | 463 | 109 | 463 | 109 | 70,1 | 1,49 | 1,49 | 23,6 | 0,14 | 3,9 | 0,02 | 14 | - | 0,15 | 0,52 | - | 18 | 0,37 | 49 | 670 | - | 8 | 0,2 | 0 | 0,34 | 2,3 | - | 12,1 | 0,095 | 0,028 | 0,55 | 0,31 | 0,23 | 16 | 0 |
| Igname, épluchée, crue | 469 | 111 | 469 | 111 | 69,6 | 1,53 | 1,53 | 23,8 | 0,17 | 4,1 | 0,023 | 17 | - | 0,18 | 0,54 | - | 21 | 0,4 | 55 | 816 | - | 9 | 0,24 | 0 | 0,35 | 2,3 | - | 17,1 | 0,11 | 0,032 | 0,55 | 0,31 | 0,29 | 23 | 0 |
| Jacques, pulpe, cru, prélevé à la Martinique | - | - | - | - | 69,9 | 1,98 | 1,98 | - | 0,21 | - | - | - | - | - | - | - | - | - | - | - | - | - | - | - | - | - | - | 13,6 | - | - | - | - | - | - | - |
| Julienne ou brunoise de légumes, surgelée, crue | - | - | - | - | - | 1,19 | 1,19 | 4,08 | 0,25 | 2,28 | 0,1 | 43,5 | - | - | 0,55 | - | - | - | - | - | - | 54,5 | - | - | - | - | - | 9 | - | - | - | - | - | - | - |
| Jus de carambole, prélevé à la Martinique, jus filtré | - | - | - | - | 91,6 | 4,52 | 4,52 | - | 0,15 | - | - | - | - | - | - | - | - | - | - | - | - | - | - | - | - | - | - | 7 | - | - | - | - | - | - | - |
| Jus de carambole, prélevée à la Martinique, jus non filtré | - | - | - | - | 90,4 | 0,063 | 0,063 | - | 0,03 | - | - | - | - | - | - | - | - | - | - | - | - | - | - | - | - | - | - | 14,2 | - | - | - | - | - | - | - |
| Jus de citron punch (variété petit calibre), pur jus, prélevé à la Martinique | - | - | - | - | 92 | 0,48 | 0,48 | - | 0,14 | - | - | - | - | - | - | - | - | - | - | - | - | - | - | - | - | - | - | 23,2 | - | - | - | - | - | - | - |
| Jus de citron vert (variété gros calibre), pur jus, prélevé à la Martinique | - | - | - | - | 91,4 | 0,69 | 0,69 | - | 0,46 | - | - | - | - | - | - | - | - | - | - | - | - | - | - | - | - | - | - | 23,7 | - | - | - | - | - | - | - |
| Jus de corossol, prélevé à la Martinique | - | - | - | - | 84 | 0,65 | 0,65 | - | 0,43 | - | - | - | - | - | - | - | - | - | - | - | - | - | - | - | - | - | - | 16,3 | - | - | - | - | - | - | - |
| Jus de fruit de la passion ou maracudja, prélevé à la Martinique, pur jus | - | - | - | - | 84,4 | 1,08 | 1,08 | - | 0,37 | - | - | - | - | - | - | - | - | - | - | - | - | - | - | - | - | - | - | 54,3 | - | - | - | - | - | - | - |
| Jus de grenade, prélevée à la Martinique | - | - | - | - | 87 | 0,69 | 0,69 | - | 0 | - | - | - | - | - | - | - | - | - | - | - | - | - | - | - | - | - | - | 19,8 | - | - | - | - | - | - | - |
| Jus de lime, pur jus, prélevé à la Martinique | - | - | - | - | 92,4 | 0,35 | 0,35 | - | 0,57 | - | - | - | - | - | - | - | - | - | - | - | - | - | - | - | - | - | - | 18,5 | - | - | - | - | - | - | - |
| Jus de mandarine commune, pur jus filtré pasteurisé, prélevée à la Martinique | - | - | - | - | - | - | - | - | - | < 0,5 | - | - | - | - | - | - | - | - | - | - | - | - | - | - | - | - | - | 27,7 | - | - | - | - | - | 59,5 | - |
| Jus de mandarine commune, pur jus filtré, prélevée à la Martinique | - | - | - | - | 90 | 0,61 | 0,61 | 8,63 | 0,57 | < 0,5 | - | - | - | - | - | - | - | - | - | 102 | - | - | - | - | - | - | - | 26,7 | - | - | - | - | - | 39,9 | - |
| Jus de mandarine macaque, pur jus filtré, prélevée à la Martinique | - | - | - | - | 88,2 | 0,6 | 0,6 | 10,8 | 0 | < 0,5 | - | - | - | - | - | - | - | - | - | 199 | - | - | - | - | - | - | - | 6 | - | - | - | - | - | 31,3 | - |
| Jus de moubin, pur jus, prélevé à la Martinique | - | - | - | - | 85,5 | 0,88 | 0,88 | - | 0,19 | - | - | - | - | - | - | - | - | - | - | - | - | - | - | - | - | - | - | 10,1 | - | - | - | - | - | - | - |
| Jus de pamplemousse (variété locale), pur jus, prélevé à la Martinique | - | - | - | - | 92,3 | 0,42 | 0,42 | - | 0,19 | - | - | - | - | - | - | - | - | - | - | - | - | - | - | - | - | - | - | 35,6 | - | - | - | - | - | - | - |
| Jus de prune de cythère verte "géante" (variété locale), fruit mûr, pur jus filtré, prélevée à la Martinique | - | - | - | - | 91,1 | 0,29 | 0,29 | 7,13 | 0,19 | 1 | - | - | - | - | - | - | - | - | - | - | - | - | - | - | - | - | - | 16,5 | - | - | - | - | - | - | - |
| Jus de prune de cythère verte "géante" (variété locale), fruit vert, pur jus filtré, prélevée à la Martinique | - | - | - | - | 91,7 | 0,44 | 0,44 | 6,55 | 0,28 | 0,8 | - | - | - | - | - | - | - | - | - | - | - | - | - | - | - | - | - | 21,1 | - | - | - | - | - | - | - |
| Jus de prune de cythère verte "naine" (variété hybride), fruit mûr, pur jus filtré, prélevée à la Martinique | - | - | - | - | 87,8 | 0,69 | 0,69 | 9,97 | 0,21 | 0,9 | - | - | - | - | - | - | - | - | - | - | - | - | - | - | - | - | - | 34,3 | - | - | - | - | - | - | - |
| Jus de prune de cythère verte "naine" (variété hybride), fruit vert, pur jus filtré, prélevée à la Martinique | - | - | - | - | 91,6 | 0,67 | 0,67 | 6,34 | 0,26 | 0,8 | - | - | - | - | - | - | - | - | - | - | - | - | - | - | - | - | - | 45,5 | - | - | - | - | - | - | - |
| Kaki, pulpe, cru | 290 | 68,6 | 290 | 68,6 | 81,8 | 0,88 | 0,88 | 14,3 | < 0,3 | 3,4 | < 0,013 | 6,9 | 33 | 0,03 | 0,06 | < 20 | 7,2 | 0,13 | 13 | 160 | < 20 | < 5 | < 0,05 | < 0,25 | 0,12 | < 0,8 | - | 3,41 | < 0,015 | < 0,01 | < 0,1 | 0,14 | < 0,01 | 22,1 | 0 |
| Kamanioc, pulpe, cuit à la vapeur, prélevé à la Martinique | - | - | - | - | 62,3 | 0,6 | 0,6 | - | 0 | - | - | - | - | - | - | - | - | - | - | - | - | - | - | - | - | - | - | 21,3 | - | - | - | - | - | - | - |
| Kiwi, pulpe et graines, cru | 255 | 60,5 | 255 | 60,5 | 83,5 | 0,88 | 0,88 | 11 | 0,6 | 2,4 | < 0,013 | 29 | 34,7 | 0,15 | 0,16 | < 20 | 12 | 0,05 | 26 | 290 | < 20 | < 5 | 0,12 | 0 | 0,96 | 16,6 | - | 81,9 | < 0,015 | 0,021 | 0,23 | 0,24 | 0,036 | 22,2 | 0 |
| Kumquat, sans pépin, cru | - | - | - | - | 81,9 | 1,57 | 1,57 | 9,6 | 1,18 | 5,45 | 0,02 | 65,5 | - | 0,095 | 0,67 | - | 17,5 | 0,14 | 19 | 158 | 0 | 8 | 0,13 | 0 | 0,15 | 0 | - | 45 | 0,037 | 0,09 | 0,43 | 0,21 | 0,036 | 17 | 0 |

| | | | | | | | | | | | | | | | | | | | | | | | | | | | | | | | | | | | |
|---|---|---|---|---|---|---|---|---|---|---|---|---|---|---|---|---|---|---|---|---|---|---|---|---|---|---|---|---|---|---|---|---|---|---|---|
| Laitue iceberg, crue | - | - | - | - | 95,6 | 1,01 | 1,01 | 2,45 | 0,18 | 1,15 | 0,024 | 17,4 | - | 0,029 | 0,43 | 0,6 | 8,73 | 0,19 | 23,6 | 179 | 0,15 | 9,48 | 0,21 | 0 | 0,22 | 56,4 | - | 4,72 | 0,048 | 0,038 | 0,19 | 0,11 | 0,048 | 48 | 0 |
| Laitue romaine, crue | - | - | - | - | 94,6 | 1,24 | 1,24 | 1,2 | 0,3 | 2,1 | 0,02 | 33 | 48 | 0,048 | 0,97 | 1,25 | 14 | 0,16 | 30 | 247 | 1 | 8 | 0,23 | 0 | 0,13 | 103 | - | 14 | 0,072 | 0,067 | 0,31 | 0,14 | 0,074 | 136 | 0 |
| Laitue, crue | - | - | - | - | 95,3 | 1,3 | 1,3 | 1,33 | 0,2 | 1,2 | 0,021 | 64,6 | - | 0,036 | 0,98 | 1,22 | 14,9 | 0,4 | 29,5 | 200 | 2,33 | 8,47 | 0,2 | 0 | 0,18 | 123 | - | 11,8 | 0,067 | 0,077 | 0,36 | 0,14 | 0,09 | 43,5 | 0 |
| Légume cuit (aliment moyen) | 183 | 43,5 | 183 | 43,5 | 88,3 | 2,11 | 2,11 | 5,81 | 0,3 | 2,71 | 0,082 | 30,3 | | 0,097 | 0,46 | 5,51 | 17 | 0,16 | 51,8 | 261 | 6,06 | 18,6 | 0,31 | 0,017 | 0,44 | 20,7 | | 8,84 | 0,045 | 0,051 | 0,53 | 0,25 | 0,16 | 28,9 | 0 |
| Légume sec, cuit (aliment moyen) | | | | | 69,1 | 8,38 | 8,38 | 12,1 | 0,62 | 9,34 | 0,012 | 52,7 | 16,4 | 0,23 | 1,55 | 5,47 | 34,9 | 0,52 | 159 | 235 | 6,73 | 4,63 | 0,9 | 0,038 | 0,099 | 2,63 | | 1,05 | 0,12 | 0,056 | 0,81 | 0,44 | 0,13 | 131 | 0 |
| Légumes (3-4 sortes en mélange), purée | 257 | 61,8 | 257 | 61,8 | 86 | 1,94 | 1,94 | 4,25 | 3,4 | 3,3 | 0,63 | 47 | 408 | 0,12 | 0,67 | < 20 | 18 | 0,18 | 42 | 260 | < 20 | 252 | 0,33 | < 0,25 | 0,43 | 25,9 | - | < 0,5 | 0,025 | 0,023 | 1,1 | 0,16 | 0,048 | 19 | 0,019 |
| Légumes pour couscous, cuits | - | - | - | - | 91,7 | 1,53 | 1,53 | 2,55 | 0,43 | 2,6 | 1,06 | 39 | - | 0,1 | 0,7 | 1,2 | 15 | 0,2 | 38 | 208 | 0,6 | 45 | 0,3 | 0 | 0,38 | - | - | 18 | 0,05 | 0,04 | 0,41 | 0,17 | 0,12 | 26 | 0 |
| Légumes pour couscous, surgelés, crus | 195 | 46,5 | 195 | 46,5 | - | 2,05 | 2,05 | 6,75 | 0,7 | 2,53 | 0,12 | 35,6 | - | - | 0,8 | - | - | - | - | - | - | 39 | - | - | - | - | - | 27,4 | - | - | - | - | - | - | - |
| Légumes pour potages, surgelés, crus | 143 | 34,1 | 143 | 34,1 | 89,2 | 1,44 | 1,44 | 4,69 | 0,44 | 2,86 | 0,052 | 37 | - | - | 1 | - | - | - | - | - | - | 16,9 | - | - | - | - | - | 25 | - | - | - | - | - | - | - |
| Légumes pour ratatouille, surgelés | - | - | - | - | 94,2 | 1,01 | 1,01 | 3,23 | 0,089 | 1,67 | 0,11 | - | - | - | - | - | - | - | - | - | - | 34 | - | - | - | - | - | - | - | - | - | - | - | - | - |
| Légumes, mélange surgelé, crus | 271 | 65,1 | 271 | 65,1 | 85,2 | 2,26 | 2,26 | 5,71 | 2,65 | 4,7 | 0,6 | 33 | - | 0,093 | 0,98 | < 5 | 24 | 0,24 | 59 | 212 | - | 39 | 0,45 | 0 | - | - | - | 13,7 | 0,12 | 0,085 | 1,25 | 0,16 | 0,096 | 29 | < 0,08 |
| Lentille blonde, bouillie/cuite à l'eau | - | - | - | - | 66,4 | 9,7 | 9,7 | 16,3 | 0,7 | 6 | < 0,13 | 32 | 17,5 | 0,3 | 2,5 | < 20 | 31 | 0,46 | 120 | 230 | < 20 | < 5 | 1,1 | < 0,25 | - | 0,93 | - | < 0,5 | 0,087 | < 0,01 | 0,57 | 0,35 | 0,13 | 43,1 | - |
| Lentille blonde, sèche | - | - | - | - | 8,8 | 26,1 | 26,1 | 48,3 | 1,9 | 12,1 | < 0,13 | 54 | 75,2 | 0,79 | 7,4 | < 20 | 91 | 1,2 | 360 | 780 | 50 | < 5 | 3 | < 0,25 | - | 9,82 | - | < 0,5 | 0,32 | 0,029 | 1,92 | 1,62 | 0,44 | 103 | - |
| Lentille corail, bouillie/cuite à l'eau | - | - | - | - | 65,1 | 10,6 | 10,6 | 15 | 0,5 | 8,2 | < 0,13 | 21 | 15 | 0,29 | 2,2 | < 20 | 25 | 0,39 | 110 | 190 | < 20 | < 5 | 1,3 | < 0,25 | - | 2,56 | - | < 0,5 | 0,12 | < 0,01 | 0,47 | 0,24 | 0,082 | 23 | - |
| Lentille corail, sèche | - | - | - | - | 8,8 | 27,7 | 27,7 | 44,9 | 0,8 | 15,4 | < 0,13 | 29 | 78,2 | 0,82 | 6,3 | < 20 | 74 | 1,2 | 330 | 750 | < 20 | < 5 | 3,9 | < 0,25 | - | 9,25 | - | < 0,5 | 0,32 | 0,018 | 1,42 | 1,05 | 0,41 | 124 | 0 |
| Lentille verte, bouillie/cuite à l'eau | 535 | 127 | 535 | 127 | 64,8 | 10,1 | 10,1 | 16,2 | 0,58 | 8,45 | 0,015 | 39,5 | 17 | 0,25 | 2,45 | < 20 | 34 | 0,44 | 160 | 215 | < 20 | 5,8 | 1,25 | < 0,25 | 0,53 | 2,65 | - | < 0,5 | 0,094 | 0,022 | 0,52 | 0,24 | 0,11 | 50,5 | - |
| Lentille verte, sèche | - | - | - | - | 9,5 | 25,1 | 25,1 | 44,5 | 1,8 | 16,4 | < 0,13 | 64 | 81 | 0,6 | 6,3 | < 20 | 97 | 1 | 480 | 940 | < 20 | < 5 | 3,4 | < 0,25 | - | 11,9 | - | 1,63 | 0,29 | 0,046 | 1,98 | 1,34 | 0,35 | 117 | - |
| Lentille, bouillie/cuite à l'eau | - | - | - | - | 69,6 | 9,02 | 9,02 | 12,2 | 0,38 | 7,9 | 0,005 | 26,2 | 20 | 0,24 | 1,44 | 3 | 35,6 | 0,43 | 180 | 222 | < 10 | 2 | 0,98 | 0 | 0,11 | 1,7 | - | 1,5 | 0,17 | 0,073 | 1,06 | 0,64 | 0,18 | 181 | 0 |
| Lentille, cuisinée, appertisée, égouttée | - | - | - | - | 73,5 | 6,28 | 6,28 | 11,8 | 0,64 | 4 | 0,59 | 21 | - | 0,12 | 1,67 | 3 | 15,8 | 0,4 | 82,7 | 178 | 2,8 | 230 | 0,55 | 0 | 0,19 | - | - | traces | 0,083 | 0,043 | 0,43 | 0,2 | 0,12 | 11 | 0 |
| Lentille, germée | - | - | - | - | 67,3 | 8,86 | 8,86 | 19 | 0,58 | 3,1 | 0,028 | 25 | - | 0,35 | 3,21 | - | 37 | 0,51 | 173 | 322 | - | 11 | 1,51 | 0 | - | - | - | 16,5 | 0,23 | 0,13 | 1,13 | 0,58 | 0,19 | 100 | 0 |
| Lentille, sèche | - | - | - | - | 8,96 | 25,4 | 25,4 | 50,6 | 1,34 | 7,63 | 0,21 | 45,4 | 64 | 0,71 | 6,51 | 0,7 | 62 | 1,39 | 326 | 674 | 12 | 74 | 3,17 | 0 | 0,49 | 5 | - | 4,5 | 0,69 | 0,21 | 2,3 | 1,75 | 0,55 | 257 | 0 |
| Lin, brun, graine | - | - | - | - | 5,8 | 18 | 21,3 | 3,56 | 42,2 | 26,9 | 0,095 | 210 | < 100 | 1,3 | 5,7 | < 20 | 300 | 2 | 560 | 750 | < 20 | 38 | 5,1 | < 0,25 | 0,5 | 4,55 | - | - | 0,47 | < 0,01 | < 0,1 | 0,81 | 0,17 | 71,6 | - |
| Lin, graine | - | - | - | - | 5,93 | 20,2 | 23,9 | 6,6 | 36,6 | 27,3 | 0,051 | 228 | - | 1,22 | 10,2 | 0,5 | 372 | 2,48 | 595 | 641 | 28 | 20,5 | 6,05 | 0 | 0,31 | 4,3 | - | 0,6 | 1,22 | 0,2 | 3,08 | 0,99 | 0,63 | 93,7 | 0 |
| Litchi, pulpe, cru | 344 | 81 | 344 | 81 | 80,5 | 1,13 | 1,13 | 16,1 | < 0,5 | 2 | < 0,013 | 3,6 | < 20 | 0,2 | 0,26 | < 20 | 18 | 0,1 | 27 | 200 | < 20 | < 5 | 0,23 | 0 | < 0,08 | < 0,8 | - | 19,2 | < 0,015 | < 0,01 | 1,06 | 0,055 | 0,12 | 28,3 | 0 |
| Longan, pulpe, cru | - | - | - | - | 82,8 | 1,31 | 1,31 | 14 | 0,1 | 1,1 | 0 | 1 | - | 0,17 | 0,13 | - | 10 | 0,052 | 21 | 266 | - | 0 | 0,05 | - | - | - | - | 84 | 0,031 | 0,14 | 0,3 | - | - | 50 | 0 |
| Lupin, graine crue | - | - | - | - | 10,4 | 36,2 | 36,2 | 21,5 | 9,74 | 18,9 | 0,038 | 176 | - | 1,02 | 4,36 | - | 198 | 2,38 | 440 | 1010 | - | 15 | 4,75 | 0 | - | - | - | 4,8 | 0,64 | 0,22 | 2,19 | 0,75 | 0,36 | 355 | 0 |
| Luzerne, graine | - | - | - | - | 7,4 | 29,7 | 35 | 38,3 | 12,6 | 7,9 | - | 136 | - | - | 12,9 | - | - | - | - | - | 90 | - | 6,9 | - | - | - | - | 26 | 1,08 | 0,58 | 1,8 | - | - | - | - |
| Luzerne, graine germée | - | - | - | - | 92,8 | 3,28 | 3,87 | 0,91 | 0,7 | 1,9 | 0,015 | 32 | - | 0,16 | 0,96 | - | 27 | 0,19 | 70 | 79 | - | 6 | 0,92 | 0 | 0,02 | 30,5 | - | 8,2 | 0,076 | 0,13 | 0,48 | 0,56 | 0,034 | 36 | 0 |
| Macédoine de légumes, appertisée, égouttée | 170 | 40,4 | 170 | 40,4 | 87 | 2,32 | 2,32 | 5,59 | 0,28 | 3,16 | 0,61 | 40,7 | - | 0,072 | 1,05 | 0,2 | 10,6 | 0,15 | 42 | 291 | < 2,2 | 243 | 0,28 | 0 | 0,29 | 18,2 | - | 5 | 0,046 | 0,048 | 0,58 | 0,14 | 0,079 | 24 | 0 |
| Macédoine de légumes, surgelée, pré-cuite (à recuire) | 228 | 54,5 | 228 | 54,5 | 86,4 | 3,48 | 3,48 | 6,8 | 0,28 | 5,45 | 0,045 | - | - | - | - | - | - | - | - | - | - | 12,7 | - | - | - | - | - | - | - | - | - | - | - | - | - |
| Macédoine ou cocktail ou salade de fruits, au sirop (sans précision sur léger ou classique), appertisé, égoutté (aliment moyen) | | | | | 81,6 | 0,25 | 0,25 | 16,2 | 0,25 | 2,02 | 0,022 | 11,6 | 5,24 | 0,07 | 0,11 | | 4,24 | 0,084 | 9,04 | 70,3 | 10 | 2,5 | 0,025 | | | 0,64 | | 2,91 | 0,035 | 0,005 | 0,05 | 0,037 | 0,017 | 10,8 | |
| Macédoine ou cocktail ou salade de fruits, au sirop léger, appertisé, égoutté | - | - | - | - | 83,5 | < 0,5 | < 0,5 | 14,4 | < 0,5 | 1,9 | 0,0083 | 9,5 | 4,14 | 0,07 | 0,15 | - | 4,6 | 0,09 | 8,5 | 63 | < 20 | < 5 | < 0,05 | - | - | 1,01 | - | 5,87 | 0,025 | < 0,01 | < 0,1 | 0,034 | < 0,01 | 10,6 | - |
| Macédoine ou cocktail ou salade de fruits, au sirop léger, appertisé, non égoutté | - | - | - | - | 84 | < 0,5 | < 0,5 | 14,5 | 0,061 | 1,18 | 0,0026 | 8,18 | 4,05 | 0,055 | 0,13 | - | 4,37 | 0,082 | 7,74 | 61,1 | < 20 | 1,02 | < 0,05 | 0 | - | 0,63 | - | 6,1 | 0,016 | < 0,01 | < 0,1 | 0,033 | < 0,01 | 9,96 | 0 |
| Macédoine ou cocktail ou salade de fruits, au sirop, appertisé, égoutté | - | - | - | - | 80,4 | < 0,5 | < 0,5 | 17,4 | < 0,5 | 2,1 | 0,032 | 13 | 5,95 | 0,07 | 0,08 | - | 4 | 0,08 | 9,4 | 75 | < 20 | < 5 | < 0,05 | - | - | < 0,8 | - | 0,98 | 0,041 | < 0,01 | < 0,1 | 0,038 | 0,025 | 10,9 | - |
| Macédoine ou cocktail ou salade de fruits, au sirop, appertisé, non égoutté | - | - | - | - | 79,9 | < 0,5 | < 0,5 | 18,3 | < 0,5 | 1,26 | 0,003 | 11 | 6,25 | 0,05 | 0,072 | - | 3,76 | 0,076 | 8,36 | 67 | < 20 | 1,2 | < 0,05 | - | - | < 0,8 | - | 0,59 | 0,039 | < 0,01 | < 0,1 | 0,031 | 0,024 | 11 | - |
| Mâche, crue | 70,1 | 16,8 | 70,1 | 16,8 | 93,7 | 2 | 2 | 0,5 | < 0,5 | 2,3 | < 0,013 | 41 | 45,4 | 0,1 | 0,4 | < 20 | 19 | 0,28 | 30 | 330 | < 20 | < 5 | 0,14 | 0 | 0,22 | 48,9 | - | 2,38 | 0,042 | 0,066 | 0,37 | 0,25 | 0,096 | 45,5 | 0 |
| Maïs doux, appertisé, égoutté | 447 | 106 | 447 | 106 | 75,4 | 2,82 | 2,82 | 18,4 | 1,68 | 3,1 | 0,59 | 3,51 | 364 | 0,036 | 0,35 | 0,5 | 26,9 | 0,11 | 53,8 | 224 | < 10 | 236 | 0,39 | 0 | 0,29 | 0 | - | 3,55 | 0,033 | 0,048 | 1,04 | 0,29 | 0,13 | 75 | 0 |
| Maïs doux, en épis, cuit | - | - | - | - | 73,4 | 3,41 | 3,41 | 18,6 | 1,5 | 2,4 | 0,0025 | 3 | - | 0,049 | 0,45 | 1 | 26 | 0,17 | 77 | 218 | traces | 1 | 0,62 | 0 | 0,09 | 0,4 | - | 5,5 | 0,093 | 0,057 | 1,68 | 0,79 | 0,14 | 23 | 0 |
| Maïs doux, en épis, surgelé, cru | - | - | - | - | 73,3 | 3,36 | 3,36 | 17 | 1,19 | 2,65 | 0,01 | 4,18 | - | 0,15 | 0,59 | 0,5 | 30,3 | 0,25 | 108 | 276 | 4 | 4 | 1,05 | 0 | 0,37 | 0,35 | - | 6,9 | 0,11 | 0,082 | 1,72 | 0,39 | 0,18 | 36,5 | 0 |
| Maïs doux, surgelé, cru | - | - | - | - | 71,8 | 3,28 | 3,28 | 19,8 | 0,78 | 2,8 | 0,013 | 4 | - | 0,051 | 0,68 | - | 32 | 0,16 | 87 | 294 | - | 5 | 0,7 | 0 | 0,09 | 0,4 | - | 7,2 | 0,1 | 0,088 | 1,68 | 0,29 | 0,18 | 40 | 0 |
| Mandarine commune, pulpe, au sirop, appertisée, non égouttée, prélevée à la Martinique | - | - | - | - | - | - | - | - | - | 0,7 | - | - | - | - | - | - | - | - | - | - | - | - | - | - | - | - | - | 0,2 | - | - | - | - | - | 33,5 | - |
| Mango bassignac, pulpe, cru, prélevé à la Martinique | - | - | - | - | 77,3 | 0,32 | 0,32 | 18,5 | traces | 3,2 | - | - | - | - | - | - | - | - | - | - | - | - | - | - | - | - | - | 23,3 | - | - | - | - | - | - | - |
| Mango moussache, pulpe, cru, prélevé à la Martinique | - | - | - | - | 75,5 | 1,28 | 1,28 | 20,9 | traces | 2 | - | - | - | - | - | - | - | - | - | - | - | - | - | - | - | - | - | 8,1 | - | - | - | - | - | - | - |
| Mangue José, pulpe, crue, prélevée à La Réunion (Mangifera indica L.) | 376 | 88,8 | 376 | 88,8 | 77,1 | 0,75 | 0,75 | 19,3 | 0,5 | 1,8 | < 0,013 | 14 | < 20 | 0,08 | 0,12 | < 20 | 14 | 0,07 | 13 | 150 | < 20 | < 5 | 0,08 | - | 1,96 | 1,58 | - | 2,9 | 0,033 | 0,015 | 0,74 | 0,2 | 0,13 | 9,95 | - |
| Mangue julie, pulpe, crue, prélevée à la Martinique | - | - | - | - | 77,8 | 0,71 | 0,71 | 18,3 | traces | 2,8 | - | - | - | - | - | - | - | - | - | - | - | - | - | - | - | - | - | 22 | - | - | - | - | - | - | - |
| Mangue verte, pulpe, crue, prélevée à la Martinique | - | - | - | - | 82,1 | 1,31 | 1,31 | 12,2 | traces | 3,4 | - | - | - | - | - | - | - | - | - | - | - | - | - | - | - | - | - | 17,2 | - | - | - | - | - | - | - |
| Mangue, pulpe, crue | 314 | 73,9 | 314 | 73,9 | 81,1 | 0,63 | 0,63 | 14,3 | < 0,5 | 1,6 | < 0,013 | 12 | 21,8 | 0,07 | 0,09 | < 20 | 11 | 0,07 | 12 | 150 | < 20 | < 5 | 0,11 | 0 | 2,05 | 1,12 | - | 25 | < 0,015 | < 0,01 | 0,72 | 0,18 | 0,1 | 70,2 | 0 |
| Manioc, racine crue | 664 | 157 | 664 | 157 | 59,7 | 1,31 | 1,31 | 36,3 | 0,29 | 1,8 | 0,035 | 16 | - | 0,1 | 0,27 | - | 21 | 0,38 | 27 | 271 | - | 14 | 0,34 | 0 | 0,19 | 1,9 | - | 20,6 | 0,087 | 0,048 | 0,85 | 0,11 | 0,088 | 27 | 0 |
| Manioc, racine cuite | 561 | 132 | 561 | 132 | 65,2 | 0,69 | 0,69 | 32 | 0,043 | 0,4 | - | 12,1 | - | 0,018 | 0,6 | - | - | 0,038 | 40,5 | - | - | - | 0,35 | - | - | - | - | 13,1 | 0,023 | 0,018 | 0,47 | - | - | - | - |
| Massissi, pulpe, cru, prélevé à la Martinique | - | - | - | - | 83,8 | 2,58 | 2,58 | - | 0,8 | - | - | - | - | - | - | - | - | - | - | - | - | - | - | - | - | - | - | 73,2 | - | - | - | - | - | - | - |
| Mélange apéritif de fruits exotiques, sec | 1640 | 389 | 1640 | 389 | 11,7 | 2,19 | 2,19 | 69,5 | 10,5 | 3,4 | 0,22 | 43 | 75,9 | 0,26 | 1,1 | 3 | 39 | 0,45 | 67 | 530 | < 20 | 88 | 0,32 | < 0,25 | 0,69 | 7,52 | - | 0,69 | 0,039 | < 0,01 | < 0,1 | 0,25 | 0,077 | 19,1 | 0 |
| Mélange apéritif de graines (non salées) et fruits séchés | - | - | - | - | 25,6 | 8,29 | 9,78 | 32,2 | 26,3 | 6,36 | 0,3 | 88 | - | - | 4 | - | 120 | - | - | 712 | - | 117 | 2 | - | - | - | - | - | - | - | - | - | - | - | - |
| Mélange apéritif de graines (non salées) et raisins secs | 2250 | 541 | 2210 | 532 | 6,1 | 12,5 | 14,8 | 31,8 | 37,2 | 9,6 | 0,018 | 90 | < 20 | 1 | 3,1 | < 5 | 160 | 2 | 320 | 750 | < 20 | 7 | 2,4 | < 0,25 | 8,88 | 8,8 | - | < 0,5 | 0,15 | 0,036 | 2,35 | 0,94 | 0,092 | 34,4 | 0 |
| Mélange apéritif de graines salées et raisins secs | 2330 | 560 | 2280 | 548 | 5,3 | 16,6 | 19,6 | 27,1 | 39,6 | 8,4 | 0,42 | 72 | 285 | 0,82 | 2,1 | < 20 | 160 | 1,4 | 340 | 700 | < 20 | 169 | 2,3 | < 0,25 | 8,47 | 5,47 | - | < 0,5 | 0,062 | 0,044 | 4,08 | 1,1 | 0,17 | 49,5 | - |
| Melon cantaloup (par ex.: Charentais, de Cavaillon) pulpe, cru | 265 | 62,7 | 265 | 62,7 | 84,2 | 1,13 | 1,13 | 14,8 | < 0,5 | 1,3 | 0,06 | 11 | 25,8 | 0,06 | 0,21 | < 20 | 16 | 0,04 | 17 | 380 | < 20 | 24 | 0,18 | 0 | < 0,08 | < 0,8 | - | 8,14 | < 0,015 | < 0,01 | 0,6 | 0,15 | 0,055 | 58,9 | 0 |
| Melon miel ou melon honeydew, pulpe, cru | - | - | - | - | 91,9 | 0,58 | 0,58 | 4,3 | 0,17 | 0,75 | 0,06 | 6 | - | 0,033 | 0,34 | 0 | 9,08 | 0,034 | 15,8 | 237 | 0 | 24,1 | 0,086 | 0 | 0,035 | 2,9 | - | 16,8 | 0,032 | 0,016 | 0,58 | 0,13 | 0,072 | 17 | 0 |
| Melon, pulpe, cru, prélevé à la Martinique | - | - | - | - | 87 | 1,08 | 1,08 | - | 0,047 | - | - | - | - | - | - | - | - | - | - | - | - | - | - | - | - | - | - | 27,2 | - | - | - | - | - | - | - |
| Mesclun ou salade, mélange de jeunes pousses | - | - | - | - | - | 2 | 2 | 2,5 | 0 | 2 | 0,048 | - | - | - | - | - | - | - | - | - | - | 19 | - | - | - | - | - | - | - | - | - | - | - | - | - |

| | | | | | | | | | | | | | | | | | | | | | | | | | | | | | | | | | | | |
|---|---|---|---|---|---|---|---|---|---|---|---|---|---|---|---|---|---|---|---|---|---|---|---|---|---|---|---|---|---|---|---|---|---|---|---|
| Mirabelle, crue | 325 | 76,9 | 325 | 76,9 | 78,1 | 0,63 | 0,63 | 18 | < 0,5 | 2,2 | < 0,013 | 11 | < 20 | 0,08 | 0,13 | < 20 | 8,2 | 0,07 | 21 | 240 | < 20 | < 5 | 0,11 | - | 1,52 | 1,82 | - | 5,29 | - | 0,015 | < 0,1 | 0,27 | 0,074 | 12,5 | - |
| Mûre (de ronce), crue | 198 | 47,3 | 198 | 47,3 | 86,1 | 1,13 | 1,13 | 6,53 | 0,7 | 5,2 | < 0,013 | 31 | < 20 | 0,07 | 0,4 | < 20 | 20 | 1,1 | 25 | 200 | < 20 | < 5 | 0,18 | 0 | 1,28 | 14,1 | - | 10,1 | < 0,015 | 0,02 | 0,33 | 0,31 | 0,015 | 17 | 0 |
| Mûre (de ronce), surgelée, crue | - | - | - | - | 82,2 | 1,22 | 1,22 | 10,7 | 0,42 | 5 | 0,0025 | 29 | - | 0,12 | 0,8 | - | 22 | 1,22 | 30 | 140 | - | 1 | 0,25 | 0 | 1,17 | 19,8 | - | 3,1 | 0,029 | 0,046 | 1,21 | 0,15 | 0,061 | 34 | 0 |
| Mûre noire (du mûrier), crue | - | - | - | - | 87,7 | 1,44 | 1,44 | 8,1 | 0,39 | 1,7 | 0,025 | 39 | - | 0,06 | 1,85 | 0,4 | 18 | 1,1 | 38 | 194 | - | 10 | 0,12 | 0 | 0,87 | 7,8 | - | 36,4 | 0,029 | 0,1 | 0,62 | 0,26 | 0,05 | 6 | 0 |
| Myrtille, crue | 244 | 57,7 | 244 | 57,7 | 84,2 | 0,87 | 0,87 | 10,6 | 0,33 | 2,4 | 0,0025 | 6 | - | 0,057 | 0,28 | 0,5 | 6 | 0,34 | 12 | 77 | traces | 1 | 0,16 | 0 | 0,57 | 19,3 | - | 9,7 | 0,037 | 0,041 | 0,42 | 0,12 | 0,052 | 6 | 0 |
| Myrtille, surgelée, crue | - | - | - | - | 84,3 | 0,68 | 0,68 | 8,9 | 0,43 | 2,81 | 0,0026 | 10 | - | 0,035 | 0,39 | - | 4,25 | 0,18 | 10 | 55,5 | - | 1 | 0,068 | 0 | 0,5 | 17,1 | - | 8,08 | 0,029 | 0,041 | 0,45 | 0,13 | 0,059 | 7 | 0 |
| Navet, bouilli/cuit à l'eau | 88,3 | 21,1 | 88,3 | 21,1 | 94,2 | 0,75 | 0,75 | 3,23 | < 0,3 | 2,1 | 0,025 | 32 | < 20 | 0,06 | 0,08 | < 20 | 5,8 | 0,04 | 20 | 170 | < 20 | 10 | 0,06 | - | < 0,08 | < 0,8 | - | 7,25 | < 0,015 | < 0,01 | 0,26 | 0,097 | 0,029 | 17,1 | - |
| Navet, cuit | - | - | - | - | 93,4 | 0,9 | 0,9 | 3,8 | 0,2 | 2,2 | 0,073 | 37,3 | - | 0,045 | 0,11 | < 10 | 8,75 | 0,045 | 26 | 237 | < 10 | 29 | 0,099 | 0 | 0,02 | 0,1 | - | 5,6 | < 0,05 | < 0,05 | 0,05 | 0,21 | 0,06 | 9 | 0 |
| Navet, pelé, cru | - | - | - | - | 91,9 | 0,76 | 0,76 | 4,7 | 0,1 | 1,9 | 0,11 | 40 | - | 0,063 | 0,4 | 0,5 | 9,5 | 0,1 | 38,5 | 237 | 1,25 | 42 | 0,25 | 0 | 0,03 | 0,1 | - | 18,5 | 0,04 | 0,038 | 0,7 | 0,2 | 0,085 | 17,5 | 0 |
| Navet, surgelé, cru | - | - | - | - | 95,7 | 0,82 | 0,82 | 2 | 0,18 | 1,85 | 0,066 | 23 | - | 0,045 | 0,7 | - | 10 | 0,071 | 20 | 137 | - | 25 | 0,14 | 0 | - | - | - | 4,4 | 0,03 | 0,02 | 0,4 | 0,11 | 0,048 | 8 | 0 |
| Nectarine ou brugnon, blanche, pulpe et peau, crue | 220 | 51,8 | 220 | 51,8 | 86,5 | 0,81 | 0,81 | 11,4 | < 0,5 | < 0,5 | < 0,013 | 5 | < 20 | 0,07 | 0,13 | < 20 | 8,1 | 0,06 | 18 | 210 | < 20 | < 5 | 0,1 | - | 1,04 | < 0,8 | - | 3,01 | < 0,015 | 0,02 | < 0,1 | 0,22 | < 0,01 | 11,9 | - |
| Nectarine ou brugnon, jaune, pulpe et peau, crue | 218 | 51,3 | 218 | 51,3 | 86,6 | 0,69 | 0,69 | 11,3 | < 0,5 | 0,6 | < 0,013 | 4,8 | < 20 | 0,06 | 0,13 | < 20 | 8 | 0,05 | 19 | 220 | < 20 | < 5 | 0,11 | - | < 0,08 | < 0,8 | - | 3,38 | < 0,015 | < 0,01 | 0,56 | 0,22 | < 0,01 | 8,4 | - |
| Nectarine ou brugnon, pulpe et peau, crue | - | - | - | - | 87,6 | 1,16 | 1,16 | 8,9 | 0,31 | 1,7 | 0,0027 | 3,38 | - | 0,078 | 0,28 | 0,3 | 8,31 | 0,061 | 26,7 | 201 | < 2,2 | 1,09 | 0,091 | 0 | 0,77 | 2,2 | - | 5,4 | 0,034 | 0,027 | 1,13 | 0,19 | 0,025 | 5 | 0 |
| Noisette | - | - | - | - | 7,3 | 14,4 | 17 | 7,16 | 56,9 | 11,6 | < 0,013 | 120 | 30,8 | 1,7 | 3 | < 20 | 160 | 3,3 | 340 | 860 | < 20 | < 5 | 2,3 | < 0,25 | 16,3 | < 0,8 | - | < 0,5 | 0,35 | 0,088 | 0,71 | 1,64 | 0,21 | 65,2 | 0 |
| Noisette grillée | - | - | - | - | 2,52 | 14,4 | 17 | 5,21 | 66 | 9,4 | - | 123 | 3,08 | 1,75 | 4,38 | - | 173 | 5,55 | 310 | 755 | - | - | 2,5 | 0 | 15,3 | - | - | 3,8 | 0,34 | 0,12 | 2,05 | 0,92 | 0,62 | 88 | 0 |
| Noisette grillée, salée | - | - | - | - | - | 13,4 | 15,8 | 7,67 | 61,2 | 9,43 | 0,83 | - | - | - | - | - | - | - | - | - | - | 333 | - | - | - | - | - | - | - | - | - | - | - | - | - |
| Noix de cajou, grillée à sec, non salée | 2610 | 630 | 2560 | 617 | 1,3 | 17,4 | 20,5 | 23,5 | 49 | 5,7 | 0,02 | 38 | 23,9 | 2,7 | 5,7 | - | 280 | 1,8 | 530 | 680 | 20 | 8 | 5,6 | < 0,25 | 0,58 | 31,3 | - | < 0,5 | 0,36 | 0,049 | 0,56 | 0,98 | 0,17 | 25,6 | - |
| Noix de cajou, grillée, non salée | 2560 | 618 | 2510 | 606 | 1,7 | 17,4 | 20,5 | 21,3 | 48,1 | 8,4 | 0,015 | 39 | 22,5 | 2,2 | 5,4 | - | 260 | 1,6 | 510 | 610 | < 20 | 6 | 5 | < 0,25 | 1,33 | 34,7 | - | < 0,5 | 0,4 | 0,031 | 1,03 | 0,83 | 0,11 | 30,2 | - |
| Noix de cajou, grillée, salée | - | - | - | - | 1,51 | 15,2 | 18 | 26,7 | 49,5 | 3,85 | 1,15 | 40,3 | 364 | 2 | 3,9 | 6,2 | 223 | 1,1 | 452 | 546 | < 5 | 455 | 5,4 | 0 | 0,75 | 34,7 | - | < 0,5 | 0,36 | 0,15 | 1,14 | 0,49 | 0,38 | 51,4 | 0 |
| Noix de coco, amande immature, fraîche | - | - | - | - | 47 | 3,33 | 3,93 | 6,22 | 33,5 | 9 | 0,05 | 14 | - | 0,44 | 2,43 | - | 32 | 1,5 | 113 | 356 | - | 20 | 1,1 | 0 | 0,24 | 0,2 | - | 3,3 | 0,066 | 0,02 | 0,54 | 0,3 | 0,054 | 26 | 0 |
| Noix de coco, amande mûre, fraîche | - | - | - | - | 47 | 3,33 | 3,93 | 6,22 | 33,5 | 9 | 0,05 | 14 | - | 0,44 | 2,43 | 1,2 | 32 | 1,5 | 113 | 356 | - | 20 | 1,1 | 0 | 0,24 | 0,2 | - | 3,3 | 0,066 | 0,02 | 0,54 | 0,3 | 0,054 | 26 | - |
| Noix de coco, amande, sèche | - | - | - | - | 2,65 | 6,62 | 7,8 | 8,55 | 66,3 | 14 | 0,064 | 18,4 | - | 0,67 | 3,46 | 0,3 | 90 | 2,75 | 197 | 647 | 6,8 | 26,6 | 1,26 | 0 | 0,44 | 0,3 | - | 1,5 | 0,048 | 0,07 | 0,6 | 0,5 | 0,17 | 16,5 | 0 |
| Noix de macadamia | - | - | - | - | 1,36 | 7,91 | 9,33 | 5,22 | 75,8 | 8,6 | 0,013 | 85 | - | 0,76 | 3,69 | 2 | 130 | 4,13 | 188 | 368 | - | 5 | 1,3 | 0 | 0,54 | - | - | 1,2 | 1,2 | 0,16 | 2,47 | 0,76 | 0,28 | 11 | 0 |
| Noix de macadamia, grillée, salée | - | - | - | - | 1,61 | 7,04 | 8,3 | 9,95 | 70,2 | 9,5 | 0,9 | 70 | - | 0,57 | 2,65 | - | 118 | 3,04 | 198 | 363 | - | 360 | 1,29 | 0 | 0,57 | 0 | - | 0,7 | 0,71 | 0,087 | 2,27 | 0,6 | 0,36 | 10 | 0 |
| Noix de pécan | - | - | - | - | 2,57 | 9,57 | 11,3 | 5,43 | 72,6 | 8,33 | 0,0025 | 69,8 | - | 1,2 | 2,57 | 2 | 123 | 4,27 | 277 | 409 | 5,2 | 1 | 4,61 | 0 | 1,4 | 3,5 | - | 0,97 | 0,59 | 0,12 | 1,17 | 0,82 | 0,2 | 19,8 | 0 |
| Noix de pécan, salées | - | - | - | - | - | 10,2 | 12 | 5 | 73 | 3 | 1 | - | - | - | - | - | - | - | - | - | - | 400 | - | - | - | - | - | - | - | - | - | - | - | - | - |
| Noix du Brésil | - | - | - | - | 3,01 | 14,8 | 16,9 | 6,17 | 66,1 | 6,4 | 0,0075 | 150 | - | 1,75 | 2,47 | 0,05 | 367 | 2 | 658 | 591 | 103 | 3 | 4,13 | 0 | 5,33 | 0 | - | 0,7 | 0,87 | 0,035 | 0,25 | 0,21 | 0,1 | 13 | 0 |
| Noix, fraîche | - | - | - | - | 41,9 | 11 | 12,9 | 9,94 | 31,8 | 3,9 | < 0,013 | 94 | 28,3 | 0,93 | 1,5 | < 20 | 120 | 2,7 | 260 | 390 | < 20 | < 5 | 2,2 | < 0,25 | 1,6 | 2,04 | - | < 0,5 | 0,21 | 0,12 | 0,38 | 1,06 | 0,16 | 108 | 0 |
| Noix, séchée, cerneaux | - | - | - | - | 4 | 13,3 | 15,7 | 6,88 | 67,3 | 6,7 | < 0,13 | 75 | 35,1 | 1,1 | 2,2 | < 20 | 140 | 2,9 | 360 | 450 | < 20 | < 5 | 2,7 | < 0,25 | 1,67 | 2,4 | - | 0,77 | 0,3 | 0,05 | 0,4 | 0,67 | 0,19 | 120 | < 0,08 |
| Oignon blanc ou jaune, sauté/poêlé sans matière grasse | 169 | 40,2 | 169 | 40,2 | 88,1 | 1,56 | 1,56 | 6,73 | < 0,3 | 2,8 | < 0,013 | 32 | 24,5 | 0,08 | 0,24 | < 20 | 9,3 | 0,13 | 35 | 270 | < 20 | < 5 | 0,21 | - | 0,11 | < 0,8 | - | 2,31 | 0,04 | 0,022 | 0,17 | 0,24 | 0,066 | 6,75 | - |
| Oignon jaune, cru | 159 | 37,8 | 159 | 37,8 | 89 | 1,19 | 1,19 | 6,39 | < 0,5 | 2,6 | < 0,013 | 13 | 21,8 | 0,04 | 0,1 | < 20 | 5,4 | 0,07 | 28 | 150 | < 20 | < 5 | 0,1 | - | < 0,08 | < 0,8 | - | 2,08 | 0,026 | < 0,01 | 0,13 | 0,14 | 0,064 | 13 | - |
| Oignon nouveau ou oignon frais ou cébette, sauté/poêlé sans matière grasse | 115 | 27,6 | 115 | 27,6 | 91,2 | 1,13 | 1,13 | 3,56 | 0,3 | 3,1 | < 0,013 | 55 | 28,6 | 0,03 | 0,85 | < 20 | 11 | 0,36 | 26 | 260 | < 20 | < 5 | 0,21 | - | < 0,08 | 4,03 | - | 7,38 | < 0,015 | < 0,01 | 0,41 | 0,14 | 0,074 | 40,5 | - |
| Oignon rouge, cru | 153 | 36,3 | 153 | 36,3 | 90,3 | 1,31 | 1,31 | 5,63 | 0,4 | 2,5 | < 0,013 | 22 | 20,7 | 0,05 | 0,19 | < 20 | 9 | 0,12 | 29 | 190 | < 20 | < 5 | 0,21 | - | 0,12 | < 0,8 | - | 4,09 | 0,029 | < 0,01 | 0,18 | 0,15 | 0,049 | 10,9 | - |
| Oignon rouge, sauté/poêlé sans matière grasse | 179 | 42,4 | 179 | 42,4 | 87,9 | 1,69 | 1,69 | 7,55 | < 0,3 | 2,1 | < 0,013 | 26 | 22,1 | 0,07 | 0,26 | < 20 | 10 | 0,15 | 33 | 220 | < 20 | < 5 | 0,26 | - | < 0,08 | < 0,8 | - | 5,68 | 0,031 | < 0,01 | 0,15 | 0,19 | 0,062 | 33,3 | - |
| Oignon, cru | - | - | - | - | 89,6 | 1,1 | 1,1 | 6,25 | 0,62 | 1,7 | 0,098 | 25 | 61 | < 0,0001 | 0,00019 | < 10 | 9 | 0,1 | 37 | 190 | 2,3 | 39 | 0,23 | 0 | 0,045 | 0,35 | - | 3,9 | < 0,05 | < 0,05 | < 0,05 | 0,15 | 0,1 | 29,6 | 0 |
| Oignon, cuit | - | - | - | - | 90,4 | 1,3 | 1,3 | 6,2 | 0,2 | 1,4 | 0,021 | 29,2 | - | 0,077 | 0,21 | 1 | 13 | 0,12 | 44 | 220 | < 10 | 8,4 | 0,19 | 0 | 0,015 | 0,4 | - | 2,1 | < 0,05 | < 0,05 | < 0,05 | 0,13 | 0,09 | 21 | 0 |
| Oignon, séché | - | - | - | - | 3,93 | 8,95 | 8,95 | 75 | 0,46 | 9,2 | 0,053 | 257 | - | 0,42 | 1,55 | - | 92 | 1,39 | 303 | 1620 | - | 170 | 1,89 | 0 | 0,18 | 3,8 | - | 75 | 0,5 | 0,1 | 0,99 | 1,38 | 1,6 | 166 | 0 |
| Oignon, surgelé, cru | - | - | - | - | 91,1 | 1 | 1 | 5,39 | 0,12 | 2,65 | 0,02 | 26,5 | - | 0,035 | 0,4 | - | 8,5 | 0,1 | 22,5 | 133 | - | 9,43 | 0,095 | 0 | 0,045 | 0,4 | - | 5,65 | 0,028 | 0,026 | 0,16 | 0,1 | 0,084 | 19 | 0 |
| Orange (variété locale), pulpe, prélevée à la Martinique | - | - | - | - | 90,2 | 0,6 | 0,6 | - | 0,67 | - | - | - | - | - | - | - | - | - | - | - | - | - | - | - | - | - | - | 41,2 | - | - | - | - | - | - | - |
| Orange amère, pulpe, crue, prélevée à la Martinique | - | - | - | - | 91,6 | 0,73 | 0,73 | - | 0,43 | - | - | - | - | - | - | - | - | - | - | - | - | - | - | - | - | - | - | 44,5 | - | - | - | - | - | - | - |
| Orange, pulpe, crue | 192 | 45,5 | 192 | 45,5 | 87,3 | 0,75 | 0,75 | 8,03 | < 0,5 | 2,7 | < 0,013 | 66 | < 20 | 0,04 | 0,57 | < 20 | 15 | 0,02 | 38 | 180 | < 20 | < 5 | 0,25 | 0 | 0,19 | < 0,8 | - | 47,5 | 0,045 | < 0,01 | 0,37 | 0,16 | < 0,01 | 25,9 | 0 |
| Oseille, crue | - | - | - | - | 93 | 1,84 | 1,84 | 1,6 | 0,65 | 1,8 | 0,009 | 44 | - | 0,13 | 2,4 | 0,4 | 103 | 0,35 | 63 | 390 | 0,9 | 4 | 0,2 | 0 | - | - | - | 48 | 0,04 | 0,1 | 0,5 | 0,041 | 0,12 | 13 | 0 |
| Pamplemousse chinois, pulpe, cru | - | - | - | - | 89,1 | 0,76 | 0,76 | 8,62 | 0,04 | 1 | 0,0025 | 4 | - | 0,048 | 0,11 | - | 6 | 0,017 | 17 | 216 | - | 1 | 0,08 | - | - | - | - | 61 | 0,034 | 0,027 | 0,22 | - | 0,036 | - | 0 |
| Panais, cru | 245 | 58,3 | 245 | 58,3 | 80,6 | 1,54 | 1,54 | 10,1 | 0,4 | 4,7 | 0,02 | 45,5 | - | 0,13 | 0,7 | 0,25 | 29,5 | 0,48 | 82,5 | 505 | 0,2 | 8 | 0,72 | 0 | 1,9 | 22,5 | - | 11,5 | 0,09 | 0,12 | 1,95 | 0,6 | 0,1 | 67 | 0 |
| Panais, cuit | 286 | 67,8 | 286 | 67,8 | 80,2 | 1,32 | 1,32 | 13,4 | 0,3 | 3,6 | 0,025 | 37 | - | 0,14 | 0,58 | - | 29 | 0,29 | 69 | 367 | - | 10 | 0,26 | 0 | 1 | 1 | - | 13 | 0,083 | 0,051 | 0,72 | 0,59 | 0,093 | 58 | 0 |
| Panais, cuit à l'étouffée | 383 | 90,8 | 383 | 90,8 | 73,6 | 1,94 | 1,94 | 16,6 | 0,5 | 5,6 | 0,018 | 44 | 95,5 | 0,14 | 0,37 | < 20 | 25 | 0,16 | 74 | 510 | < 20 | 7 | 0,35 | - | 1,83 | < 0,8 | - | 7,92 | 0,089 | 0,056 | 1,53 | 0,41 | 0,079 | 80,1 | - |
| Papaye Colombo (fruit mûr), pulpe sans pépin, crue, prélevée à La Réunion (Carica papaya L.) | 169 | 40 | 169 | 40 | 88,6 | 0,56 | 0,56 | 7,88 | < 0,3 | 2,2 | < 0,013 | 18 | 53,2 | 0,03 | 0,16 | < 20 | 12 | < 0,01 | 11 | 240 | < 20 | < 5 | 0,09 | - | 0,22 | < 0,8 | - | 68 | < 0,015 | < 0,01 | 0,32 | 0,26 | < 0,01 | 50,1 | - |
| Papaye, pulpe, crue | 178 | 42,2 | 178 | 42,2 | 88,8 | 0,75 | 0,75 | 8,53 | < 0,3 | 1,8 | < 0,013 | 22 | 63 | < 0,01 | 0,16 | < 20 | 18 | 0,04 | 8,7 | 200 | < 20 | < 5 | < 0,05 | 0 | < 0,08 | < 0,8 | - | 65,3 | 0,017 | 0,017 | 0,29 | 0,18 | 0,02 | 55,3 | 0 |
| Papaye, pulpe, crue, prélevée à la Martinique | - | - | - | - | 90,4 | 0,67 | 0,67 | - | 0,0033 | - | - | - | - | - | - | - | - | - | - | - | - | - | - | - | - | - | - | 39,3 | - | - | - | - | - | - | - |
| Papaye, pulpe, cuite à la vapeur, prélevée à la Martinique | - | - | - | - | 90,5 | 0,63 | 0,63 | - | 0 | - | - | - | - | - | - | - | - | - | - | - | - | - | - | - | - | - | - | 25,7 | - | - | - | - | - | - | - |
| Pastèque, pulpe, crue | 165 | 38,9 | 165 | 38,9 | 91 | 0,69 | 0,69 | 8,33 | < 0,5 | 0,5 | < 0,013 | 6 | < 20 | 0,03 | 0,16 | < 20 | 11 | 0,02 | 9,7 | 100 | < 20 | < 5 | 0,09 | 0 | < 0,08 | < 0,8 | - | 4,26 | 0,015 | 0,01 | 0,3 | 0,2 | 0,064 | 36,9 | 0 |
| Pastèque, pulpe, crue, prélevée à la Martinique | - | - | - | - | 91,5 | 0,77 | 0,77 | - | 0,26 | - | - | - | - | - | - | - | - | - | - | - | - | - | - | - | - | - | - | 7,87 | - | - | - | - | - | - | - |
| Patate douce, crue | 365 | 86,3 | 365 | 86,3 | 78,8 | 1,51 | 1,51 | 18,3 | 0,15 | 2,87 | 0,098 | 37,5 | - | 0,15 | 0,76 | 0,2 | 20 | 0,26 | 47 | 373 | 0,6 | 39,3 | 0,25 | 0 | 0,26 | 1,8 | - | 2,4 | 0,072 | 0,061 | 0,56 | 0,8 | 0,19 | 11 | 0 |
| Patate douce, cuite | 265 | 62,8 | 265 | 62,8 | 78 | 1,69 | 1,69 | 12,2 | 0,15 | 2,9 | 0,079 | 32,5 | - | 0,13 | 0,71 | 3 | 22,5 | 0,38 | 43 | 353 | - | 31,5 | 0,26 | 0 | 0,83 | 2,2 | - | 16,2 | 0,082 | 0,077 | 1,01 | 0,73 | 0,23 | 6 | 0 |
| Patate douce, pulpe, blanchie, surgelée, prélevée à la Martinique | - | - | - | - | 63,1 | - | - | 19,3 | - | 4,9 | - | - | - | - | - | - | - | - | - | - | - | - | - | - | - | - | - | 4 | - | - | - | - | - | - | - |
| Patate douce, pulpe, cuite à la vapeur, prélevée à la Martinique | - | - | - | - | 63,3 | 1,48 | 1,48 | 25,2 | 0,067 | 9,2 | - | - | - | 0,26 | - | - | - | 0,026 | - | 425 | - | - | - | - | 0,24 | - | - | 12,5 | - | - | - | - | - | 92,2 | - |
| Patate douce, purée, cuisinée à la crème | 334 | 79,8 | 334 | 79,8 | - | 1,1 | 1,1 | 12,1 | 2 | 4,5 | 0,56 | - | - | - | - | - | - | - | - | - | - | 225 | - | - | - | - | - | - | - | - | - | - | - | - | - |

| | | | | | | | | | | | | | | | | | | | | | | | | | | | | | | | | | | | |
|---|---|---|---|---|---|---|---|---|---|---|---|---|---|---|---|---|---|---|---|---|---|---|---|---|---|---|---|---|---|---|---|---|---|---|---|
| Pâte d'amande, préemballée | 1820 | 432 | 1800 | 428 | 6,6 | 5,8 | 7 | 70,3 | 13,5 | 2,8 | 0,018 | 91 | < 20 | 0,27 | 0,88 | 0,5 | 74 | 0,49 | 150 | 190 | < 20 | 7 | 0,86 | < 0,25 | 5,52 | < 0,8 | - | < 0,5 | 0,029 | 0,02 | 0,34 | 0,15 | 0,019 | 13,5 | - |
| Pêche au sirop léger, appertisée, égouttée | - | - | - | - | 85,9 | 0,5 | 0,5 | 13,8 | < 0,5 | 1,5 | 0,033 | 10 | 13,1 | 0,05 | 0,13 | < 20 | 5 | 0,02 | 11 | 100 | < 20 | < 5 | 0,05 | - | - | < 0,8 | - | 2,78 | < 0,01 | < 0,01 | 0,37 | 0,071 | < 0,01 | 8,4 | - |
| Pêche au sirop léger, appertisée, non égouttée | - | - | - | - | 84,3 | 0,31 | 0,31 | 14,1 | 0,038 | 0,92 | 0,0022 | 9,57 | 13,2 | 0,038 | 0,13 | < 20 | 4,88 | 0,02 | 9,79 | 100 | < 20 | 0,9 | 0,054 | 0 | 0,49 | < 0,8 | - | 3,32 | < 0,01 | < 0,01 | 0,35 | 0,063 | < 0,01 | 8,4 | 0 |
| Pêche blanche, pulpe et peau, crue | 199 | 47,2 | 199 | 47,2 | 87,6 | 0,69 | 0,69 | 9,48 | < 0,5 | 2,1 | < 0,013 | 4,1 | < 20 | 0,06 | 0,11 | < 20 | 7,1 | 0,05 | 18 | 180 | < 20 | < 5 | 0,11 | - | 0,19 | < 0,8 | - | 3,85 | < 0,015 | < 0,01 | 0,2 | 0,18 | < 0,01 | 7,92 | - |
| Pêche blanche, pulpe, crue | 195 | 46,1 | 195 | 46,1 | 87,6 | 0,63 | 0,63 | 9,63 | < 0,5 | 1,1 | < 0,013 | 3,9 | < 20 | 0,06 | 0,12 | < 20 | 6,8 | 0,05 | 18 | 170 | < 20 | < 5 | 0,1 | - | 0,28 | < 0,8 | - | 4,09 | < 0,015 | < 0,01 | < 0,1 | 0,15 | < 0,01 | 10,5 | - |
| Pêche jaune, pulpe, crue | 196 | 46,3 | 196 | 46,3 | 87,8 | 0,69 | 0,69 | 9,8 | < 0,5 | 1 | < 0,013 | 4,2 | < 20 | 0,06 | 0,11 | < 20 | 6,7 | 0,04 | 17 | 190 | < 20 | < 5 | 0,09 | - | 0,5 | < 0,8 | - | 3,43 | < 0,015 | < 0,01 | 0,31 | 0,16 | 0,015 | 12 | - |
| Pêche, pulpe et peau, crue | - | - | - | - | 88,6 | 1,08 | 1,08 | 9 | 0,33 | 1,6 | 0,0075 | 7,32 | - | 0,075 | 0,15 | 0,3 | 11,2 | 0,064 | 21 | 215 | 0,16 | < 1,11 | 0,12 | 0 | 1,27 | 2,6 | - | 6,6 | 0,022 | 0,046 | 0,7 | 0,16 | 0,028 | 4,1 | 0 |
| Pêche, sèche | - | - | - | - | 7,5 | 5 | 5 | 68,9 | 1 | 14,3 | 0,2 | 25 | - | 0,63 | 6,8 | 2,8 | 80 | 0,67 | 120 | 945 | - | 79 | 0,57 | 0 | - | - | - | 4,8 | 0,082 | 0,065 | 1,3 | 0,3 | 0,15 | 14 | 0 |
| Petits pois et carottes, appertisés, égouttés | 206 | 49 | 206 | 49 | 87,4 | 2,51 | 2,51 | 7,39 | 0,33 | 3,27 | 0,61 | 23 | - | 0,1 | 0,75 | 2 | 14 | 0,36 | 46 | 100 | - | 244 | 0,58 | 0 | traces | - | - | 6,6 | 0,074 | 0,053 | 0,58 | 0,12 | 0,088 | 18 | 0 |
| Petits pois et carottes, surgelés, crus | - | - | - | - | 82,3 | 3,89 | 3,89 | 7,1 | 0,37 | 4,3 | 0,082 | 29,5 | - | 0,089 | 1,2 | - | 18 | 0,24 | 60 | 194 | - | 31,3 | 0,52 | 0 | - | - | - | 13,6 | 0,19 | 0,081 | 1,41 | 0,2 | 0,1 | 36 | 0 |
| Petits pois et carottes, surgelés, cuits | - | - | - | - | 85,8 | 3,09 | 3,09 | 7,02 | 0,42 | 3,1 | 0,17 | 23 | - | 0,076 | 0,94 | - | 16 | 0,2 | 49 | 158 | - | 68 | 0,45 | 0 | 0,52 | 18,8 | - | 8,1 | 0,23 | 0,064 | 1,15 | 0,16 | 0,087 | 26 | 0 |
| Petits pois, appertisés, égouttés | 342 | 81,5 | 342 | 81,5 | 79,5 | 5,12 | 5,12 | 10,7 | 0,8 | 5,75 | 0,64 | 20,6 | 510 | 0,079 | 1,29 | 0,9 | 16,3 | 0,17 | 73,1 | 71,5 | < 3,35 | 257 | 0,42 | 0 | 0,21 | 53,4 | - | 6,9 | 0,12 | 0,059 | 0,89 | 0,2 | 0,053 | 26,5 | 0 |
| Petits pois, bouillis/cuits à l'eau | 338 | 80,3 | 338 | 80,3 | 77,4 | 6,38 | 6,38 | 9,93 | 0,6 | 4,8 | 0,023 | 38 | < 20 | 0,14 | 1,2 | < 20 | 29 | 0,29 | 100 | 190 | < 20 | 9 | 0,72 | - | < 0,08 | 9,9 | - | 11,4 | 0,14 | < 0,01 | 1,4 | 0,24 | 0,044 | 38,2 | - |
| Petits pois, crus | - | - | - | - | 79,2 | 5,84 | 5,84 | 7 | 0,55 | 5,5 | 0,0088 | 27,5 | 37,5 | 0,14 | 1,64 | 0,15 | 30,5 | 0,4 | 119 | 272 | 1,02 | 3,5 | 1 | 0 | 0,075 | 47,4 | - | 41,5 | 0,29 | 0,13 | 2,15 | 0,43 | 0,16 | 45 | 0 |
| Petits pois, cuits | - | - | - | - | 82,9 | 5,8 | 5,8 | 4,7 | 0,87 | 5,8 | 0,018 | 32,7 | 243 | 0,14 | 1,17 | < 5 | 31,9 | 0,32 | 96 | 135 | < 10 | 7,1 | 0,8 | 0 | < 0,1 | 25,9 | - | 1,8 | 0,13 | 0,05 | 0,4 | < 0,05 | < 0,05 | 65,6 | 0 |
| Petits pois, purée | - | - | - | - | - | 4,8 | 4,8 | 8 | 2,8 | 4,8 | 0,74 | - | - | - | - | - | - | - | - | - | - | - | - | - | - | - | - | - | - | - | - | - | - | - | - |
| Petits pois, surgelés, crus | - | - | - | - | 80,2 | 5,86 | 5,86 | 8,77 | 0,44 | 6,45 | 0,09 | 27,3 | 20 | 0,15 | 1,53 | 0,15 | 25,9 | 0,35 | 86,9 | 158 | 1,02 | 21,7 | 0,86 | 0 | 0,025 | 49 | - | 20,8 | 0,27 | 0,099 | 1,72 | 0,43 | 0,08 | 87 | 0 |
| Petits pois, surgelés, cuits | - | - | - | - | 79,5 | 5,15 | 5,15 | 7,95 | 0,27 | 5,5 | 0,18 | 24 | 12 | 0,11 | 1,52 | - | 22 | 0,28 | 77 | 110 | 1 | 72 | 0,67 | 0 | 0,03 | 24 | - | 9,9 | 0,28 | 0,1 | 1,48 | 0,14 | 0,11 | 59 | 0 |
| Pignon de pin | - | - | - | - | 2,45 | 13,7 | 16,2 | 6,31 | 65 | 10 | 0,023 | 6,2 | 61 | 1,2 | 4,6 | 11,2 | 227 | 8,2 | 527 | 662 | < 5 | 9 | 5,6 | 0 | 8,47 | 53,9 | - | < 0,5 | 0,57 | 0,11 | 3,29 | 0,26 | 0,14 | 80,2 | 0 |
| Piment, cru | - | - | - | - | 87,9 | 1,87 | 1,87 | 7,7 | 0,32 | 1,65 | 0,02 | 14 | - | 0,15 | 1,12 | - | 24 | 0,21 | 55,3 | 331 | - | 8 | 0,28 | 0 | 0,69 | 14 | - | 155 | 0,081 | 0,088 | 1,07 | 0,13 | 0,39 | 23 | 0 |
| Pissenlit, cru | - | - | - | - | 85,3 | 2,91 | 2,91 | 6,1 | 0,85 | 2,7 | 0,22 | 62,3 | - | 0,067 | 3,1 | 0,4 | 13,2 | 0,16 | 66 | 397 | < 2,2 | 87,8 | 0,21 | 0 | 3,44 | 778 | - | 37,5 | 0,19 | 0,2 | 0,8 | 0,084 | 0,25 | 27 | 0 |
| Pistache, grillée | - | - | - | - | 2,43 | 18,4 | 21,7 | 18,6 | 47,4 | 10,1 | 0,015 | 108 | - | 1,3 | 4,1 | - | 115 | 1,24 | 478 | 1020 | 4,7 | 6 | 2,3 | 0 | 2,42 | 13,2 | - | 3,33 | 0,78 | 0,2 | 1,37 | 0,51 | 1,3 | 50,8 | 0 |
| Pistache, grillée, salée | - | - | - | - | 2,09 | 18,9 | 22,3 | 15,9 | 49,5 | 9,32 | 1,67 | 98,5 | 728 | 1,1 | 2,4 | 6,2 | 105 | 0,65 | 437 | 655 | < 5 | 664 | 2,3 | 0 | 1,09 | 13,2 | - | < 0,5 | 0,67 | 0,17 | 1,2 | 0,43 | 1,41 | 94,2 | 0 |
| Poire au sirop léger, appertisée, égouttée | 274 | 64,9 | 274 | 64,9 | 82,3 | < 0,5 | < 0,5 | 13,9 | 0,4 | 2,6 | < 0,013 | 8,8 | < 20 | 0,05 | < 0,05 | - | 4 | 0,01 | 7,9 | 69 | < 20 | < 5 | 0,06 | < 0,25 | 0,31 | < 0,8 | - | 0,85 | < 0,015 | < 0,01 | < 0,1 | 0,038 | 0,011 | < 5 | - |
| Poire au sirop léger, appertisée, non égouttée | - | - | - | - | 83,7 | 0,3 | 0,3 | 14,5 | 0,12 | 1,23 | 0,013 | 5 | - | 0,049 | 0,28 | - | 4 | 0,033 | 7 | 66 | - | 5 | 0,08 | 0 | 0,08 | 0,3 | - | 0,7 | 0,01 | 0,016 | 0,15 | 0,022 | 0,014 | 1 | 0 |
| Poire Conférence, pulpe, crue | 223 | 53,1 | 223 | 53,1 | 85,3 | < 0,5 | < 0,5 | 11,4 | < 0,5 | 3,1 | < 0,013 | 3,9 | < 20 | 0,05 | 0,09 | < 20 | 5,3 | 0,02 | 9,3 | 99 | < 20 | < 5 | 0,07 | - | < 0,08 | < 0,8 | - | 1,39 | < 0,015 | < 0,01 | < 0,1 | 0,065 | < 0,01 | 12,6 | - |
| Poire Williams, pulpe, crue | 228 | 54,1 | 228 | 54,1 | 84,7 | < 0,5 | < 0,5 | 11,5 | < 0,5 | 3,1 | < 0,013 | 6,1 | < 20 | 0,05 | 0,07 | < 20 | 5,3 | 0,02 | 9,8 | 120 | < 20 | < 5 | 0,09 | - | < 0,08 | < 0,8 | - | 2,57 | < 0,015 | < 0,01 | < 0,1 | 0,06 | < 0,01 | 18,4 | - |
| Poire, pulpe et peau, crue | - | - | - | - | 83,5 | 0,49 | 0,49 | 10,9 | 0,27 | 2,9 | 0,0045 | 6,46 | 61 | 0,071 | 0,072 | 0,4 | 8,23 | 0,03 | 15,4 | 132 | < 10 | 1,8 | 0,097 | 0 | 0,41 | 4,4 | - | 4,62 | 0,015 | 0,021 | 0,23 | 0,06 | 0,022 | 11,5 | 0 |
| Poire, pulpe, crue | - | - | - | - | 85,1 | 0,3 | 0,3 | 10,4 | 0,1 | 2,47 | 0,0075 | 11 | - | 0,06 | 0,2 | 1 | 7 | 0 | 13 | 150 | 1 | 3 | 0,1 | 0 | - | - | - | 6 | 0,2 | 0,03 | 0,2 | 0,07 | 0,02 | 2 | 0 |
| Poireau, bouilli/cuit à l'eau | 114 | 27,4 | 114 | 27,4 | 91,6 | 2,56 | 2,56 | 2,23 | < 0,3 | 3,4 | < 0,013 | 35 | 33 | 0,08 | 0,45 | < 20 | 6,5 | 0,11 | 26 | 120 | < 20 | < 5 | 0,14 | - | 0,75 | 3,6 | - | 3,79 | 0,028 | < 0,01 | 0,3 | 0,088 | 0,055 | 22,8 | - |
| Poireau, cru | 136 | 32,3 | 136 | 32,3 | 87,6 | 1,49 | 1,49 | 4,9 | 0,25 | 2,27 | 0,036 | 50,7 | - | 0,087 | 1,5 | 1,1 | 19 | 0,34 | 40,3 | 208 | 0,56 | 14,5 | 0,21 | 0 | 0,74 | 47 | - | 18,5 | 0,065 | 0,065 | 0,5 | 0,13 | 0,24 | 73 | 0 |
| Poireau, cuit | - | - | - | - | 92 | 1,1 | 1,1 | 3 | 0,2 | 3,2 | 0,015 | 25,6 | - | 0,12 | 0,34 | < 10 | 8,45 | 0,075 | 20 | 151 | < 10 | 5,8 | 0,1 | 0 | 0,5 | 25,4 | - | 1,3 | < 0,05 | < 0,05 | < 0,05 | < 0,05 | 0,06 | 21,2 | 0 |
| Poireau, surgelé, cru | 112 | 26,7 | 112 | 26,7 | 93,2 | 1,42 | 1,42 | 3,35 | 0,29 | 2,55 | 0,016 | 63 | - | - | 0,8 | - | - | - | - | - | - | 6,75 | - | - | - | - | - | 26 | - | - | - | - | - | - | - |
| Pois cassé, bouilli/cuit à l'eau | - | - | - | - | 64,9 | 8,6 | 8,6 | 16,3 | 1,49 | 7,95 | 0,011 | 24,7 | - | 0,19 | 2,09 | 2 | 20,9 | 0,28 | 134 | 327 | < 2,2 | 4,5 | 0,65 | 0 | 0,19 | 4,5 | - | 0,85 | 0,15 | 0,06 | 0,71 | 0,44 | 0,094 | 119 | 0 |
| Pois cassé, sec | - | - | - | - | 8,33 | 22,8 | 22,8 | 52 | 1,44 | 15,8 | 0,051 | 37,4 | 56 | 0,75 | 5,16 | 0,5 | 65,5 | 1,21 | 364 | 969 | 14 | 15,4 | 3,68 | 0 | 0,09 | 47,8 | - | 1,4 | 0,77 | 0,2 | 2,69 | 1,88 | 0,12 | 154 | 0 |
| Pois chiche, appertisé, égoutté | - | - | - | - | 72,1 | 6,74 | 6,74 | 15 | 2,68 | 5,45 | 0,54 | 41,2 | - | 0,21 | 1,15 | - | 27,5 | 0,67 | 82,5 | 135 | 23,3 | 216 | 0,93 | 0 | 0,29 | 3,4 | - | 0,1 | 0,03 | 0,015 | 0,14 | 0,3 | 0,29 | 36,5 | 0 |
| Pois chiche, bouilli/cuit à l'eau | 618 | 147 | 618 | 147 | 62 | 8,31 | 8,31 | 17,7 | 3 | 8,2 | 0,027 | 72 | < 20 | 0,24 | 1,3 | < 20 | 44 | 0,86 | 140 | 170 | < 20 | 10,9 | 1,1 | < 0,25 | 1,22 | 2,53 | - | < 0,5 | 0,06 | 0,027 | 0,21 | 0,14 | 0,09 | 84,4 | - |
| Pois chiche, sec | - | - | - | - | 8,99 | 20,5 | 20,5 | 47,5 | 5,85 | 13,3 | 0,08 | 90,5 | 60 | 0,71 | 5,36 | 0,6 | 120 | 2,2 | 276 | 759 | 2 | 23,2 | 1,88 | 0 | 3,1 | 9 | - | 3,5 | 0,49 | 0,18 | 1,52 | 1,59 | 0,48 | 369 | 0 |
| Pois d'angole, entier, cuit à la vapeur, prélevé à la Martinique | - | - | - | - | 49,1 | 16,3 | 16,3 | - | 0,74 | - | - | - | - | - | - | - | - | - | - | - | - | - | - | - | - | - | - | - | - | - | - | - | - | - | - |
| Pois mange-tout ou pois gourmand, bouilli/cuit à l'eau | 128 | 30,5 | 128 | 30,5 | 90,9 | 2,25 | 2,25 | 3,39 | 0,3 | 2,8 | 0,013 | 35 | < 20 | 0,08 | 0,46 | < 20 | 14 | 0,18 | 40 | 100 | < 20 | 5 | 0,28 | < 0,25 | 0,59 | 3,87 | - | 20,6 | 0,073 | < 0,01 | 0,39 | 0,22 | 0,022 | 56 | 0 |
| Pois mange-tout ou pois gourmand, cru | - | - | - | - | 88,3 | 3,08 | 3,08 | 6,2 | 0,27 | 2,63 | 0,01 | 40,7 | - | 0,079 | 2,08 | - | 24 | 0,24 | 56,8 | 200 | - | 4 | 0,27 | 0 | 0,39 | 25 | - | 60 | 0,15 | 0,08 | 0,6 | 0,75 | 0,16 | 42 | 0 |
| Pois mange-tout ou pois gourmands, cuits | - | - | - | - | 86,6 | 3,5 | 3,5 | 5,92 | 0,38 | 3,1 | 0,013 | 59 | - | 0,09 | 2,4 | - | 28 | 0,28 | 58 | 217 | - | 5 | 0,49 | 0 | 0,47 | 30,2 | - | 22 | 0,064 | 0,12 | 0,56 | 0,86 | 0,17 | 35 | 0 |
| Poivron jaune, cru | 130 | 30,7 | 130 | 30,7 | 91,9 | 0,94 | 0,94 | 4,73 | 0,3 | 1,2 | < 0,013 | 8,6 | 29,6 | 0,07 | 0,21 | < 20 | 12 | 0,11 | 23 | 220 | < 20 | < 5 | 0,23 | 0 | 1,76 | 1,18 | - | 121 | 0,03 | < 0,01 | 0,28 | 0,14 | 0,079 | 43,3 | 0 |
| Poivron jaune, sauté/poêlé sans matière grasse | 150 | 35,8 | 150 | 35,8 | 90,5 | 1 | 1 | 5,33 | 0,6 | 2,2 | < 0,013 | 7,3 | 25,4 | 0,06 | 0,23 | < 20 | 11 | 0,09 | 23 | 220 | < 20 | < 5 | 0,14 | - | 2,12 | < 0,8 | - | 126 | 0,03 | 0,047 | 0,56 | 0,16 | 0,24 | 57,1 | - |
| Poivron rouge, appertisé, égoutté | - | - | - | - | 90,3 | 0,91 | 0,91 | 4,42 | 0,77 | 1,7 | 1,14 | 25 | - | 0,098 | 0,63 | - | 12,5 | 0,12 | 19 | 156 | - | 440 | 0,15 | 0 | 1,65 | 5,1 | - | 109 | 0,042 | 0,03 | 0,51 | 0,079 | 0,23 | 34 | 0 |
| Poivron rouge, cru | 153 | 36,6 | 153 | 36,6 | 90,2 | 1,06 | 1,06 | 5,98 | < 0,5 | 3,2 | < 0,013 | 4,8 | 29,7 | 0,04 | 0,21 | < 20 | 8,2 | 0,08 | 19 | 180 | < 20 | < 5 | 0,14 | 0 | 3,45 | 3,16 | - | 121 | 0,036 | 0,11 | 0,34 | 0,18 | 0,31 | 81,8 | 0 |
| Poivron rouge, cuit | - | - | - | - | 90,2 | 1,4 | 1,4 | 7 | 0,3 | 2,2 | 0,006 | 9 | - | 0,01 | - | 1 | 14 | 0,1 | 26 | 180 | traces | - | 0,2 | - | 0,9 | - | - | 81 | 0,01 | 0,03 | 0,9 | 0,06 | 0,31 | 11 | - |
| Poivron rouge, sauté/poêlé sans matière grasse | 151 | 35,7 | 151 | 35,7 | 89,8 | 0,94 | 0,94 | 6,53 | < 0,3 | 2 | < 0,013 | 7 | 28,1 | 0,06 | 0,27 | < 20 | 12 | 0,11 | 25 | 240 | < 20 | < 5 | 0,14 | - | 3,81 | 5,61 | - | 144 | 0,033 | 0,1 | 0,83 | 0,15 | 0,31 | 69,7 | - |
| Poivron vert, cru | 109 | 26 | 109 | 26 | 93 | 0,81 | 0,81 | 3,43 | < 0,5 | 3,2 | < 0,013 | 6,2 | < 20 | 0,04 | 0,22 | < 20 | 7,7 | 0,07 | 17 | 160 | < 20 | < 5 | 0,11 | 0 | 0,28 | < 0,8 | - | 26,9 | < 0,015 | < 0,01 | 0,3 | 0,11 | 0,21 | 55,1 | 0 |
| Poivron vert, cuit | - | - | - | - | 91,9 | 0,92 | 0,92 | 2,36 | 0,2 | 1,2 | 0,005 | 9 | - | 0,065 | 0,46 | 1 | 10 | 0,12 | 18 | 166 | traces | 2 | 0,12 | 0 | 0,5 | 9,8 | - | 74,4 | 0,059 | 0,03 | 0,48 | 0,079 | 0,23 | 16 | 0 |
| Poivron vert, sauté/poêlé sans matière grasse | 120 | 28,6 | 120 | 28,6 | 92,4 | 1,25 | 1,25 | 3,53 | 0,3 | 3 | < 0,013 | 6,7 | 26,6 | 0,05 | 0,21 | < 20 | 9,5 | 0,08 | 20 | 200 | < 20 | < 5 | 0,12 | - | 0,76 | 4,1 | - | 64,6 | 0,017 | 0,021 | 0,35 | 0,097 | 0,26 | 27,8 | - |
| Poivron, vert, jaune ou rouge, cru | - | - | - | - | 92,1 | 0,8 | 0,8 | 4,55 | 0,27 | 1,5 | 0,021 | 7,7 | 61 | < 0,1 | 0,4 | < 5 | 11,9 | 0,1 | 22,5 | 155 | 0,1 | 8,4 | 0,13 | < 0,5 | 1,44 | - | - | 121 | 0,041 | 0,041 | 0,74 | 0,12 | 0,38 | 40,7 | - |
| Pomelo (dit Pamplemousse) jaune, pulpe, cru | - | - | - | - | 90,5 | 0,69 | 0,69 | 7,31 | 0,1 | 1,1 | 0 | 12 | - | 0,05 | 0,06 | - | 9 | 0,013 | 8 | 148 | - | 0 | 0,07 | 0 | 0,13 | 0 | - | 33,3 | 0,037 | 0,02 | 0,27 | 0,28 | 0,043 | 10 | 0 |
| Pomelo (dit Pamplemousse) rose, pulpe, cru | - | - | - | - | 88,1 | 0,77 | 0,77 | 6,2 | 0,14 | 1,6 | 0 | 22 | - | 0,032 | 0,08 | - | 9 | 0,022 | 18 | 135 | - | 0 | 0,07 | 0 | 0,13 | 0 | - | 31,2 | 0,043 | 0,031 | 0,2 | 0,26 | 0,053 | 13 | 0 |
| Pomelo (dit Pamplemousse), pulpe, cru | 169 | 39,8 | 169 | 39,8 | 89,3 | < 0,5 | < 0,5 | 8,02 | < 0,5 | 0,8 | < 0,013 | 14 | < 20 | 0,02 | 0,06 | < 20 | 7,2 | 0,02 | 17 | 140 | < 20 | < 5 | 0,07 | 0 | < 0,08 | < 0,8 | - | 42,4 | 0,043 | < 0,01 | < 0,1 | 0,19 | 0,36 | 52,9 | 0 |
| Pomme cajou, pulpe, crue, prélevée à la Martinique | - | - | - | - | 88,5 | 0,79 | 0,79 | - | 0,11 | - | - | - | - | - | - | - | - | - | - | - | - | - | - | - | - | - | - | 556 | - | - | - | - | - | - | - |
| Pomme Canada, pulpe, crue | - | - | - | - | 84,4 | 0,4 | 0,4 | 11,6 | 0,1 | 1,6 | 0,0092 | 3,59 | - | 0,035 | 0,14 | 1,12 | 3,76 | 0,03 | 8,61 | 154 | 0,045 | 3,69 | 0,09 | 0 | 0,27 | - | - | 4 | 0,03 | 0,02 | 0,1 | traces | 0,06 | 1 | 0 |
| Pomme cannelle, pulpe, crue, prélevée à la Martinique | - | - | - | - | 67,2 | 1,48 | 1,48 | - | 0,57 | - | - | - | - | - | - | - | - | - | - | - | - | - | - | - | - | - | - | 11,5 | - | - | - | - | - | - | - |

| | | | | | | | | | | | | | | | | | | | | | | | | | | | | | | | | | | | |
|---|---|---|---|---|---|---|---|---|---|---|---|---|---|---|---|---|---|---|---|---|---|---|---|---|---|---|---|---|---|---|---|---|---|---|---|
| Pomme Chantecler, pulpe, crue | 219 | 51,8 | 219 | 51,8 | 86,4 | < 0,5 | < 0,5 | 11,2 | < 0,5 | 1,9 | < 0,013 | 4 | < 20 | 0,04 | 0,06 | < 20 | 3,6 | 0,05 | 10 | 120 | < 20 | < 5 | 0,05 | - | < 0,28 | < 0,8 | - | 2,85 | < 0,015 | < 0,01 | < 0,1 | 0,078 | 0,039 | < 5 | - |
| Pomme de terre dauphine, surgelée, crue | 1190 | 285 | 1190 | 285 | 49 | 4,62 | 4,62 | 25,6 | 17,7 | 2,34 | 1,17 | 22,2 | - | 0,09 | 0,99 | - | 9,25 | 0,23 | 41,8 | 138 | - | 468 | 0,68 | - | - | - | - | - | - | - | - | - | - | - | - |
| Pomme de terre dauphine, surgelée, cuite | 1260 | 302 | 1260 | 302 | 43,5 | 4,5 | 4,5 | 30,2 | 17,7 | 2 | 1,32 | 120 | 524 | 0,06 | 0,8 | < 20 | 14 | 0,18 | 210 | 180 | < 20 | 527 | 0,38 | < 0,25 | 6,6 | < 0,8 | - | - | 0,08 | 0,071 | 0,31 | 0,75 | 0,11 | 15,1 | 0,39 |
| Pomme de terre de conservation, sans peau, bouillie/cuite à l'eau | 322 | 76,1 | 322 | 76,1 | 79,7 | 1,81 | 1,81 | 15,7 | < 0,5 | 1,9 | < 0,01 | 6,5 | 76,3 | 0,07 | 0,31 | < 20 | 17 | 0,1 | 40 | 390 | < 20 | < 5 | 0,23 | - | 0,1 | < 0,8 | - | 3,78 | 0,062 | < 0,01 | 0,61 | 0,37 | 0,17 | 7,89 | 0,074 |
| Pomme de terre duchesse, surgelée, crue | 745 | 178 | 745 | 178 | 65,5 | 2,8 | 2,8 | 23,2 | 7,72 | 2,13 | 0,93 | - | - | - | - | - | - | - | - | - | - | 368 | - | - | - | - | - | - | - | - | - | - | - | - | - |
| Pomme de terre duchesse, surgelée, cuite | 913 | 218 | 913 | 218 | 51,5 | 3,44 | 3,44 | 29,5 | 8,5 | 4,8 | 0,99 | 51 | 622 | 0,12 | 0,81 | < 20 | 28 | 0,19 | 81 | 530 | < 20 | 395 | 0,53 | < 0,25 | 2,33 | < 0,8 | - | < 0,5 | 0,092 | 0,051 | 0,34 | 0,42 | 0,26 | 18,5 | - |
| Pomme de terre noisette, surgelée, crue | 760 | 181 | 760 | 181 | 61 | 2,76 | 2,76 | 24,9 | 7,22 | 2,82 | 0,93 | 25,2 | - | 0,1 | 1 | 3 | 27,9 | 0,18 | 111 | 550 | < 2,2 | 373 | 0,58 | 0 | 0,2 | - | - | 8 | 0,13 | 0,03 | 2,2 | 0,5 | 0,2 | 28 | - |
| Pomme de terre noisette, surgelée, cuite | 804 | 191 | 804 | 191 | 56,9 | 2,81 | 2,81 | 28,5 | 6,6 | 3,4 | 0,87 | 27 | 592 | 0,09 | 0,61 | < 20 | 22 | 0,14 | 59 | 420 | < 20 | 347 | 0,39 | - | 1,29 | 0,89 | - | - | 0,18 | 0,033 | 0,66 | 0,46 | 0,17 | 12,4 | 0,07 |
| Pomme de terre nouvelle, bouillie/cuite à l'eau | - | - | - | - | 81,1 | 1,44 | 1,44 | - | 0,3 | 1,6 | 0,025 | 13 | 43 | 0,06 | 1,6 | 2,25 | 18 | 0,2 | 54 | 430 | 1 | 10 | 0,3 | 0 | 0,06 | - | - | 15 | 0,09 | 0,06 | 0,4 | 0,38 | 0,36 | 7,5 | 0 |
| Pomme de terre nouvelle, crue | 324 | 76,4 | 324 | 76,4 | 79,9 | 1,88 | 1,88 | 15,9 | 0,3 | 1,3 | 0,058 | 10 | - | - | 0,6 | 0,9 | - | - | 54 | 367 | - | 23 | 0,5 | 0 | 0,06 | - | - | 22 | 0,12 | 0,03 | 2,4 | - | 0,44 | 25 | 0 |
| Pomme de terre poêlée, avec matière grasse | 574 | 137 | 574 | 137 | 69,9 | 2,5 | 2,5 | 17,3 | 5,7 | 2,6 | 0,51 | 18 | 370 | 0,15 | 0,73 | < 20 | 28 | 0,17 | 46 | 680 | < 20 | 204 | 0,35 | - | 0,37 | 2,66 | - | 1,97 | 0,043 | < 0,01 | 0,72 | 0,41 | 0,15 | 26 | - |
| Pomme de terre primeur, sans peau, bouillie/cuite à l'eau | 303 | 71,6 | 303 | 71,6 | 79,9 | 1,84 | 1,84 | 14,9 | < 0,1 | 2,13 | - | - | - | 0,17 | 0,7 | - | 19,7 | - | - | 291 | - | - | 0,34 | - | - | - | - | 18,9 | - | - | - | - | 0,21 | 53,8 | - |
| Pomme de terre sautée/ rissolée, pré-frites, surgelée, crue | 569 | 135 | 569 | 135 | 69,1 | 2,32 | 2,32 | 20,9 | 4,2 | 2,41 | 0,17 | - | - | - | - | - | - | - | - | - | - | 66 | - | - | - | - | - | 6 | - | - | - | - | - | - | - |
| Pomme de terre sautée/poêlée à la graisse de canard | 889 | 213 | 889 | 213 | - | 2,54 | 2,54 | 21,9 | 12,1 | 2,89 | 0,67 | - | - | - | - | - | - | - | - | - | - | 266 | - | - | - | - | - | - | - | - | - | - | - | - | - |
| Pomme de terre sautée/rissolée, pré-frite, surgelée, cuite | 654 | 156 | 654 | 156 | 63,3 | 2,81 | 2,81 | 23 | 4,8 | 4 | 0,19 | 11 | 117 | 0,15 | 0,63 | < 20 | 26 | 0,13 | 100 | 570 | < 20 | 77 | 0,42 | - | 1,28 | 2,29 | - | 1,25 | 0,086 | < 0,01 | 0,54 | 0,51 | 0,14 | 14 | - |
| Pomme de terre vapeur, sous vide | 312 | 73,6 | 312 | 73,6 | - | 1,7 | 1,7 | 15,4 | 0,2 | 1,62 | 0,24 | - | - | - | - | - | - | - | - | - | - | 96,2 | - | - | - | - | - | 14,7 | - | - | - | - | - | - | - |
| Pomme de terre, appertisée, égouttée | 253 | 59,9 | 253 | 59,9 | 84,3 | 1,47 | 1,47 | 11,7 | 0,34 | 2,2 | 0,15 | 25,4 | - | 0,051 | 1,26 | - | 9,4 | 0,071 | 28,8 | 229 | < 3,35 | 60,6 | 0,18 | 0 | - | - | - | 5,1 | 0,068 | 0,013 | 0,92 | 0,35 | 0,19 | 6 | 0 |
| Pomme de terre, bouillie/cuite à l'eau | 341 | 80,5 | 341 | 80,5 | 78,3 | 1,8 | 1,8 | 16,7 | 0,34 | 1,8 | 0,052 | 5,83 | 61 | 0,076 | 0,27 | < 5 | 17,3 | 0,074 | 37,2 | 363 | < 10 | 20,6 | 0,15 | 0 | 0,01 | 2,1 | - | 2,92 | 0,079 | < 0,04 | 1,73 | 0,57 | 0,34 | 31,1 | 0 |
| Pomme de terre, cuite (aliment moyen) | 393 | 93,2 | 393 | 93,2 | 76,4 | 2,01 | 2,01 | 17,2 | 1,37 | 1,96 | 0,12 | 9,62 | 150 | 0,095 | 0,43 | 6,78 | 20,7 | 0,12 | 43,7 | 450 | 7,07 | 48,8 | 0,24 | | 0,12 | 1,34 | | 5,05 | 0,07 | 0,013 | 1,08 | 0,46 | 0,24 | 18,4 | 0,03 |
| Pomme de terre, flocons déshydratés, au lait ou à la crème | 1570 | 371 | 1570 | 372 | 6,21 | 10,6 | 10,5 | 69,4 | 3,97 | 8 | 0,2 | 106 | 189 | 0,47 | 2,55 | 39 | 69,5 | 0,61 | 238 | 1510 | 0,5 | 78,1 | 1,3 | 0 | 0,03 | 8,7 | - | 22,3 | 0,19 | 0,24 | 6,35 | 1,31 | 1,03 | 30 | 0 |
| Pomme de terre, flocons déshydratés, nature | 1530 | 361 | 1530 | 361 | 5,09 | 8,04 | 8,04 | 76,1 | 0,46 | 10,3 | 0,55 | 48,5 | - | 0,27 | 2,26 | 0,5 | 70,7 | 0,45 | 223 | 1130 | 0,5 | 220 | 1,05 | 0 | 0,03 | 8,7 | - | 59 | 0,59 | 0,21 | 5,23 | 1,21 | 0,8 | 38 | 0 |
| Pomme de terre, purée (aliment moyen) | 385 | 91,8 | 385 | 91,8 | 78,7 | 2,11 | 2,11 | 12,7 | 3,25 | 1,58 | 0,35 | 35,4 | | 0,049 | 0,26 | | 15,5 | 0,086 | 51,7 | 238 | | 140 | 0,24 | 0,17 | 0,17 | | | 5,98 | 0,077 | 0,065 | 0,97 | 0,32 | 0,16 | 6,68 | 0,089 |
| Pomme de terre, purée à base de flocons, reconstituée avec lait demi-écrémé et eau, non salée | 284 | 67,3 | 284 | 67,3 | 82,8 | 2,63 | 2,63 | 12,1 | 0,53 | 1,8 | 0,23 | 36,5 | - | 0,062 | 0,39 | 5,15 | 15 | 0,082 | 57,3 | 270 | 0,96 | 91,5 | 0,31 | traces | 0,13 | - | - | 3,2 | 0,033 | 0,091 | 1,06 | 0,28 | 0,16 | 4,7 | 0,091 |
| Pomme de terre, purée à base de flocons, reconstituée avec lait entier, matière grasse | 409 | 97,9 | 409 | 97,9 | 77,8 | 1,92 | 1,92 | 10,2 | 5,18 | 1,47 | 0,56 | 38,7 | - | 0,025 | 0,19 | - | 16,3 | 0,11 | 52,3 | 184 | - | 223 | 0,2 | 0,3 | 0,33 | 3,9 | - | 8,63 | 0,11 | 0,062 | 0,77 | 0,2 | 0,089 | 7,33 | 0,11 |
| Pomme de terre, purée, avec lait et beurre, non salée | 374 | 88,8 | 374 | 88,9 | 77 | 1,89 | 1,87 | 14,2 | 2,4 | 1,5 | 0,11 | 31,1 | - | 0,061 | 0,21 | 5 | 15,2 | 0,066 | 45,5 | 260 | < 10 | 44,9 | 0,22 | 0,2 | 0,07 | 1,9 | - | 6,1 | 0,088 | 0,041 | 1,1 | 0,48 | 0,23 | 8 | 0,07 |
| Pomme de terre, rôtie/cuite au four | 379 | 89,4 | 379 | 89,4 | 74,9 | 2,5 | 2,5 | 18,5 | 0,13 | 2,2 | 0,025 | 15 | - | 0,12 | 1,08 | - | 28 | 0,22 | 70 | 535 | - | 10 | 0,36 | 0 | 0,04 | 2 | - | 9,6 | 0,064 | 0,048 | 1,41 | 0,38 | 0,31 | 28 | 0 |
| Pomme de terre, sans peau, crue | 341 | 80,5 | 341 | 80,5 | 79 | 2,16 | 2,16 | 16,2 | 0,18 | 1,8 | 0,016 | 14,3 | 50 | 0,079 | 0,91 | 1,2 | 22 | 0,11 | 56,2 | 418 | 22,9 | < 0,44 | 0,35 | 0 | 0,055 | 8,95 | - | 18,9 | 0,068 | 0,048 | 1,33 | 0,34 | 0,25 | 26 | 0 |
| Pomme de terre, sans peau, rôtie/cuite au four | 390 | 91,9 | 390 | 91,9 | 75,4 | 1,96 | 1,96 | 20,1 | 0,1 | 1,5 | 0,0036 | 9,48 | - | 0,094 | 0,35 | 3 | 21,7 | 0,15 | 50 | 391 | < 2,2 | 1,43 | 0,24 | 0 | 0,04 | 0,3 | - | 12,8 | 0,11 | 0,021 | 1,4 | 0,56 | 0,3 | 9 | 0 |
| Pomme d'eau ou pomme malaca, pulpe, crue, prélevée à la Martinique | - | - | - | - | 91,1 | 0,44 | 0,44 | - | 0,023 | - | - | - | - | - | - | - | - | - | - | - | - | - | - | - | - | - | - | 0,33 | - | - | - | - | - | - | - |
| Pomme Gala, pulpe, crue | 230 | 54,5 | 230 | 54,5 | 85,5 | < 0,5 | < 0,5 | 11,9 | < 0,5 | 1,9 | < 0,013 | 3,6 | < 20 | 0,03 | 0,06 | < 20 | 3 | 0,02 | 7,8 | 91 | < 20 | < 5 | < 0,05 | - | < 0,08 | < 0,8 | - | 1,33 | 0,02 | < 0,01 | < 0,1 | 0,1 | 0,035 | 6,98 | - |
| Pomme Golden, pulpe et peau, crue | 242 | 57,1 | 242 | 57,1 | 85,1 | < 0,5 | < 0,5 | 12,8 | < 0,5 | 1,4 | < 0,013 | 4 | < 20 | 0,04 | 0,07 | < 20 | 3,8 | 0,03 | 10 | 150 | < 20 | < 5 | 0,07 | < 0,25 | < 0,08 | < 0,8 | - | < 0,5 | < 0,015 | < 0,01 | < 0,1 | 0,096 | 0,046 | < 5 | - |
| Pomme Golden, pulpe, crue | 232 | 54,9 | 232 | 54,9 | 85,5 | < 0,5 | < 0,5 | 11,7 | < 0,5 | 2,5 | < 0,013 | 2,8 | < 20 | 0,03 | 0,06 | < 20 | 3,5 | 0,03 | 8,8 | 110 | < 20 | < 5 | < 0,05 | - | < 0,08 | < 0,8 | - | 1,29 | < 0,015 | < 0,01 | < 0,1 | 0,078 | 0,052 | 8,93 | - |
| Pomme Granny Smith, pulpe et peau, crue | 217 | 51,5 | 217 | 51,5 | 85,8 | < 0,5 | < 0,5 | 10,7 | < 0,5 | 2,8 | < 0,013 | 3,2 | < 20 | 0,04 | 0,06 | < 20 | 2,8 | 0,02 | 7,6 | 110 | < 20 | < 5 | < 0,05 | - | < 0,08 | < 0,8 | - | 2,31 | < 0,015 | < 0,01 | < 0,1 | 0,07 | 0,03 | 5,76 | - |
| Pomme Granny Smith, pulpe, crue | 227 | 53,9 | 227 | 53,9 | 85,3 | < 0,5 | < 0,5 | 11,4 | < 0,5 | 2,6 | < 0,013 | 3,2 | < 20 | 0,03 | 0,08 | < 20 | 2,9 | 0,02 | 7,6 | 110 | < 20 | < 5 | < 0,05 | - | < 0,08 | < 0,8 | - | 2,21 | < 0,015 | < 0,01 | < 0,1 | 0,061 | 0,029 | 7,16 | - |
| Pomme liane, pulpe, crue, prélevée à la Martinique | - | - | - | - | 82,6 | 0,88 | 0,88 | - | 0,22 | - | - | - | - | - | - | - | - | - | - | - | - | - | - | - | - | - | - | 41,4 | - | - | - | - | - | - | - |
| Pomme Pink lady, pulpe, crue | 254 | 60,1 | 254 | 60,1 | 83,6 | < 0,5 | < 0,5 | 12,8 | < 0,5 | 2,9 | < 0,013 | 2,8 | < 20 | 0,03 | 0,07 | < 20 | 3 | 0,02 | 7,5 | 110 | < 20 | < 5 | < 0,05 | - | < 0,08 | < 0,8 | - | 2,13 | < 0,015 | < 0,01 | < 0,1 | 0,093 | 0,058 | < 5 | - |
| Pomme, pulpe et peau, crue | - | - | - | - | 85,4 | 0,25 | 0,25 | 11,6 | 0,25 | 1,4 | 0,0037 | 5,34 | 61 | 0,041 | 0,099 | 0,2 | 6,47 | 0,036 | 14,4 | 119 | < 10 | 1,5 | 0,031 | 0 | 0,37 | 2,39 | - | 6,25 | 0,016 | 0,019 | 0,091 | 0,079 | 0,048 | 6 | 0 |
| Pomme, pulpe, bouillie/cuite à l'eau | - | - | - | - | 85,5 | 0,26 | 0,26 | 11,2 | 0,36 | 2,4 | 0,0025 | 5 | - | 0,035 | 0,19 | - | 3 | 0,12 | 8 | 88 | - | 1 | 0,04 | 0 | 0,05 | 0,6 | - | 0,2 | 0,016 | 0,012 | 0,095 | 0,046 | 0,044 | 1 | 0 |
| Pomme, pulpe, crue | - | - | - | - | 86,7 | 0,27 | 0,27 | 10,7 | 0,13 | 1,3 | 0 | 5 | - | 0,031 | 0,07 | 0,1 | 4 | 0,038 | 11 | 90 | 0,15 | 0 | 0,05 | 0 | 0,05 | 0,6 | - | 4 | 0,019 | 0,028 | 0,091 | 0,071 | 0,037 | 0 | 0 |
| Pomme, pulpe, rôtie/cuite au four | - | - | - | - | 75,6 | < 0,5 | < 0,5 | 21 | < 0,5 | 3 | < 0,01 | 4,7 | < 3,6 | 0,06 | 0,1 | < 20 | 5,2 | 0,05 | 12 | 150 | < 20 | < 5 | 0,06 | - | 0,47 | < 0,8 | - | 3,18 | < 0,01 | < 0,01 | 0,11 | 0,16 | 0,06 | 9,55 | - |
| Pomme, sèche | - | - | - | - | 31,8 | 0,78 | 0,78 | 57,2 | 0,31 | 8,7 | 0,22 | 22 | - | 0,12 | 1,4 | 0,8 | 10,5 | 0,15 | 38 | 450 | 0,5 | 87 | 0,2 | 0 | 0,53 | 3 | - | 3,9 | 0,065 | 0,13 | 0,76 | 0,27 | 0,13 | 0 | 0 |
| Potatoes ou Wedges ou Quartiers de pommes de terre épicés, surgelées, cuites | 732 | 174 | 732 | 174 | 59,9 | 3,25 | 3,25 | 25,3 | 5,8 | 3,4 | 0,76 | 17 | 499 | 0,13 | 0,7 | < 20 | 24 | 0,13 | 84 | 550 | < 20 | 302 | 0,43 | < 0,25 | 1,39 | 2,08 | - | 1,08 | 0,091 | < 0,01 | < 0,1 | 0,6 | 0,14 | 24,3 | - |
| Potimarron, pulpe, bouilli/cuit à l'eau | 161 | 38,2 | 161 | 38,2 | 88,9 | 1,06 | 1,06 | 6,88 | 0,3 | 1,8 | < 0,013 | 18 | 62 | 0,08 | 0,19 | < 20 | 8,5 | 0,06 | 27 | 330 | < 20 | < 5 | 0,24 | - | 1,52 | 3,17 | - | 8,65 | < 0,015 | < 0,01 | 0,74 | 0,29 | 0,06 | 9,28 | - |
| Potimarron, pulpe, cru | - | - | - | - | 94,6 | 0,63 | 0,63 | 3,1 | 0,1 | 1,1 | 0,074 | 18,5 | - | 0,029 | 0,27 | 0,15 | 7,13 | 0,038 | 33 | 243 | < 3,35 | 29,7 | 0,11 | 0 | 0,1 | - | - | 5 | 0,05 | 0,02 | 0,2 | 0,4 | 0,05 | 36 | 0 |
| Potimarron, pulpe, cuit à l'étouffée | 188 | 44,8 | 188 | 44,8 | 87,3 | 1,31 | 1,31 | 7,05 | 0,5 | 3,3 | < 0,013 | 18 | 67,5 | 0,08 | 0,24 | < 20 | 10 | 0,05 | 33 | 430 | < 20 | < 5 | 0,27 | < 0,25 | 1,25 | 1,56 | - | 10,9 | < 0,015 | < 0,01 | 0,6 | 0,27 | 0,15 | 24,7 | - |
| Potiron, appertisé, égoutté | - | - | - | - | 90 | 1,1 | 1,1 | 5,19 | 0,28 | 2,9 | 0,6 | 26 | - | 0,11 | 1,39 | 0,7 | 23 | 0,15 | 35 | 206 | - | 241 | 0,17 | 0 | 1,06 | 16 | - | 4,2 | 0,024 | 0,054 | 0,37 | 0,4 | 0,056 | 12 | 0 |
| Potiron, cru | - | - | - | - | 91,6 | 1 | 1 | 3,5 | 0,1 | 0,5 | 0,0025 | 21 | - | 0,13 | 0,8 | - | 12 | 0,13 | 44 | 340 | - | 1 | 0,32 | 0 | 1,06 | 1,1 | - | 9 | 0,05 | 0,11 | 0,6 | 0,3 | 0,061 | 16 | 0 |
| Potiron, cuit | - | - | - | - | 92 | 0,81 | 0,81 | 4,5 | 0,07 | 2 | < 0,01 | 13 | 21,2 | 0,04 | 0,18 | < 20 | 6,9 | 0,04 | 21 | 230 | < 20 | < 5 | 0,1 | < 0,25 | 0,24 | < 0,8 | - | 0,84 | 0,02 | < 0,01 | 0,22 | 0,14 | 0,058 | < 5 | 0 |
| Potiron, rôti/cuit au four | 127 | 30,2 | 127 | 30,2 | 90,7 | < 0,5 | < 0,5 | 5,66 | < 0,3 | 2 | < 0,013 | 31 | 45,4 | 0,05 | 0,11 | < 20 | 8,6 | 0,02 | 14 | 330 | < 20 | < 5 | 0,1 | - | 0,5 | < 0,8 | - | < 0,5 | < 0,015 | < 0,01 | 0,56 | 0,4 | 0,02 | 10 | - |
| Printanière de légumes, surgelée, crue (haricots verts, carottes, pomme de terre, petits pois, oignons) | - | - | - | - | 85,7 | 2,4 | 2,4 | 8,48 | 0,078 | 3,62 | 0,04 | - | - | - | - | - | - | - | - | - | - | 16,7 | - | - | - | - | - | 16 | - | - | - | - | - | - | - |
| Prune de cythère, mûre, fruit entier, crue, prélevée à la Martinique | - | - | - | - | 86,5 | 0,54 | 0,54 | - | 0,03 | - | - | - | - | - | - | - | - | - | - | - | - | - | - | - | - | - | - | 20 | - | - | - | - | - | - | - |
| Prune de cythère, verte ou immature, fruit entier, crue, prélevée à la Martinique | - | - | - | - | 89,6 | 0,6 | 0,6 | - | 0,07 | - | - | - | - | - | - | - | - | - | - | - | - | - | - | - | - | - | - | 23,9 | - | - | - | - | - | - | - |
| Prune Reine-Claude, crue | 301 | 71,3 | 301 | 71,3 | 80,1 | 0,94 | 0,94 | 16,4 | < 0,5 | 1,7 | < 0,013 | 13 | < 20 | 0,1 | 0,16 | < 20 | 9,5 | 0,08 | 20 | 250 | < 20 | < 5 | 0,12 | 0 | 1,2 | 0,8 | - | 4,16 | < 0,015 | < 0,01 | 0,23 | 0,25 | 0,049 | 18,4 | 0 |
| Prune, crue | - | - | - | - | 87 | 0,66 | 0,66 | 9,92 | 0,29 | 1,5 | 0,0075 | 7,29 | - | 0,06 | 0,16 | 0,4 | 6 | 0,076 | 17,8 | 149 | 0,11 | 3 | 0,096 | 0 | 0,33 | 6,4 | - | 7,25 | 0,024 | 0,026 | 0,41 | 0,14 | 0,037 | 4 | 0 |

| | | | | | | | | | | | | | | | | | | | | | | | | | | | | | | | | | | | |
|---|---|---|---|---|---|---|---|---|---|---|---|---|---|---|---|---|---|---|---|---|---|---|---|---|---|---|---|---|---|---|---|---|---|---|---|
| Pruneau, sec | 969 | 229 | 969 | 229 | 34,9 | 1,63 | 1,63 | 55,4 | 0,4 | 5,1 | < 0,013 | 50 | < 20 | 0,23 | 0,38 | 0,8 | 30 | 0,25 | 66 | 610 | < 20 | < 5 | 0,28 | < 0,25 | 0,41 | 12,8 | - | < 0,5 | 0,029 | 0,034 | 0,43 | 0,38 | 0,25 | < 5 | 0 |
| Purée de fruits, tout type de fruits, type "compote sans sucres ajoutés" | 249 | 58,9 | 249 | 58,9 | 84,2 | < 0,5 | < 0,5 | 13,4 | < 0,3 | 1,7 | < 0,013 | 6,2 | < 20 | 0,05 | < 0,05 | <20 | 6 | 0,06 | 11 | 140 | < 20 | < 5 | < 0,05 | < 0,25 | 0,4 | < 0,8 | - | 16,7 | < 0,015 | < 0,01 | 0,18 | 0,11 | < 0,01 | 8,24 | - |
| Purée de pommes, type "compote sans sucres ajoutés" | 239 | 56,4 | 239 | 56,4 | 85,1 | 1,13 | 1,13 | 11,7 | < 0,3 | 1,7 | < 0,013 | 4,3 | < 20 | 0,04 | 0,1 | - | 4,2 | 0,03 | 9,5 | 110 | < 20 | < 5 | < 0,05 | < 0,25 | 0,4 | < 0,8 | - | 9,86 | < 0,015 | < 0,01 | < 0,1 | 0,085 | 0,035 | 6,29 | - |
| Radis noir, cru | 123 | 29,2 | 123 | 29,2 | 91,4 | 0,94 | 0,94 | 5,52 | < 0,3 | 1,2 | 0,058 | 39 | 42,1 | 0,01 | 0,23 | <20 | 10 | 0,06 | 32 | 300 | < 20 | 23 | 0,15 | 0 | < 0,08 | < 0,8 | - | 9,58 | 0,021 | 0,016 | 0,23 | 0,14 | 0,077 | 17,3 | - |
| Radis rouge, cru | 60,7 | 14,5 | 60,7 | 14,5 | 95,5 | 0,94 | 0,94 | 1,53 | < 0,5 | 1,4 | 0,04 | 15 | 32,2 | < 0,01 | 0,4 | <20 | 6,2 | 0,04 | 20 | 250 | < 20 | 16 | 0,08 | 0 | < 0,08 | < 0,8 | - | 12,2 | < 0,015 | < 0,01 | < 0,1 | 0,09 | 0,063 | 95,4 | 0 |
| Raisin blanc, à gros grain (type Italia ou Dattier), cru | 311 | 73,4 | 311 | 73,4 | 80,9 | 0,75 | 0,75 | 16,6 | < 0,5 | 1 | < 0,013 | 12 | < 20 | 0,19 | 0,2 | <20 | 7,6 | 0,08 | 16 | 200 | < 20 | < 5 | < 0,05 | < 0,25 | 0,31 | 5,44 | - | 1,07 | 0,043 | < 0,01 | < 0,1 | 0,077 | 0,043 | 16,3 | 0 |
| Raisin Chasselas, cru | 335 | 79,1 | 335 | 79,1 | 79,4 | 0,75 | 0,75 | 16,9 | 0,5 | 2 | < 0,013 | 16 | < 20 | 0,08 | 0,14 | <20 | 7,5 | 0,07 | 21 | 150 | < 20 | < 5 | 0,07 | - | 0,35 | 2,18 | - | 4,14 | 0,031 | 0,056 | < 0,1 | 0,1 | 0,054 | 8,34 | - |
| Raisin noir Muscat, cru | 381 | 90,1 | 381 | 90,1 | 76,7 | 0,69 | 0,69 | 20 | < 0,5 | 2,7 | < 0,013 | 13 | < 20 | 0,12 | 0,22 | <20 | 7,3 | 0,09 | 20 | 210 | < 20 | < 5 | 0,06 | - | 0,99 | 5,7 | - | 3,11 | 0,057 | < 0,01 | < 0,1 | 0,09 | 0,034 | 11 | - |
| Raisin noir, cru | - | - | - | - | 84 | 0,63 | 0,63 | 15,6 | 0,4 | 1,4 | 0,0075 | 8,98 | - | 0,078 | 0,2 | 0,4 | 6,78 | 0,051 | 21,8 | 152 | < 2,97 | < 0,39 | 0,038 | 0 | 0,4 | - | - | 10,8 | 0,045 | 0,01 | 0,15 | 0,075 | 0,04 | 5 | 0 |
| Raisin, cru | - | - | - | - | 80,5 | 0,72 | 0,72 | 15,7 | 0,16 | 0,9 | 0,005 | 10 | - | 0,13 | 0,36 | 0,75 | 7 | 0,071 | 20 | 191 | 0,8 | 2 | 0,07 | 0 | 0,19 | 14,6 | - | 3,2 | 0,069 | 0,07 | 0,19 | 0,05 | 0,086 | 2 | 0 |
| Raisin, sec | 1360 | 321 | 1360 | 321 | 16 | 3 | 3 | 73,2 | 0,9 | 4,2 | 0,013 | 54 | < 20 | 0,38 | 1,7 | < 20 | 35 | 0,28 | 82 | 960 | < 20 | 5 | 0,24 | < 0,25 | 1,67 | 12,6 | - | < 0,5 | 0,075 | < 0,01 | 0,53 | 0,084 | 0,061 | 82 | 0,012 |
| Ramboutan, pulpe, crue | - | - | - | - | 78 | 0,65 | 0,65 | 20 | 0,21 | 0,9 | 0,028 | 22 | - | 0,066 | 0,35 | - | 7 | 0,34 | 9 | 42 | - | 11 | 0,08 | - | - | - | - | 4,9 | 0,013 | 0,022 | 1,35 | 0,018 | 0,02 | 8 | 0 |
| Rhubarbe, tige, crue | - | - | - | - | 92 | 0,78 | 0,78 | 1,47 | 0,23 | 2,47 | 0,0095 | 129 | - | 0,039 | 0,28 | 0,1 | 13,5 | 0,16 | 18,6 | 239 | 0,37 | 3 | 0,17 | 0 | 0,31 | 29,3 | - | 9,7 | 0,028 | 0,03 | 0,25 | 0,08 | 0,032 | 7,25 | 0 |
| Rhubarbe, tige, cuite, sucrée | - | - | - | - | 67,8 | 0,39 | 0,39 | 29,2 | 0,05 | 2 | 0,0025 | 145 | - | 0,027 | 0,21 | 0,3 | 12 | 0,073 | 8 | 96 | - | 1 | 0,08 | 0 | 0,19 | 21,1 | - | 3,3 | 0,018 | 0,023 | 0,2 | 0,05 | 0,02 | 5 | 0 |
| Roquette, crue | - | - | - | - | 91,7 | 2,58 | 2,58 | 2,1 | 0,66 | 1,6 | 0,068 | 160 | - | 0,076 | 1,46 | - | 47 | 0,32 | 52 | 369 | - | 27 | 0,47 | 0 | 0,43 | 109 | - | 15 | 0,044 | 0,086 | 0,31 | 0,44 | 0,073 | 97 | 0 |
| Rostis ou Galette de pomme de terre | 760 | 182 | 760 | 182 | 61,4 | 2,38 | 2,38 | 22,5 | 8,3 | 3,3 | 0,82 | 18 | 515 | 0,13 | 0,55 | < 20 | 22 | 0,11 | 70 | 460 | < 20 | 328 | 0,35 | < 0,25 | 2,92 | 2,31 | - | < 0,5 | 0,061 | < 0,01 | 0,68 | 0,44 | 0,12 | 9,13 | - |
| Rutabaga, cru | - | - | - | - | 89,5 | 1,17 | 1,17 | 5,7 | 0,13 | 2,6 | 0,028 | 45,2 | - | 0,031 | 0,35 | 0,5 | 17 | 0,14 | 47,8 | 349 | 0 | 11 | 0,21 | 0 | 0,3 | 0,3 | - | 33 | 0,068 | 0,04 | 1,25 | 0,14 | 0,11 | 24 | 0 |
| Rutabaga, cuit | - | - | - | - | 91,5 | 0,93 | 0,93 | 4,26 | 0,18 | 1,8 | 0,013 | 18 | - | 0,029 | 0,18 | - | 10 | 0,097 | 41 | 216 | - | 5 | 0,12 | 0 | 0,24 | 0,2 | - | 18,8 | 0,082 | 0,041 | 0,72 | 0,16 | 0,1 | 15 | 0 |
| Salade de fruits, crue | - | - | - | - | 78 | 0,5 | 0,5 | 11,4 | 0,15 | 1,23 | 0,0037 | 14,5 | - | 0,071 | 0,29 | 1,18 | 16,4 | 0,34 | 16,7 | 203 | 4,9 | 0,81 | 0,094 | 0 | 0,37 | - | - | 20 | 0,043 | 0,032 | 0,32 | 0,15 | 0,17 | 20,2 | 0 |
| Salade feuille de chêne, crue | 59,3 | 14,3 | 59,3 | 14,3 | 95,8 | 1,13 | 1,13 | 1,05 | < 0,5 | 1,8 | 0,023 | 33 | 72,1 | 0,04 | 0,82 | <20 | 12 | 0,19 | 20 | 280 | < 20 | 9 | 0,18 | - | < 0,08 | 34,8 | - | 2,16 | 0,047 | < 0,01 | 0,2 | 0,11 | 0,069 | 36,7 | - |
| Salade ou chicorée frisée, crue | - | - | - | - | 92,9 | 1,48 | 1,48 | 2,4 | 0,25 | 2,45 | 0,084 | 76 | - | 0,2 | 0,85 | 0,4 | 22,5 | 0,42 | 37,5 | 367 | 1 | 33,5 | 0,61 | 0 | 2,26 | 298 | - | 18,5 | 0,07 | 0,088 | 0,45 | 1,03 | 0,063 | 126 | 0 |
| Salade sucrine, crue | 75,5 | 17,9 | 75,5 | 17,9 | 94,8 | 1,13 | 1,13 | 2,58 | < 0,3 | 0,8 | < 0,013 | 23 | 51,2 | 0,05 | 0,27 | - | 9,6 | 0,12 | 30 | 230 | < 20 | < 5 | 0,18 | - | 0,11 | 9,88 | - | 2,32 | 0,048 | 0,027 | 0,16 | 0,17 | 0,068 | 57,2 | - |
| Salade verte, crue, sans assaisonnement | - | - | - | - | 96,1 | 1,01 | 1,01 | 1,5 | 0,1 | 1,3 | 0,046 | 35,3 | 61 | 0,061 | 0,64 | < 5 | 15,3 | 0,22 | 23,5 | 260 | 3,23 | 18,3 | 0,18 | < 0,5 | 0,17 | - | - | 1,95 | 0,041 | < 0,04 | 0,14 | < 0,16 | 0,052 | 55,3 | 0 |
| Salicorne (Salicornia sp.), fraîche | - | - | - | - | 92,2 | 0,67 | 0,67 | 1,1 | 0,24 | 2,46 | 2,56 | 34,3 | - | 0,07 | 4,89 | 0 | 75,3 | 0,7 | 19,5 | 119 | - | 1020 | 0,48 | - | 0,51 | - | - | - | - | - | - | - | - | - | - |
| Salsifis noir, ou scorsonère d'Espagne, cru | - | - | - | - | 77 | 3,13 | 3,13 | 15,3 | 0,3 | 3,3 | 0,011 | 24,1 | - | 0,12 | 0,7 | 0,2 | 32,1 | - | 94 | 387 | - | 4,5 | - | 0 | 0,8 | - | - | 3 | 0,15 | 0,05 | 0,3 | - | 0,07 | - | 0 |
| Salsifis, appertisé, égoutté | - | - | - | - | 91,5 | 1,25 | 1,25 | 4,25 | 0,15 | 2,5 | 0,61 | 21 | - | - | - | 0,2 | 8,9 | - | - | 173 | 0,6 | 235 | - | - | - | - | - | - | - | - | - | - | - | - | - |
| Salsifis, bouilli/cuit à l'eau | 243 | 57,9 | 243 | 57,9 | 82,8 | 2,63 | 2,63 | 9,29 | < 0,3 | 4,5 | 0,02 | 38 | 23,4 | 0,2 | 0,56 | <20 | 23 | 0,21 | 88 | 170 | < 20 | 8 | 0,49 | - | 1,9 | < 0,8 | - | 0,74 | 0,07 | < 0,01 | 0,11 | 0,16 | 0,042 | 39,6 | - |
| Salsifis, cuit | - | - | - | - | - | 2,73 | 2,73 | 12,3 | 0,17 | 3,1 | 0,04 | 47 | - | 0,07 | 0,55 | 0,2 | 18 | 0,21 | 56 | 283 | - | 16 | 0,3 | 0 | 0,19 | 0,3 | - | 4,6 | 0,056 | 0,17 | 0,39 | 0,28 | 0,22 | 15 | 0 |
| Salsifis, surgelé, cru | - | - | - | - | 85,8 | 2,65 | 2,65 | 4,15 | 0,15 | 5,8 | 0,043 | - | - | - | - | - | - | - | - | - | - | 16 | - | - | - | - | - | - | - | - | - | - | - | - | - |
| Scarole, crue | - | - | - | - | 95,8 | 1,25 | 1,25 | 0,3 | 0,2 | 3,1 | 0,055 | 52 | - | 0,099 | 0,83 | 0,4 | 15 | 0,42 | 28 | 314 | 0,6 | 22 | 0,79 | traces | traces | - | - | 6,5 | 0,08 | 0,075 | 0,4 | 0,9 | 0,02 | 142 | 0 |
| Sésame, graine | - | - | - | - | 3,48 | 17,6 | 20,8 | 9,85 | 49,7 | 14,9 | < 0,1 | 962 | - | 1,58 | 14,6 | < 5 | 324 | 1,24 | 605 | 468 | 26,5 | 2,31 | 5,74 | 0 | 0,25 | 0 | - | 0 | 0,79 | 0,25 | 4,52 | 0,05 | 0,79 | 97 | 0 |
| Sésame, graine décortiquée | - | - | - | - | 3,58 | 21,1 | 24,9 | 4,5 | 56,1 | 11,9 | 0,073 | 62,4 | - | 1,4 | 6,36 | 0,2 | 348 | 1,95 | 703 | 419 | 24,4 | 29 | 7,24 | 0 | 1,68 | 0 | - | 0 | 0,75 | 0,17 | 5,16 | 0,17 | 0,6 | 106 | 0 |
| Sésame, grillé, graine décortiquée | - | - | - | - | 5 | 17 | 20 | 9,14 | 48 | 16,9 | 0,098 | 131 | - | 1,46 | 7,78 | - | 346 | 1,43 | 774 | 406 | - | 39 | 10,2 | 0 | 0,25 | 0 | - | 0 | 1,21 | 0,47 | 5,44 | 0,68 | 0,15 | 96 | 0 |
| Soja, graine entière | 1810 | 432 | 1750 | 419 | 7,77 | 34,5 | 37,8 | 20,8 | 19,2 | 13 | 0,0075 | 220 | - | 0,88 | 15,7 | 0,6 | 253 | 2,26 | 586 | 1740 | 8,8 | 3 | 2,95 | 0 | 0,85 | 47 | - | 6 | 0,87 | 0,87 | 1,62 | 1,36 | 0,4 | 328 | 0 |
| Spécialité de fruits divers, sucrée (mélange pulpes et/ou purées de fruits, mais toujours avec autre ingrédient) | - | - | - | - | - | 0,41 | 0,41 | 16,9 | 0,56 | 1,53 | 0,0084 | 5 | < 3,6 | 0,03 | 0,18 | < 20 | 4,9 | 0,04 | 12 | 120 | < 20 | 3,33 | < 0,05 | - | 1,67 | < 0,8 | - | 10,5 | 0,23 | 0,27 | 3 | 0,095 | 0,33 | < 5 | - |
| Sureau, baie, crue | - | - | - | - | 79,8 | 0,64 | 0,64 | 11,4 | 0,5 | 7 | 0,018 | 42,7 | - | 0,062 | 1,25 | 2,5 | 18,5 | - | 47,4 | 301 | 0,6 | 7 | 0,16 | 0 | 1 | - | - | 32,5 | 0,07 | 0,06 | 1,1 | 0,14 | 0,21 | 9 | 0 |
| Tahin ou Purée de sésame | - | - | - | - | 3,03 | 17,7 | 20,3 | 13,8 | 53,4 | 7 | 0,19 | 284 | - | 1,61 | 6,69 | - | 95 | 1,46 | 761 | 437 | - | 75 | 4,62 | 0 | 0,25 | 0 | - | 4,2 | 1,41 | 0,3 | 5,55 | 0,69 | 0,15 | 98 | 0 |
| Tamarin, fruit mûr, pulpe, cru | - | - | - | - | 31,4 | 2,65 | 2,65 | 57,4 | 0,6 | 5,1 | 0,07 | 74 | - | 0,086 | 2,8 | - | 92 | - | 113 | 628 | - | 28 | 0,1 | 0 | 0,1 | 2,8 | - | 3,5 | 0,43 | 0,15 | 1,94 | 0,14 | 0,066 | 14 | 0 |
| Tapioca ou Perles du Japon, cru | 1500 | 354 | 1500 | 354 | 11 | 0,19 | 0,19 | 87,8 | 0,02 | 0,9 | 0,0025 | 20 | - | 0,02 | 1,58 | 2,7 | 1 | 0,11 | 7 | 11 | 0,3 | 1 | 0,12 | 0 | 0 | 0 | - | 0 | 0,004 | 0 | 0 | 0,14 | 0,008 | 4 | 0 |
| Taro, tubercule, cru | 453 | 107 | 453 | 107 | 70,6 | 1,5 | 1,5 | 22,8 | 0,2 | 4,1 | 0,028 | 43 | - | 0,17 | 0,55 | - | 33 | 0,38 | 84 | 591 | - | 11 | 0,23 | 0 | 2,38 | 1 | - | 4,5 | 0,095 | 0,025 | 0,6 | 0,3 | 0,28 | 22 | 0 |
| Taro, tubercule, cuit | 555 | 131 | 555 | 131 | 63,8 | 0,52 | 0,52 | 29,5 | 0,11 | 5,1 | 0,038 | 18 | - | 0,2 | 0,72 | - | 30 | 0,45 | 76 | 484 | - | 15 | 0,27 | 0 | 2,93 | 1,2 | - | 5 | 0,11 | 0,028 | 0,51 | 0,34 | 0,33 | 19 | 0 |
| Tétragone cornue, cuite | - | - | - | - | 94,8 | 1,3 | 1,3 | 1 | 0,17 | 1,4 | 0,27 | 48 | - | 0,077 | 0,66 | - | 32 | 0,53 | 22 | 102 | - | 107 | 0,31 | 0 | 1,23 | 292 | - | 16 | 0,03 | 0,11 | 0,39 | 0,26 | 0,24 | 8 | 0 |
| Ti nain, pulpe, cuit à la vapeur, prélevé à la Martinique | - | - | - | - | 72,1 | 1,13 | 1,13 | - | 0,05 | - | - | - | - | - | - | - | - | - | - | - | - | - | - | - | - | - | - | 9,23 | - | - | - | - | - | - | - |
| Tomate cerise, crue | 142 | 33,7 | 142 | 33,7 | 90,1 | 1,31 | 1,31 | 5,62 | < 0,5 | 1,2 | < 0,013 | 6,6 | 52,3 | 0,05 | 0,37 | < 20 | 11 | 0,16 | 31 | 330 | < 20 | < 5 | 0,18 | 0 | 0,4 | 2,83 | - | 21,8 | 0,048 | 0,018 | 0,92 | 0,079 | 0,096 | 17,6 | 0 |
| Tomate côtelée ou coeur de boeuf, crue | 83,4 | 19,8 | 83,4 | 19,8 | 94,4 | 0,5 | 0,5 | 3,23 | < 0,5 | 0,9 | < 0,013 | 7,2 | 41,4 | 0,03 | 0,14 | < 20 | 6,9 | 0,12 | 19 | 210 | < 20 | < 5 | 0,09 | - | 0,36 | 0,82 | - | 11,7 | 0,035 | < 0,01 | 0,44 | 0,06 | 0,057 | 65,4 | - |
| Tomate grappe, crue | 84,3 | 20,1 | 84,3 | 20,1 | 94,6 | 0,5 | 0,5 | 3,03 | < 0,5 | 1,5 | < 0,013 | 6,6 | 31,9 | 0,02 | 0,18 | < 20 | 6,5 | 0,12 | 20 | 200 | < 20 | < 5 | 0,08 | - | 0,45 | 0,97 | - | 18,2 | 0,036 | < 0,01 | 0,47 | 0,06 | 0,07 | 15 | - |
| Tomate ronde, crue | 85,6 | 20,3 | 85,6 | 20,3 | 94,1 | 0,5 | 0,5 | 3,59 | < 0,5 | 1 | < 0,013 | 6,8 | 34,3 | 0,02 | 0,17 | < 20 | 7 | 0,1 | 20 | 210 | < 20 | < 5 | 0,08 | - | 0,38 | 1,13 | - | 16,3 | - | < 0,01 | 0,45 | 0,061 | 0,07 | 13,5 | - |
| Tomate verte, crue | - | - | - | - | 93 | 1,2 | 1,2 | 4 | 0,2 | 1,1 | 0,033 | 13 | - | 0,09 | 0,51 | - | 10 | 0,1 | 28 | 204 | - | 13 | 0,07 | 0 | 0,38 | 10,1 | - | 23,4 | 0,06 | 0,04 | 0,5 | 0,5 | 0,081 | 9 | 0 |
| Tomate, concentré, appertisé | 419 | 99,2 | 419 | 99,2 | 72 | 4,4 | 4,4 | 17,1 | 0,53 | 4,2 | 0,27 | 45,6 | traces | < 0,1 | 1,2 | 4,4 | 63,3 | < 0,1 | 103 | 1010 | 0,4 | 108 | < 0,1 | 0 | 5,7 | 11,4 | - | 2,6 | 0,08 | 0,1 | 2,2 | 0,2 | 0,09 | 58 | 0 |
| Tomate, coulis, appertisé (purée de tomates mi-réduite à 11%) | - | - | - | - | 88,2 | 2,05 | 2,05 | 8,53 | 0,2 | 2,6 | 0,88 | - | - | - | - | - | - | - | - | - | - | 350 | - | - | - | - | - | - | - | - | - | - | - | - | - |
| Tomate, crue | 81,1 | 19,3 | 81,1 | 19,3 | 94,1 | 0,86 | 0,86 | 2,49 | 0,26 | 1,2 | 0,0081 | 8,14 | 51 | 0,029 | 0,12 | 0,2 | 10,1 | 0,066 | 26,6 | 256 | < 10 | 3,22 | 0,087 | 0 | 0,66 | 7,9 | - | 15,5 | 0,039 | 0,019 | 0,65 | 0,21 | 0,082 | 22,7 | 0 |
| Tomate, double concentré, appertisé | 392 | 92,8 | 392 | 92,8 | 74,3 | 3,73 | 3,73 | 17 | 0,29 | 3,62 | 0,93 | - | - | - | - | - | - | - | - | - | - | 329 | - | - | - | - | - | - | - | - | - | - | - | - | - |
| Tomate, entière, crue, prélevée à la Martinique | - | - | - | - | 94,1 | 1,06 | 1,06 | - | 0,06 | - | - | - | - | - | - | - | - | - | - | - | - | - | - | - | - | - | - | 15,2 | - | - | - | - | - | - | - |
| Tomate, pelée, appertisée, égouttée | 76,9 | 18,3 | 76,9 | 18,3 | 94,4 | 1,07 | 1,07 | 1,89 | 0,28 | 1,27 | 0,28 | 21,1 | - | 0,06 | 0,77 | 0,2 | 9,2 | 0,073 | 18 | 190 | < 2,2 | 112 | 0,11 | 0 | 0,67 | 2,6 | - | 12 | 0,31 | 0,055 | 0,64 | 0,17 | 0,11 | 16 | 0 |
| Tomate, pulpe et peau, bouillie/cuite à l'eau | - | - | - | - | 94,3 | 0,95 | 0,95 | 2,8 | 0,11 | 0,7 | 0,028 | 11 | - | 0,075 | 0,68 | 1,5 | 9 | 0,11 | 28 | 218 | 0,18 | 11 | 0,14 | 0 | 0,56 | 2,8 | - | 22,8 | 0,036 | 0,022 | 0,53 | 0,13 | 0,079 | 13 | 0 |
| Tomate, pulpe et peau, rôtie/cuite au four | 130 | 31,1 | 130 | 31,1 | 92 | 0,88 | 0,88 | 4,13 | 0,6 | 2,2 | < 0,013 | 10 | 58,2 | 0,05 | 0,25 | < 20 | 13 | 0,23 | 30 | 300 | < 20 | < 5 | 0,15 | < 0,25 | 0,98 | 1,61 | - | 14,5 | < 0,015 | < 0,01 | 0,45 | 0,14 | 0,1 | 30 | - |
| Tomate, pulpe, appertisée | 110 | 26,1 | 110 | 26,1 | 93 | 1,2 | 1,2 | 3,63 | < 0,5 | 1,8 | 0,068 | 20 | 117 | 0,08 | 0,39 | < 20 | 11 | 0,08 | 20 | 260 | < 20 | 27,1 | 0,09 | < 0,25 | 1,3 | < 0,8 | - | 10,5 | 0,036 | 0,023 | 0,65 | 0,054 | 0,083 | 30,2 | 0 |
| Tomate, purée, appertisée | 199 | 47,1 | 199 | 47,1 | 85,6 | 1,4 | 1,4 | 7,54 | 0,4 | 2,38 | 0,51 | 19,9 | 283 | 0,15 | 0,6 | 2,2 | 30 | 0,16 | 41,9 | 798 | 0,4 | 239 | 0,2 | 0 | 2,3 | - | - | 9 | 0,1 | 0,05 | 1,3 | 0,25 | 0,18 | 72 | 0 |

| | | | | | | | | | | | | | | | | | | | | | | | | | | | | | | | | | | | |
|---|---|---|---|---|---|---|---|---|---|---|---|---|---|---|---|---|---|---|---|---|---|---|---|---|---|---|---|---|---|---|---|---|---|---|---|
| Tomate, rôtie/cuite au four | 106 | 25,3 | 106 | 25,3 | 92,6 | 1 | 1 | 3,53 | 0,4 | 1,9 | < 0,013 | 12 | 37 | 0,03 | 0,22 | < 20 | 9,2 | 0,18 | 29 | 310 | < 20 | < 5 | 0,11 | - | 1,09 | 1,12 | - | 14,9 | < 0,015 | < 0,01 | 0,57 | 0,089 | 0,066 | 20,1 | - |
| Tomate, séchée | - | - | - | - | 14,6 | 14,2 | 14,2 | 43,3 | 2,99 | 12,3 | 0,62 | 110 | - | 1,42 | 9,09 | - | 194 | 1,85 | 356 | 3430 | 5,5 | 247 | 1,99 | 0 | 0,01 | 43 | - | 39,2 | 0,53 | 0,49 | 9,05 | 2,09 | 0,33 | 68 | 0 |
| Tomate, séchée, à l'huile | 779 | 187 | 779 | 187 | 60,4 | 4,27 | 4,27 | 13,4 | 11,7 | 5,63 | 2,21 | 47 | - | 0,47 | 2,68 | - | 81 | 0,47 | 139 | 1570 | - | 1020 | 0,78 | 0 | - | - | - | 102 | 0,19 | 0,38 | 3,63 | 0,48 | 0,32 | 23 | 0 |
| Topinambour, cru | 257 | 60,7 | 257 | 60,7 | 80,1 | 1,94 | 1,94 | 11,5 | 0,31 | 2,1 | 0,0088 | 21 | - | 0,13 | 2 | 0,1 | 16,5 | 0,06 | 75,2 | 495 | 0,03 | 3,5 | 0,11 | 0 | 0,17 | 0,1 | - | 5 | 0,14 | 0,058 | 1,3 | 0,39 | 0,084 | 24,5 | 0 |
| Topinambour, cuit | 346 | 81,9 | 346 | 81,9 | 80,2 | 1,8 | 1,8 | 16 | 0,7 | 2,2 | 0,09 | 32,9 | - | < 0,1 | 0,2 | < 10 | 14,1 | < 0,1 | 64 | 452 | - | 35,9 | 0,34 | 0 | 0,19 | 0,1 | - | 3,2 | < 0,05 | 0,03 | 0,05 | 0,38 | 0,04 | 44,3 | 0 |
| Topinambour, pulpe, cuit à la vapeur, prélevé à la Martinique | - | - | - | - | 82,7 | 1,23 | 1,23 | - | 0,18 | - | - | - | - | - | - | - | - | - | - | - | - | - | - | - | - | - | - | 9,57 | - | - | - | - | - | - | - |
| Tournesol, graine | - | - | - | - | 3,38 | 21,3 | 25,1 | 10,1 | 55,5 | 6,4 | 0,014 | 86,5 | - | 1,5 | 4,9 | 5 | 364 | 1,95 | 477 | 578 | 8,41 | 5,5 | 3,8 | 0 | 42,3 | 0 | - | 1,4 | 1,98 | 0,16 | 4,8 | 0,83 | 1,24 | 254 | 0 |
| Tournesol, graine, grillé, salé | - | - | - | - | 1,2 | 19,3 | 22,8 | 15,1 | 49,8 | 9 | 1,64 | 70 | - | 1,83 | 3,8 | - | 129 | 2,11 | 1160 | 850 | - | 655 | 5,29 | 0 | 26,1 | 2,7 | - | 1,4 | 0,11 | 0,25 | 7,04 | 7,04 | 0,8 | 237 | 0 |

# Appendix 5

| eaux et boissons | Energie, Règlement UE N° 1169/2011 (kJ/100 g) | Energie, Règlement UE N° 1169/2011 (kcal/100 g) | Energie, N x facteur Jones, avec fibres (kJ/100 g) | Energie, N x facteur Jones, avec fibres (kcal/100 g) | Eau (g/100 g) | Protéines, N x facteur de Jones (g/100 g) | Protéines, N x 6.25 (g/100 g) | Glucides (g/100 g) | Lipides (g/100 g) | Fibres alimentaires (g/100 g) | Sel chlorure de sodium (g/100 g) | Calcium (mg/100 g) | Chlorure (mg/100 g) | Cuivre (mg/100 g) | Fer (mg/100 g) | Iode (µg/100 g) | Magnésium (mg/100 g) | Manganèse (mg/100 g) | Phosphore (mg/100 g) | Potassium (mg/100 g) | Sélénium (µg/100 g) | Sodium (mg/100 g) | Zinc (mg/100 g) | Vitamine D (µg/100 g) | Vitamine E (mg/100 g) | Vitamine K1 (µg/100 g) | Vitamine K2 (µg/100 g) | Vitamine C (mg/100 g) | Vitamine B1 ou Thiamine (mg/100 g) | Vitamine B2 ou Riboflavine (mg/100 g) | Vitamine B3 ou PP ou Niacine (mg/100 g) | Vitamine B5 ou Acide pantothénique (mg/100 g) | Vitamine B6 (mg/100 g) | Vitamine B9 ou Folates totaux (µg/100 g) | Vitamine B12 (µg/100 g) |
|---|---|---|---|---|---|---|---|---|---|---|---|---|---|---|---|---|---|---|---|---|---|---|---|---|---|---|---|---|---|---|---|---|---|---|---|
| Boisson à base d'avoine, nature, préemballée | 179 | 42,6 | 179 | 42,5 | 90,5 | < 0,46 | < 0,5 | 7,8 | 1,1 | < 0,5 | 0,066 | 1 | 50 | 0,01 | 0,02 | < 20 | 2,3 | 0,04 | 11 | 32 | < 20 | 26,4 | < 0,05 | < 0,25 | 0,37 | < 0,8 | - | < 0,5 | 0,029 | < 0,01 | < 0,1 | 0,079 | < 0,01 | 6,39 | - |
| Boisson à la châtaigne, nature, préemballée | 291 | 68,9 | 290 | 68,8 | 83,9 | < 0,42 | < 0,5 | 13,5 | 1,4 | 0,61 | 0,077 | 77 | 52,9 | 0,02 | 0,22 | < 20 | 9,3 | 0,17 | 11 | 42 | 20 | 30,6 | 0,12 | < 0,25 | 0,51 | < 0,8 | - | < 0,5 | 0,016 | < 0,01 | 0,25 | 0,16 | < 0,01 | < 5 | - |
| Boisson à la noix de coco, nature, préemballée | 131 | 31,4 | 130 | 31,3 | 94,6 | < 0,42 | < 0,5 | 2,75 | 2,1 | < 0,5 | 0,089 | 2,7 | 55 | 0,02 | < 0,05 | < 20 | 2,1 | 0,03 | 3,9 | 14 | < 20 | 35,6 | < 0,05 | < 0,25 | < 0,08 | < 0,8 | - | < 0,5 | < 0,015 | < 0,01 | < 0,1 | < 0,007 | < 0,01 | < 5 | - |
| Boisson à l'amande, nature, non sucrée, non enrichie, préemballée | 150 | 36,3 | 150 | 36,3 | 94,6 | 1,06 | 1,06 | 0,68 | 3,2 | < 0,5 | 0,051 | 12 | - | 0,05 | 0,1 | < 20 | 8,3 | 0,05 | 16 | 31 | < 20 | 20,5 | 0,11 | < 0,25 | 0,96 | < 0,8 | - | < 0,5 | < 0,015 | 0,016 | 0,16 | 0,028 | < 0,01 | < 5 | - |
| Boisson à l'amande, sucrée, enrichie en calcium, préemballée | 185 | 44,4 | 185 | 44,4 | 92 | 0,69 | 0,69 | 3,96 | 2,8 | < 0,5 | 0,095 | 54 | 36,2 | 0,04 | 0,35 | < 20 | 12 | 0,05 | 20 | 32 | < 20 | 38,1 | 0,12 | < 0,25 | 0,67 | < 0,8 | - | < 0,5 | < 0,015 | 0,012 | 0,11 | 0,02 | < 0,01 | < 5 | - |
| Boisson à l'eau minérale ou de source, aromatisée, non sucrée, avec édulcorants | - | - | - | - | - | 0,086 | 0,086 | 0,28 | 0,086 | 0,067 | 0,012 | - | - | - | - | - | - | - | - | - | - | 4,67 | - | - | - | - | - | - | - | - | - | - | - | - | - |
| Boisson à l'eau minérale ou de source, aromatisée, non sucrée, sans édulcorant | - | - | - | - | - | 0,06 | 0,06 | 0,14 | 0 | 0 | 0,019 | - | - | - | - | - | 5 | - | - | 3 | - | 7,17 | - | - | - | - | - | - | - | - | - | - | - | - | - |
| Boisson à l'eau minérale ou de source, aromatisée, sucrée | - | - | - | - | - | 0,086 | 0,086 | 3,48 | 0,086 | 0,044 | 0,012 | 11,7 | - | 0,005 | traces | 1 | 2,3 | traces | 0,001 | 0,33 | 0,5 | 4,74 | traces | 0 | 0 | - | - | 0 | 0 | 0 | 0 | 0 | 0 | 0 | 0 |
| Boisson au jus de fruit et au lait | 201 | 47,5 | 201 | 47,5 | 87,5 | 0,56 | 0,56 | 9,63 | 0,3 | < 3 | 0,03 | 57,4 | < 60 | < 0,1 | 0,2 | 1 | 8,1 | 0,1 | 19 | 68,4 | - | 12 | 0,1 | < 0,5 | 0,23 | - | - | 9 | < 0,01 | 0,035 | 0,11 | 0,085 | < 0,04 | 9,2 | < 0,08 |
| Boisson au riz, nature, préemballée | 227 | 53,7 | 226 | 53,6 | 87,6 | < 0,46 | < 0,5 | 10,8 | 1 | < 0,5 | 0,074 | 5 | 51 | < 0,01 | 0,01 | < 20 | 3,3 | 0,03 | 10 | 16 | < 20 | 29,5 | < 0,05 | < 0,25 | 0,48 | < 0,8 | - | < 0,5 | < 0,015 | < 0,01 | 0,18 | 0,15 | < 0,01 | < 5 | - |
| Boisson au soja et jus de fruits concentrés, préemballée | 238 | 56,2 | 238 | 56,2 | - | 2,3 | 2,3 | 9,5 | 1 | traces | 0,22 | - | - | - | - | - | - | - | - | - | - | 87 | - | - | - | - | - | - | - | - | - | - | - | - | - |
| Boisson au soja, aromatisée, sucrée, enrichie en calcium, préemballée | 179 | 42,8 | 174 | 41,6 | 90,6 | 3,01 | 3,3 | 2,89 | 1,8 | 1,1 | 0,12 | 53 | 32 | 0,1 | 0,4 | < 20 | 21 | 0,1 | 48 | 137 | < 20 | 47,9 | 0,2 | < 0,25 | 0,16 | 2,52 | - | < 0,5 | < 0,015 | 0,055 | 0,12 | 0,11 | 0,064 | 7,31 | - |
| Boisson au soja, aromatisée, sucrée, non enrichie, préemballée | 254 | 60,5 | 249 | 59,3 | 86,8 | 2,96 | 3,25 | 7,24 | 1,92 | 0,6 | 0,12 | 12 | 48,8 | 0,11 | 0,51 | < 20 | 18 | 0,18 | 49 | 140 | < 50 | 48,7 | 0,3 | < 0,5 | 0,11 | 3,53 | - | 0 | 0,025 | 0,01 | < 0,1 | 0,068 | 0,021 | 5,4 | 0,014 |
| Boisson au soja, nature, enrichie en calcium, préemballée | 185 | 44,2 | 179 | 42,9 | 90,8 | 3,42 | 3,75 | 2,18 | 2,05 | 1 | 0,097 | 98 | 26 | 0,09 | 0,45 | < 20 | 13 | 0,17 | 81 | 86 | < 50 | 38,8 | 0,24 | 1,16 | - | 3,55 | - | - | 0,025 | 0,01 | < 0,1 | 0,033 | 0,022 | 22,3 | - |
| Boisson au soja, nature, non enrichie, préemballée | 155 | 37,1 | 149 | 35,8 | 93 | 3,31 | 3,63 | 0,7 | 2,07 | 0,6 | 0,061 | 12 | 7 | 0,11 | 0,41 | < 20 | 16 | 0,19 | 50 | 110 | < 50 | 24,3 | 0,29 | < 0,5 | 0,11 | 3,81 | - | 1 | 0,025 | 0,01 | 0,18 | 0,058 | 0,034 | 26,1 | 0 |
| Boisson au thé, aromatisée, à teneur réduite en sucres | - | - | - | - | 94,5 | 0 | 0 | - | 0 | 0 | - | - | - | - | - | - | - | - | - | - | - | - | - | - | - | - | - | - | - | - | - | - | - | - | - |
| Boisson au thé, aromatisée, non sucrée, avec édulcorants | - | - | - | - | 99,5 | 0,014 | 0,014 | 0,34 | 0 | 0 | 0,099 | 0 | - | 0 | 0 | 0 | 2,3 | 0 | 0 | 0 | 0 | 40,3 | 0 | 0 | 0 | - | - | 0 | 0 | 0 | 0 | 0 | 0 | 0 | 0 |
| Boisson au thé, aromatisée, sucrée | 116 | 27,3 | 116 | 27,3 | 93,1 | 0,033 | 0,033 | 6,76 | < 0,022 | 0 | 0,058 | 2 | - | 0,0027 | 0,0099 | 0,4 | 1,2 | 0,072 | 0 | 12,9 | < 10 | 22,8 | 0,015 | 0 | 0 | - | - | 0 | 0 | 0 | 0 | 0 | 0 | 0 | 0 |
| Boisson au thé, aromatisée, sucrée, avec édulcorants | - | - | - | - | - | 0 | 0 | 5,3 | 0 | 0 | traces | - | - | - | - | - | - | - | - | - | - | traces | - | - | - | - | - | - | - | - | - | - | - | - | - |
| Boisson au thé, aromatisée, teneur en sucre et édulcorant inconnue (aliment moyen) | 116 | 27,3 | 116 | 27,3 | 93,8 | 0,028 | 0,028 | 6,11 | 0,0092 | 0 | 0,061 | 1,8 | | 0,0024 | 0,0089 | 0,36 | 1,31 | 0,065 | 0 | 11,6 | 4,5 | 24,5 | 0,013 | 0 | 0 | | | 0 | 0 | 0 | 0 | 0 | 0 | 0 | 0 |
| Boisson cacaotée ou au chocolat, instantanée, sucrée, enrichie en vitamines, prête à boire (reconstituée avec du lait demi-écrémé standard) | 303 | 71,9 | 303 | 71,9 | 83,2 | 4,08 | 4,08 | 9,91 | 1,71 | < 0,5 | - | - | - | - | - | | - | - | - | - | - | - | - | - | - | - | - | - | - | - | - | - | - | - | 0,07 |
| Boisson cacaotée ou au chocolat, instantanée, sucrée, prête à boire (reconstituée avec du lait demi-écrémé standard) | 292 | 69,3 | 292 | 69,3 | 84,7 | 3,78 | 3,78 | 9,5 | 1,42 | 1,7 | 0,12 | 97,1 | - | 0,11 | 0,87 | 24,3 | 27,5 | 0,1 | - | 239 | < 10 | 46,2 | 0,5 | 0,083 | 0,046 | - | - | 0 | 0,037 | 0,16 | 0,15 | 0,34 | 0,03 | 3,2 | 0,085 |
| Boisson concentrée à diluer, sans sucres ajoutés, avec édulcorants, type "sirop 0%" | 92 | 21,7 | 92 | 21,7 | 93,8 | < 0,5 | < 0,5 | 2,76 | < 0,5 | 0,7 | 0,028 | 6,6 | < 1 | < 0,01 | 0,04 | < 20 | 2 | 0,02 | 2,2 | 57 | < 20 | 11 | < 0,05 | - | < 0,08 | < 0,8 | - | < 0,5 | 0,027 | < 0,01 | 0,49 | 0,17 | < 0,01 | < 5 | - |
| Boisson énergisante, non sucrée, avec édulcorants | - | - | - | - | - | 0,073 | 0,073 | 0,95 | 0,1 | 0 | 0,17 | - | - | - | - | - | - | - | - | - | - | 68,3 | - | - | - | - | - | - | - | 0,65 | 8,2 | 2 | 1,64 | - | - |
| Boisson énergisante, sucrée | - | - | - | - | 88 | 0,2 | 0,2 | 11,9 | 0 | 0,1 | 0,11 | 0 | - | 0 | 0 | 0 | 0 | 0 | 0 | 0 | 0 | 46,2 | 0 | 0 | 1,8 | - | - | 40 | 0,21 | 0,52 | 6,08 | 1,76 | 1,05 | 0 | 0,63 |
| Boisson gazeuse à la pomme (de 50 à 99% de fruits), non sucrée | - | - | - | - | - | 0 | 0 | 7,5 | 0 | 0 | 0,00025 | 2,92 | - | 0,01 | 0,1 | 0,7 | 1,94 | - | 3 | 60,3 | 0,05 | 0,1 | 0,01 | 0 | 0 | - | - | 0 | 0,01 | 0,006 | 0,01 | 0,02 | 0,01 | 4 | 0 |
| Boisson gazeuse aux fruits (à moins de 10% de jus), non sucrée, avec édulcorants | - | - | - | - | 99,2 | 0,063 | 0,063 | 0,45 | 0,13 | traces | 0,022 | 0 | - | 0 | 0 | 0 | 0 | 0 | 0 | 8,3 | 0 | 10 | 0 | 0 | 0 | - | - | 0 | 0 | 0 | 0 | 0 | 0 | 0 | 0 |
| Boisson gazeuse aux fruits (à moins de 10% de jus), non sucrée, sans édulcorant | - | - | - | - | - | 0 | 0 | 4,17 | 0 | 0 | 0,002 | - | - | - | - | - | - | - | - | - | - | 5 | - | - | - | - | - | - | - | - | - | - | - | - | - |
| Boisson gazeuse aux fruits (à moins de 10% de jus), sucrée | - | - | - | - | 90,5 | 0,081 | 0,081 | 9,97 | 0,051 | 0,005 | 0,025 | - | - | - | - | - | 2 | - | - | - | - | 9,73 | - | - | - | - | - | - | - | - | - | - | - | - | - |
| Boisson gazeuse aux fruits (à moins de 10% de jus), sucrée, avec édulcorants | - | - | - | - | - | traces | traces | 7 | traces | 0 | 0,0075 | - | - | - | - | - | - | - | - | - | - | 3 | - | - | - | - | - | - | - | - | - | - | - | - | - |
| Boisson gazeuse aux fruits (de 10 à 50% de jus), non sucrée, avec édulcorants | - | - | - | - | 98 | 0,073 | 0,073 | 1,39 | 0,014 | traces | 0,017 | 6,37 | - | - | - | - | 1,99 | - | - | 63,7 | - | 6,28 | - | - | - | - | - | 3,89 | 0,01 | - | 0,03 | - | 0,04 | - | 0 |
| Boisson gazeuse aux fruits (de 10 à 50% de jus), sucrée | 198 | 46,6 | 198 | 46,6 | 88,2 | 0,067 | 0,067 | 11,3 | 0,028 | traces | 0,033 | 4,63 | - | 0,0065 | 0,025 | 0,4 | 2,67 | 0,0043 | 1 | 26,5 | < 10 | 13,5 | 0,014 | 0 | 0 | - | - | 0 | 0 | 0 | 0 | 0 | 0 | 0 | 0 |
| Boisson gazeuse aux fruits (de 10 à 50% de jus), sucrée, avec édulcorants | - | - | - | - | - | 0,067 | 0,067 | 3,83 | 0 | 0 | 0,058 | - | - | - | - | - | - | - | - | - | - | 23,4 | - | - | - | - | - | 12 | - | - | - | - | - | - | - |
| Boisson gazeuse aux fruits (teneur en jus non spécifiée), sucrée (aliment moyen) | 198 | 46,6 | 198 | 46,6 | 88,4 | 0,068 | 0,068 | 11,2 | 0,03 | 0,00043 | 0,032 | 4,63 | | 0,0065 | 0,025 | 0,4 | 2,61 | 0,0043 | 1 | 26,5 | 5 | 13,2 | 0,014 | 0 | 0 | | | 0 | 0 | 0 | 0 | 0 | 0 | 0 | 0 |
| Boisson gazeuse, sans jus de fruit, non sucrée, avec édulcorants | - | - | - | - | 99,9 | 0,05 | 0,05 | 0,16 | traces | 0 | 0,031 | 20 | - | 0,005 | 0,05 | 1 | 0 | 0,002 | 3 | 6 | 0,001 | 12,5 | 0,013 | 0 | 0 | 0 | - | 0 | 0 | 0 | 0 | 0 | 0 | 0 | 0 |

| | | | | | | | | | | | | | | | | | | | | | | | | | | | | | | | | | | | |
|---|---|---|---|---|---|---|---|---|---|---|---|---|---|---|---|---|---|---|---|---|---|---|---|---|---|---|---|---|---|---|---|---|---|---|---|
| Boisson gazeuse, sans jus de fruit, sucrée | 158 | 37,1 | 158 | 37,1 | 90,6 | 0,14 | 0,14 | 8,67 | 0,078 | 0 | 0,091 | 11,5 | - | 0,012 | 0,12 | 2,9 | 1 | 0,015 | 3 | 3,5 | 0,029 | 36,4 | 0,032 | 0 | 0 | 0 | - | 0 | 0 | 0 | 0 | 0 | 0 | 0 | 0 |
| Boisson gazeuse, sans jus de fruit, sucrée, avec édulcorants | - | - | - | - | - | 0 | 0 | 6,8 | 0 | 0 | 0 | - | - | - | - | - | - | - | - | - | - | 0 | - | - | - | - | - | - | - | - | - | - | - | - | - |
| Boisson lactée aromatisée (arôme inconnu), sucrée, au lait partiellement écrémé, enrichie et/ou restaurée en vitamines et/ou minéraux (aliment moyen) | | | | | | 2,93 | 2,87 | 12,1 | 1,08 | 0,045 | 0,098 | 122 | | | | | | | | | | 38,9 | | 0,77 | | | | | | 0,2 | | | | | 0,34 |
| Boisson lactée aromatisée à la fraise, sucrée, au lait partiellement écrémé, enrichie à la vitamine D | 223 | 52,6 | 224 | 52,7 | 86,9 | 2,17 | 2,13 | 10 | 0,45 | 0 | 0,14 | 76,6 | 106 | 0,06 | < 1 | 66 | 9,3 | 0,05 | 72 | 125 | 1,9 | 56 | 0,4 | 0,5 | < 0,1 | - | - | 0,96 | 0,04 | 0,16 | 0,16 | 0,71 | < 0,05 | < 5 | 0,25 |
| Boisson lactée aromatisée au chocolat, sucrée, au lait partiellement écrémé, enrichie et/ou restaurée en vitamines et/ou minéraux | 270 | 63,8 | 271 | 64,1 | 84,7 | 2,75 | 2,7 | 10,8 | 0,99 | < 1 | 0,095 | 85 | - | 0,11 | 0,3 | 15,4 | 17 | 0,036 | 99 | 185 | 1,34 | 38,1 | 0,44 | 0,64 | 0,042 | 0,2 | - | 0 | 0,041 | 0,17 | 0,15 | 0,44 | 0,027 | 2,6 | 0,21 |
| Boisson plate aux fruits (10 à 50% de jus de jus), sucrée, avec édulcorants | - | - | - | - | - | 0 | 0 | 6,17 | 0 | 0 | 0,033 | - | - | - | - | - | - | - | - | - | - | 13,3 | - | - | - | - | - | - | - | - | - | - | - | - | - |
| Boisson plate aux fruits (10 à 50% de jus), à teneur réduite en sucres | - | - | - | - | 94,5 | 0,014 | 0,014 | 5,16 | 0,01 | 0 | 0,019 | 0 | - | 0 | 0 | 0 | 0 | 0 | 0 | 1,6 | 0 | 7,45 | 0 | 0 | 0 | - | - | 0 | 0 | 0 | 0 | 0 | 0 | 0 | 0 |
| Boisson plate aux fruits (10 à 50% de jus), non sucrée, avec édulcorants | - | - | - | - | - | 0,1 | 0,1 | 1,4 | 0,1 | 0,1 | 0,033 | - | - | - | - | - | - | - | - | - | - | 13,3 | - | - | - | - | - | - | - | - | - | - | - | - | - |
| Boisson plate aux fruits (10 à 50% de jus), sucrée | - | - | - | - | 89,3 | 0,13 | 0,13 | 10,1 | 0,06 | 0,041 | 0,067 | 2,5 | < 60 | < 0,1 | 0,1 | 1 | 2,4 | < 0,1 | < 10 | 20,1 | < 2,2 | 28 | < 0,1 | 0 | 0 | - | - | 12,8 | < 0,01 | < 0,01 | < 0,04 | < 0,04 | < 0,04 | 6,2 | 0 |
| Boisson plate aux fruits (à moins de 10% de jus), sucrée | - | - | - | - | - | 0,1 | 0,1 | 4,4 | 0,1 | 0,1 | 0,017 | - | - | - | - | - | - | - | - | - | - | 6,86 | - | - | - | - | - | - | - | - | - | - | - | - | - |
| Boisson plate aux fruits (teneur en jus non spécifiée), sucrée | - | - | - | - | 88,3 | 0,1 | 0,1 | 11,2 | 0,07 | 0 | 0,0018 | 13 | - | 0,02 | 0,1 | 1 | 4,6 | 0,007 | 5 | 48 | 0 | 0,7 | 0,06 | 0 | 0 | - | - | 4,33 | 0,043 | 0,01 | 0,9 | traces | 0,055 | 10 | 0 |
| Boisson plate aux fruits, (à moins de 10% de jus), non sucrée, avec édulcorants | - | - | - | - | - | 0,06 | 0,06 | 0,96 | 0,13 | 0,05 | 0,05 | - | - | - | - | - | - | - | - | - | - | 19,9 | - | - | - | - | - | - | - | - | - | - | - | - | - |
| Boisson préparée à partir de boisson concentrée à diluer, non sucrée, avec édulcorants, type "sirop 0%", diluée dans l'eau | - | - | - | - | 93,4 | < 0,5 | < 0,5 | 2,42 | 0 | 0 | 0,074 | - | - | - | - | - | - | - | - | 61,3 | - | 29,7 | - | - | - | - | - | - | - | - | - | - | - | - | - |
| Boisson préparée à partir de sirop à diluer type menthe, fraise, etc, sucré, dilué dans l'eau | - | - | - | - | 92,2 | 0,055 | 0,055 | 7,7 | 0 | 0 | 0,012 | 7,55 | - | 0,004 | 0,025 | 0 | 2,71 | 0,001 | 0,5 | 5,06 | traces | 4,94 | 0,01 | 0 | 0 | - | - | 1,3 | 0,001 | 0,0011 | 0,0071 | 0,0033 | 0,0025 | 1,51 | 0 |
| Boisson rafraîchissante sans alcool (aliment moyen) | | | | | 92,1 | 0,11 | 0,11 | 7,42 | 0,042 | 0,025 | 0,028 | 6,22 | | 0,01 | 0,066 | 1,16 | 2,5 | 0,012 | 6,02 | 11,3 | 1,21 | 11,3 | 0,02 | 0,0099 | 0,0092 | | | 1,78 | 0,0034 | 0,0081 | 0,047 | 0,018 | 0,0085 | 1,3 | 0,0054 |
| Cacao, non sucré, poudre soluble | 1610 | 387 | 1610 | 387 | 3,5 | 22,4 | 22,4 | 11,6 | 20,6 | 29,5 | 0,11 | 140 | < 20 | 3,9 | 48,5 | < 20 | 500 | 4,1 | 690 | 3900 | < 20 | 45 | 6,4 | 2,73 | 0,61 | 3,9 | - | < 0,5 | 0,076 | 0,12 | 1,15 | 0,82 | < 0,01 | 107 | 0 |
| Café au lait ou cappuccino au chocolat, poudre soluble | 1690 | 402 | 1700 | 402 | - | 11,2 | 11 | 67,1 | 9,13 | 3,51 | 1,29 | - | - | - | - | - | - | - | - | - | - | 516 | - | - | - | - | - | - | - | - | - | - | - | - | - |
| Café au lait ou cappuccino, poudre soluble | 1750 | 416 | 1760 | 417 | 1 | 12,2 | 11,9 | 70,1 | 8,8 | 3,3 | 0,79 | - | - | - | - | - | - | - | - | - | - | 316 | - | - | - | - | - | - | - | - | - | - | - | - | - |
| Café au lait, café crème ou cappuccino, instantané ou non, non sucré, prêt à boire | 90,5 | 21,5 | 91 | 21,6 | 95,3 | 1,39 | 1,36 | 2,31 | 0,76 | 0 | 0,059 | 62,2 | - | 0,007 | 0,13 | 5,6 | 8,52 | 0,016 | 42,6 | 109 | 0,6 | 23,4 | 0,22 | 0,025 | 0,02 | - | - | - | 0,032 | 0,12 | 0,15 | 0,31 | 0,015 | 2,81 | 0,03 |
| Café décaféiné, instantané, non sucré, prêt à boire | 8,91 | 2,1 | 8,91 | 2,1 | 98,9 | 0,12 | 0,12 | 0,4 | 0,002 | 0 | 0,01 | 4 | - | 0,01 | 0,04 | 1 | 4 | 0,012 | 3 | 36 | - | 4 | 0,01 | 0 | 0 | 0 | - | 0 | 0 | 0,014 | 0,28 | 0,001 | 0 | 0 | 0 |
| Café décaféiné, non instantané, non sucré, prêt à boire | 18,6 | 4,42 | 18,6 | 4,42 | 98,6 | 0,1 | 0,1 | 0,6 | 0,18 | 0 | 0,02 | 2 | - | 0,029 | 0,09 | 1 | 42,5 | 0,039 | 4 | 84,5 | - | 8 | 0,035 | 0 | 0 | 0,1 | - | 0 | 0,001 | 0,18 | 2,71 | 0,015 | 0 | 1 | 0 |
| Café expresso, non instantané, non sucré, prêt à boire | 32 | 7,64 | 32 | 7,64 | 97,8 | < 0,5 | < 0,5 | 1,16 | 0,2 | < 0,5 | 0,0071 | 3,5 | 5,1 | < 0,01 | < 0,05 | < 20 | 7,9 | 0,05 | 6,9 | 150 | < 20 | 2,82 | < 0,05 | < 0,25 | < 0,08 | < 0,8 | - | < 0,5 | < 0,015 | < 0,01 | 1,73 | 0,028 | 0,74 | < 5 | 0 |
| Café, décaféiné, poudre soluble | 1510 | 357 | 1510 | 357 | 3,2 | 12,7 | 12,7 | 76 | 0,2 | 0 | 0,058 | 140 | - | 0,069 | 3,8 | - | 311 | 1,22 | 286 | 3500 | 17,3 | 23 | 0,11 | 0 | 0 | 1,9 | - | 0 | 0 | 1,36 | 28,1 | 0,097 | 0 | 0 | 0 |
| Café, instantané, non sucré, prêt à boire | 6,95 | 1,64 | 6,95 | 1,64 | 99,2 | 0,1 | 0,1 | 0,3 | 0,004 | 0 | 0,01 | 4 | - | 0,006 | 0,04 | 1 | 4 | 0,015 | 3 | 30 | 0,03 | 4 | 0,008 | 0 | 0 | 0 | - | 0 | 0 | 0,001 | 0,24 | 0,001 | 0 | 0 | 0 |
| Café, moulu | 1660 | 397 | 1660 | 397 | 5,5 | 14,4 | 14,4 | 40,2 | 15,4 | 19,8 | 0,19 | 120 | - | 1,55 | 4,1 | 0,5 | 240 | 2,13 | 160 | 2020 | 10 | 74 | 0,79 | 0 | 2,7 | 0 | - | 0 | 0,07 | 0,2 | 15 | 0,23 | 0,001 | 22 | 0 |
| Café, non instantané, non sucré, prêt à boire | 29 | 6,88 | 29 | 6,88 | 97,8 | < 0,5 | < 0,5 | 1,35 | 0,018 | < 0,5 | 0,0067 | 4,3 | 5,16 | < 0,01 | < 0,05 | < 20 | 9,2 | 0,07 | 7,4 | 150 | < 20 | 2,7 | < 0,05 | < 0,25 | < 0,08 | < 0,8 | - | < 0,5 | < 0,015 | < 0,01 | 1,26 | 0,013 | 0,73 | < 5 | 0 |
| Café, poudre soluble | 1250 | 296 | 1250 | 296 | 4,95 | 19,4 | 19,4 | 42,6 | 1,1 | 19,1 | 0,098 | 151 | - | 0,12 | 4,41 | 0,5 | 359 | 1,71 | 327 | 3770 | 10,7 | 39 | 0,46 | 0 | 0 | 1,9 | - | 0 | 0,008 | 0,092 | 25,1 | 0,25 | 0,03 | 22 | 0 |
| Chicorée et café, instantané, non sucré, prêt à boire (reconstituée avec de l'eau) | 121 | 28,7 | 121 | 28,7 | 92,2 | 0,56 | 0,56 | 6,06 | < 0,3 | 0,6 | 0,022 | 11 | 12,6 | 0,02 | 0,2 | < 20 | 35 | 0,1 | 15 | 160 | < 20 | 8,93 | < 0,05 | < 0,25 | < 0,08 | < 0,8 | - | < 0,5 | < 0,015 | < 0,01 | 1,53 | 0,055 | 0,86 | 5,5 | - |
| Chicorée et café, instantané, non sucré, prête à boire (reconstituée avec du lait demi-écrémé standard) | 226 | 53,7 | 227 | 54,1 | 87 | 3,74 | 3,66 | 5,9 | 1,62 | 0,46 | 0,13 | 117 | - | 0 | 0,37 | 10,4 | 13,2 | 0 | - | 162 | 0,9 | 52,2 | 0,5 | 0,01 | 0,16 | - | - | - | 0,054 | 0,18 | 0,93 | 0,35 | 0,025 | 2,6 | 0,26 |
| Chicorée et café, poudre soluble | 1400 | 332 | 1400 | 332 | 4,1 | 9,3 | 9,3 | 66,9 | 0,35 | 12 | 0,72 | 103 | - | 0,05 | 4,76 | 0,5 | 213 | 1,2 | 271 | 3400 | - | 289 | 0,37 | 0 | traces | - | - | 0 | 0 | 0,29 | 21,7 | 0,097 | 0,029 | 0 | 0 |
| Chicorée, instantanée, non sucrée, prête à boire (reconstituée avec du lait demi-écrémé standard) | 388 | 93,5 | 388 | 93,5 | - | traces | traces | 7,1 | 7,1 | 0,59 | 0,13 | - | - | - | - | - | - | - | - | - | - | 50,9 | - | - | - | - | - | - | - | - | - | - | - | - | - |
| Chicorée, poudre soluble | 1350 | 323 | 1350 | 323 | 2,7 | 3,3 | 3,3 | 57,2 | 2,5 | 29,2 | 0,39 | 109 | - | < 0,1 | 5,5 | 0,5 | 65 | < 0,1 | 223 | - | < 5 | 156 | < 0,1 | 0 | 0 | - | - | 0 | 0,04 | 0,12 | 10,1 | < 0,05 | 0,28 | 19,1 | 0 |
| Citron ou Lime, spécialité à diluer pour boissons, sans sucres ajoutés | 84,9 | 19,7 | 84,9 | 19,7 | 93,8 | < 0,5 | < 0,5 | 1,31 | 0 | 0 | 0,0035 | 6,1 | 2,8 | 0,01 | 0,05 | < 20 | 3,2 | < 0,01 | 4,7 | 44 | < 20 | 1,4 | < 0,05 | - | 0 | - | - | 5,56 | 0,27 | < 0,01 | < 0,1 | 0,025 | < 0,01 | 6,44 | - |
| Cola, non sucré, avec édulcorants | 5,47 | 1,3 | 5,47 | 1,3 | 99,7 | 0,081 | 0,081 | 0,11 | 0,053 | 0,003 | 0,024 | 2,26 | - | 0,0068 | 0,065 | 0,8 | 2,38 | < 0,0012 | 10 | 6 | < 2,2 | 9,88 | < 0,016 | 0 | 0 | 0 | - | 0 | 0,005 | 0,023 | 0 | 0 | 0 | 0 | 0 |
| Cola, non sucré, avec édulcorants, sans caféine | 3,7 | 0,87 | 3,7 | 0,87 | 99,7 | 0 | 0 | 0,2 | 0 | 0 | 0,018 | 3 | - | 0,002 | 0,02 | - | 0 | 0 | 10 | 7 | - | 7 | 0,01 | 0 | 0 | 0 | - | 0 | 0,005 | 0,023 | 0 | 0 | 0 | 0 | 0 |
| Cola, sucré | 178 | 41,8 | 178 | 41,8 | 90,3 | 0,093 | 0,093 | 10,2 | 0,062 | 0 | 0,017 | 1,92 | - | 0,0044 | 0,11 | 0,8 | 2,44 | < 0,0012 | 10 | 2 | < 2,2 | 6,71 | < 0,016 | 0 | 0 | 0 | - | 0 | 0 | 0 | 0 | 0 | 0 | 0 | 0 |
| Cola, sucré, avec édulcorants | - | - | - | - | 95 | 0,0093 | 0,0093 | 6,65 | 0,0033 | 0,0033 | 0,011 | 0 | - | 0 | 0 | 0 | 0 | 0 | 17 | 10 | 0 | 4,2 | 0 | 0 | 0 | - | - | 0 | 0 | 0 | 0 | 0 | 0 | 0 | 0 |
| Cola, sucré, sans caféine | 186 | 43,8 | 186 | 43,8 | 89,6 | traces | traces | 11 | traces | 0 | 0,025 | 2 | - | 0 | 0,02 | - | 0 | 0 | 11 | 3 | - | 10 | 0,01 | 0 | 0 | 0 | - | 0 | 0 | 0 | 0 | 0 | 0 | 0 | 0 |
| Cola, teneur en sucre et édulcorant inconnue (aliment moyen) | 131 | 30,7 | 131 | 30,7 | 92,9 | 0,089 | 0,089 | 7,45 | 0,059 | 0,00084 | 0,019 | 2 | | 0,005 | 0,097 | 0,79 | 2,4 | 0,0006 | 10,1 | 3,14 | 1,09 | 7,55 | 0,0081 | 0 | 0 | 0 | | 0 | 0,0014 | 0,0062 | 0 | 0 | 0 | 0 | 0 |
| Diabolo (limonade et sirop) | - | - | - | - | - | 0,075 | 0,075 | 10,6 | traces | 0,049 | 0,021 | - | - | - | - | - | - | - | - | - | - | 8,5 | - | - | - | - | - | - | - | - | - | - | - | - | - |
| Eau de coco | 64,2 | 15,2 | 63,5 | 15,1 | 95,7 | < 0,42 | < 0,5 | 3,33 | < 0,3 | < 0,5 | 0,051 | 14 | 140 | < 0,01 | < 0,05 | < 20 | 8,3 | 0,15 | 7,9 | 200 | < 20 | 20,4 | < 0,05 | < 0,25 | < 0,08 | < 0,8 | - | < 0,5 | < 0,015 | < 0,01 | < 0,1 | 0,028 | 0,031 | < 5 | 0 |
| Eau de source Cristaline, embouteillée, non gazeuse | 0 | 0 | 0 | 0 | - | 0 | 0 | 0 | 0 | 0 | 0,0026 | 12,4 | 1,6 | - | - | - | 2,5 | - | - | 0,35 | - | 1,1 | - | - | - | - | - | - | - | - | - | - | - | - | - |
| Eau de source Ogeu, embouteillée, faiblement minéralisée (Ogeu, 64) | 0 | 0 | 0 | 0 | 100 | 0 | 0 | 0 | 0 | 0 | - | - | - | - | - | 0 | - | - | - | - | - | - | - | 0 | 0 | - | - | 0 | 0 | 0 | 0 | 0 | 0 | 0 | 0 |
| Eau de source, embouteillée (aliment moyen) | 0 | 0 | 0 | 0 | | 0 | 0 | 0 | 0 | 0 | | | | | | | | | | | | | | | | | | | | | | | | | |
| Eau du robinet | 0 | 0 | 0 | 0 | 100 | 0 | 0 | 0 | 0 | 0 | 0,0076 | 7,13 | - | 0,017 | 0,03 | - | 0,99 | 0,00082 | 0 | 0,73 | - | 2,4 | 0,011 | 0 | 0 | 0 | - | 0 | 0 | 0 | 0 | 0 | 0 | 0 | 0 |
| Eau embouteillée de source | 0 | 0 | 0 | 0 | - | 0 | 0 | 0 | 0 | 0 | - | 16,9 | - | 0,008 | 0,023 | - | 1,81 | 0,00089 | - | 0,99 | < 10 | 2,17 | 0,011 | - | - | - | - | - | - | - | - | - | - | - | - |
| Eau minérale (aliment moyen) | 0 | 0 | 0 | 0 | 99,9 | 0 | 0 | 0 | 0 | 0 | 0,0042 | 19,3 | 2,68 | 0,0012 | 0,0017 | 0 | 4,01 | 0,0091 | 0,0034 | 0,58 | 2,83 | 5,02 | 0,0067 | 0 | 0 | | | 0 | 0 | 0 | 0 | 0 | 0 | 0 | 0 |
| Eau minérale Abatilles, embouteillée, non gazeuse, faiblement minéralisée (Arcachon, 33) | 0 | 0 | 0 | 0 | 100 | 0 | 0 | 0 | 0 | 0 | 0,023 | 1,9 | 13,7 | < 0,0005 | 0,01 | - | 0,9 | 0,0001 | < 0,003 | 0,4 | < 1 | 10 | < 0,0005 | - | - | - | - | - | - | - | - | - | - | - | - |
| Eau minérale Aix-les-Bains, embouteillée, non gazeuse, faiblement minéralisée (Aix-les-Bains, 73) | 0 | 0 | 0 | 0 | 100 | 0 | 0 | 0 | 0 | 0 | 0,00035 | 8,23 | 0,21 | < 0,0005 | < 0,001 | 0 | 2,25 | < 0,0001 | < 0,003 | 0,08 | < 1 | 0,44 | < 0,0005 | 0 | 0 | - | - | 0 | 0 | 0 | 0 | 0 | 0 | 0 | 0 |
| Eau minérale Aizac, embouteillée, gazeuse, faiblement minéralisée (Aizac, 07) | 0 | 0 | 0 | 0 | 100 | 0 | 0 | 0 | 0 | 0 | 0,00033 | 8,01 | 1,4 | < 0,0005 | < 0,001 | 0 | 2,77 | 0,0001 | < 0,003 | 0,12 | < 1 | 0,13 | 0,0016 | 0 | 0 | - | - | 0 | 0 | 0 | 0 | 0 | 0 | 0 | 0 |
| Eau minérale Amanda, embouteillée, non gazeuse, fortement minéralisée (St-Amand, 59) | 0 | 0 | 0 | 0 | 99,9 | 0 | 0 | 0 | 0 | 0 | 0,01 | 24 | 6,2 | < 0,0005 | 0,04 | 0 | 6,6 | 0,0004 | < 0,003 | 0,61 | < 1 | 5,5 | < 0,0005 | 0 | 0 | - | - | 0 | 0 | 0 | 0 | 0 | 0 | 0 | 0 |
| Eau minérale Appollinaris, embouteillée, non gazeuse, fortement minéralisée (Allemagne) | 0 | 0 | 0 | 0 | 99,8 | 0 | 0 | 0 | 0 | 0 | 0,017 | 9 | 10 | 0 | 0 | 0 | 11 | - | 0 | 3 | 0 | 38 | traces | 0 | 0 | - | - | 0 | 0 | 0 | 0 | 0 | 0 | 0 | 0 |

| | | | | | | | | | | | | | | | | | | | | | | | | | | | | | | | | | | | |
|---|---|---|---|---|---|---|---|---|---|---|---|---|---|---|---|---|---|---|---|---|---|---|---|---|---|---|---|---|---|---|---|---|---|---|---|
| Eau minérale Arcens, embouteillée, gazeuse, moyennement minéralisée (Arcens, 07) | 0 | 0 | 0 | 0 | 99,9 | 0 | 0 | 0 | 0 | 0 | 0,004 | 0,9 | 2,4 | < 0,0005 | 0,0006 | 0 | 1,6 | 0,015 | < 0,003 | 0,53 | < 1 | 25,2 | < 0,0005 | 0 | 0 | - | - | 0 | 0 | 0 | 0 | 0 | 0 | 0 | 0 |
| Eau minérale Ardesy, embouteillée, gazeuse, fortement minéralisée (Ardes, 63) | 0 | 0 | 0 | 0 | 99,7 | 0 | 0 | 0 | 0 | 0 | 0,064 | 17 | 38,7 | < 0,0005 | 0,0006 | 0 | 9,2 | 0,039 | < 0,003 | 13 | < 1 | 65 | 0,001 | 0 | 0 | - | - | 0 | 0 | 0 | 0 | 0 | 0 | 0 | 0 |
| Eau minérale Avra, embouteillée, non gazeuse, faiblement minéralisée (Grèce) | 0 | 0 | 0 | 0 | 100 | 0 | 0 | 0 | 0 | 0 | 0,0013 | 11,1 | 0,8 | 0 | 0 | 0 | 1 | - | 0 | 0,07 | 0 | 8,4 | traces | 0 | 0 | - | - | 0 | 0 | 0 | 0 | 0 | 0 | 0 | 0 |
| Eau minérale Badoit, embouteillée, gazeuse, moyennement minéralisée (St-Galmier, 42) | 0 | 0 | 0 | 0 | 99,9 | 0 | 0 | 0 | 0 | 0 | 0,0089 | 15,3 | 5,4 | < 0,0023 | < 0,001 | 0 | 8 | < 0,0015 | < 0,003 | 1,1 | < 10 | 18 | < 0,01 | 0 | 0 | - | - | 0 | 0 | 0 | 0 | 0 | 0 | 0 | 0 |
| Eau minérale Beckerich, embouteillée, non gazeuse, faiblement minéralisée (Luxembourg) | 0 | 0 | 0 | 0 | 100 | 0 | 0 | 0 | 0 | 0 | 0,00075 | 9,7 | 0,6 | 0 | 0 | 0 | 0,5 | - | 0 | 0,06 | 0 | 0,3 | traces | 0 | 0 | - | - | 0 | 0 | 0 | 0 | 0 | 0 | 0 | 0 |
| Eau minérale Biovive, embouteillée, non gazeuse, faiblement minéralisée (Dax, 40) | 0 | 0 | 0 | 0 | 100 | 0 | 0 | 0 | 0 | 0 | 0,0028 | 4,2 | 1,71 | - | - | - | 0,38 | - | - | 0,24 | - | 1,86 | - | - | - | - | - | - | - | - | - | - | - | - | - |
| Eau minérale Carola, embouteillée, gazeuse ou non gazeuse, moyennement minéralisée (Ribeauville, 68) | 0 | 0 | 0 | 0 | - | 0 | 0 | 0 | 0 | 0 | - | 7,3 | - | - | - | - | 2 | - | - | - | - | 13,1 | - | - | - | - | - | - | - | - | - | - | - | - | - |
| Eau minérale Celtic, embouteillée, gazeuse ou non gazeuse, très faiblement minéralisée (Niederbronn, 67) | 0 | 0 | 0 | 0 | 100 | 0 | 0 | 0 | 0 | 0 | 0,00028 | 1,05 | < 0,5 | < 0,0005 | < 0,001 | 0 | 0,4 | 0,0001 | < 0,003 | 0,19 | < 1 | 0,11 | < 0,0005 | 0 | 0 | - | - | 0 | 0 | 0 | 0 | 0 | 0 | 0 | 0 |
| Eau minérale Chambon, embouteillée, non gazeuse, faiblement minéralisée (Chambon, 45) | 0 | 0 | 0 | 0 | 100 | 0 | 0 | 0 | 0 | 0 | 0,0027 | 9,6 | 2,26 | < 0,0005 | < 0,001 | 0 | 0,61 | < 0,0001 | < 0,003 | 0,37 | < 1 | 1,06 | < 0,0005 | 0 | 0 | - | - | 0 | 0 | 0 | 0 | 0 | 0 | 0 | 0 |
| Eau minérale Chantemerle, embouteillée, non gazeuse, faiblement minéralisée (Le Pestrin, 07) | 0 | 0 | 0 | 0 | 100 | 0 | 0 | 0 | 0 | 0 | 0,00028 | 2,62 | 0,17 | 0,0016 | < 0,001 | 0 | 0,96 | 0,06 | < 0,003 | 0,13 | < 1 | 1,34 | 0,007 | 0 | 0 | - | - | 0 | 0 | 0 | 0 | 0 | 0 | 0 | 0 |
| Eau minérale Chateauneuf, embouteillée, gazeuse, fortement minéralisée (Chateauneuf, 63) | 0 | 0 | 0 | 0 | 99,8 | 0 | 0 | 0 | 0 | 0 | 0,038 | 11,9 | 22,9 | < 0,0005 | < 0,001 | 0 | 3 | 0,004 | < 0,003 | 4,4 | < 1 | 70,3 | < 0,0005 | 0 | 0 | - | - | 0 | 0 | 0 | 0 | 0 | 0 | 0 | 0 |
| Eau minérale Chateldon, embouteillée, gazeuse, fortement minéralisée (Chateldon, 63) | 0 | 0 | 0 | 0 | 99,8 | 0 | 0 | 0 | 0 | 0 | 0,0012 | 34 | 0,7 | < 0,0005 | < 0,001 | 0 | 4,9 | 0,1 | < 0,003 | 4,1 | < 1 | 24 | 0,004 | 0 | 0 | - | - | 0 | 0 | 0 | 0 | 0 | 0 | 0 | 0 |
| Eau minérale Chaudfontaine, embouteillée, non gazeuse, faiblement minéralisée (Belgique) | 0 | 0 | 0 | 0 | 100 | 0 | 0 | 0 | 0 | 0 | 0,0058 | 6,5 | 3,5 | 0 | 0 | 0 | 1,8 | - | 0 | 0,25 | 0 | 4,4 | traces | 0 | 0 | - | - | 0 | 0 | 0 | 0 | 0 | 0 | 0 | 0 |
| Eau minérale Christinen Brunnen, embouteillée, non gazeuse, moyennement minéralisée (Allemagne) | 0 | 0 | 0 | 0 | 99,9 | 0 | 0 | 0 | 0 | 0 | 0,047 | 3,3 | 28,2 | 0 | 0 | 0 | 0 | - | 0 | 0 | 0 | 37 | traces | 0 | 0 | - | - | 0 | 0 | 0 | 0 | 0 | 0 | 0 | 0 |
| Eau minérale Cilaos, embouteillée, gazeuse, fortement minéralisée (Cilaos, 974) | 0 | 0 | 0 | 0 | - | 0 | 0 | 0 | 0 | 0 | 0,0005 | 11,1 | 0,3 | - | - | - | 7,05 | - | - | 0,53 | - | 23,8 | - | - | - | - | - | - | - | - | - | - | - | - | - |
| Eau minérale Clos de l'Abbaye, embouteillée, non gazeuse, moyennement minéralisée (St-Amand, 59) | 0 | 0 | 0 | 0 | 99,9 | 0 | 0 | 0 | 0 | 0 | 0,006 | 16,4 | 3,66 | < 0,0005 | 0,0005 | 0 | 4,15 | 0,0013 | < 0,003 | 0,36 | < 1 | 2,94 | 0,0005 | 0 | 0 | - | - | 0 | 0 | 0 | 0 | 0 | 0 | 0 | 0 |
| Eau minérale Contrex, embouteillée, non gazeuse, fortement minéralisée (Contrexéville, 88) | 0 | 0 | 0 | 0 | 99,8 | 0 | 0 | 0 | 0 | 0 | 0,0013 | 46,8 | 0,76 | 0,0042 | 0,0005 | 0 | 7,45 | 0,0015 | < 0,003 | 0,28 | < 10 | 0,94 | 0,0095 | 0 | 0 | - | - | 0 | 0 | 0 | 0 | 0 | 0 | 0 | 0 |
| Eau minérale Courmayeur, embouteillée, non gazeuse, fortement minéralisée (Italie) | 0 | 0 | 0 | 0 | 99,8 | 0 | 0 | 0 | 0 | 0 | 0,00025 | 51,7 | - | 0 | 0 | 0 | 6,7 | - | 0 | 0,2 | 0 | 0,1 | traces | 0 | 0 | - | - | 0 | 0 | 0 | 0 | 0 | 0 | 0 | 0 |
| Eau minérale Dax, embouteillée, non gazeuse, moyennement minéralisée (Dax, 40) | 0 | 0 | 0 | 0 | 99,9 | 0 | 0 | 0 | 0 | 0 | 0,031 | - | - | - | - | 0 | - | - | - | - | - | - | - | 0 | 0 | - | - | 0 | 0 | 0 | 0 | 0 | 0 | 0 | 0 |
| Eau minérale Didier, embouteillée non gazeuse, fortement minéralisée (Martinique) | 0 | 0 | 0 | 0 | 99,9 | 0 | 0 | 0 | 0 | 0 | 0,0041 | 17,2 | 2,48 | < 0,0005 | < 0,001 | 0 | 10,7 | 0,016 | < 0,003 | 1,31 | < 1 | 13 | < 0,0005 | 0 | 0 | - | - | 0 | 0 | 0 | 0 | 0 | 0 | 0 | 0 |
| Eau minérale Didier, embouteillée, gazeuse, fortement minéralisée (Martinique) | 0 | 0 | 0 | 0 | 99,9 | 0 | 0 | 0 | 0 | 0 | 0,032 | 13,7 | 2,31 | < 0,0005 | 0,0005 | 0 | 11,3 | 0,026 | < 0,003 | 1,4 | < 1 | 12,6 | < 0,0005 | 0 | 0 | - | - | 0 | 0 | 0 | 0 | 0 | 0 | 0 | 0 |
| Eau minérale Eden (La Goa), embouteillée, non gazeuse, faiblement minéralisée (Suisse) | 0 | 0 | 0 | 0 | - | 0 | 0 | 0 | 0 | 0 | - | 3,45 | < 0,01 | - | - | - | 0,31 | - | - | 0,07 | - | 0,4 | - | - | - | - | - | - | - | - | - | - | - | - | - |
| Eau minérale Evian, embouteillée, non gazeuse, faiblement minéralisée (Evian, 74) | 0 | 0 | 0 | 0 | 100 | 0 | 0 | 0 | 0 | 0 | 0,0011 | 8 | 0,68 | < 0,002 | < 0,0083 | 0 | 2,6 | < 0,0005 | < 0,003 | 0,1 | < 1 | 0,65 | < 0,005 | 0 | 0 | - | - | 0 | 0 | 0 | 0 | 0 | 0 | 0 | 0 |
| Eau minérale Hépar, embouteillée, non gazeuse, fortement minéralisée (Vittel, 88) | 0 | 0 | 0 | 0 | 99,7 | 0 | 0 | 0 | 0 | 0 | 0,0031 | 54,9 | 1,88 | < 0,0023 | < 0,001 | 0 | 11,9 | < 0,0015 | < 0,003 | 0,41 | < 10 | 1,42 | < 0,01 | 0 | 0 | - | - | 0 | 0 | 0 | 0 | 0 | 0 | 0 | 0 |
| Eau minérale Hydroxydase, embouteillée, gazeuse, fortement minéralisée (Le Breuil sur Couze, 63) | 0 | 0 | 0 | 0 | 99,1 | 0 | 0 | 0 | 0 | 0 | 0,058 | 28,1 | 35 | < 0,0005 | 0,018 | 0 | 24 | 0,07 | 0,003 | 18,2 | < 1 | 184 | 0,0005 | 0 | 0 | - | - | 0 | 0 | 0 | 0 | 0 | 0 | 0 | 0 |
| Eau minérale La Cairolle, embouteillée, non gazeuse, fortement minéralisée (Les Aires, 34) | 0 | 0 | 0 | 0 | 99,8 | 0 | 0 | 0 | 0 | 0 | 0,0021 | 34 | 1,3 | - | - | - | 4 | - | - | 1,25 | - | 2,7 | - | - | - | - | - | - | - | - | - | - | - | - | - |
| Eau minérale La Française, embouteillée, non gazeuse, fortement minéralisée (Propiac, 26) | 0 | 0 | 0 | 0 | 99,7 | 0 | 0 | 0 | 0 | 0 | 0,16 | 35,4 | 98,2 | - | - | - | 8,3 | - | - | 2,2 | - | 68 | - | - | - | - | - | - | - | - | - | - | - | - | - |
| Eau minérale Levissima, embouteillée, non gazeuse, faiblement minéralisée (Italie) | 0 | 0 | 0 | 0 | 100 | 0 | 0 | 0 | 0 | 0 | 0,00005 | 2 | 0,03 | 0 | 0 | 0 | 0,2 | - | 0 | 0,2 | 0 | 0,2 | traces | 0 | 0 | - | - | 0 | 0 | 0 | 0 | 0 | 0 | 0 | 0 |
| Eau minérale Luchon, embouteillée, non gazeuse, faiblement minéralisée (Luchon, 31) | 0 | 0 | 0 | 0 | 100 | 0 | 0 | 0 | 0 | 0 | 0,0002 | 2,65 | 0,23 | < 0,0005 | < 0,001 | 0 | 0,1 | < 0,0001 | < 0,003 | 0,02 | < 1 | 0,08 | < 0,0005 | 0 | 0 | - | - | 0 | 0 | 0 | 0 | 0 | 0 | 0 | 0 |
| Eau minérale Luso, embouteillée, non gazeuse, très faiblement minéralisée (Portugal) | 0 | 0 | 0 | 0 | 100 | 0 | 0 | 0 | 0 | 0 | 0,0015 | 0,1 | 0,9 | 0 | 0 | 0 | 0,2 | - | 0 | 0 | 0 | 0,6 | traces | 0 | 0 | - | - | 0 | 0 | 0 | 0 | 0 | 0 | 0 | 0 |
| Eau minérale Mont-Blanc, embouteillée, non gazeuse, faiblement minéralisée (Italie) | 0 | 0 | 0 | 0 | - | 0 | 0 | 0 | 0 | 0 | - | 2,9 | < 0,1 | - | - | - | 0,24 | - | - | 0,19 | - | 0,16 | - | - | - | - | - | - | - | - | - | - | - | - | - |
| Eau minérale Montcalm, embouteillée, non gazeuse, très faiblement minéralisée (Auzat, 09) | 0 | 0 | 0 | 0 | 100 | 0 | 0 | 0 | 0 | 0 | 0,000099 | 0,3 | 0,06 | - | - | - | 0,07 | - | - | 0,06 | - | 0,22 | - | - | - | - | - | - | - | - | - | - | - | - | - |
| Eau minérale Montclar, embouteillée, non gazeuse, faiblement minéralisée (Montclar, 04) | 0 | 0 | 0 | 0 | 100 | 0 | 0 | 0 | 0 | 0 | 0,0005 | 4,1 | 0,3 | - | - | - | 0,3 | - | - | - | - | 0,2 | - | - | - | - | - | - | - | - | - | - | - | - | - |
| Eau minérale Mont-Roucous, embouteillée, très faiblement minéralisée (Lacaune, 81) | 0 | 0 | 0 | 0 | 100 | 0 | 0 | 0 | 0 | 0 | 0,0005 | 0,24 | 0,3 | < 0,0005 | < 0,001 | 0 | 0,05 | < 0,0001 | < 0,003 | 0,04 | < 1 | 0,31 | < 0,0005 | 0 | 0 | - | - | 0 | 0 | 0 | 0 | 0 | 0 | 0 | 0 |
| Eau minérale Néro, embouteillée, non gazeuse, faiblement minéralisée (Grèce) | 0 | 0 | 0 | 0 | 100 | 0 | 0 | 0 | 0 | 0 | 0,01 | 5,9 | 6,6 | 0 | 0 | 0 | 3,5 | - | 0 | 0,2 | 0 | 4,1 | traces | 0 | 0 | - | - | 0 | 0 | 0 | 0 | 0 | 0 | 0 | 0 |
| Eau minérale Nessel, embouteillée, gazeuse, moyennement minéralisée (Soultzmatt, 68) | 0 | 0 | 0 | 0 | 99,9 | 0 | 0 | 0 | 0 | 0 | 0,0068 | 11,4 | 4,1 | - | - | - | 5,42 | - | - | 4 | - | 23,1 | - | - | - | - | - | - | - | - | - | - | - | - | - |
| Eau minérale Ogeu, embouteillée, gazeuse, faiblement minéralisée (Ogeu-les-Bains, 64) | 0 | 0 | 0 | 0 | 100 | 0 | 0 | 0 | 0 | 0 | 0,0078 | 4,8 | 4,8 | - | - | - | 1,2 | - | - | 0,1 | - | 3,1 | - | - | - | - | - | - | - | - | - | - | - | - | - |
| Eau minérale Ogeu, embouteillée, non gazeuse, faiblement minéralisée (Ogeu-les-Bains, 64) | 0 | 0 | 0 | 0 | 100 | 0 | 0 | 0 | 0 | 0 | 0,0058 | 4,9 | 2,3 | - | - | - | 1,7 | - | - | 0,1 | - | 2,7 | - | - | - | - | - | - | - | - | - | - | - | - | - |
| Eau minérale Orée du bois, embouteillée, non gazeuse, moyennement minéralisée (St-Amand, 59) | 0 | 0 | 0 | 0 | 99,9 | 0 | 0 | 0 | 0 | 0 | 0,01 | 24 | 6,06 | < 0,0005 | 0,0044 | 0 | 6,12 | 0,0006 | < 0,003 | 0,64 | < 1 | 4,92 | < 0,0005 | 0 | 0 | - | - | 0 | 0 | 0 | 0 | 0 | 0 | 0 | 0 |
| Eau minérale Orezza, embouteillée, gazeuse, moyennement minéralisée (Rapaggio, 20B) | 0 | 0 | 0 | 0 | 99,9 | 0 | 0 | 0 | 0 | 0 | 0,0017 | 18,5 | 1 | < 0,0005 | 0,035 | 0 | 1,65 | 0,41 | < 0,003 | 0,16 | 0,1 | 0,7 | < 0,0005 | 0 | 0 | - | - | 0 | 0 | 0 | 0 | 0 | 0 | 0 | 0 |
| Eau minérale Parot, embouteillée, gazeuse, moyennement minéralisée (St-Romain-le-Puy, 42) | 0 | 0 | 0 | 0 | 99,9 | 0 | 0 | 0 | 0 | 0 | 0,016 | 11 | 9,9 | < 0,0005 | 0,025 | 0 | 9,4 | 0,031 | 0,003 | 11 | < 1 | 101 | < 0,0005 | 0 | 0 | - | - | 0 | 0 | 0 | 0 | 0 | 0 | 0 | 0 |
| Eau minérale Penacova, embouteillée, non gazeuse, très faiblement minéralisée (Portugal) | 0 | 0 | 0 | 0 | 100 | 0 | 0 | 0 | 0 | 0 | 0,0015 | 0,1 | 0,9 | 0 | 0 | 0 | 0,1 | - | 0 | 0 | 0 | 0,6 | traces | 0 | 0 | - | - | 0 | 0 | 0 | 0 | 0 | 0 | 0 | 0 |
| Eau minérale Perrier, embouteillée, gazeuse, faiblement minéralisée (Vergèse, 30) | 0 | 0 | 0 | 0 | 100 | 0 | 0 | 0 | 0 | 0 | 0,0024 | 16 | 2,2 | < 0,0023 | < 0,001 | 0 | 0,42 | < 0,0015 | < 0,003 | < 0,1 | < 10 | 0,95 | 0,011 | 0 | 0 | - | - | 0 | 0 | 0 | 0 | 0 | 0 | 0 | 0 |
| Eau minérale Plancoet, embouteillée, gazeuse ou non gazeuse, faiblement minéralisée (Plancoet, 22) | 0 | 0 | 0 | 0 | 100 | 0 | 0 | 0 | 0 | 0 | 0,0059 | 2,3 | 3,6 | < 0,0005 | < 0,001 | 0 | 1,4 | 0,0018 | < 0,003 | 0,45 | < 1 | 3,1 | < 0,0005 | 0 | 0 | - | - | 0 | 0 | 0 | 0 | 0 | 0 | 0 | 0 |

| | | | | | | | | | | | | | | | | | | | | | | | | | | | | | | | | | | | |
|---|---|---|---|---|---|---|---|---|---|---|---|---|---|---|---|---|---|---|---|---|---|---|---|---|---|---|---|---|---|---|---|---|---|---|---|
| Eau minérale Prince Noir, embouteillée, non gazeuse, fortement minéralisée (St-Antonin-Noble-Val, 82) | 0 | 0 | 0 | 0 | - | 0 | 0 | 0 | 0 | 0 | 0,0015 | 52,8 | 0,9 | - | - | - | 7,8 | - | - | 0,3 | - | 0,9 | - | - | - | - | - | - | - | - | - | - | - | - | - |
| Eau minérale Propiac, embouteillée, non gazeuse, fortement minéralisée (Propiac, 26) | 0 | 0 | 0 | 0 | 99,6 | 0 | 0 | 0 | 0 | 0 | 0,17 | - | - | - | - | 0 | - | - | - | - | - | - | - | 0 | 0 | - | - | 0 | 0 | 0 | 0 | 0 | 0 | 0 | 0 |
| Eau minérale Puits St-Georges, embouteillée, gazeuse, moyennement minéralisée (St-Romain-le-Puy, 42) | 0 | 0 | 0 | 0 | 99,9 | 0 | 0 | 0 | 0 | 0 | 0,0063 | 4,5 | 3,8 | < 0,0005 | 0,0032 | 0 | 3,3 | 0,025 | < 0,003 | 1,8 | < 1 | 43 | < 0,0005 | 0 | 0 | - | - | 0 | 0 | 0 | 0 | 0 | 0 | 0 | 0 |
| Eau minérale Quézac, embouteillée, gazeuse, moyennement minéralisée (Quézac, 48) | 0 | 0 | 0 | 0 | 99,9 | 0 | 0 | 0 | 0 | 0 | 0,003 | 17 | 1,8 | < 0,0005 | < 0,001 | 0 | 6,9 | 0,18 | < 0,003 | 3,8 | < 1 | 11 | 0,0007 | 0 | 0 | - | - | 0 | 0 | 0 | 0 | 0 | 0 | 0 | 0 |
| Eau minérale Reine des basaltes, embouteillée, gazeuse, moyennement minéralisée (Asperjoc, 07) | 0 | 0 | 0 | 0 | 99,9 | 0 | 0 | 0 | 0 | 0 | 0,0019 | 12,5 | 1,18 | 0,003 | < 0,001 | 0 | 6,45 | 0,074 | < 0,003 | 2,57 | < 1 | 24,6 | 0,0021 | 0 | 0 | - | - | 0 | 0 | 0 | 0 | 0 | 0 | 0 | 0 |
| Eau minérale Rozana, embouteillée, gazeuse, fortement minéralisée (Beauregard, 63) | 0 | 0 | 0 | 0 | 99,7 | 0 | 0 | 0 | 0 | 0 | 0,11 | 30,1 | 64,9 | 0,0011 | < 0,001 | 0 | 16 | 0,0011 | < 0,003 | 5,2 | < 1 | 49,3 | < 0,0005 | 0 | 0 | - | - | 0 | 0 | 0 | 0 | 0 | 0 | 0 | 0 |
| Eau minérale Sail-les-Bains, embouteillée, non gazeuse, faiblement minéralisée (Sail-les-Bains, 42) | 0 | 0 | 0 | 0 | 100 | 0 | 0 | 0 | 0 | 0 | 0,0057 | 2,1 | 3,47 | < 0,0005 | < 0,001 | 0 | 0,2 | 0,0004 | < 0,003 | 0,48 | < 1 | 8,5 | 0,0007 | 0 | 0 | - | - | 0 | 0 | 0 | 0 | 0 | 0 | 0 | 0 |
| Eau minérale Salvetat, embouteillée, gazeuse, moyennement minéralisée (La Salvetat, 34) | 0 | 0 | 0 | 0 | 99,9 | 0 | 0 | 0 | 0 | 0 | 0,00086 | 21 | 0,52 | < 0,0005 | < 0,001 | 0 | 0,95 | 0,08 | < 0,003 | 0,23 | < 1 | 0,7 | < 0,0005 | 0 | 0 | - | - | 0 | 0 | 0 | 0 | 0 | 0 | 0 | 0 |
| Eau minérale San Bernardo, embouteillée, très faiblement minéralisée (Italie) | 0 | 0 | 0 | 0 | 100 | 0 | 0 | 0 | 0 | 0 | 0,000083 | 1,1 | 0,05 | 0 | 0 | 0 | 0,05 | - | 0 | 0,04 | 0 | 0,05 | traces | 0 | 0 | - | - | 0 | 0 | 0 | 0 | 0 | 0 | 0 | 0 |
| Eau minérale San Pellegrino, embouteillée, gazeuse, moyennement minéralisée (Italie) | 0 | 0 | 0 | 0 | 99,9 | 0 | 0 | 0 | 0 | 0 | 0,011 | 20,8 | 7,4 | 0 | 0 | 0 | 5,1 | - | 0 | 0,3 | 0 | 4,4 | traces | 0 | 0 | - | - | 0 | 0 | 0 | 0 | 0 | 0 | 0 | 0 |
| Eau minérale Spa-Reine, embouteillée, gazeuse ou non non gazeuse, moyennement minéralisée (Belgique) | 0 | 0 | 0 | 0 | 100 | 0 | 0 | 0 | 0 | 0 | 0,00075 | 0,45 | 0,5 | 0 | 0 | 0 | 0,13 | - | 0 | 0,05 | 0 | 0,3 | traces | 0 | 0 | - | - | 0 | 0 | 0 | 0 | 0 | 0 | 0 | 0 |
| Eau minérale St-Alban, embouteillée, gazeuse, moyennement minéralisée (St-Alban, 42) | 0 | 0 | 0 | 0 | 99,9 | 0 | 0 | 0 | 0 | 0 | 0,0033 | 18 | 2 | 0,0006 | 0,002 | 0 | 5 | 0,072 | < 0,003 | 2,8 | < 1 | 25 | 0,001 | 0 | 0 | - | - | 0 | 0 | 0 | 0 | 0 | 0 | 0 | 0 |
| Eau minérale St-Amand, embouteillée, gazeuse ou non gazeuse, moyennement minéralisée (St-Amand, 59) | 0 | 0 | 0 | 0 | 99,9 | 0 | 0 | 0 | 0 | 0 | 0,0061 | 17,6 | 3,7 | < 0,0005 | 0,0008 | 0 | 4,6 | 0,0006 | < 0,003 | 0,5 | < 1 | 2,8 | < 0,0005 | 0 | 0 | - | - | 0 | 0 | 0 | 0 | 0 | 0 | 0 | 0 |
| Eau minérale St-Antonin, embouteillée, non gazeuse, fortement minéralisée (St-Antonin-Noble-Val, 82) | 0 | 0 | 0 | 0 | 99,8 | 0 | 0 | 0 | 0 | 0 | 0,0015 | 56,8 | 0,9 | < 0,0005 | 0,009 | 0 | 8,9 | 0,0003 | < 0,003 | 0,3 | < 1 | 0,79 | 0,0008 | 0 | 0 | - | - | 0 | 0 | 0 | 0 | 0 | 0 | 0 | 0 |
| Eau minérale St-Diéry, embouteillée, gazeuse, fortement minéralisée (St-Diéry, 63) | 0 | 0 | 0 | 0 | 99,8 | 0 | 0 | 0 | 0 | 0 | 0,019 | 8,5 | 28,5 | < 0,0005 | < 0,001 | 0 | 8 | 0,019 | < 0,003 | 6,5 | < 1 | 7,7 | < 0,0005 | 0 | 0 | - | - | 0 | 0 | 0 | 0 | 0 | 0 | 0 | 0 |
| Eau minérale Ste-Marguerite, embouteillée, gazeuse, moyennement minéralisée (St-Maurice, 63) | 0 | 0 | 0 | 0 | 99,9 | 0 | 0 | 0 | 0 | 0 | 0,038 | 7,1 | 23 | 0,0019 | < 0,001 | 0 | 4 | < 0,0001 | < 0,003 | 3,3 | < 1 | 30,2 | 0,0012 | 0 | 0 | - | - | 0 | 0 | 0 | 0 | 0 | 0 | 0 | 0 |
| Eau minérale St-Géron, embouteillée, gazeuse, moyennement minéralisée (St-Géron, 43) | 0 | 0 | 0 | 0 | 99,9 | 0 | 0 | 0 | 0 | 0 | 0,0073 | 7,91 | 4,42 | - | - | - | 5,37 | - | - | 1,84 | - | 25,6 | - | - | - | - | - | - | - | - | - | - | - | - | - |
| Eau minérale St-Michel-de-Mourcairol, embouteillée, gazeuse, moyennement minéralisée (Les Aires, 34) | 0 | 0 | 0 | 0 | 99,9 | 0 | 0 | 0 | 0 | 0 | 0,0036 | 23 | 2,2 | - | - | - | 10 | - | - | 5,5 | - | 17,5 | - | - | - | - | - | - | - | - | - | - | - | - | - |
| Eau minérale St-Yorre, embouteillée, gazeuse, fortement minéralisée (Saint-Yorre, 03) | 0 | 0 | 0 | 0 | 99,5 | 0 | 0 | 0 | 0 | 0 | 0,053 | 9 | 32,2 | < 0,0005 | 0,0015 | 0 | 1,1 | 0,0013 | < 0,003 | 11 | < 1 | 171 | < 0,0005 | 0 | 0 | - | - | 0 | 0 | 0 | 0 | 0 | 0 | 0 | 0 |
| Eau minérale Thonon, embouteillée, non gazeuse, faiblement minéralisée (Thonon, 74) | 0 | 0 | 0 | 0 | 100 | 0 | 0 | 0 | 0 | 0 | 0,0014 | 9,2 | 1,1 | < 0,0005 | < 0,001 | 0 | 1,9 | < 0,0001 | < 0,003 | < 0,1 | < 1 | 0,57 | < 0,0005 | 0 | 0 | - | - | 0 | 0 | 0 | 0 | 0 | 0 | 0 | 0 |
| Eau minérale Treignac, embouteillée, non gazeuse, très faiblement minéralisée (Treignac, 19) | 0 | 0 | 0 | 0 | 100 | 0 | 0 | 0 | 0 | 0 | 0,00053 | 0,12 | 0,32 | - | - | - | - | - | - | < 0,05 | - | 0,28 | - | - | - | - | - | - | - | - | - | - | - | - | - |
| Eau minérale Vals, embouteillée, gazeuse, moyennement minéralisée (Vals-les-Bains, 07) | 0 | 0 | 0 | 0 | 99,9 | 0 | 0 | 0 | 0 | 0 | 0,0041 | 2,22 | 2,46 | < 0,0005 | < 0,001 | 0 | 1,35 | 0,06 | 0,0023 | 3,38 | < 1 | 38,1 | 0,0031 | 0 | 0 | - | - | 0 | 0 | 0 | 0 | 0 | 0 | 0 | 0 |
| Eau minérale Valvert, embouteillée, non gazeuse, faiblement minéralisée (Belgique) | 0 | 0 | 0 | 0 | 100 | 0 | 0 | 0 | 0 | 0 | 0,00049 | 6,78 | 0,4 | 0 | 0 | 0 | 0,2 | - | 0 | 0,045 | 0 | 0,2 | traces | 0 | 0 | - | - | 0 | 0 | 0 | 0 | 0 | 0 | 0 | 0 |
| Eau minérale Vauban, embouteillée, non gazeuse, moyennement minéralisée (St-Amand-les-Eaux, 59) | 0 | 0 | 0 | 0 | 99,9 | 0 | 0 | 0 | 0 | 0 | 0,0096 | 23 | 5,8 | - | - | - | 6,6 | - | - | 0,8 | - | 4 | - | - | - | - | - | - | - | - | - | - | - | - | - |
| Eau minérale Ventadour, embouteillée, gazeuse, faiblement minéralisée (Le Pestrin, 07) | 0 | 0 | 0 | 0 | 100 | 0 | 0 | 0 | 0 | 0 | 0,00033 | 4,5 | 0,2 | 0,0027 | < 0,001 | 0 | 1,05 | 0,04 | < 0,003 | 0,21 | < 1 | 1,34 | 0,006 | 0 | 0 | - | - | 0 | 0 | 0 | 0 | 0 | 0 | 0 | 0 |
| Eau minérale Vernet, embouteillée, gazeuse, faiblement minéralisée (Prades, 07) | 0 | 0 | 0 | 0 | 100 | 0 | 0 | 0 | 0 | 0 | 0,00086 | 3,2 | 0,52 | < 0,0005 | < 0,001 | 0 | 1,7 | 0,05 | < 0,003 | 2,2 | < 1 | 12,9 | 0,004 | 0 | 0 | - | - | 0 | 0 | 0 | 0 | 0 | 0 | 0 | 0 |
| Eau minérale Vernière, embouteillée, gazeuse, moyennement minéralisée (Les Aires, 34) | 0 | 0 | 0 | 0 | 99,9 | 0 | 0 | 0 | 0 | 0 | 0,0023 | 18 | 1,4 | < 0,0005 | < 0,001 | 0 | 7,3 | 0,04 | < 0,003 | 4 | < 1 | 11 | 0,016 | 0 | 0 | - | - | 0 | 0 | 0 | 0 | 0 | 0 | 0 | 0 |
| Eau minérale Vichy Célestins, embouteillée, gazeuse, fortement minéralisée (Saint-Yorre, 03) | 0 | 0 | 0 | 0 | 99,7 | 0 | 0 | 0 | 0 | 0 | 0,039 | 10,3 | 23,5 | < 0,0005 | < 0,001 | 0 | 1 | 0,0014 | < 0,003 | 6,6 | < 1 | 117 | 0,0008 | 0 | 0 | - | - | 0 | 0 | 0 | 0 | 0 | 0 | 0 | 0 |
| Eau minérale Vittel, embouteillée, non gazeuse, moyennement minéralisée (Vittel, 88) | 0 | 0 | 0 | 0 | 99,9 | 0 | 0 | 0 | 0 | 0 | 0,0013 | 24 | 0,8 | < 0,0023 | < 0,001 | 0 | 4,2 | < 0,0015 | < 0,003 | 0,19 | < 10 | 0,52 | 0,02 | 0 | 0 | - | - | 0 | 0 | 0 | 0 | 0 | 0 | 0 | 0 |
| Eau minérale Volvic active, embouteillée, gazeuse, faiblement minéralisée (Volvic, 63) | 0 | 0 | 0 | 0 | 100 | 0 | 0 | 0 | 0 | 0 | 0,0019 | 1,07 | 1,17 | < 0,0005 | < 0,001 | 0 | 0,75 | < 0,0001 | 0,02 | 0,58 | < 1 | 1,17 | < 0,0005 | 0 | 0 | - | - | 0 | 0 | 0 | 0 | 0 | 0 | 0 | 0 |
| Eau minérale Volvic, embouteillée, non gazeuse, faiblement minéralisée (Volvic, 63) | 0 | 0 | 0 | 0 | 100 | 0 | 0 | 0 | 0 | 0 | 0,0025 | 1,2 | 1,5 | < 0,0023 | < 0,001 | 0 | 0,8 | < 0,0015 | 0,02 | 0,6 | < 10 | 1,2 | < 0,01 | 0 | 0 | - | - | 0 | 0 | 0 | 0 | 0 | 0 | 0 | 0 |
| Eau minérale Wattwiller, embouteillée, gazeuse ou non gazeuse, faiblement minéralisée (Wattwiller, 68) | 0 | 0 | 0 | 0 | 100 | 0 | 0 | 0 | 0 | 0 | 0,00075 | 3,5 | 0,32 | < 0,0005 | < 0,001 | 0 | 1,1 | 0,005 | < 0,003 | 0,11 | < 1 | 0,3 | 0,011 | 0 | 0 | - | - | 0 | 0 | 0 | 0 | 0 | 0 | 0 | 0 |
| Eau minérale, embouteillée, faiblement minéralisée (aliment moyen) | 0 | 0 | 0 | 0 | 100 | 0 | 0 | 0 | 0 | 0 | 0,0019 | 6,54 | 1,2 | 0,0012 | 0,0033 | 0 | 1,64 | 0,00069 | 0,0061 | 0,22 | 2,23 | 0,92 | 0,0047 | 0 | 0 | | | 0 | 0 | 0 | 0 | 0 | 0 | 0 | 0 |
| Eau minérale, gazeuse (aliment moyen) | 0 | 0 | 0 | 0 | 99,9 | 0 | 0 | 0 | 0 | 0 | 0,011 | 15,9 | 6,89 | 0,00057 | 0,00058 | 0 | 3,78 | 0,033 | 0,0014 | 1,49 | 1,84 | 16,3 | 0,0056 | 0 | 0 | | | 0 | 0 | 0 | 0 | 0 | 0 | 0 | 0 |
| Eau minérale, plate (aliment moyen) | 0 | 0 | 0 | 0 | 99,9 | 0 | 0 | 0 | 0 | 0 | 0,0019 | 20,5 | 1,2 | 0,0014 | 0,0021 | 0 | 4,16 | 0,0008 | 0,0041 | 0,26 | 3,07 | 0,96 | 0,0077 | 0 | 0 | | | 0 | 0 | 0 | 0 | 0 | 0 | 0 | 0 |
| Jus d'ananas, à base de concentré | 222 | 52,3 | 222 | 52,3 | 87,1 | 0,41 | 0,41 | 11,9 | 0,096 | 0,18 | 0,011 | 14,3 | - | 0,03 | 0,15 | 1 | 17 | 1,46 | 9,12 | 144 | < 10 | 4,62 | 0,066 | 0 | 0,02 | 0,3 | - | 15 | 0,058 | 0,021 | 0,2 | 0,056 | 0,1 | 18 | 0 |
| Jus d'ananas, pur jus | 225 | 52,9 | 225 | 52,9 | 86,3 | 0,41 | 0,41 | 12,1 | 0,074 | 0,22 | 0,013 | 8,08 | - | 0,04 | 0,18 | 1 | 13,6 | 1,2 | 9,8 | 134 | 1,1 | 5,2 | 0,073 | 0 | 0,02 | - | - | 14 | 0,055 | 0,02 | 0,3 | 0,15 | 0,1 | 23 | 0 |
| Jus de carotte, pur jus | - | - | - | - | 88,9 | 0,4 | 0,4 | 6,55 | 0,1 | 0,33 | 0,1 | 13,6 | 45,6 | 0,051 | 0,46 | 2 | 8,75 | 0,071 | 25,6 | 210 | < 2,2 | 42,5 | 0,086 | 0 | 1,16 | 15,5 | - | 8,5 | 0,092 | 0,035 | 0,39 | 0,23 | 0,22 | 4 | 0 |
| Jus de citron vert, maison | - | - | - | - | 90,8 | 0,42 | 0,42 | 1,69 | 0,07 | 0,4 | 0,005 | 14 | - | 0,027 | 0,09 | - | 8 | 0,018 | 14 | 117 | - | 2 | 0,08 | 0 | 0,22 | 0,6 | - | 30 | 0,025 | 0,015 | 0,14 | 0,12 | 0,038 | 10 | 0 |
| Jus de citron vert, pur jus | - | - | - | - | 92,5 | 0,25 | 0,25 | 1,24 | 0,23 | 0,45 | 0,026 | 12 | - | 0,03 | 0,23 | 1,2 | 7 | 0,008 | 10 | 75 | - | 10,5 | 0,06 | 0 | 0,12 | 0,5 | - | 6,4 | 0,033 | 0,003 | 0,16 | 0,066 | 0,027 | 8 | 0 |
| Jus de citron, maison | - | - | - | - | 91,5 | 0,49 | 0,49 | 2,41 | 0,24 | 0,3 | 0,0065 | 12,4 | - | 0,042 | 0,24 | 0,6 | 5,99 | 0,008 | 14,5 | 132 | < 2,2 | 2,62 | 0,049 | 0 | 0,095 | 0 | - | 42,4 | 0,022 | 0,013 | 0,096 | 0,12 | 0,048 | 13,5 | 0 |
| Jus de citron, pur jus | - | - | - | - | 92,5 | 0,4 | 0,4 | 6,1 | 0,29 | 0,4 | 0,053 | 11 | - | 0,037 | 0,13 | - | 8 | 0,02 | 9 | 102 | - | 21 | 0,06 | 0 | 0,15 | 0 | - | 24,8 | 0,041 | 0,009 | 0,2 | 0,091 | 0,043 | 10 | 0 |
| Jus de clémentine ou mandarine, pur jus | - | - | - | - | 88 | 0,66 | 0,66 | 10,6 | 0,14 | 0,15 | 0,0094 | 2,8 | < 60 | < 0,1 | 0,3 | 1 | 9,3 | < 0,1 | 16 | 123 | 0,5 | 3,79 | < 0,1 | 0 | 0,09 | - | - | 17,6 | 0,077 | 0,018 | 0,25 | 0,13 | 0,066 | 19 | 0 |
| Jus de fruit de la passion ou maracudja, frais | - | - | - | - | 84,2 | 0,67 | 0,67 | 14,3 | 0,18 | 0,2 | 0,015 | 4 | - | 0,05 | 0,36 | - | 17 | - | 25 | 278 | - | 6 | 0,06 | 0 | 0,01 | 0,4 | - | 18,2 | 0 | 0,1 | 2,24 | - | 0,06 | 8 | 0 |
| Jus de fruit(s) et de légume(s), pur jus | - | - | - | - | - | 0,85 | 0,85 | 9,5 | traces | 0,7 | traces | - | - | - | - | - | - | - | - | - | - | 0,1 | - | - | - | - | - | - | - | - | - | - | - | - | - |
| Jus de fruits (aliment moyen) | 200 | 47 | 200 | 47 | 88,5 | 0,52 | 0,52 | 9,58 | 0,14 | 0,29 | 0,012 | 5 | | 0,029 | 0,12 | 2,02 | 9,22 | 0,068 | 14,3 | 152 | | 5,18 | 0,041 | 0,0001 | 0,18 | 0,14 | | 32,5 | 0,064 | 0,023 | 0,29 | 0,19 | 0,086 | 22,8 | 0 |
| Jus de fruits, à base de concentré (aliment moyen) | 200 | 47,1 | 200 | 47,1 | 88,8 | 0,54 | 0,54 | 10,5 | 0,13 | 0,35 | 0,012 | 8,62 | | 0,028 | 0,097 | 3,22 | 9,82 | 0,12 | 14 | 147 | 3,93 | 4,88 | 0,03 | 0 | 0,24 | 0,027 | | 31,6 | 0,079 | 0,043 | 0,44 | 0,23 | 0,11 | 23,9 | 0 |
| Jus de fruits, pur jus (aliment moyen) | 199 | 46,9 | 199 | 46,9 | 88,4 | 0,51 | 0,51 | 9,32 | 0,14 | 0,27 | 0,012 | 3,93 | | 0,029 | 0,12 | 1,66 | 9,05 | 0,054 | 14,3 | 154 | | 5,26 | 0,044 | 0,00013 | 0,16 | 0,16 | | 32,7 | 0,059 | 0,018 | 0,25 | 0,18 | 0,08 | 22,4 | 0 |

| | | | | | | | | | | | | | | | | | | | | | | | | | | | | | | | | | | | |
|---|---|---|---|---|---|---|---|---|---|---|---|---|---|---|---|---|---|---|---|---|---|---|---|---|---|---|---|---|---|---|---|---|---|---|---|
| Jus de grenade, frais | - | - | - | - | 85,4 | 0,2 | 0,2 | 11,6 | 0 | traces | 0,0025 | 3 | - | - | 0,2 | - | - | - | 8 | 200 | - | 1 | 0,1 | 0 | 0,55 | - | - | 8 | 0,02 | 0,03 | 0,2 | - | 0,31 | 6 | - |
| Jus de grenade, pur jus | - | - | - | - | 86 | 0,15 | 0,15 | 13 | 0,29 | 0,1 | 0,023 | 11 | - | 0,021 | 0,1 | - | 7 | 0,095 | 11 | 214 | - | 9 | 0,09 | 0 | 0,38 | 10,4 | - | 0,1 | 0,015 | 0,015 | 0,23 | 0,29 | 0,04 | 24 | 0 |
| Jus de légumes, pur jus (aliment moyen) | 91,7 | 21,7 | 91,7 | 21,7 | 92,4 | 0,72 | 0,72 | 4,66 | 0,13 | 0,37 | 0,33 | 8,89 | 134 | 0,068 | 0,45 | 0,87 | 11,7 | 0,08 | 23 | 258 | 0,7 | 131 | 0,16 | 0 | 0,85 | 10,3 | | 24,3 | 0,07 | 0,043 | 0,53 | 0,24 | 0,16 | 8,5 | 0 |
| Jus de mangue, frais | - | - | - | - | 86 | 0,19 | 0,19 | 9,5 | 0 | traces | 0,028 | - | - | 0,02 | - | 3 | 14,3 | 0,02 | - | - | 0,1 | - | 0,02 | 0 | 1,05 | - | - | - | 0,01 | - | - | 0,14 | 0,12 | 27,1 | 0 |
| Jus de pamplemousse (pomélo), à base de concentré | 174 | 40,9 | 174 | 40,9 | 89,9 | 0,52 | 0,52 | 8,41 | 0,1 | 0,21 | 0,01 | 10,6 | - | 0,031 | 0,15 | 0,6 | 7,91 | 0,016 | 12,5 | 115 | <2,2 | 3,97 | 0,037 | 0 | 0,04 | 0 | - | 23,7 | 0,041 | 0,015 | 0,055 | 0,13 | 0,016 | 11,8 | 0 |
| Jus de pamplemousse (pomélo), maison | - | - | - | - | 90 | 0,5 | 0,5 | 9,1 | 0,1 | 0,1 | 0,0025 | 9 | - | 0,033 | 0,2 | 0,6 | 12 | 0,02 | 15 | 162 | - | 1 | 0,05 | 0 | 0,22 | 0 | - | 38 | 0,04 | 0,02 | 0,2 | 0,19 | 0,044 | 10 | 0 |
| Jus de pamplemousse (pomelo), pur jus | 181 | 42,5 | 181 | 42,5 | 89,9 | 0,49 | 0,49 | 8,89 | 0,083 | 0,15 | 0,008 | 4,83 | <60 | <0,1 | 0,15 | 1 | 8,63 | <0,1 | 15 | 128 | - | 3,05 | <0,1 | 0 | 0,19 | - | - | 22,5 | 0,043 | 0,017 | 0,12 | 0,13 | 0,048 | 12,6 | - |
| Jus de pomme, à base de concentré | 206 | 48,5 | 206 | 48,5 | 88,5 | 0,24 | 0,24 | 11,3 | 0,098 | 0,3 | 0,014 | 6,45 | - | 0,011 | 0,085 | 0,7 | 7,07 | 0,026 | 6 | 128 | <10 | 3,98 | 0,01 | 0 | 0,01 | 0 | - | 13,2 | 0,021 | 0,017 | 0,16 | 0,063 | 0,03 | 4,95 | 0 |
| Jus de pomme, pur jus | 208 | 48,9 | 208 | 48,9 | 88,4 | 0,17 | 0,17 | 11,4 | 0,14 | 0,23 | 0,022 | 4,32 | <60 | <0,1 | 0,3 | 1 | 4,56 | <0,1 | 6,59 | 113 | 0,1 | 13,3 | <0,1 | 0 | 0,01 | 0 | - | 14,5 | 0,013 | <0,01 | 0,043 | 0,045 | 0,042 | 6,75 | 0 |
| Jus de pruneau | - | - | - | - | 81,2 | 0,37 | 0,37 | 18,7 | 0,039 | 0,45 | 0,025 | 12 | - | 0,068 | 1,18 | - | 14 | 0,15 | 25 | 276 | - | 10 | 0,21 | 0 | 0,12 | 3,4 | - | 4,1 | 0,016 | 0,07 | 0,79 | 0,11 | 0,22 | - | 0 |
| Jus de raisin, à base de concentré | 293 | 68,8 | 293 | 68,8 | 84,2 | 0,31 | 0,31 | 16 | 0,11 | 0,15 | 0,0074 | 16 | 6,5 | <0,01 | 0,43 | <20 | 7,8 | 0,06 | 17 | 62 | <20 | 2,83 | <0,05 | - | 0,67 | <0,8 | - | 9 | 0,046 | 0,021 | 0,23 | 0,063 | 0,073 | 43 | - |
| Jus de raisin, pur jus | 294 | 69,2 | 294 | 69,2 | 82,4 | 0,25 | 0,25 | 16,3 | 0,057 | 0,14 | 0,0089 | 8,29 | - | 0,031 | 0,47 | 0,5 | 8,35 | 0,089 | 12,5 | 114 | <10 | 3,81 | 0,056 | 0 | 0,4 | 0,4 | - | - | 0,022 | 0,026 | 0,29 | 0,045 | 0,049 | <5 | 0 |
| Jus de tomate, pur jus (aliment moyen) | 91,7 | 21,7 | 91,7 | 21,7 | 94,1 | 0,88 | 0,88 | 3,72 | 0,14 | 0,4 | 0,45 | 6,54 | | | | 0,3 | 13,1 | | 21,8 | 283 | | 175 | | | | | | | | | | | | | |
| Jus de tomate, pur jus, salé à 3 g/L | 98,4 | 23,2 | 98,4 | 23,2 | 94,3 | 0,88 | 0,88 | 4,13 | 0,17 | 0,23 | 0,3 | 7,98 | 222 | - | - | 0,4 | 12,5 | - | 23,2 | 263 | - | 119 | - | - | - | - | - | - | - | - | - | - | - | - | - |
| Jus de tomate, pur jus, salé à 6g/L | 85 | 20,1 | 85 | 20,1 | 93,9 | 0,88 | 0,88 | 3,31 | 0,11 | 0,56 | 0,6 | 5,1 | - | 0,086 | 0,43 | 0,2 | 13,7 | 0,09 | 20,3 | 302 | 0,3 | 231 | 0,23 | 0 | 0,53 | 5,1 | - | 40 | 0,047 | 0,031 | 0,67 | 0,25 | 0,11 | 13 | 0 |
| Jus d'orange sanguine, pur jus | - | - | - | - | - | 0,63 | 0,63 | 10,2 | 0,1 | 0,68 | 0,01 | - | - | - | - | - | - | - | - | - | - | 4 | - | 0 | - | - | - | 39,8 | - | - | - | - | - | - | 0 |
| Jus d'orange, à base de concentré | 194 | 45,6 | 194 | 45,6 | 89,7 | 0,63 | 0,63 | 9,59 | 0,13 | 0,25 | 0,0084 | 7,25 | - | 0,023 | 0,08 | 0,4 | 10,9 | 0,02 | 16,9 | 154 | <10 | 3,44 | 0,022 | 0 | 0,15 | 0 | - | 44,2 | 0,08 | 0,034 | 0,3 | 0,24 | 0,1 | 30 | 0 |
| Jus d'orange, maison | - | - | - | - | 88,3 | 0,7 | 0,7 | 9,4 | 0,2 | 0,2 | 0,0025 | 10,5 | - | 0,025 | 0,079 | 1 | 13,2 | 0,022 | 17 | 200 | <10 | <1,11 | 0,027 | 0 | 0,04 | 0,1 | - | 50 | 0,09 | 0,03 | 0,4 | 0,19 | 0,04 | 30 | 0 |
| Jus d'orange, pur jus | 193 | 45,4 | 193 | 45,4 | 88,2 | 0,61 | 0,61 | 9,61 | 0,11 | 0,28 | 0,012 | 0,41 | - | 0,022 | 0,068 | 0,7 | 10 | 0,021 | 16,5 | 165 | - | 4,81 | 0,045 | 0 | 0,2 | 0,1 | - | 37 | 0,08 | 0,021 | 0,3 | 0,24 | 0,1 | 30 | 0 |
| Jus multifruit - base orange, multivitaminé | - | - | - | - | 87,6 | <0,5 | <0,5 | 11,7 | <0,8 | <0,5 | 0,008 | 11 | 11,2 | 0,02 | 0,1 | <20 | 8,7 | 0,27 | 11 | 150 | <20 | 3,2 | <0,05 | - | - | <0,8 | - | 34,7 | 0,34 | 0,077 | 4,76 | 1,19 | 0,45 | 90 | 1 |
| Jus multifruit - base orange, standard | - | - | - | - | 88,6 | 0,7 | 0,7 | 10,4 | 0,16 | 0,44 | 0,069 | - | - | - | - | - | 11,9 | - | - | 152 | - | 1,55 | - | - | - | - | - | 20,8 | - | - | 0,2 | - | - | 19,2 | - |
| Jus multifruit - base pomme, multivitaminé | - | - | - | - | 88,2 | 0,33 | 0,33 | 11 | <0,1 | <0,5 | - | - | - | - | - | - | 5,9 | - | - | 69,5 | - | - | - | - | - | - | - | 26,2 | - | - | 0,36 | - | - | 20,7 | - |
| Jus multifruit - base pomme, standard | - | - | - | - | 87,8 | 0,33 | 0,33 | 12,2 | <0,1 | <0,5 | 0 | - | - | - | - | - | 5,94 | - | - | 99,1 | - | 0 | - | - | - | - | - | 25,4 | - | - | 0,38 | - | - | 16,2 | - |
| Jus multifruit - base raisin, standard | - | - | - | - | - | 0,5 | 0,5 | 13 | 0 | 0,3 | 0,0025 | - | - | - | - | - | - | - | - | - | - | 1 | - | - | - | - | - | - | - | - | - | - | - | - | - |
| Jus multifruit, à base de concentré, multivitaminé | - | - | - | - | - | 0,29 | 0,29 | 10,8 | 0,091 | 0,27 | 0,019 | - | - | - | - | - | - | - | - | - | - | 7,5 | - | - | 2,15 | - | - | 16,2 | 0,3 | 0,36 | 3,84 | 1,28 | 0,42 | 45,6 | - |
| Jus multifruit, à base de concentré, standard | - | - | - | - | 87,8 | 0,52 | 0,52 | 11,2 | 0,15 | 0,62 | 0,02 | 11,1 | - | 0,05 | 0,1 | 10 | 8,2 | 0,2 | 12,1 | 148 | 0,5 | 8,57 | 0,05 | 0 | 0,53 | - | - | 17,8 | 0,11 | 0,08 | 0,93 | 0,31 | 0,17 | 18,8 | 0 |
| Jus multifruit, à base de jus et purée de fruits | - | - | - | - | - | 0,46 | 0,46 | 11,5 | 0,11 | 0,21 | 0,01 | - | - | - | - | - | - | - | - | - | - | 3,99 | - | - | 1,12 | - | - | 19,4 | 0,13 | 0,13 | 1,45 | 0,52 | 0,18 | 23 | - |
| Jus multifruit, pur jus, multivitaminé | - | - | - | - | 86,5 | 0,48 | 0,48 | 11,3 | 0,1 | 0,25 | 0,0038 | 9,6 | <60 | 0,018 | 0,12 | 1 | 9,98 | 0,14 | 10 | 171 | <10 | 1,5 | 0,036 | <0,5 | 2 | - | - | 24 | 0,24 | 0,32 | 0,29 | 0,095 | 0,33 | 30,5 | 0 |
| Jus multifruit, pur jus, standard | 208 | 48,9 | 208 | 48,9 | 87,3 | <0,5 | <0,5 | 11,2 | <0,3 | <0,5 | 0,0023 | 5,4 | 6,5 | 0,03 | 0,08 | <20 | 7 | 0,09 | 10 | 140 | <20 | 0,92 | <0,05 | 0 | 0,2 | <0,8 | - | 11,5 | <0,015 | <0,01 | 0,17 | 0,099 | 0,083 | <5 | 0 |
| Kombucha, préemballé | 40 | 9,46 | 40 | 9,46 | 97,6 | <0,5 | <0,5 | 1,45 | <0,3 | <0,5 | 0,0053 | 5,7 | 6,5 | <0,01 | <0,05 | <20 | 1,5 | 0,07 | 0,5 | 7,7 | <20 | 2,1 | <0,05 | <0,25 | <0,08 | <0,8 | - | <0,5 | <0,015 | <0,01 | <0,1 | 0,0077 | <0,01 | <5 | <0,01 |
| Lait de coco ou Crème de coco | 775 | 188 | 775 | 188 | 75,1 | 1,77 | 1,77 | 3,4 | 18,4 | 0,57 | 0,075 | 18 | - | 0,22 | 3,3 | 1 | 46 | 0,77 | 96 | 220 | 3 | 30 | 0,56 | 0 | 0,7 | - | - | 1 | 0,022 | 0 | 0,64 | 0,15 | 0,028 | 14 | 0 |
| Limonade, non sucrée, avec édulcorants | - | - | - | - | 99 | 0,067 | 0,067 | 0,1 | 0 | 0 | 0,022 | 0 | - | 0 | 0 | 0 | 0 | 0 | 0 | 0 | 0 | 8,56 | 0 | 0 | 0 | - | - | 0 | 0 | 0 | 0 | 0 | 0 | 0 | 0 |
| Limonade, sucrée | - | - | - | - | 90,5 | 0,04 | 0,04 | 8,45 | 0,02 | traces | 0,015 | 6,52 | - | 0,0029 | 0,0092 | 0,5 | <0,58 | <0,0015 | - | 3,51 | <10 | 5,72 | <0,01 | 0 | 0 | - | - | 0 | 0 | 0 | 0 | 0 | 0 | 0 | 0 |
| Limonade, sucrée, avec édulcorants | - | - | - | - | - | 0 | 0 | 3 | 0 | 0 | - | - | - | - | - | - | - | - | - | - | - | - | - | - | - | - | - | - | - | - | - | - | - | - | - |
| Nectar d'abricot | - | - | - | - | 84,9 | 0,21 | 0,21 | 13,7 | 0,03 | 0,08 | 0,0042 | 11,5 | - | 0,039 | 0,38 | 0,2 | 5,28 | 0,022 | 9 | 114 | <2,2 | 1,68 | 0,05 | 0 | 0,31 | 1,2 | - | 17,6 | 0,009 | 0,014 | 0,26 | 0,096 | 0,022 | 1 | 0 |
| Nectar d'ananas | - | - | - | - | 88,1 | 0,1 | 0,1 | 13 | 0 | 0 | 0 | 13 | - | - | - | - | 8 | - | - | 90 | - | 0 | - | - | - | - | - | - | 0,02 | 0,01 | - | - | 0,05 | - | - |
| Nectar de banane | - | - | - | - | 86,7 | 0,23 | 0,23 | 13,6 | 0,048 | 0,51 | 0,0046 | 5,8 | 19 | 0,02 | 0,07 | <20 | 8 | 0,06 | 5,7 | 91 | <20 | 1,82 | <0,05 | - | - | <0,8 | - | 11,6 | 0,013 | <0,01 | 1,24 | 0,053 | 0,059 | 14,8 | - |
| Nectar de fruit de la passion ou maracuja | - | - | - | - | 85,5 | 0,3 | 0,3 | 15,2 | 0,1 | traces | 0,025 | 1,7 | - | 0,02 | - | 1,1 | 3,8 | - | - | 72,8 | 0,1 | 3 | 0,02 | - | traces | - | - | 17,4 | - | - | - | 0,05 | 0,02 | 3 | - |
| Nectar de goyave | - | - | - | - | 88 | 0,14 | 0,14 | 11,3 | 0,16 | 0,14 | 0,0051 | 7,8 | 8,4 | 0,02 | 0,07 | <20 | 2,6 | 0,03 | 4,5 | 61 | <20 | 2,02 | <0,05 | 0 | - | <0,8 | - | 36 | 0,011 | <0,01 | 1,39 | 0,034 | 0,011 | 10,3 | 0 |
| Nectar de mangue | - | - | - | - | 87 | 0,2 | 0,2 | 12,3 | 0,083 | 0,15 | 0,011 | 8,1 | 4,8 | 0,02 | 0,08 | <20 | 3,7 | 0,03 | 3,8 | 39 | <20 | 4,59 | <0,05 | - | - | <0,8 | - | 18 | 0,024 | 0,011 | 2,1 | 0,17 | 0,039 | 17,1 | - |
| Nectar de papaye | - | - | - | - | 85 | 0,17 | 0,17 | 13,9 | 0,15 | 0,6 | 0,013 | 10 | - | 0,013 | 0,34 | - | 3 | 0,013 | 0 | 31 | - | 5 | 0,15 | 0 | 0,24 | 0,8 | - | 3 | 0,006 | 0,004 | 0,15 | 0,054 | 0,009 | 2 | 0 |
| Nectar de pêche | - | - | - | - | 85,6 | 0,22 | 0,22 | 12,1 | 0,05 | 0,33 | 0,011 | 5 | - | 0,069 | 0,19 | - | 4 | 0,019 | 6 | 40 | - | 4,37 | 0,08 | 0 | 0,29 | 1 | - | 5,3 | 0,003 | 0,014 | 0,29 | 0,068 | 0,007 | 1 | 0 |
| Nectar de poire | - | - | - | - | 84 | 0,19 | 0,19 | 14,8 | 0,043 | 0,45 | 0,0033 | 5 | - | 0,067 | 0,26 | 0,7 | 3 | 0,03 | 3 | 13 | - | 1,3 | 0,07 | 0 | 0,05 | 1,8 | - | 1,1 | 0,002 | 0,013 | 0,13 | 0,022 | 0,014 | 1 | 0 |
| Nectar de pomme | - | - | - | - | 90,2 | 0,14 | 0,14 | 10,9 | 0,075 | 0,1 | 0,01 | 4,9 | - | - | - | - | 3,94 | - | 4,33 | 64,9 | - | 4 | - | - | - | - | - | - | - | - | 0,08 | - | - | 12,2 | - |
| Nectar d'orange | - | - | - | - | 89,7 | 0,29 | 0,29 | 8,79 | 0,08 | 0,18 | 0,018 | 7,69 | - | 0,023 | - | 1,1 | 5,96 | 0,022 | 7,7 | 82,8 | <2,2 | 6,51 | 0,041 | - | - | - | - | 21,2 | 0,04 | 0,01 | 0,17 | - | 0,01 | 10,8 | - |
| Nectar multifruit - base orange, multivitaminé | - | - | - | - | 88,1 | 0,4 | 0,4 | 11,1 | <0,1 | <0,5 | - | - | - | - | - | - | 6,3 | - | - | 85,7 | - | - | - | - | - | - | - | 36,2 | - | - | 1,64 | - | - | 23 | - |
| Nectar multifruit - base orange, standard | - | - | - | - | 87,4 | 0,3 | 0,3 | 12,2 | 0,12 | 0,5 | 0,025 | - | - | - | - | - | 6,7 | - | - | 86,8 | - | 10 | - | - | - | - | - | 13,2 | - | - | 0,21 | - | - | 13,8 | - |
| Nectar multifruit - base pomme, multivitaminé | - | - | - | - | 88,7 | <0,1 | <0,1 | 9,1 | <0,1 | traces | - | - | - | - | - | - | 4,4 | - | - | 43,2 | - | - | - | - | - | - | - | 15,1 | - | - | 0,28 | - | - | 17 | - |
| Nectar multifruit - base pomme, standard | - | - | - | - | 88,7 | 0,28 | 0,28 | 10 | <0,1 | traces | - | - | - | - | - | - | 4,7 | - | - | 78,5 | - | - | - | - | - | - | - | - | - | - | 0,27 | - | - | 35 | - |
| Nectar multifruit, multivitaminé | - | - | - | - | 87,4 | 0,24 | 0,24 | 11,3 | 0,1 | 0,15 | 0,016 | 3,1 | <60 | <0,1 | 0,2 | 1 | 5,4 | <0,1 | <10 | 79,1 | 0,3 | 6,36 | <0,1 | <0,5 | 1,05 | - | - | 16 | 0,14 | 0,2 | 1,63 | 0,62 | 0,22 | 22,9 | 0 |
| Nectar multifruit, standard | 211 | 49,7 | 211 | 49,7 | 87,2 | 0,3 | 0,3 | 11,5 | 0,089 | 0,3 | 0,009 | 9,17 | - | 0,05 | traces | 10 | 5,35 | 0,05 | 6,17 | 87,4 | 0,3 | 3,43 | traces | 0 | 0,97 | - | - | 20,6 | 0,095 | 0,025 | 1,09 | 0,6 | 0,14 | 22,1 | 0 |
| Nectar, avec édulcorants, allégé en sucres | - | - | - | - | - | 0,18 | 0,18 | 5,47 | 0,037 | 0,11 | 0,0087 | - | - | - | - | - | - | - | - | - | - | 3,48 | - | - | 1,5 | - | - | 10,8 | 0,21 | 0,24 | 2,7 | 0,9 | 0,3 | 30 | - |
| Poudre cacaotée ou au chocolat pour boisson, sucrée | 1609 | 379 | 1600 | 379 | 1,67 | 6,65 | 6,65 | 75,6 | 3,29 | 10,5 | 0,21 | 116 | - | 0,069 | 13 | 1,2 | 166 | 0,041 | 244 | 605 | <2,2 | 85 | 0,7 | 0 | 0,1 | - | - | 0 | 0,06 | 0,04 | 0,5 | 0,041 | 0,02 | 23 | 0 |
| Poudre cacaotée ou au chocolat pour boisson, sucrée, enrichie en vitamines | 1640 | 386 | 1640 | 386 | 2 | 4,29 | 4,29 | 85 | 2,04 | 5,3 | 0,31 | 36 | - | - | 2,1 | - | 71 | - | 120 | - | - | 124 | 1,7 | - | - | - | - | 25 | 0,48 | 0,66 | 5,23 | 2,2 | 0,76 | 66,7 | 0,33 |
| Poudre cacaotée ou au chocolat sucrée pour boisson, enrichie en vitamines et minéraux | 1590 | 376 | 1590 | 376 | - | 4,68 | 4,68 | 79,3 | 3,04 | 6,48 | 0,43 | 523 | - | - | 2,1 | - | 113 | - | 120 | - | - | 169 | - | - | 9 | - | - | 49,2 | 0,52 | 0,7 | 8,61 | 1,61 | 0,93 | 122 | 0,25 |
| Poudre maltée, cacaotée ou au chocolat pour boisson, sucrée, enrichie en vitamines et minéraux | - | - | - | - | 2,25 | 9,9 | 9,9 | 79,4 | 5,03 | - | 0,55 | 253 | 420 | 0,6 | 4,7 | 6 | 248 | 0,6 | 350 | 1100 | - | 219 | 1 | <0,5 | 11,8 | 2,5 | - | 48 | 0,98 | 1,57 | 16,8 | 6,02 | 1,84 | 219 | 2 |
| Sirop à diluer, sucré | 1080 | 255 | 1080 | 255 | 36,2 | <0,5 | <0,5 | 62,6 | <0,5 | <0,5 | 0,017 | 2,7 | <5 | <0,01 | 0,06 | <20 | 1,2 | 0,03 | 2,5 | 17 | <20 | 6,76 | <0,05 | - | - | <0,8 | - | <0,5 | <0,015 | <0,01 | <0,1 | 0,0074 | <0,01 | <5 | - |
| Smoothie | - | - | - | - | 84,2 | 0,62 | 0,62 | 12,4 | 0,26 | 1,08 | 0,01 | 9,65 | 17,6 | 0,045 | 0,24 | <20 | 11,5 | 0,27 | 15,5 | 165 | <20 | 3,84 | 0,07 | - | - | 0,88 | - | 20,2 | 0,059 | 0,012 | 0,2 | 0,079 | 0,064 | 28,6 | - |
| Thé infusé, non sucré | 0,26 | 0,063 | 0,26 | 0,063 | 99,8 | 0 | 0 | 0 | 0,007 | 0 | 0,0041 | 0,2 | - | 0,018 | 0,015 | 0,22 | 2 | 0,19 | 1 | 25,5 | 0,05 | 1,65 | 0,025 | 0 | 0 | 0 | - | 0 | 0 | 0,015 | 0,1 | 0,011 | 0 | 4,95 | 0 |
| Thé noir, infusé, non sucré | 5,36 | 1,26 | 5,36 | 1,26 | 99,7 | 0 | 0 | 0,3 | 0,007 | 0 | 0,0075 | 0 | - | 0,01 | 0,02 | - | 3 | 0,22 | 1 | 37 | - | 3 | 0,02 | 0 | 0 | 0 | - | 0 | 0 | 0,014 | 0 | 0,011 | 0 | 5 | 0 |
| Thé oolong, infusé, non sucré | 0 | 0 | 0 | 0 | - | 0 | 0 | 0 | 0 | 0 | - | - | - | - | - | - | - | - | - | - | - | - | - | - | - | - | - | - | - | - | - | - | - | - | - |

| | | | | | | | | | | | | | | | | | | | | | | | | | | | | | | | | | | | |
|---|---|---|---|---|---|---|---|---|---|---|---|---|---|---|---|---|---|---|---|---|---|---|---|---|---|---|---|---|---|---|---|---|---|---|---|
| Thé vert, infusé, non sucré | 21,7 | 5,1 | 21,7 | 5,1 | 98,5 | 0 | 0 | 1,09 | 0 | 0 | 0,011 | 3,78 | 8,27 | 0,03 | 0,06 | < 20 | 3,49 | 1,49 | 3,68 | 53,8 | < 20 | 4,25 | < 0,05 | - | - | < 0,8 | - | < 0,5 | < 0,015 | < 0,01 | 0,21 | 0,13 | < 0,01 | 6,99 | - |
| Thé, feuille | 957 | 232 | 957 | 232 | 9,3 | 19,4 | 19,4 | 6,5 | 2 | 55,8 | 0,11 | 430 | - | 2,5 | 18 | 16 | 250 | 71 | 630 | 2160 | - | 45 | 2,9 | 0 | 2,1 | - | - | 0 | 0,14 | 0,95 | 7,5 | 1,3 | 0,3 | - | 0 |
| Tisane infusée, non sucrée | 3,7 | 0,87 | 3,7 | 0,87 | 99,7 | 0 | 0 | 0,2 | 0,008 | 0 | 0,0025 | 2 | - | 0,015 | 0,08 | 1 | 1 | 0,044 | 0 | 9 | - | 1 | 0,04 | 0 | 0 | 0 | - | 0 | 0,01 | 0,004 | 0 | 0,011 | 0 | 1 | 0 |
| Tonic ou bitter, non sucré, avec édulcorants | - | - | - | - | - | 0,05 | 0,05 | 0,4 | traces | 0,1 | 0,13 | - | - | - | - | - | - | - | - | - | - | 50 | - | - | - | - | - | - | - | - | - | - | - | - | - |
| Tonic ou bitter, sucré | - | - | - | - | 88,9 | 0,068 | 0,068 | 8,28 | 0,066 | 0,05 | 0,014 | 7,48 | - | 0,0029 | 0,03 | 1 | 2,45 | 0,0011 | 0 | 1 | < 2,2 | 6,83 | 0,017 | 0 | 0 | 0 | - | 0 | 0 | 0 | 0 | 0 | 0 | 0 | 0 |
| Tonic ou bitter, sucré, avec édulcorants | - | - | - | - | - | 0 | 0 | 6,45 | 0 | 0,1 | 0 | - | - | - | - | - | - | - | - | - | - | 0 | - | - | - | - | - | - | - | - | - | - | - | - | - |

# Appendix 6

| aides culinaires et ingrédients divers | Energie, Règlement UE N° 1169/2011 (kJ/100 g) | Energie, Règlement UE N° 1169/2011 (kcal/100 g) | Energie, N x facteur Jones, avec fibres (kJ/100 g) | Energie, N x facteur Jones, avec fibres (kcal/100 g) | Eau (g/100 g) | Protéines, N x facteur de Jones (g/100 g) | Protéines, N x 6.25 (g/100 g) | Glucides (g/100 g) | Lipides (g/100 g) | Fibres alimentaires (g/100 g) | Sel chlorure de sodium (g/100 g) | Calcium (mg/100 g) | Chlorure (mg/100 g) | Cuivre (mg/100 g) | Fer (mg/100 g) | Iode (µg/100 g) | Magnésium (mg/100 g) | Manganèse (mg/100 g) | Phosphore (mg/100 g) | Potassium (mg/100 g) | Sélénium (µg/100 g) | Sodium (mg/100 g) | Zinc (mg/100 g) | Vitamine D (µg/100 g) | Vitamine E (mg/100 g) | Vitamine K1 (µg/100 g) | Vitamine K2 (µg/100 g) | Vitamine C (mg/100 g) | Vitamine B1 ou Thiamine (mg/100 g) | Vitamine B2 ou Riboflavine (mg/100 g) | Vitamine B3 ou PP ou Niacine (mg/100 g) | Vitamine B5 ou Acide pantothénique (mg/100 g) | Vitamine B6 (mg/100 g) | Vitamine B9 ou Folates totaux (µg/100 g) | Vitamine B12 (µg/100 g) |
|---|---|---|---|---|---|---|---|---|---|---|---|---|---|---|---|---|---|---|---|---|---|---|---|---|---|---|---|---|---|---|---|---|---|---|---|
| Cornichon, au vinaigre | 81,9 | 19,4 | 81,9 | 19,4 | 93,7 | 1,06 | 1,06 | 0,78 | < 0,6 | 1,5 | 1,72 | 65 | 1310 | 0,05 | 0,25 | < 20 | 23 | 0,06 | 26 | 120 | < 20 | 689 | 0,13 | 0 | 0,48 | 42,8 | - | < 0,5 | 0,13 | < 0,01 | 0,2 | 0,11 | < 0,01 | 11,5 | 0 |
| Moutarde | 632 | 152 | 632 | 152 | 67,9 | 6,92 | 6,92 | 4,33 | 11,2 | < 1 | 6,3 | 86,3 | - | 0,082 | 1,83 | 3 | 48,3 | 0,39 | 130 | 150 | 6 | 1860 | 0,67 | 0 | 0,36 | 1,4 | - | 0,3 | 0,11 | 0,048 | 0,57 | 0,19 | 0,071 | 7,5 | 0 |
| Vinaigre | 97,4 | 22,6 | 97,4 | 22,6 | 92,9 | 0,04 | 0,04 | 0,62 | 0,1 | 0 | 0,033 | 9 | 0,1 | 0,0075 | 0,37 | < 5 | 12,3 | 0,15 | 19 | 54,8 | 0,1 | 13,1 | 0,015 | 0 | 0 | 0 | - | 0,5 | 0 | 0 | 0 | 0 | 0,001 | 0 | 0 |
| Moutarde à l'ancienne | 591 | 142 | 591 | 142 | 67,8 | 6,6 | 6,6 | 2,2 | 9,9 | 4,5 | 4,9 | 110 | 2920 | 0,16 | 1,7 | < 20 | 95 | 0,57 | 210 | 200 | 70 | 1960 | 1,1 | < 0,25 | 1,18 | 2,63 | - | < 0,5 | 0,22 | 0,044 | 0,94 | 0,21 | 0,018 | 18 | - |
| Câpres, au vinaigre | - | - | - | - | 83,9 | 2,18 | 2,18 | 3,5 | 0,86 | 3,6 | 4,06 | 40 | - | 0,37 | 1,67 | - | 33 | 0,078 | 10 | 40 | - | 1620 | 0,32 | 0 | 0,88 | 24,6 | - | 4,3 | 0,018 | 0,14 | 0,65 | 0,027 | 0,023 | 23 | 0 |
| Tapenade | - | - | - | - | 62,3 | 1,3 | 1,3 | 2,6 | 23 | 3,7 | 3,2 | 63 | - | 0,15 | 4,3 | 4,5 | 20 | 0,05 | 23 | 258 | 2,2 | 1280 | 0,4 | 0 | 4,49 | - | - | 1,05 | 0 | 0,09 | 1,1 | 0,02 | 0,02 | 0 | 0 |
| Oignon au vinaigre | - | - | - | - | 93 | 0,56 | 0,56 | 2,5 | 0,2 | 1,8 | 2,36 | 24 | - | - | 0,4 | - | 12 | - | 29 | 110 | - | 908 | - | - | - | - | - | 7 | 0,03 | 0,03 | 0,2 | - | 0,22 | 20,7 | 0 |
| Vinaigre de cidre | - | - | - | - | 93,8 | 0 | 0 | 0,93 | 0 | 0 | 0,013 | 7 | - | 0,008 | 0,2 | - | 5 | 0,25 | 8 | 73 | - | 5 | 0,04 | 0 | 0 | 0 | - | 0 | 0 | 0 | 0 | 0 | 0 | 0 | 0 |
| Vinaigre balsamique | 531 | 125 | 531 | 125 | 70,3 | 0,69 | 0,69 | 25,8 | < 0,6 | 0,5 | 0,068 | 18 | < 20 | 0,03 | 0,95 | < 20 | 12 | 0,19 | 24 | 160 | < 20 | 27 | 0,12 | < 0,25 | - | < 0,8 | - | < 0,5 | < 0,015 | < 0,01 | < 0,1 | 0,072 | 1,71 | 5,5 | - |
| Cornichon, aigre-doux | 152 | 36 | 152 | 36 | 91,5 | 1,25 | 1,25 | 5 | 0,35 | 1,5 | 1,1 | - | - | - | - | - | - | - | - | - | - | 386 | - | - | - | - | - | - | - | - | - | - | - | - | - |
| Vinaigre de vin rouge | 93,3 | 21,6 | 93,3 | 21,6 | 91,5 | < 0,5 | < 0,5 | traces | 0 | 0 | 0,013 | 10 | < 5,1 | < 0,01 | 0,41 | - | 5,9 | 0,06 | 11 | 74 | < 20 | 5,1 | 0,08 | < 0,25 | < 0,08 | < 0,8 | - | < 0,5 | < 0,015 | < 0,01 | < 0,1 | 0,037 | 0,015 | < 5 | - |
| Olive noire, en saumure, égouttée | - | - | - | - | 74,9 | 1,38 | 1,38 | < 0,1 | 17,2 | 6,2 | 1,77 | 80 | 1070 | 0,18 | 8,5 | < 20 | 16 | 0,07 | 7,9 | 19 | < 50 | 706 | 0,14 | 0 | 6,1 | 1,4 | - | < 0,5 | 0,017 | < 0,01 | < 0,1 | 0,038 | 0,025 | < 5 | 0 |
| Olive verte, en saumure, égouttée | 637 | 155 | 637 | 155 | 75,8 | 1,31 | 1,31 | traces | 15,7 | 3,6 | 3,15 | 58 | 1760 | 0,11 | 0,16 | < 20 | 24 | 0,03 | 6,7 | 31 | < 20 | 1260 | 0,07 | < 0,25 | 5,23 | 13,9 | - | < 0,5 | < 0,015 | < 0,01 | < 0,1 | 0,035 | 0,018 | < 5 | 0 |
| Olive noire, à l'huile (à la grecque) | 1360 | 332 | 1360 | 332 | 45,7 | 2,19 | 2,19 | < 0,1 | 33,3 | 11,6 | 8,45 | 110 | 4610 | 0,27 | 1,7 | < 20 | 29 | 0,28 | 60 | 320 | < 50 | 3380 | 0,62 | 0 | 2,57 | - | - | 0,62 | 0,058 | 0,05 | 0,19 | 0,17 | 0,048 | 5,83 | 0,028 |
| Olives vertes, fourrées ou farcies (anchois, poivrons, etc.) | - | - | - | - | 75 | 2,09 | 2,09 | 0,61 | 15,2 | 4,33 | 2,78 | 66,8 | 1880 | 0,23 | 0,3 | 27,7 | 11,2 | < 0,1 | 14,8 | 32,9 | < 5 | 1110 | 0,11 | < 0,5 | 4,18 | - | - | 38,6 | < 0,04 | 0,021 | 0,11 | < 0,08 | < 0,04 | 13,8 | 0,08 |
| Olive (aliment moyen) | 667 | 162 | 667 | 162 | 74,6 | 1,39 | 1,39 | 0,043 | 16,5 | 4,38 | 3,03 | 64,2 | 1720 | 0,13 | 1,79 | 10,9 | 22 | 0,046 | 9,02 | 37,9 | 12,9 | 1210 | 0,1 | 0,1 | 5,26 | 11,3 | | 2,27 | 0,012 | 0,0072 | 0,057 | 0,04 | 0,02 | 3,2 | 0,0051 |
| Olive noire (aliment moyen) | | | | | 70,7 | 1,49 | 1,49 | 0,05 | 19,5 | 6,97 | 2,72 | 84,3 | 1580 | 0,19 | 7,53 | 10 | 17,9 | 0,1 | 15,3 | 62 | 25 | 1090 | 0,21 | 0 | 5,6 | 1,4 | | 0,3 | 0,023 | 0,011 | 0,07 | 0,056 | 0,028 | 2,98 | 0,004 |
| Sel blanc alimentaire, non iodé, non fluoré (marin, ignigène ou gemme) | 0 | 0 | 0 | 0 | 0,027 | 0 | 0 | 0 | 0 | 0 | 97,8 | 13,3 | 60800 | 0,15 | 0,29 | 1,8 | 3,15 | 0,12 | 8 | 16,9 | 0,5 | 39100 | 0,088 | 0 | 0 | 0 | - | 0 | 0 | 0 | 0 | 0 | 0 | 0 | 0 |
| Sel au céleri | 204 | 49 | 204 | 49 | - | 1,81 | 1,81 | 4,14 | 2,53 | 1,19 | 87,5 | - | - | - | - | - | - | - | - | - | - | 35000 | - | - | - | - | - | - | - | - | - | - | - | - | - |
| Sel blanc alimentaire, iodé, non fluoré (marin, ignigène ou gemme) | 0 | 0 | 0 | 0 | 0,027 | 0 | 0 | 0 | 0 | 0 | 97,8 | 13,3 | 60800 | - | - | 1860 | 3,15 | - | - | 16,9 | - | 39100 | - | - | - | - | - | - | - | - | - | - | - | - | - |
| Fleur de sel, non iodée, non fluorée | 0 | 0 | 0 | 0 | 3,75 | 0 | 0 | 0 | 0 | 0 | 94,3 | 171 | 58700 | - | - | - | 424 | - | - | 103 | - | 37700 | - | - | - | - | - | - | - | - | - | - | - | - | - |
| Sel marin gris, non iodé, non fluoré | 0 | 0 | 0 | 0 | 5,09 | 0 | 0 | 0 | 0 | 0 | 80,5 | 181 | 53200 | - | - | < 200 | 503 | - | - | 99,3 | - | 32200 | - | - | - | - | - | - | - | - | - | - | - | - | - |
| Sel blanc alimentaire, iodé, fluoré à 25 mg /100 g (marin, ignigène ou gemme) | 0 | 0 | 0 | 0 | 0,027 | 0 | 0 | 0 | 0 | 0 | 97,8 | 13,3 | 60800 | - | - | 1860 | 3,15 | - | - | 16,9 | - | 39100 | - | - | - | - | - | - | - | - | - | - | - | - | - |
| Curry, poudre | 1230 | 301 | 1230 | 301 | 8,8 | 14,5 | 14,5 | 2,63 | 14 | 53,2 | 0,13 | 525 | - | 1,2 | 19,1 | 0,5 | 255 | 8,3 | 367 | 1170 | - | 52 | 4,7 | 0 | 25,2 | 99,8 | - | 0,7 | 0,18 | 0,2 | 3,26 | 1,07 | 0,11 | 56 | 0 |
| Gingembre, poudre | 1410 | 335 | 1410 | 335 | 9,94 | 8,98 | 8,98 | 58,3 | 4,24 | 14,1 | 0,068 | 114 | - | 0,48 | 19,8 | - | 214 | 33,3 | 168 | 1320 | - | 27 | 3,64 | 0 | 0 | 0,8 | - | 0,7 | 0,046 | 0,17 | 9,62 | 0,48 | 0,63 | 13 | 0 |
| Poivre noir, poudre | 1380 | 330 | 1380 | 330 | 9 | 13,3 | 13,3 | 39,5 | 7,5 | 25,7 | 0,036 | 480 | 300 | 1 | 17 | < 20 | 190 | 15 | 170 | 1600 | < 20 | 14,2 | 0,99 | < 0,25 | 1,93 | 144 | - | < 0,5 | 0,12 | < 0,01 | 0,46 | 1,82 | 0,23 | 21,8 | 0 |
| Poivre blanc, poudre | 1300 | 310 | 1300 | 310 | 11,4 | 11,4 | 11,4 | 48,3 | 2,11 | 26,2 | 0,013 | 265 | - | 0,91 | 14,3 | - | 90 | 4,3 | 176 | 73 | 3 | 5 | 1,13 | 0 | - | 0 | - | 21 | 0,022 | 0,13 | 0,21 | - | 0,1 | 10 | 0 |
| Cannelle, poudre | 1000 | 243 | 1000 | 243 | 10,6 | 3,87 | 3,87 | 27,5 | 1,22 | 53,1 | 0,025 | 1000 | - | 0,34 | 8,32 | - | 60 | 17,5 | 64 | 431 | 3,1 | 10 | 1,83 | 0 | 2,32 | 31,2 | - | 3,8 | 0,022 | 0,041 | 1,33 | 0,36 | 0,16 | 6 | 0 |
| Coriandre, graine | 1430 | 346 | 1430 | 346 | 8,88 | 12,4 | 12,4 | 13 | 17,8 | 41,9 | 0,088 | 709 | - | 0,98 | 16,3 | - | 330 | 1,9 | 409 | 1270 | - | 35 | 4,7 | 0 | - | - | - | 21 | 0,24 | 0,29 | 2,13 | - | - | 0 | 0 |
| Safran | 1490 | 352 | 1490 | 352 | 11,9 | 11,4 | 11,4 | 61,5 | 5,85 | 3,9 | 0,37 | 111 | - | 0,33 | 11,1 | - | 264 | 28,4 | 252 | 1720 | - | 148 | 1,09 | 0 | 1,69 | - | - | 80,8 | 0,12 | 0,27 | 1,46 | - | 1,01 | 93 | 0 |
| Cumin, graine | 1780 | 427 | 1780 | 427 | 8,06 | 17,8 | 17,8 | 33,7 | 22,3 | 10,5 | 0,42 | 931 | - | 0,87 | 66,4 | - | 366 | 3,33 | 499 | 1790 | - | 168 | 4,8 | 0 | 3,33 | 5,4 | - | 7,7 | 0,63 | 0,33 | 4,58 | - | 0,44 | 10 | 0 |
| Noix de muscade | 2100 | 507 | 2080 | 504 | 6,22 | 5,3 | 6,26 | 28,5 | 36,3 | 20,8 | 0,04 | 184 | - | 1,03 | 3,04 | - | 183 | 2,9 | 213 | 350 | 1,6 | 16 | 2,15 | 0 | 0 | 0 | - | 3 | 0,35 | 0,057 | 1,3 | - | 0,16 | 76 | 0 |
| Paprika | 1320 | 319 | 1320 | 319 | 11,2 | 14,1 | 14,1 | 19,1 | 12,9 | 34,9 | 0,17 | 229 | - | 0,71 | 21,1 | - | 178 | 1,59 | 314 | 2280 | - | 68 | 4,33 | 0 | 29,1 | 80,3 | - | 0,9 | 0,33 | 1,23 | 10,1 | 2,51 | 2,14 | 49 | 0 |
| Clou de girofle | 1390 | 335 | 1390 | 335 | 9,87 | 5,97 | 5,97 | 31,6 | 13 | 33,9 | 0,69 | 632 | - | 0,37 | 11,8 | - | 259 | 60,1 | 104 | 1020 | - | 277 | 2,32 | 0 | 8,82 | 142 | - | 0,2 | 0,16 | 0,22 | 1,56 | 0,51 | 0,39 | 25 | 0 |
| Laurier, feuille | 1480 | 353 | 1480 | 353 | 5,44 | 7,61 | 7,61 | 48,6 | 8,36 | 26,3 | 0,058 | 834 | - | 0,42 | 43 | - | 120 | 8,17 | 113 | 529 | - | 23 | 3,7 | 0 | 1,79 | - | - | 46,5 | 0,009 | 0,42 | 2,01 | - | 1,74 | 180 | 0 |
| Quatre épices | 1460 | 348 | 1460 | 348 | 8,46 | 6,09 | 6,09 | 50,5 | 8,69 | 21,6 | 0,19 | 661 | - | 0,55 | 7,06 | - | 135 | 2,94 | 113 | 1040 | - | 77 | 1,01 | 0 | - | - | - | 39,2 | 0,1 | 0,063 | 2,86 | - | 0,21 | 36 | 0 |
| Vanille, gousse | - | - | - | - | - | - | - | - | - | - | 0 | - | - | - | 0 | - | - | - | 0 | - | - | 0 | - | - | - | - | - | 0 | 0 | 0 | 0 | - | 0 | 0 | 0 |
| Pavot, graine | 2280 | 551 | 2280 | 551 | 5,17 | 19,7 | 19,7 | 13,7 | 43,1 | 14,8 | 0,059 | 1440 | - | 1,63 | 9,58 | - | 339 | 6,71 | 860 | 710 | - | 23,5 | 9,05 | 0 | 1,14 | 0 | - | 1 | 0,85 | 0,14 | 0,94 | 0,32 | 0,4 | 82 | 0 |
| Carvi, graine | 1380 | 334 | 1380 | 334 | 9,87 | 19,8 | 19,8 | 11,9 | 14,6 | 38 | 0,043 | 689 | - | 0,91 | 16,2 | - | 258 | 1,3 | 568 | 1350 | - | 17 | 5,5 | 0 | 2,5 | 0 | - | 21 | 0,38 | 0,38 | 3,61 | - | 0,36 | 10 | 0 |
| Vanille, extrait alcoolique | 996 | 240 | 996 | 240 | 64,5 | 0,05 | 0,05 | 2,41 | 0 | 0 | 0,01 | - | - | 0,023 | 0,13 | - | 5 | 0,48 | 22 | 98 | - | 4 | 0,06 | 0 | - | - | - | 0 | 0,008 | 0,071 | 0,32 | 0,026 | 0,019 | 0 | 0 |
| Fenouil, graine | 1350 | 326 | 1350 | 326 | 8,81 | 15,7 | 15,7 | 12,5 | 14,9 | 39,8 | 0,22 | 1200 | - | 1,07 | 18,5 | - | 385 | 6,53 | 487 | 1690 | - | 88 | 3,7 | 0 | 2,72 | 0 | - | 21 | 0,41 | 0,55 | 6,05 | - | 0,47 | 0 | 0 |
| Gingembre, racine crue | 137 | 32,9 | 137 | 32,9 | 90,9 | 1,1 | 1,1 | 3,4 | 1,1 | 2,7 | 0,016 | 11 | 17,1 | 0,05 | 0,4 | < 20 | 26 | 7 | 26 | 320 | < 20 | 6,3 | 0,24 | < 0,25 | 0,75 | < 0,8 | - | < 0,5 | 0,018 | < 0,01 | < 0,1 | 0,13 | 0,1 | 11,8 | 0 |
| Cardamome, poudre | 1340 | 321 | 1340 | 321 | 8,28 | 10,8 | 10,8 | 40,5 | 6,7 | 28 | 0,045 | 383 | - | 0,38 | 14 | - | 229 | 28 | 178 | 1120 | - | 18 | 7,47 | 0 | - | - | - | 21 | 0,2 | 0,18 | 1,1 | - | 0,23 | - | 0 |

| | | | | | | | | | | | | | | | | | | | | | | | | | | | | | | | | | | | |
|---|---|---|---|---|---|---|---|---|---|---|---|---|---|---|---|---|---|---|---|---|---|---|---|---|---|---|---|---|---|---|---|---|---|---|---|
| Fenugrec, graine | 1470 | 350 | 1470 | 350 | 7,72 | 27,1 | 27,1 | 33,8 | 6,41 | 24,6 | 0,17 | 176 | - | 1,11 | 33,5 | - | 191 | 1,23 | 296 | 770 | - | 67 | 2,5 | 0 | - | - | - | 3 | 0,32 | 0,37 | 1,64 | - | 0,6 | 57 | 0 |
| Epice (aliment moyen) | 1440 | 346 | 1440 | 345 | 8,86 | 12,3 | 12,4 | 37,7 | 10,4 | 25,9 | 0,048 | 450 | 300 | 0,97 | 17 | 9,75 | 189 | 13,1 | 186 | 1430 | 8,86 | 19,2 | 1,44 | 0,096 | 2,99 | 118 | | 2,61 | 0,15 | 0,064 | 1,05 | 1,77 | 0,29 | 30,2 | 0 |
| Poivre de Cayenne ou piment de Cayenne | 1560 | 376 | 1560 | 376 | 8,05 | 12 | 12 | 29,4 | 17,3 | 27,2 | 0,075 | 148 | - | 0,37 | 7,8 | - | 152 | 2 | 293 | 2010 | - | 30 | 2,48 | 0 | 29,8 | 80,3 | - | 76,4 | 0,33 | 0,92 | 8,7 | - | 2,45 | 106 | 0 |
| Curcuma, poudre | 1220 | 291 | 1220 | 291 | 12,9 | 9,68 | 9,68 | 44,4 | 3,25 | 22,7 | 0,068 | 168 | - | 1,3 | 55 | - | 208 | 19,8 | 299 | 2080 | - | 27 | 4,5 | 0 | 4,43 | 13,4 | - | 0,7 | 0,058 | 0,15 | 1,35 | 0,54 | 0,11 | 20 | 0 |
| Vanille, extrait aqueux | 245 | 57,7 | 245 | 57,7 | 85,6 | 0,03 | 0,03 | 14,4 | 0 | 0 | 0,0075 | 3 | - | 0,003 | 0,05 | - | 1 | 0,001 | 0 | - | - | 3 | 0,01 | 0 | 0 | 0 | - | 0 | 0,003 | 0,029 | 0,13 | 0,011 | 0,008 | 0 | 0 |
| Ail, cru | 470 | 111 | 470 | 111 | 67,8 | 5,31 | 5,31 | 18,6 | <0,5 | 5,8 | 0,023 | 11 | 30,3 | 0,19 | 0,63 | <20 | 20 | 0,2 | 130 | 530 | <20 | 9 | 0,75 | <0,25 | <0,08 | <0,8 | - | <0,5 | 0,2 | <0,01 | <0,1 | 0,36 | 1,1 | 26,1 | 0 |
| Cerfeuil, frais | 167 | 39,9 | 167 | 39,9 | 87,4 | 3,72 | 3,72 | 3,63 | 0,6 | 2,55 | 0,038 | 273 | - | 0,073 | 1,6 | 2,8 | 34 | 1,7 | 40,6 | 597 | 0,1 | 10 | 1,1 | traces | 2,9 | - | - | 37 | 0,13 | 0,34 | 1,5 | 0,3 | 0,026 | 220 | 0 |
| Ciboule ou Ciboulette, fraîche | 124 | 29,7 | 124 | 29,7 | 90,6 | 2,57 | 2,57 | 2,1 | 0,52 | 3,19 | 0,028 | 92,8 | - | 0,1 | 1,31 | 2,2 | 23,8 | 0,85 | 46,3 | 275 | 0,35 | 11,5 | 0,45 | 0 | 0,99 | 260 | - | 39,7 | 0,065 | 0,12 | 0,57 | 0,19 | 0,12 | 78,3 | 0 |
| Persil, frais | 180 | 43 | 180 | 43 | 85,5 | 3,71 | 3,71 | 3,48 | 0,63 | 4,5 | 0,59 | 218 | 364 | 0,12 | 4,67 | <5 | 39,5 | 1,7 | 65 | 598 | 0,1 | 54,3 | 0,77 | 0 | 1,73 | 1220 | - | 177 | 0,1 | 0,2 | 1,51 | 0,35 | 0,15 | 134 | 0 |
| Raifort, cru | - | - | - | - | 69,3 | 7,5 | 7,5 | 13 | 0,7 | 7,5 | 0,015 | 93,7 | - | 0,23 | 1,3 | 0,9 | 25 | 0,46 | 80,5 | 587 | 0,2 | 6 | 1,4 | 0 | 0 | - | - | 152 | 0,14 | 0,11 | 0,6 | - | 0,18 | - | 0 |
| Menthe, fraîche | 240 | 57,6 | 240 | 57,6 | 82,1 | 3,52 | 3,52 | 5,3 | 0,84 | 7,4 | 0,076 | 221 | - | 0,28 | 8,48 | - | 71,5 | 1,15 | 66,5 | 514 | 0,5 | 30,5 | 1,1 | 0 | 5 | - | - | 22,6 | 0,08 | 0,22 | 1,33 | 0,29 | 0,14 | 110 | 0 |
| Basilic, frais | 145 | 34,8 | 145 | 34,8 | 91,7 | 3,35 | 3,35 | 2,55 | 0,47 | 3,47 | 0,028 | 273 | - | 0,39 | 5,24 | - | 64 | 1,15 | 56 | 295 | - | 12 | 0,81 | 0 | 0,8 | 415 | - | 14,5 | 0,034 | 0,076 | 0,9 | 0,21 | 0,16 | 68 | 0 |
| Romarin, frais | 498 | 121 | 498 | 121 | 67,8 | 3,31 | 3,31 | 6,6 | 5,86 | 14,1 | 0,065 | 317 | - | 0,3 | 6,65 | - | 91 | 0,96 | 66 | 668 | - | 26 | 0,93 | 0 | - | - | - | 21,8 | 0,036 | 0,15 | 0,91 | 0,8 | 0,34 | 109 | 0 |
| Sauge, fraîche | 152 | 36,7 | 152 | 36,7 | 85,6 | 4 | 4 | 0,97 | <0,5 | 7 | 0,028 | 230 | 107 | 0,13 | 4,6 | <20 | 54 | 1,9 | 57 | 510 | <20 | 11,1 | 0,28 | <0,25 | 1,46 | 44,9 | - | <0,5 | 0,14 | 0,09 | <0,1 | 0,68 | 0,14 | 72,1 | - |
| Thym, frais | 446 | 107 | 446 | 107 | 65,1 | 5,56 | 5,56 | 10,5 | 1,68 | 14 | 0,023 | 405 | - | 0,56 | 17,5 | - | 160 | 1,72 | 106 | 609 | - | 9 | 1,81 | 0 | 1,7 | - | - | 160 | 0,048 | 0,47 | 1,82 | 0,41 | 0,35 | 45 | 0 |
| Herbes aromatiques fraîches (aliment moyen) | 206 | 49,4 | 206 | 49,4 | 84,2 | 3,58 | 3,58 | 4,15 | 0,8 | 5,64 | 0,28 | 218 | | 0,22 | 6,06 | | 56,7 | 1,37 | 64,8 | 508 | 0,28 | 34,7 | 0,88 | 0,00018 | 2,23 | 870 | | 103 | 0,08 | 0,2 | 1,29 | 0,32 | 0,16 | 105 | 0 |
| Estragon, frais | 184 | 44 | 184 | 44 | - | 3,8 | 3,8 | 4,1 | traces | 6,2 | 0,025 | - | - | - | - | - | - | - | - | - | - | - | - | - | - | - | - | - | - | - | - | - | - | - | - |
| Aneth, frais | 202 | 48,2 | 202 | 48,2 | 86 | 3,93 | 3,93 | 3,9 | 1,1 | 3,5 | 0,15 | 202 | - | 0,22 | 5,5 | 3,9 | 28 | 2,7 | 51,9 | 647 | 2,7 | 27 | 1,8 | 0 | 1,7 | - | - | 70 | 0,19 | 0,43 | 2,4 | 0,3 | 0,3 | 116 | 0 |
| Coriandre, fraîche | 92,6 | 22,3 | 92,6 | 22,3 | 92,2 | 2,13 | 2,13 | 0,87 | 0,52 | 2,8 | 0,12 | 67 | - | 0,23 | 1,77 | - | 26 | 0,43 | 48 | 521 | - | 46 | 0,5 | 0 | 2,5 | 310 | - | 27 | 0,067 | 0,16 | 1,11 | 0,57 | 0,15 | 62 | 0 |
| Ail, rôti/cuit au four | 538 | 127 | 538 | 127 | 65,4 | 5,56 | 5,56 | 23,8 | <0,5 | 2,7 | <0,013 | 11 | 37,5 | 0,18 | 0,64 | <20 | 21 | 0,22 | 140 | 490 | <20 | <5 | 0,73 | <0,25 | 0,44 | <0,8 | - | <0,5 | 0,16 | 0,043 | <0,1 | 0,34 | 1,37 | 37,4 | - |
| Ail, sauté/poêlé, sans matière grasse | 548 | 130 | 548 | 130 | 63,5 | 6 | 6 | 22 | 0,4 | 5,8 | <0,013 | 12 | 20,7 | 0,17 | 0,77 | <20 | 22 | 0,23 | 150 | 580 | <20 | <5 | 0,96 | - | 0,39 | <0,8 | - | 4,09 | 0,14 | 0,032 | 0,75 | 0,3 | 0,94 | 19,3 | - |
| Ail séché, poudre | 1460 | 344 | 1460 | 344 | 6,48 | 16,7 | 16,7 | 62,8 | 0,77 | 9,45 | 0,11 | 79,5 | - | 0,53 | 4,2 | - | 67,5 | 0,98 | 416 | 1150 | - | 43 | 2,81 | 0 | 0,67 | 0,4 | - | 2,39 | 0,45 | 0,15 | 0,74 | 0,74 | 1,65 | 47 | 0 |
| Persil, séché | 1210 | 291 | 1210 | 291 | 3,95 | 29 | 29 | 16,8 | 5,34 | 29,7 | 1,05 | 658 | - | 0,62 | 38 | - | 386 | 5,57 | 492 | 4490 | - | 422 | 5,78 | 0 | 8,96 | 1360 | - | 137 | 0,62 | 2,32 | 10,2 | 1,79 | 1,14 | 187 | 0 |
| Menthe, séchée | 1180 | 283 | 1180 | 283 | 11,3 | 19,9 | 19,9 | 22,2 | 6,03 | 29,8 | 0,86 | 1490 | - | 1,54 | 87,5 | - | 602 | 11,5 | 276 | 1920 | - | 344 | 2,41 | 0 | - | - | - | 0 | 0,29 | 1,42 | 6,56 | 1,4 | 2,58 | 530 | 0 |
| Basilic, séché | 1010 | 244 | 1010 | 244 | 10,4 | 23 | 23 | 10,1 | 4,07 | 37,7 | 0,19 | 2240 | - | 2,1 | 89,8 | - | 711 | 9,8 | 274 | 2630 | - | 76 | 7,1 | 0 | 10,7 | 1710 | - | 0,8 | 0,08 | 1,2 | 4,9 | 0,84 | 1,34 | 310 | 0 |
| Marjolaine, séchée | 1140 | 276 | 1140 | 276 | 7,64 | 12,7 | 12,7 | 20,3 | 7,04 | 40,3 | 0,19 | 1990 | - | 1,13 | 82,7 | - | 346 | 5,43 | 306 | 1520 | - | 77 | 3,6 | 0 | 1,69 | 622 | - | 51,4 | 0,29 | 0,32 | 4,12 | - | 1,19 | 274 | 0 |
| Origan, séché | 1100 | 265 | 1100 | 265 | 9,93 | 9 | 9 | 26,4 | 4,28 | 42,5 | 0,063 | 1600 | - | 0,63 | 36,8 | - | 270 | 4,99 | 148 | 1260 | - | 25 | 2,69 | 0 | 18,3 | 622 | - | 2,3 | 0,18 | 0,53 | 4,64 | 0,92 | 1,04 | 237 | 0 |
| Romarin, séché | 1350 | 328 | 1350 | 328 | 9,31 | 4,88 | 4,88 | 21,5 | 15,2 | 42,6 | 0,13 | 1280 | - | 0,55 | 29,3 | - | 220 | 1,87 | 70 | 955 | - | 50 | 3,23 | 0 | - | - | - | 61,2 | 0,51 | 0,43 | 1 | - | 1,74 | 307 | 0 |
| Sauge, séchée | 1320 | 320 | 1320 | 320 | 7,96 | 10,6 | 10,6 | 20,4 | 12,8 | 40,3 | 0,028 | 1650 | - | 0,76 | 28,1 | - | 428 | 3,13 | 91 | 1070 | - | 11 | 4,7 | 0 | 7,48 | 1710 | - | 32,4 | 0,75 | 0,34 | 5,72 | - | 2,69 | 274 | 0 |
| Thym, séché | 1180 | 285 | 1180 | 285 | 7,79 | 9,11 | 9,11 | 26,9 | 7,43 | 37 | 0,14 | 1890 | - | 0,86 | 124 | - | 220 | 7,87 | 201 | 814 | - | 55 | 6,18 | 0 | 7,48 | 1710 | - | 50 | 0,51 | 0,4 | 4,94 | traces | 0,55 | 274 | 0 |
| Herbes de Provence, séchées | 1170 | 283 | 1170 | 283 | 7,1 | 11,5 | 11,5 | 23 | 7,2 | 40,1 | 0,088 | 1860 | - | 1,06 | 69,8 | 15 | 304 | 5,2 | 297 | 1970 | 4,4 | 35 | 5,47 | 0 | 1,69 | - | - | 54 | 0,33 | 0,34 | 6,05 | 0 | 1,58 | 274 | 0 |
| Sarriette, séchée | 1090 | 264 | 1090 | 264 | 9 | 6,73 | 6,73 | 25 | 5,91 | 45,7 | 0,06 | 2130 | - | 0,85 | 37,9 | - | 377 | 6,1 | 140 | 1050 | - | 24 | 4,3 | 0 | - | - | - | 50 | 0,37 | - | 4,08 | - | 1,81 | - | 0 |
| Meloukhia, feuilles de corète séchées, en poudre | 994 | 239 | 994 | 239 | 8,4 | 23 | 23 | 12,7 | 1,8 | 40 | 0,17 | 2000 | 963 | 1,6 | 87 | 321 | 609 | 12,5 | 307 | 3580 | <5 | 67,9 | 3 | - | 2,8 | - | - | - | 0,09 | 0,36 | 2,6 | 2,4 | 0,55 | 28,3 | - |
| Agar (algue), cru | - | - | - | - | 91,3 | 0,54 | 0,54 | 6,25 | 0,03 | 0,5 | 0,023 | 54 | - | 0,061 | 1,86 | - | 67 | 0,37 | 5 | 226 | - | 9 | 0,58 | 0 | 0,87 | 2,3 | - | 0 | 0,005 | 0,022 | 0,055 | 0,3 | 0,032 | 85 | 0 |
| Agar (algue), séché | - | - | - | - | 14,3 | 4,36 | 4,36 | - | 0,3 | - | 0,13 | 658 | - | 0,56 | 15,2 | 36000 | 770 | 4,3 | 33 | 577 | 0 | 53 | 3,4 | 0 | 5 | 24,4 | - | 0 | 0,01 | 0,22 | 0,2 | 3,02 | 0,3 | 580 | 0 |
| Spiruline (Spirulina sp.), séchée ou déshydratée | - | - | - | - | 4,68 | 57,5 | 57,5 | 20,3 | 7,72 | 3,6 | 2,62 | 120 | - | 6,1 | 28,5 | - | 195 | 1,9 | 118 | 1360 | - | 1050 | 2 | 0 | 5 | 25,5 | - | 10,1 | 2,38 | 3,67 | 12,8 | 3,48 | 0,36 | 94 | 0 |
| Wakamé (Undaria pinnatifida), séchée ou déshydratée | 760 | 184 | 760 | 184 | 8,54 | 14,1 | 14,1 | 5,7 | 2,51 | 41,4 | 12,9 | 1000 | - | 0,23 | 17,2 | 19100 | 1110 | 0,72 | 319 | 7140 | 72,5 | 5170 | 2,03 | 0 | 1,13 | 732 | - | 28 | 0,34 | 0,78 | 6,39 | 0,16 | 0,18 | 237 | 0,23 |
| Laitue de mer (Ulva sp.), séchée ou déshydratée | 856 | 206 | 856 | 206 | 10,9 | 15,9 | 15,9 | 13,8 | 2,03 | 34,4 | 4,93 | 1200 | - | 1,31 | 78,9 | 9190 | 2780 | 3,87 | 181 | 1950 | 14,9 | 1970 | 3,65 | 1,31 | 1,95 | - | - | 54,6 | 0,13 | 0,25 | 8,59 | 0,15 | - | 53 | 9,55 |
| Kombu royal (Saccharina latissima), séchée ou déshydratée | 849 | 204 | 849 | 204 | 8,63 | 10,3 | 10,3 | 23,6 | 1,07 | 29,3 | 9,08 | 801 | - | 0,24 | 23,1 | 341000 | 834 | 0,96 | 208 | 6250 | 521 | 3630 | 2,3 | 1,28 | 0,57 | - | - | 11,4 | 0,44 | 0,34 | - | - | - | - | - |
| Nori (Porphyra sp.), séchée ou déshydratée | 1060 | 255 | 1060 | 255 | 6,54 | 31,5 | 31,5 | 10,5 | 1,63 | 36,3 | 4,96 | 318 | - | 1,06 | 37,2 | 5100 | 486 | 3,94 | 518 | 1730 | 51,2 | 1980 | 4,52 | 0,56 | 2,87 | - | - | 57,3 | 0,58 | 1,91 | 5,82 | 0,25 | 0,53 | 21,7 | 38,8 |
| Dulse (Palmaria palmata), séchée ou déshydratée | 949 | 227 | 949 | 227 | 6,72 | 17,2 | 17,2 | 22,6 | 1,33 | 28 | 4,15 | 547 | - | 1,1 | 34,8 | 32500 | 241 | 12,1 | 280 | 6810 | 8,95 | 1660 | 4,17 | 0,93 | 3,45 | 420 | - | 83,6 | 0,37 | 0,48 | 4,31 | 0,4 | 0,013 | 92 | 9,81 |
| Kombu ou kombu japonais (Laminaria japonica), séchée ou déshydratée | 931 | 224 | 931 | 224 | 9,5 | 8,38 | 8,38 | 24,2 | 2,65 | 34,8 | 6,41 | 805 | - | 0,16 | 13,3 | 236000 | 1050 | 0,49 | 761 | 10600 | 137 | 2560 | 1,22 | 0 | - | - | - | 15,1 | 0,21 | 0,48 | 1,5 | - | 0,081 | 127 | 1,52 |
| Kombu breton (Laminaria digitata), séchée ou déshydratée | 905 | 217 | 905 | 217 | 5,31 | 9,51 | 9,51 | 25,6 | 1,13 | 33,2 | 7,87 | 847 | - | 0,48 | 9,29 | 486000 | 800 | 0,46 | 857 | 4590 | - | 3150 | 4,89 | 2,27 | 0,32 | - | - | 0,024 | - | - | 3,03 | - | - | - | - |
| Haricot de mer (Himanthalia elongata), séchée ou déshydratée | 998 | 239 | 998 | 239 | 8,09 | 10,1 | 10,1 | 28,3 | 2,63 | 31 | 9,22 | 712 | - | 0,23 | 2,53 | 14400 | 1620 | 2,76 | 104 | 5970 | 6,25 | 3690 | 4,55 | 0,28 | 5,5 | - | - | 62,8 | 0,27 | 0,46 | - | - | - | 60,7 | - |
| Gracilaire ou ogonori (Gracilaria verrucosa), séchée ou déshydratée | 1010 | 243 | 1010 | 243 | 8,4 | 16,5 | 16,5 | 18,8 | 3,54 | 35 | 7,8 | 1290 | - | 0,42 | 19,7 | 494000 | 412 | 104 | 177 | 5850 | 6,41 | 3120 | 3,33 | - | - | - | - | - | - | - | - | - | - | - | - |
| Fucus vésiculeux (Fucus serratus ou Fucus vesiculosus), séché ou déshydraté | 799 | 194 | 799 | 194 | 11,6 | 7,41 | 7,41 | 15,7 | 1,33 | 44,6 | 10,1 | 1170 | - | 0,4 | 14,7 | 40000 | 885 | 8,34 | - | 3270 | 88,4 | 4020 | 8,19 | - | 12,2 | - | - | - | - | - | 1,67 | - | - | - | - |
| Ao-nori (Enteromorpha sp.), séchée ou déshydratée | 932 | 224 | 932 | 224 | 7,65 | 13,7 | 13,7 | 18,8 | 2,47 | 36 | 11,9 | 1610 | - | 1,21 | 234 | 9390 | 2440 | 24,7 | 1040 | 1850 | 10,7 | 4760 | 6,08 | 0 | 11 | - | - | 35,4 | 0,29 | 1,1 | 4,77 | - | - | 39,6 | 30,6 |
| Lichen de mer ou pioca ou goémon rouge (Chondrus crispus), séché ou déshydraté | 978 | 234 | 978 | 234 | 10,7 | 16,6 | 16,6 | 21,5 | 2,3 | 30,6 | 5,17 | 362 | - | 0,45 | 21,1 | 34600 | 1230 | 1,47 | 750 | 1570 | 3,35 | 2070 | 7,86 | 0 | - | - | - | 14,3 | 0,071 | 2,23 | 2,84 | 0,84 | 0,33 | 866 | - |
| Ascophylle noueux ou goémon noir (Ascophyllum nodosum), séché ou déshydraté | 874 | 211 | 874 | 211 | 10,8 | 7,22 | 7,22 | 18,5 | 2,78 | 41,8 | 7,15 | 1650 | - | 0,72 | 21,8 | 68200 | 836 | 2,52 | 162 | 2270 | 6,69 | 2860 | 6,45 | 0,89 | 14 | 1020 | - | 94,8 | 0,31 | 0,96 | 2,74 | 0,018 | 5,61 | 22,7 | 2,11 |
| Wakamé atlantique (Alaria esculenta), séchée ou déshydratée | 769 | 187 | 769 | 187 | 10,3 | 12,3 | 12,3 | 6,28 | 1,53 | 49,5 | 4,81 | 235 | - | 0,24 | 61,5 | 34600 | - | 2,06 | 220 | 2180 | - | 1920 | 2,09 | - | - | - | - | - | - | - | - | - | - | - | - |
| Son de blé | 1160 | | | | | | | | | | | | | | | | | | | | | | | | | | | | | | | | | | |
| Son de blé | | 279 | 1160 | 279 | 9,05 | 15,4 | 15,2 | 23,6 | 4,35 | 42 | 0,033 | 73,9 | - | 1,05 | 14,8 | 2,4 | 546 | 13,3 | 1030 | 1260 | 6,8 | 13,3 | 7,49 | 0 | 1,55 | 42,5 | - | 0 | 0,71 | 0,47 | 21,6 | 2,29 | 1,34 | 110 | 0 |
| Son d'avoine | 1510 | 359 | 1510 | 359 | 7,53 | 15,8 | 15,8 | 51 | 6,48 | 16,7 | 0,016 | 62,6 | - | 0,4 | 5,41 | - | 235 | 5,63 | 627 | 566 | 45,2 | 6,5 | 3,11 | 0 | 1,01 | 3,2 | - | 0 | 1,17 | 0,22 | 0,93 | 1,49 | 0,17 | 52 | 0 |
| Son de maïs | 921 | 226 | 921 | 226 | 4,71 | 8,36 | 8,36 | 6,65 | 0,92 | 79 | 0,018 | 42 | - | 0,25 | 2,79 | - | 64 | 0,14 | 72 | 44 | - | 7 | 1,56 | 0 | 0,42 | 0,3 | - | 0 | 0,01 | 0,1 | 2,74 | 0,64 | 0,15 | 4 | 0 |
| Son de riz | 1650 | 398 | 1650 | 398 | 6,12 | 13,9 | 13,9 | 28,1 | 20,9 | 21 | 0,013 | 57 | - | 0,73 | 18,5 | - | 781 | 14,2 | 1680 | 1490 | 15,6 | 5 | 6,04 | 0 | 4,92 | 1,9 | - | 0 | 2,75 | 0,28 | 34 | 7,39 | 4,07 | 63 | 0 |
| Son (aliment moyen) | 1450 | 344 | 1450 | 344 | 7,8 | 15,7 | 15,7 | 46 | 6,09 | 21,3 | 0,019 | 64,6 | | 0,52 | 7,11 | | 291 | 7,02 | 701 | 692 | 38,2 | 7,74 | 3,91 | 0 | 1,11 | 10,3 | | 0 | 1,09 | 0,27 | 4,69 | 1,64 | 0,38 | 62,5 | 0 |

| | | | | | | | | | | | | | | | | | | | | | | | | | | | | | | | | | | | |
|---|---|---|---|---|---|---|---|---|---|---|---|---|---|---|---|---|---|---|---|---|---|---|---|---|---|---|---|---|---|---|---|---|---|---|---|
| Germe de blé | 1580 | 375 | 1540 | 368 | 7,5 | 27,2 | 29,2 | 35,1 | 9,5 | 16,3 | 0,018 | 54 | 107 | 0,78 | 8,9 | < 20 | 250 | 17 | 1000 | 990 | < 20 | 7,2 | 14 | < 0,25 | 10,2 | 2,87 | - | < 0,5 | 1,32 | 0,14 | 1,68 | 1,55 | 0,83 | 143 | 0 |
| Gélatine, sèche | 1480 | 348 | 1480 | 348 | 11,7 | 86,9 | 86,9 | 0 | 0,1 | 0 | 0,75 | 280 | - | - | 1,2 | 5 | 11 | - | 0 | 12 | - | 298 | - | 0 | 0 | - | - | 0 | 0 | 0,07 | 0 | 0 | 0,006 | 0 | 0 |
| Levure alimentaire | 1400 | 334 | 1400 | 334 | 5 | 40,4 | 40,4 | 21,8 | 4,5 | 22,5 | 0,35 | 130 | 251 | 0,39 | 4,1 | < 20 | 150 | 0,89 | 1100 | 1700 | < 20 | 141 | 8,4 | < 0,25 | 0,098 | < 0,8 | - | < 0,5 | 11,6 | 1 | 18,5 | 5,14 | 0,88 | 697 | 0,34 |
| Levure de boulanger, compressée | - | - | - | - | 69 | 8,23 | 8,3 | 11,9 | 1,9 | 7,15 | 0,058 | 17 | 20 | 0,15 | 2,43 | 0,6 | 35,5 | 0,26 | 338 | 606 | 1 | 23 | 7,24 | 0 | 0,2 | 0 | - | 0,1 | 1,18 | 1,27 | 9,9 | 4,2 | 0,77 | 893 | 0,01 |
| Levure de boulanger, déshydratée | - | - | - | - | 2,94 | 44,5 | 44,5 | 17,9 | 6,37 | 23,3 | 0,56 | 30 | 0 | 0,44 | 2,17 | 5 | 54 | 0,31 | 637 | 955 | - | 226 | 7,94 | 0 | 0 | 0,4 | - | 0,3 | 11 | 4 | 40,2 | 13,5 | 1,5 | 2340 | 0,07 |
| Levure chimique ou Poudre à lever | - | - | - | - | 5,4 | 1,96 | 1,96 | 33,2 | 0,2 | 1,9 | - | 8960 | - | 0,012 | 11,1 | - | 21 | 0,013 | 7240 | 30,8 | - | 14200 | 0,015 | 0 | 0 | 0 | - | 0 | 0 | 0 | 0 | 0 | 0 | 0 | 0 |
| Préparation culinaire à base de soja, type "crème de soja" | - | - | - | - | 78,5 | 3,25 | 3,25 | 2,03 | 14,7 | - | 0,063 | 16 | 225 | 0,11 | 0,47 | < 20 | 19 | 0,18 | 62 | 130 | < 20 | 25 | 0,37 | < 0,25 | 5,31 | 6,28 | - | - | 0,02 | < 0,01 | < 0,1 | 0,14 | 0,048 | 16,5 | - |
| Bicarbonate de soude | 1,7 | 0,4 | 1,7 | 0,4 | 0,2 | 0,1 | 0,1 | 0 | 0 | 0 | - | 0 | - | 0 | 0 | - | 0 | 0 | 0 | 0 | - | 27400 | 0 | 0 | 0 | 0 | - | 0 | 0 | 0 | 0 | 0 | 0 | 0 | 0 |
| Tofu soyeux, préemballé | 227 | 54,3 | 219 | 52,5 | 89,4 | 4,57 | 5 | 1,53 | 2,9 | 0,6 | < 0,013 | 21 | 111 | 0,19 | 0,64 | < 20 | 64 | 0,27 | 58 | 210 | < 20 | < 5 | 0,47 | < 0,25 | 0,41 | 5,48 | - | < 0,5 | 0,036 | < 0,01 | < 0,1 | 0,21 | 0,048 | 24,2 | - |
| Miso | - | - | - | - | 48 | 10,7 | 11,7 | 18,1 | 5,31 | 5,4 | 8,35 | 62,5 | - | 0,42 | 2,1 | - | 48 | 0,86 | 235 | 273 | - | 3340 | 2,56 | 0 | 0,01 | 29,3 | - | 0 | 0,079 | 0,17 | 0,6 | 0,34 | 0,2 | 19 | 0,08 |
| Sirop pour fruits appertisés au sirop | - | - | - | - | 80,6 | < 0,5 | < 0,5 | 18,4 | < 0,5 | < 0,5 | 0,019 | 6,4 | 4,6 | 0,025 | 0,07 | 16 | 3,75 | 0,045 | 6,8 | 74 | < 20 | 7,5 | < 0,05 | - | - | < 0,8 | - | < 0,5 | 0,37 | < 0,01 | 0,095 | 0,049 | 0,041 | 11 | - |
| Sirop léger pour fruits appertisés au sirop | - | - | - | - | 84,3 | 0,34 | 0,34 | 15 | < 0,5 | < 0,5 | 0,011 | 7,13 | 9,1 | 0,025 | 0,095 | < 20 | 6,3 | 0,3 | 5,98 | 99,5 | < 20 | 4,23 | 0,043 | - | - | < 0,8 | - | 4,5 | 0,18 | < 0,01 | 0,12 | 0,044 | 0,0082 | 9,02 | - |
| Jus d'ananas pour ananas appertisé au jus | - | - | - | - | 85,5 | < 0,5 | < 0,5 | 13,7 | < 0,5 | < 0,5 | 0,0014 | 9 | 17,9 | 0,02 | 0,19 | < 20 | 11 | 1 | 5,9 | 120 | < 20 | 0,57 | 0,1 | - | - | < 0,8 | - | 4,95 | 0,063 | < 0,01 | < 0,1 | 0,05 | 0,041 | 11,9 | - |
| Sirop léger pour poire appertisée | 258 | 60,7 | 258 | 60,7 | 84,2 | < 0,5 | < 0,5 | 14,6 | < 0,3 | < 0,5 | 0,0043 | 6,7 | 1,9 | 0,04 | < 0,05 | - | 3,6 | 0,01 | 5,2 | 70 | < 20 | 1,7 | 0,05 | < 0,25 | < 0,08 | < 0,8 | - | 0,66 | 0,14 | < 0,01 | < 0,1 | 0,025 | < 0,01 | 6,74 | - |
| Gélifiant pour confitures | 1170 | 278 | 1170 | 278 | 7,2 | 0 | 0 | 62 | 0 | 15 | - | - | - | - | - | - | - | - | - | - | - | - | - | - | - | - | - | - | - | - | - | - | - | - | - |
| Gelée royale | - | - | - | - | 66 | 13 | 13 | 14 | 4 | - | - | - | - | - | traces | - | - | traces | - | - | - | - | - | - | traces | - | - | 12 | 4,3 | 7,5 | 105 | 133 | 6,2 | 200 | 150 |
| Pollen, partiellement séché | 1510 | 358 | 1510 | 358 | 7,58 | 21,9 | 21,9 | 52,5 | 4,2 | 11,3 | 0,22 | 85,2 | - | 1,95 | - | - | 82,8 | - | 367 | 464 | - | 89,8 | 5,6 | - | - | - | - | - | 0,00092 | 0,0019 | - | 0,004 | 0,0005 | - | - |
| Pollen,frais | 1200 | 285 | 1200 | 285 | - | 17,5 | 17,5 | 37,7 | 5,17 | 9,07 | - | 105 | - | 0,51 | 11,6 | - | 63,3 | - | 273 | 411 | 41,4 | 7,57 | 3,95 | - | 4,46 | - | - | 6,34 | 0,67 | 1,08 | 4,84 | 0,87 | 0,28 | 993 | - |
| Base de pizza à la crème | 1100 | 261 | 1100 | 261 | - | 7,7 | 7,7 | 51,1 | 2,6 | 1 | 1,75 | 9 | - | - | - | - | - | - | - | - | - | 700 | - | - | - | - | - | - | - | - | - | - | - | - | - |
| Base de pizza tomatée | 983 | 232 | 983 | 232 | - | 6,9 | 6,9 | 44,6 | 2,7 | 1 | 1,75 | 12 | - | - | - | - | - | - | - | - | - | 700 | - | - | - | - | - | - | - | - | - | - | - | - | - |
| Lécithine de soja | 3210 | 779 | 3210 | 779 | 1 | 0 | 0 | 8 | 83 | traces | 0,038 | 125 | - | 0 | 2,5 | 0 | 115 | - | 2750 | 875 | - | 15 | 0 | 0 | 32,5 | - | - | 0 | 0 | 0 | 0 | 0 | 0 | 0 | 0 |

# Bibliographical references


Ainane, T. (2020). Moroccan traditional treatment for fever and influenza, similar to symptoms of coronavirus COVID-19 disease: Mini Review. *Journal of Analytical Sciences and Applied Biotechnology*, *2*(1), 1-3.

Caillavet, F., Darmon, N., Lhuissier, A., & Régnier, F. (2005). L'alimentation des populations défavorisées en France: synthèse des travaux dans les domaines économique, sociologique et nutritionnel. Les travaux de *l'Observatoire national de la pauvreté et de l'exclusion sociale*, *2006,* 279-322.

Cherroud, S., Ainane, A., El Hassan, A. B. B. A., El Yaacoubi, A., Ainane, T. Stress and confinement during the COVID 19 pandemic: Statistical study of the situation in Morocco (March-May 2020). *Journal of Analytical Sciences and Applied Biotechnology*, *2*(1), 12-15.

Dupin, H. (1992). *Human food and nutrition*. ESF editor.

Gourch, A., Zejli, H., Lfitat, A., Bousraf, F. Z., Yassine, E. L., Ainane, A., Abdellaoui, A., Ainane, T., Taleb, M. (2020). Preventive impact of traditional medicine against covid-19. *Journal of Analytical Sciences and Applied Biotechnology*, *2*(2), 78-82.

Jean, G., Bergot, P., & Kaushik, S. (1999). *Nutrition and feeding of fish and crustaceans*. Editions Quae.

Khripounoff, A. & Sibuet, M. (1980). La nutrition d'echinodermes abyssaux I. Alimentation des holothuries. *Marine Biology*, *60*(1), 17-26.

Mathé, T., Pilorin, T., Hébel, P., Denizeau, M. (2008). Du discours nutritionnel aux représentations de l'alimentation. *Cahier de recherche*, *252,* 1-74.

Peretti, N., Lambe, C., Lopez, M., Valla, F., Viala, J., & de Gastroentérologie Hépatologie, G. F. (2020). Specificities of nutritional management of pediatric patients Covid 19.

Samouh, Y., Benider, A., & Derfoufi, S. (2020). Nutrition and emerging viral disease Covid-19. *Therapeutic medicine*, *26*(2), 107-109.

Thibault, R., Quilliot, D., Seguin, P., Tamion, F., Schneider, S., Déchelotte, P. (2020). Stratégie de prise en charge nutritionnelle à l'hôpital au cours de l'épidémie virale Covid-19: avis d'experts de la Société Francophone de Nutrition Clinique et Métabolisme (SFNCM). *Clinical Nutrition and Metabolism*.

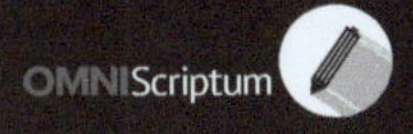

MIX
Papier aus verantwortungsvollen Quellen
Paper from responsible sources
FSC® C105338

FSC
www.fsc.org

Printed by Books on Demand GmbH, Norderstedt / Germany